Systems Analysis and Simulation in Ecology

VOLUME IV

Contributors

PETER H. ASTOR
LEWIS J. BLEDSOE
ROBERT W. BOSSERMAN
DANIEL B. BOTKIN
WILLIAM G. CALE
HAL CASWELL
MICHAEL CONRAD
NELSON T. EDWARDS
GERALD N. ESTBERG
JOHN T. FINN
W. FRANKLIN HARRIS
STEVEN O. HOWE
ROBERT A. KELLY
HERMAN E. KOENIG
ARIEL E. LUGO
JOHN McLEOD
ROBERT J. MULHOLLAND
JAMES J. O'BRIEN
HOWARD T. ODUM
BERNARD C. PATTEN
MAURICE SELL
CRAIG S. SIMS
SAMUEL C. SNEDAKER
PHILLIP SOLLINS
JACK B. WAIDE
JOHN J. WALSH
JACKSON R. WEBSTER
JOSEPH S. WROBLESWSKI
BERNARD P. ZEIGLER

SYSTEMS ANALYSIS AND SIMULATION IN ECOLOGY

Edited by BERNARD C. PATTEN
Department of Zoology
University of Georgia
Athens, Georgia

VOLUME IV

ACADEMIC PRESS New York San Francisco London 1976
A Subsidiary of Harcourt Brace Jovanovich, Publishers

ACADEMIC PRESS, INC.
111 Fifth Avenue, New York, New York 10003

United Kingdom Edition published by
ACADEMIC PRESS, INC. (LONDON) LTD.
24/28 Oval Road, London NW1

Library of Congress Cataloging in Publication Data

Main entry under title:

Systems analysis and simulation in ecology.

Vols. 3-4 based on proceedings of symposium held Mar. 1-3, 1975, at the University of Georgia.
Includes bibliographies.
1. Ecology–Mathematical models–Collected works.
2. Ecology–Data processing. I. Patten, Bernard C., Date ed.
QH541.15.M3S9 574.5′01′84 76-127695
ISBN 0–12–547204–8 (v. 4)

PRINTED IN THE UNITED STATES OF AMERICA

Contents

3. A Simulation of the Mesoscale Distribution of the Lower Marine Trophic Levels off West Florida

James J. O'Brien and Joseph S. Wroblewski

PART II

MODELS OF TERRESTRIAL ECOSYSTEMS

4. Mangrove Ecosystem Analysis

Ariel E. Lugo, Maurice Sell, and Samuel C. Snedaker

5. The Role of Species Interactions in the Response of a Forest Ecosystem to Environmental Perturbation

Daniel B. Botkin

6. Simulating the Physiology of a Temperate Deciduous Forest
Phillip Sollins, W. Franklin Harris, and Nelson T. Edwards

PART III

MODELS OF HUMAN ECOSYSTEMS

7. Human Ecosystem Design and Management: A Sociocybernetic Approach
Herman E. Koenig

8. Simulating the World Ecosystem
John McLeod

9. Macroscopic Minimodels of Man and Nature
Howard T. Odum

PART IV

SPECIAL PROBLEMS IN ECOSYSTEM MODELING

10. Linear and Nonlinear Approaches for Ecosystem Dynamic Modeling

Lewis J. Bledsoe

11. The Aggregation Problem

Bernard P. Zeigler

12. The Validation Problem

Hal Caswell

PART V

THEORY IN ECOSYSTEM ANALYSIS

13. Engineering Systems Analysis: Applicability to Ecosystems

Jack B. Waide and Jackson R. Webster

List of Contributors

Numbers in parentheses indicate the pages on which the authors' contributions begin.

PETER H. ASTOR, Envirosphere Company: A Division of Ebasco Services, New York, New York (389)

LEWIS J. BLEDSOE, Center for Quantitative Science in Forestry, Fisheries, and Wildlife, University of Washington, Seattle, Washington (283)

ROBERT W. BOSSERMAN, Department of Zoology and Institute of Ecology, University of Georgia, Athens, Georgia (457)

DANIEL B. BOTKIN, Ecosystems Center, Marine Biological Laboratory, Woods Hole, Massachusetts (147)

WILLIAM G. CALE,* Department of Zoology and Institute of Ecology, University of Georgia, Athens, Georgia (457)

HAL CASWELL,† Department of Zoology and Division of Engineering Research, Michigan State University, East Lansing, Michigan (313)

MICHAEL CONRAD,‡ Institute for Information Sciences, University of Tübingen, Tübingen, Germany (431)

NELSON T. EDWARDS, Environmental Sciences Division, Oak Ridge National Laboratory, Oak Ridge, Tennessee (173)

GERALD N. ESTBERG, Department of Physics, University of San Diego, Alcala Park, San Diego, California (389)

JOHN T. FINN, Department of Zoology and Institute of Ecology, University of Georgia, Athens, Georgia (457)

* Present address: University of Texas at Dallas, Richardson, Texas.

† Present address: Ecology Section U-42, University of Connecticut, Storrs, Connecticut.

‡ Present address: Department of Computer and Communication Sciences, University of Michigan, Ann Arbor, Michigan.

W. FRANKLIN HARRIS, Environmental Sciences Division, Oak Ridge National Laboratory, Oak Ridge, Tennessee (173)

STEVEN O. HOWE, Division of Applied Sciences, Brookhaven National Laboratory, Upton, Long Island, New York (47)

ROBERT A. KELLY,§ Resources for the Future, Inc., Washington, D.C. (3)

HERMAN E. KOENIG, Department of Electrical Engineering and Systems Science, Michigan State University, East Lansing, Michigan (221)

ARIEL E. LUGO, Department of Botany, University of Florida, Gainesville, Florida (113)

JOHN McLEOD, Society for Computer Simulation, Simulation Councils, Inc., La Jolla, California (239)

ROBERT J. MULHOLLAND, Center for Systems Science and School of Electrical Engineering, Oklahoma State University, Stillwater, Oklahoma (373)

JAMES J. O'BRIEN,¶ Departments of Meteorology and Oceanography, Florida State University, Tallahassee, Florida (63)

HOWARD T. ODUM, Department of Environmental Engineering Sciences, University of Florida, Gainesville, Florida (249)

BERNARD C. PATTEN, Department of Zoology, University of Georgia, Athens, Georgia (389, 457)

MAURICE SELL, Department of Environmental Engineering Sciences, University of Florida, Gainesville, Florida (113)

CRAIG S. SIMS, Center for Systems Science and School of Electrical Engineering, Oklahoma State University, Stillwater, Oklahoma (373)

SAMUEL C. SNEDAKER,* Resource Management Systems Program, School of Forest Resources and Conservation, University of Florida, Gainesville, Florida (113)

PHILLIP SOLLINS, Environmental Sciences Division, Oak Ridge National Laboratory, Oak Ridge, Tennessee (173)

§ Present address: Fisheries and Wildlife Division, Ministry for Conservation, Melbourne, Victoria, Australia.

¶ Present address: Office of Naval Research, Arlington, Virginia.

* Present address: Rosenstiel School of Marine and Atmospheric Sciences, University of Miami, Florida.

JACK B. WAIDE,† Institute of Ecology and Department of Zoology, University of Georgia, Athens, Georgia (329)

JOHN J. WALSH,‡ Department of Oceanography, University of Washington, Seattle, Washington (47)

JACKSON R. WEBSTER,§ Institute of Ecology and Department of Zoology, University of Georgia, Athens, Georgia (329)

JOSEPH S. WROBLESWSKI,¶ Department of Oceanography, Florida State University, Tallahassee, Florida (63)

BERNARD P. ZEIGLER, Department of Applied Mathematics, Weizmann Institute of Science, Rehovot, Israel (299)

† Present address: Department of Zoology, Clemson University, Clemson, South Carolina.

‡ Present address: Division of Oceanographic Sciences, Brookhaven National Laboratory, Upton, Long Island, New York.

§ Present address: Department of Biology, Virginia Polytechnic Institute and State University, Blacksburg, Virginia.

¶ Present address: Department of Oceanography, Dalhousie University, Halifax, Nova Scotia.

Preface

Volume IV continues the organization begun in Volume III to document a meeting, *Modeling and Analysis of Ecosystems*, held March 1-3, 1973 at the University of Georgia. Several chapters are considerably expanded over their original concept, and several others are included which were not part of the symposium.

The book has five parts. Part I contains chapters on estuarine-marine ecosystems. Part II presents models of several terrestrial ecosystems. Part III has chapters devoted to human aspects of ecology. Part IV considers special problems of ecosystem modeling, namely linear versus nonlinear models, aggregation and validation. Part V, the most extensive section, describes theory in ecosystem analysis.

These seventeen chapters amply demonstrate the current scope of systems ecology—its past and present emphasis on parts and mechanisms in simulation modeling, and its movement toward systems analysis and new, more formal consideration of wholes in theory. They make clear that although the systems approach is young in ecology, it has substantially enriched the science both methodologically and conceptually. The original expectations for it are valid and growing.

Athens, Georgia
August 1976

BERNARD C. PATTEN

Contents of Other Volumes

Volume I

Volume II

Volume III

Part I ECOSYSTEM MODELING IN THE U.S. INTERNATIONAL BIOLOGICAL PROGRAM

Part II MODELS OF FRESHWATER-ESTUARINE ECOSYSTEMS

PART I

MODELS OF ESTUARINE–MARINE ECOSYSTEMS

The three chapters of this section continue the organization begun in Volume III, whose last part treated models of freshwater-estuarine ecosystems.

Chapter 1 by Dr. Kelly describes a model of the Delaware Estuary developed as part of a regional management model for the lower Delaware Valley. The objective was to formulate an ecological model in a regional management context to predict water quality parameters and effects. The model interacts with other submodels of the regional system model. An input–output mass balance approach is used. The estuary is partitioned into longitudinal reaches within each of which eight functional compartments occur. The compartments are algae, zooplankton, bacteria, fish, oxygen, organic matter, nitrogen, and phosphorus. Exchanges of mass between both compartments and reaches are formulated by differential equations. Factors represented in these equations include temperature, light intensity, turbidity, toxic matter, advective flow, and reaeration. Numerical solutions are obtained by Euler integration. Material flows between compartments within each reach include feeding or nutrient uptake, excretion and death, respiration, and predation. Mathematical rationales are presented and formulations described for processes within reaches and across reach boundaries. Data sources are discussed and a simulation for September 1970, is described. Oxygen concentrations conform to observed, particularly in the upper reaches. The BOD pattern is realistic, although absolute levels do not match data. Phosphorus agrees with measurements except in the upper estuary. Simulated nitrogen (total) was different from observed (Kjeldahl), so results could not be directly compared. Algae were in general agreement with data, while zooplankton, bacteria and fish were probably several times too high. The need for many (hundreds) computer trials to obtain the reasonably realistic simulation reported suggests that the model lacks the controllability required to serve management objectives. The rationale should probably be reevaluated, although the author suggests several avenues for improvement within the existing rationale. These amount to better ecological understanding and more data, always welcome and any modeling exercise. The possible need for new theory is not considered.

Chapter 2 by Drs. Walsh and Howe compares simulated nitrogen and carbon productivity of the Peru upwelling ecosystem. Part of the U.S. International Biological Program, this work was motivated by considerations of future availability of food from the sea. It explores phytoplankton C/N ratios as a possible index to food chain quality. Carbon is not limiting in buffered marine systems, but nitrogen is recycled slowly and may determine regional productivity through regulation of protein synthesis. Possible reflection of nutrient limitation and other factors in phytoplankton C/N ratios is examined with nitrogen and carbon simulation models of the upwelling ecosystem. The models are two-layered spatial representations in which diel changes in nutrient uptake, grazing, and vertical zooplankton migration are simulated with periodic functions. The state variables are light, nitrate, recycled nitrogen, phosphate, silicate, phytoplankton, zooplankton, and fish. Their equations, in terms of particulate nitrogen and carbon, are nonlinear, time-varying, and differential in form. Numerical computer solutions are obtained by Adams–Bashforth integration. Predicted C/N ratios corresponded to observed values well enough to suggest that the Peru upwelling ecosystem within 40 km of the coast may be regulated by silicate and phytoplankton intrinsic growth rates. The role of diatoms versus dinoflagellates in various silicate conditions is considered, and ecosystem manipulations to increase protein yield to man (diatoms are more protein-rich than dinoflagellates) are discussed. Finally, simulation as a necessary tool for total resource management is advocated; it is the only means to experimentally couple lower trophic level dynamics and factors to upper trophic level yields.

In *Chapter 3*, Drs. O'Brien and Wroblewski present a simulation of the flow of biologically limiting nutrient through marine trophic levels over the continental shelf off west Florida. The limiting nutrient concentrations, in phytoplankton, zooplankton, pelagic fish, organic detritus, and dissolved nutrients are accounted for by biotic uptake and release, and mass transport. Biological processes are modeled within spatial blocks, and exchanges between blocks are accomplished by advective and diffusive fluxes. A scaling procedure for nondimensionalizing system equations is described. The equations are solved by numerical integration. Sensitivity analysis is used to investigate the relative importance of different biological processes. The most important parameters in overall model behavior are zooplankton grazing coefficient, zooplankton excretion coefficient, Ivlev constant, zooplankton death coefficient, and fish grazing coefficient. The model appears quite sensitive to zooplankton dynamics. Water circulation on the shelf is modeled and the biological dynamics are placed into the simulated physical system. Spatial distributions of biotic components are determined, first without advection to demonstrate transition from a one-box to a 3200-box model. Then, spatial distributions with advection are computed for phosphate- and nitrate-limited seas, and results interpreted. The major strength of the model is successful combining of physical and biological processes in a workable framework. Weaknesses are lack of knowledge of correct expressions and parameter values, which describe the real system, and inability to check results of the sensitivity analysis. Also, the model is too aggregated to simulate biological complexity, and better feedback is needed between modeling theory and field observations.

1

Conceptual Ecological Model of the Delaware Estuary

ROBERT A. KELLY*

QUALITY OF THE ENVIRONMENT PROGRAM
RESOURCES FOR THE FUTURE, INC., WASHINGTON, D.C.

I. Introduction

This chapter describes an ecological model of the Delaware Estuary developed as part of a regional residuals-environmental quality management model. The

* Present address: Fisheries and Wildlife Division, Ministry for Conservation, Melbourne, Victoria, Australia.

regional model includes: (i) submodels of residuals generating activities (industrial operations, residences, etc.); (ii) submodels of natural systems (aquatic ecosystem, atmosphere); and (iii) a submodel of the political decision-making process. The regional model and its submodels are described conceptually by Russell and Spofford (1972), and for the lower Delaware Valley Region (parts of Pennsylvania, New Jersey, and Delaware) by Spofford *et al.* (1972).

Basically, the objective in developing the model was to investigate whether an ecological model of an aquatic ecosystem (i) might be formulated as an integral part of a regional management model, (ii) could include more outputs of information of decision-making interest than water quality models typically used, and (iii) would result in more accurate prediction of certain water quality parameters and their effects than more traditional water quality models. It is important to understand these objectives because the degree of sophistication (detail, refinement) of any model should be related to the model's use. Because the ecological model is part of a larger model, how it is formulated is affected by the complexities of the other submodels and by the need to integrate it into the overall decision-making framework.

Many different types of models are currently used to predict consequences of man's activities on estuaries and other bodies of water. Most of these models address one or, at most, a few aspects of "water quality," such as dissolved oxygen (DO), total dissolved solids (TDS), or nutrients (see Tracor, 1971, for examples of several types). Even though the results of such models have long been used in decision making, two major factors limit their practical utility: irrelevance of outputs to public policy issues, and inaccuracy of outputs. These two factors are of course interrelated.

For example, if it is hypothesized that industrial discharges of oxygen-consuming materials (BOD) into a water body will have a major impact on a fishery, the agency responsible for water quality management will want to know how DO concentration will change with changes in residuals discharges, and also how fish will be directly and indirectly affected. In this case, a simple DO model (as in Thomann, 1972) which explicitly considers only oxygen-demanding materials, water temperature, and DO, and assumes that increased DO will result in more fish, would predict that decreasing industrial discharges would increase DO concentrations, and hence increase fish. However, it may be that the fishery in question would be improved by some discharges, because the fish are indirectly able to use organic oxygen-demanding materials as energy sources. To predict this result, a model which explicitly includes a larger number of relevant variables is required.

However, even if it is true that in a particular case DO is the most important water quality parameter for a management decision, current DO models may seriously over- or underestimate DO levels. For example, large changes in discharge levels of BOD imply concurrent changes in nutrient and toxic materials discharges. Since nutrients affect oxygen production by photosynthesis, and toxic materials affect, among other things, the rate of BOD decay, the net effect of reduced BOD (and associated nutrient and toxic materials) discharge could be a more severe local oxygen deficit than previously occurred. As before, to obtain a more accurate prediction of what actually happens requires a model which explicitly includes more of the relevant variables.*

The model described below represents an attempt to include both more relevant variables and their interrelationships in a generalized estuarine water quality model in order to remedy the deficiencies noted above.

* I take issue with some of the premises expressed in this introductory section. The critique of other approaches to water quality modeling leaves the impression that such approaches have not been effective. I believe the record clearly shows that management models based on DO and other quality parameters have been extremely useful in guiding management of industrial and municipal sources which discharge oxygen-demanding wastes into streams, rivers, and estuaries.

For the Delaware River system, previous investigators have provided DO models which have been reasonably successful in predicting effects of waste treatment on oxygen levels. They have been aware of the limitations of their models which, despite imperfections, have provided a rational basis for apportioning waste loads. While it is true that these models do not consider detailed effects of algae, toxic materials, etc., on water quality in the river, I believe it is fair to state that most water pollution control agencies have little concern about growth of algae in a system where average DO is 0.5–1.0 mg l^{-1} during summer, as is the case in parts of the Delaware River today. From a management point of view, there is little or no point in trying to model effects of phosphorus discharges on fish production for a system where DO is low enough to preclude all but a few hardy fish. It is similarly futile to attempt to model subtle biological effects of nutrients in a situation where the primary water quality problem is one of large amounts of floatable solids. In other words, the critical problems have to be addressed first, and refinements can come later. This is the approach taken in the past to water quality management. In a world without critical problems, Kelly's approach would certainly have merit.

A particular point concerning counterintuitive responses of systems: I know of no practical situation where it has been demonstrated that discharge of an oxygen-demanding waste would be beneficial to fish. This would certainly not be a common occurrence. If the example is inappropriate, the general idea is still valid. Unexpected things do occur when large, complex systems are subjected to perturbations. Unfortunately, most models (such as the present one) which strive for sufficient scope of ecosystem understanding to admit this possibility are data-poor and usually have little or no practical applicability to water resources management.

Meaningful modeling of water quality for management purposes will require that the modeler have a thorough understanding of the relationship between the dominant factors that influence water quality in any particular ecosystem and the management options available to a regulatory agency.

G. FRED LEE
University of Texas at Dallas, Richardson, Texas
(continued)

II. Model Development

A. General

The model is based on an input–output or mass-balance approach. The estuary is divided into a series of cells both physically (by distance along the estuary: *reaches*) and by functional groups within the physical cells (algae, fish, oxygen: *compartments*). Within each reach, the transfer of mass between each pair of compartments is accounted for so that the time rate of change of mass in each compartment is expressed as input of materials minus output. Similarly, the time rate of change of materials between reaches is also accounted for. An overall balance for each reach for each compartment may be written as

$$dx_i/dt = y_i - \omega x_i \pm f_i(X), \tag{1}$$

where x_i is the concentration of material i within the reach (mg l^{-1}), y_i is the input of material i to the reach (mg l^{-1} day^{-1}), ωx_i is the loss of material i from the reach due to advective (net) water flow [ω is the fraction of water lost from the reach per day (day^{-1})],* and $f_i(X)$ accounts for transformations of material i to other materials, such as incorporation of nitrogen into algal mass, or incorporation of algal mass into zooplankton mass.

The model accounts for the following materials (see also Table I): algae, herbivorous zooplankton, bacteria, fish, organic matter, nitrogen, phosphorus, and oxygen. These materials comprise the eight compartments or endogenous variables. For each compartment, a single differential equation is specified which describes the change in concentration of that compartment over time as a

* Tidal dispersion is ignored. See later discussion.

Previous water quality models have served well as management tools and will continue to do so in the future. However, it is obvious that many interactions between biota and their chemical and physical environments which are important in determining water quality are assumed away in the simpler models. If counterintuitive predictions, such as increased fish production from increased organic loadings (predicted in the present model for only the first few reaches of the Delaware, and logically consistent with ideas of allochthonous versus autochthonous sources of organic matter in natural systems), can come from more comprehensive models, this indicates a need to incorporate more parts of the water quality system into management models. This is a direction for research as time and money permit. For quick answers to critical problems, obviously more complete analyses cannot be done. However, quick answers are not always sufficient answers. The continued persistence and complexity of water quality problems despite the availability of previous generations of models are enough to justify attempts at more comprehensive approaches. Inadequacies of the present models should not be equated to bankruptcy of the underlying idea that pollution problems cannot be solved definitively without recourse to whole ecosystem understanding.

ROBERT A. KELLY

TABLE I

Variables and Their Associated Symbols

Compartment (Endogenous variable)	Symbol	Exogenous variable	Symbol	Units
Algae	x_1	Temperature	θ	degrees C
Zooplankton (herbivores)	x_2	Water flow[a]	ω	day^{-1}
Bacteria	x_3	Turbidity	σ	Hellige units
Fish	x_4	Toxic materials	ϕ	$mg\ l^{-1}$
Oxygen	x_5	Reach volume	ν	m^3
Organic matter	x_6	Depth	z	m
Nitrogen	x_7	Light intensity	Ω	$ly\ sec^{-1}$
Phosphorus	x_8	Material input[b]	y	$mg\ l^{-1}\ day^{-1}$
		Reaeration constant	d	day^{-1}

[a] Water flow is expressed as a turnover rate for a specific reach; i.e., $m^3\ m^{-3}\ day^{-1}$.
[b] Milligrams of input to a reach per day, divided by its volume in liters.

function of the endogenous variables and certain exogenous variables such as advective flow (ω), temperature (θ), toxic materials (ϕ), turbidity (σ), and light intensity (Ω).

Since there are several interactions among components of the model, and since most of the interactions are nonlinear, no attempt was made to solve the model analytically. Instead, Euler's numerical integration technique was used. (For various types of numerical integration techniques, see Mankin and Brooks, 1971; IBM, 1968; Patten, 1971.) The use of this technique allows analysis of the transient response of the system. But, since the model is to be used within a larger model which is time invariant (see Spofford *et al.,* 1972), a steady-state solution can also be obtained by simulating over time until each time rate of change equation equals zero, i.e.,

$$dx_i/dt = 0, \qquad i = 1, \ \ldots, \ 8, \tag{2}$$

providing a steady-state solution exists. As an alternative, the rate of change can be set to zero initially and either Newton's method or the Gause–Seidel method (Mankin and Brooks, 1971) used to solve the nonlinear simultaneous equation set. Obtaining only a steady-state solution in some ways simplifies solution of the model.

B. Ecological Basis of the Model

Several different approaches have been utilized to model biological-ecological processes, ranging from models of single species from a behavioral aspect (Holling, 1966), to two-species interactions (Gause, 1934; Lotka, 1924), to complex systems of species (R. B. Williams, 1971). The approach taken here is closer to the last where similar functional groups are lumped together, as in the analysis framework of Lindeman (1942) and Odum (1956), but including nutrient regeneration as an important feedback control mechanism (Odum *et al.*, 1969; Odum, 1972a). Using only the most basic components of an aquatic system (because the desired output of the model and lack of data did not warrant a more complete analysis), a generalized network of material flows was constructed (Fig. 1).

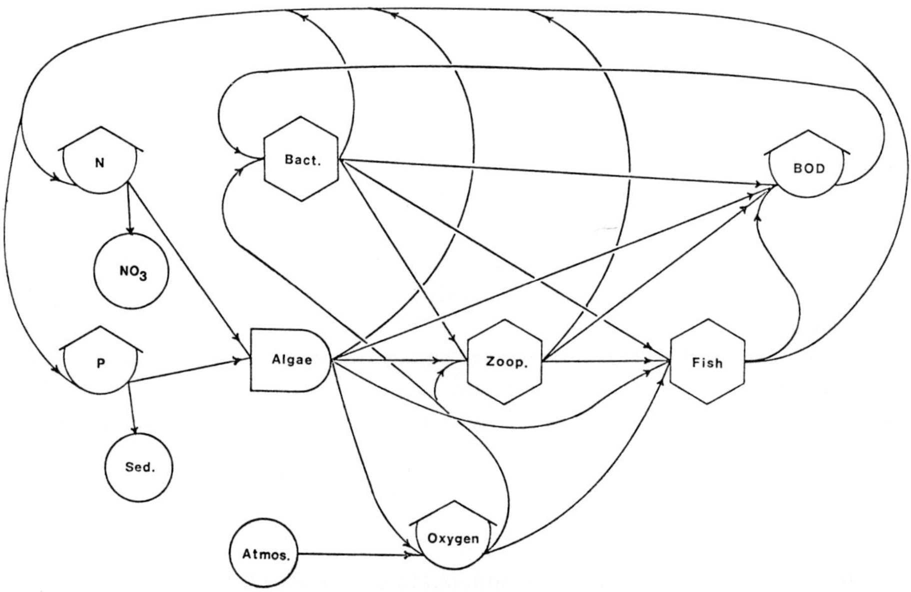

FIG. 1. Diagram of materials flows between compartments within a single reach. Source: "Ecological Modeling in a Resource Management Framework," © Resources for the Future, Inc.

The flows depicted in Fig. 1 can be divided into four classes: (i) feeding (or nutrient uptake for plants), (ii) excretion and death, (iii) respiration, and (iv) predation. Feeding is the uptake or ingestion of a material from another compartment. When the flow is described from the compartment being fed upon, it is termed predation. Death is simply loss of life, and the flows generated by death go into the organic matter compartment. Excretion (plus egestion) is the difference between material taken into a compartment and that part of the ingested material applied to growth, respiration, and death. The rate of loss of nitrogen and phosphorus from a group of organisms was assumed to be proportional to the rate of carbon loss due to respiration (see below). For phosphorus, this assumption may be valid (Hargrave and Geen, 1968); for nitrogen it probably is not.

As a final comment before discussing the mathematical forms of these processes, the transfers between all pairs of compartments, and the concentrations of the compartments themselves (with the exception of oxygen), are expressed in the total weight of nitrogen, phosphorus, and carbon per unit volume of water which is being transferred in the flow or which exists in the compartment. Thus, algae is expressed as milligrams of N plus P plus C per liter of water, as are herbivores, bacteria, fish, and organic matter. Nitrogen is expressed as milligrams of nitrogen per liter (mg N l^{-1}), phosphorus as milligrams of phosphorus per liter (mg P l^{-1}), and oxygen as milligrams of oxygen per liter (mg O_2 l^{-1}). For transfers between living materials a constant ratio of elements is assumed, based on Strickland's (1965) summary of elemental ratios for phytoplankton. The ratio assumed is 40C : 5N : 1P.* This may be the greatest weakness of the model from the standpoint of realism, but allowing elemental ratios to change would unduly complicate matters. The behavior of the model might not be greatly changed because of this assumption, for there are several compensatory tradeoffs between intracellular and extracellular storage of material in nature which affect the net balance of nutrients but not the gross balance (Odum *et al.*, 1963; F. M. Williams, 1971).

C. MATERIAL FLOWS WITHIN A REACH

In this section the nature of the functions $f_i(X)$ in Eq. (1) is discussed. These rates of transfer between compartments are due to biological activity and as such are dependent on biological and environmental factors.

It is widely recognized that there is a physical limitation to the amount of material taken up by any organism or group of organisms per unit time. This has been observed experimentally for uptake of nutrients by algae (Fuhs *et al.*, 1972; Morton *et al.*, 1972) and has been shown to be true for feeding by animals (e.g., Smith, 1969; Desert Biome, 1970). Odum (1972b) poses a theoretical argument justifying the use of a Michaelis–Menten or Monod type equation to describe this phenomenon:

$$v = k_1 S/(k_2 + S), \tag{3}$$

where v is the transfer rate, k_1 is a rate constant, k_2 is the concentration at which half the maximum rate occurs, and S is the concentration of material being eaten or taken up. Thus, the feeding rate of zooplankton, fish, and bacteria as a function of their food supply can be described by Eq. (3).

* Ratio by weight, not molecules. Ratios compare favorably with nutrient-poor Sargasso Sea zooplankton ratios (Beers, 1966).

Other factors also affect feeding rate, however. Although there are apparently few experimental data to justify the following equation types, considerations such as those presented by H. T. Odum (1967, 1968) indicate that under adverse conditions feeding rates are probably depressed and respiration rates increased because organisms must utilize their energy resources in maintaining other functions (such as detoxification in the presence of poisons). Also, since the model is designed for steady-state conditions where species replacement due to adverse conditions is not explicitly considered, the overall behavior of lumped species components of the system may be quite different from the response of individual species of organisms (see Kelly, 1971). In Fig. 2, for example, the rate of nutrient uptake by algae increases exponentially with temperature of the system, whereas the rate for each species has a definite maximum rate. There is an obvious limit to this function, but it may well be beyond the expected temperatures of the Delaware Estuary (Brock and Brock, 1966, have presented evidence that the limiting temperature may be over 50° C).

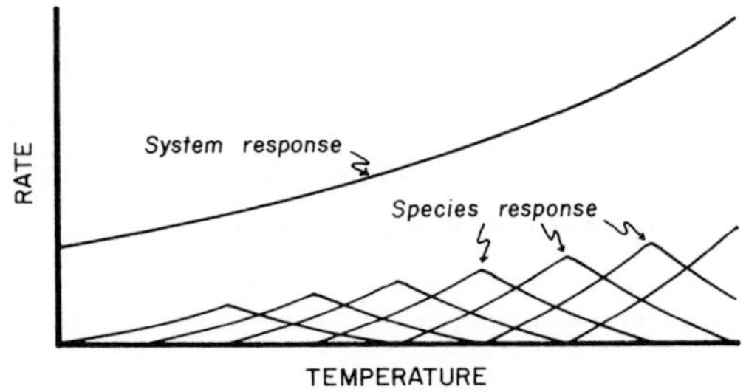

FIG. 2. System versus species response to temperature. Individual species respond to temperature changes differently. Over the restricted range where each species functions, the sum of all species' rates active at that temperature range is the system rate. The net result of species replacement along a temperature gradient (in time or space) is the curve above labeled the system rate.

The environmental variables which affect the feeding rate (as well as other rates in the model) are oxygen concentration, temperature, and toxic materials. Figure 3 shows the feeding rate as a function of each of these three factors. Notice that the y-axis is in units of a fraction of the maximum rate. Since very little work has been done on the antagonistic and synergistic effects of various environmental variables as they affect rates of transfers, for simplicity it has been assumed that simple product functions approximate the natural world relationships. The values of the functions which are depicted by these curves can be multiplied by an expression for a maximum rate to obtain the actual rate under specified conditions; that is,

$$v_{f,\, x_i,\, x_j} = a_i \cdot x_i \cdot f_1(x_j) \cdot f_2(x_5) \cdot f_3(\theta) \cdot f_4(\phi), \tag{4}$$

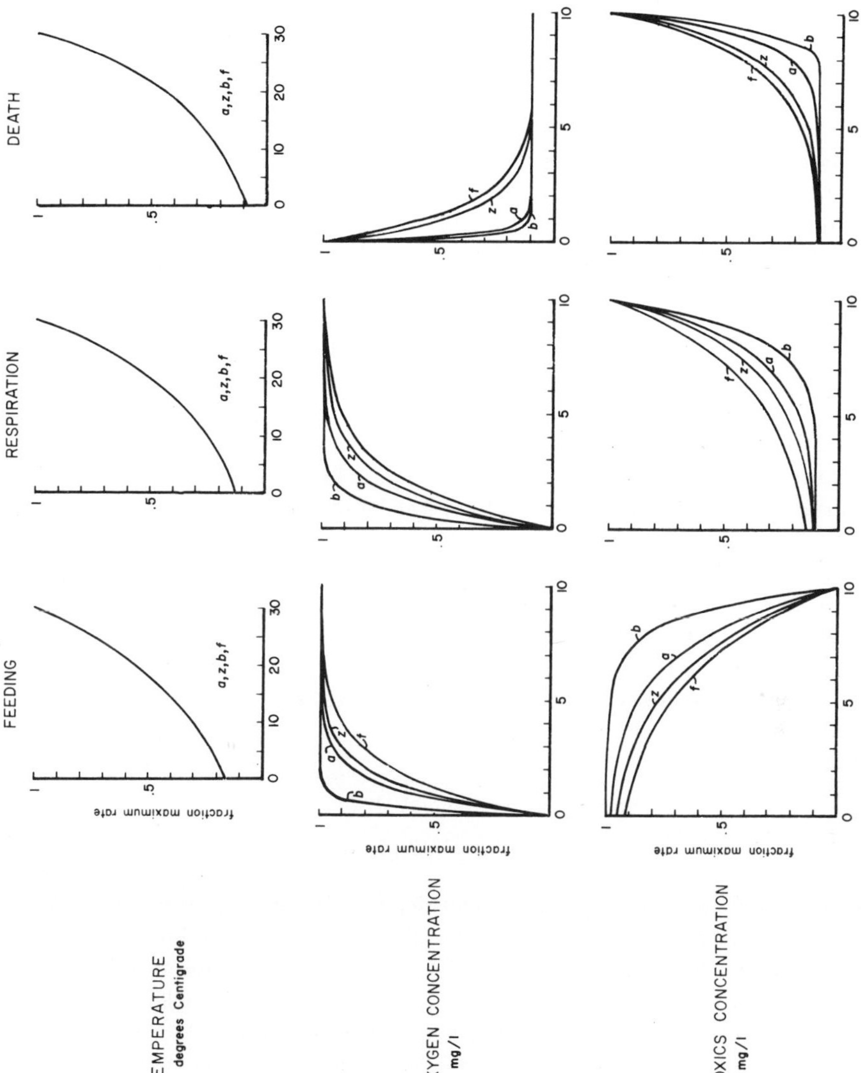

FIG. 3. Response of various rates to environmental variables. Source: "Ecological Modeling in a Resource Management Framework," © Resources for the Future, Inc.

where v_{f, x_i, x_j} is the feeding rate of species group i on species group j (mg l^{-1} day^{-1}), a_i is a feeding rate constant (day^{-1}), x_i is the concentration of species group i (mg l^{-1}), and f_1 to f_4 are the functions describing the feeding rate as modified by environmental variables (Fig. 3).

The behavior of the nutrient uptake system of algae poses a particularly difficult problem. Unlike animal feeding, where different food supplies can be exploited according to their availability (at least conceptually), plants require definite amounts of nitrogen, phosphorus, carbon, and other elements jointly in order to grow. If we assume that in the Delaware only N and P can be growth limiting, and that N and P are taken up in a specific ratio, the most straightforward approach is to express the total uptake rate as the product of two Michaelis–Menten expressions. This has been hypothesized by Chen (1970), and according to Di Toro *et al.* (1971) fits some available data adequately. The nutrient uptake rate for algae may then be expressed as

$$v_{f, x_1} = a \cdot x_1 \cdot \frac{x_7}{c_1 + x_7} \cdot \frac{x_8}{c_2 + x_8} \cdot f_2(x_5) \cdot f_3(\theta) \cdot f_4(\phi), \qquad (5)^*$$

where a is a rate constant, x_7 and x_8 are concentrations of nitrogen and phosphorus, c_1 and c_2 are half-saturation constants [as in Eq. (3)], and x_1 is the concentration of algae. However, the total uptake of inorganic materials is also determined by the rate of photosynthesis, since carbon uptake, and thus the necessity for nitrogen and phosphorus uptake, is a result of photosynthesis. Since photosynthesis is dependent upon light intensity, Eq. (5) must be expanded to include effects of turbidity, depth, and seasonal light intensity variations (Appendix I).

The other biologically mediated rate processes which are included are respiration and death. Expressions relating rates of transfer to the relevant variables are

$$v_{r, x_i} = a_i \cdot x_i \cdot r_1(x_5) \cdot r_2(\theta) \cdot r_3(\phi), \qquad (6)^\dagger$$

$$v_{d, x_i} = a_i \cdot x_i \cdot d_1(x_5) \cdot d_2(\theta) \cdot d_3(\phi), \qquad (7)$$

where v_{r, x_i} and v_{d, x_i} are respiration and death rates, respectively, a_i is a rate constant, and $r_1 - r_3$ ($d_1 - d_3$) are directly analogous to f_2 to f_4 in Eq. (4) except that they pertain to the respiration (death) rate. The graphical forms of the f, r, and d functions are presented in Fig. 3. (The mathematical expressions used to generate the curves in Fig. 3 appear in Appendix I, Table I.1.)

Since each predation term has already been described as a feeding term, only one process is left to be described: excretion. Its proper development requires that growth, as usually used in population ecology literature, also be discussed.

* The constant k_1 of Eq. (3) has been accounted for in Eq. (5) by a.

† Rate constant a_i is different for each process.

Growth is usually defined as the increase (or less often, decrease) in number of organisms or weight over time. It is a net increase or decrease, since seldom are total input–output balances attempted. Thus, the growth rate in terms of the processes already mentioned is

growth rate = feeding rate − respiration rate
− excretion (egestion) rate − death rate − predation rate.

In fact, this relationship is used for obtaining the time rate of change of each living component in the model. Ever since Lotka (1924) first explored growth from a biological point of view, a logistic or S-shaped curve has been fitted to population growth curves. The widespread use of this curve and its apparent applicability to diverse types of organisms perhaps justifies its continued use here.

The logistic curve is commonly stated in its differential form

$$dM/dt = aM[1 - (M/K)]\,, \tag{8}$$

where M is the number of individuals in (or the mass of) a population at time t, a is a rate constant, and K is the carrying capacity of the environment (the maximum number of individuals, or mass, that can be maintained by the environment under steady-state conditions (Odum, 1971; Slobodkin, 1966). Term K is usually considered as being a function of various environmental variables. It is useful to consider a new variable K^* which, instead of being variable, is constant for a given species (or species group). A physiological or behavioral significance may be attached to it in the form of biochemical inhibitors, or behaviors which inhibit reproduction, increase death, etc. Its main effect, mathematically, is to place a limit on the size any group can achieve. It is in this mathematical sense that it is used here.

Odum (1972b) treats the logistic equation in a slightly different way. In its simplest form

$$dM/dt = aMR - bM^2, \tag{9}$$

where M, t, and a are as above, and b is a/K^*. Term R now represents a variable which is related to the availability of food. It is assumed that R is equal to the last four terms of Eq. (4); that is,

$$R \equiv f_1(x_j) \cdot f_2(x_5) \cdot f_3(\theta) \cdot f_4(\phi). \tag{10}$$

By equating bM^2 in Eq. (9) with excretion, two purposes will be served: (i) food consumption will not approach zero as the population approaches its maximum size, and (ii) high excretion (egestion) rates (low assimilation rates) will be observed at high population densities. These two considerations are apparently true in real ecosystems (Lean, 1973).

In order to give a more complete picture of the nature of interactions of these transfers within a given reach, the following functions are defined:

$$f_{x_i, x_j}(x_i, x_j, x_5, \theta, \phi) = v_{f, x_i, x_j}$$
$$= p_{x_i, x_j}(x_i, x_j, x_5, \theta, \phi), \tag{11}$$
$$r_{x_i}(x_i, x_5, \theta, \phi) = v_{r, x_i}, \tag{12}$$
$$d_{x_i}(x_i, x_5, \theta, \phi) = v_{d, x_i}, \tag{13}$$
$$e_{x_i}(x_i) = b_i x_i^2, \tag{14}$$

where $p_{x_i, x_j}(x_i, x_j, x_5, \theta, \phi)$ is the predation rate on x_j by x_i, and $e_{x_i}(x_i)$ is the excretion rate of compartment i. The functional form of the model can now be written explicitly to indicate input and output terms for each compartment as a function of endogenous and exogenous variables. If interreach transfers and exogenous inputs of matter are ignored for the moment, the functional arguments can be written as in Fig. 4. The form presented corresponds exactly with Fig. 1, the diagram of material transfer pathways within a reach.

D. Material Flows across Reach Boundaries

Each reach is treated as being homogeneous. This implies that all inputs to a reach are instantaneously mixed throughout. The net effect of this assumption is to introduce a factor which partially compensates for the absence in the model of tidally influenced dispersion (see Tracor, 1971, pp. 451–456).

Inputs to each reach, both from upstream (V) and from industrial, municipal, and tributary sources (U), are initially expressed as mass per day. The vector Y [Eq. (1)] is the sum of these two sources, divided by the volume of the reach:

$$y_i^k = (u_i^k + v_i^k)/\nu^k, \qquad i = 1, \ \ldots, \ 8, \tag{15}$$

or in matrix notation

$$Y^k = (1/\nu^k)(U^k + V^k), \tag{16}$$

where $y_i^k(Y^k)$ is an input of material to reach k (mg l^{-1} day^{-1}), $u_i^k(U^k)$ is from industrial, municipal, and tributary sources (mg day^{-1}), V^k is from upstream, and ν^k is the volume of the kth reach.

The output from reach k is calculated by multiplying the endogenous variables (X mg l^{-1}) times ω^k, the fraction of water lost from reach k per day (day^{-1}). When this product is multiplied by the volume of reach k, ν^k, the vector input to the $(k + 1)$st reach, is obtained:

$$V^{k+1} = \omega^k \nu^k X^k = \omega^k \nu^k \begin{bmatrix} x_1 \\ \vdots \\ x_8 \end{bmatrix}^k . \tag{17}$$

$$\dot{x}_1 = Q_t[f_{x_1,x_8}(x_1, x_3, x_5, x_7, x_8, \theta, \sigma, \phi, z, \Omega)] - d_{x_1}(x_1, x_5, \theta, \phi) - r_{x_1}(x_1, x_5, \theta, \phi)$$
$$- e_{x_1}(x_1) - p_{x_4,x_1}(x_1, x_4, x_5, \theta, \phi) - p_{x_2,x_1}(x_1, x_2, x_5, \theta, \phi)^*$$
$$\dot{x}_2 = f_{x_2,x_1}(x_1, x_2, x_5, \theta, \phi) + f_{x_2,x_3}(x_2, x_3, x_5, \theta, \phi) - d_{x_2}(x_2, x_5, \theta, \phi) - r_{x_2}(x_2, x_5, \theta, \phi)$$
$$- e_{x_2}(x_2) - p_{x_4,x_2}(x_2, x_4, x_5, \theta, \phi)$$
$$\dot{x}_3 = f_{x_3,x_6}(x_3, x_5, x_6, \theta, \phi) - d_{x_3}(x_3, x_5, \theta, \phi) - r_{x_3}(x_3, x_5, \theta, \phi) - e_{x_3}(x_3)$$
$$- p_{x_2,x_3}(x_2, x_3, x_5, \theta, \phi)$$
$$\dot{x}_4 = f_{x_4,x_1}(x_1, x_4, x_5, \theta, \phi) + f_{x_4,x_2}(x_2, x_4, x_5, \theta, \phi) + f_{x_4,x_3}(x_3, x_4, x_5, \theta, \phi)$$
$$- d_{x_4}(x_4, x_5, \theta, \phi) - r_{x_4}(x_4, x_5, \theta, \phi) - e_{x_4}(x_4)$$
$$\dot{x}_5 = Q_o\{Q_t[f_{x_1,x_8}(x_1, x_3, x_5, x_7, x_8, \theta, \sigma, \phi, z, \Omega)] - r_{x_1}(x_1, x_5, \theta, \phi) - r_{x_2}(x_2, x_5, \theta, \phi)$$
$$- r_{x_3}(x_3, x_5, \theta, \phi) - r_{x_4}(x_4, x_5, \theta, \phi)\}^*$$
$$\dot{x}_6 = d_{x_1}(x_1, x_5, \theta, \phi) + d_{x_2}(x_2, x_5, \theta, \phi) + d_{x_3}(x_3, x_5, \theta, \phi) + d_{x_4}(x_4, x_5, \theta, \phi)$$
$$+ e_{x_1}(x_1) + e_{x_2}(x_2) + e_{x_3}(x_3) + e_{x_4}(x_4) - p_{x_3,x_6}(x_3, x_5, x_6, \theta, \phi)$$
$$\dot{x}_7 = Q_n[r_{x_1}(x_1, x_5, \theta, \phi) + r_{x_2}(x_2, x_5, \theta, \phi) + r_{x_3}(x_3, x_5, \theta, \phi) + r_{x_4}(x_4, x_5, \theta, \phi)]$$
$$- f_{x_1,x_7}(x_1, x_5, x_7, x_8, \theta, \sigma, \phi, z, \Omega)^*$$
$$\dot{x}_8 = Q_p[r_{x_1}(x_1, x_5, \theta, \phi) + r_{x_2}(x_2, x_5, \theta, \phi) + r_{x_3}(x_3, x_5, \theta, \phi) + r_{x_4}(x_4, x_5, \theta, \phi)]$$
$$- f_{x_1,x_8}(x_1, x_5, x_7, x_8, \theta, \sigma, \phi, z, \Omega)^*$$

* Total mass is in the ratio 40C : 5N : 1P, so $Q_t = 46$ [$40 + 5 + 1 = 46 \cdot$ P], $Q_n = 0.1087$ [fraction N = 5/46], $Q_p = 0.02174$ [fraction P = 1/46], $Q_o = 2.319$ [fraction C = 40/46; O_2/C = 32/12 = 2.68; $Q_o = (40/46) \cdot (32/12)$].

FIG. 4. Functional arguments of the model.

We can now specify the nearly complete model as

$$\dot{x}_i^k = \frac{\omega^{k-1} x_i^{k-1} v^{k-1} + u^k}{v^k} - \omega^k x_i^k + f_i(X)^k, \qquad \begin{matrix} i = 1, \ldots, 8, \\ k = 1, \ldots, 22, \end{matrix} \tag{18}$$

which is Eq. (1) with reaches added, and where $f_i(X)$ is the equation set presented in Fig. 4.

Fish, unlike the other components considered in the model, are capable of considerable mobility. For this reason they are not advected according to water flow. Instead, they are assumed to move to their sources of food (Chen and Orlob, 1975). For the steady-state case, the exogenous input and output terms for each reach would be equal to zero, reducing the generalized equation (18) for fish to

$$\dot{x}_4^k = f_4(X)^k. \tag{19}$$

An important factor ignored thus far is diffusion of oxygen across the air–water surface, the rate of which can be described by

$$v = d(C_S - x_5), \tag{20}$$

where v is the rate (mg l^{-1} day^{-1}), d is a (reaeration) rate constant (day^{-1}), C_s is the saturation concentration of oxygen (mg l^{-1}), and x_5 is the oxygen concentration.* This term can now be added to the equation for oxygen in Fig. 4:

$$\dot{x}_5 = Q_0\{Q_t[f_{x_1,x_8}(x_1, x_3, x_5, x_7, x_8, \theta, \sigma, \phi, z, \Omega)] - r_{x_1}(x_1, x_5, \theta, \phi) - r_{x_2}(x_2, x_5, \theta, \phi) - r_{x_3}(x_3, x_5, \theta, \phi) - r_{x_4}(x_4, x_5, \theta, \phi)\} + d(C_S - x_5). \quad (21)$$

Additionally, phosphate loss by precipitation is allowed to occur. The loss rate is proportional to concentration,

$$v = -kx_8, \quad (22)$$

where v is the rate of precipitation loss (mg l^{-1} day^{-1}), k is a (precipitation) rate constant (day^{-1}), and x_8 is the concentration of phosphorus. The equation for phosphorus in Fig. 4 thus becomes

$$\dot{x}_8 = Q_p[r_{x_1}(x_1, x_5, \theta, \phi) + r_{x_2}(x_2, x_5, \theta, \phi) + r_{x_3}(x_3, x_5, \theta, \phi) + r_{x_4}(x_4, x_5, \theta, \phi)] - f_{x_1,x_8}(x_1, x_3, x_5, x_7, x_8, \theta, \sigma, \phi, z, \Omega) - kx_8. \quad (23)$$

The equation set in Fig. 4, as modified by Eqs. (18), (19), (21), and (23), is the basic model. Further explanation and modification will follow a discussion of data requirements and sources.

III. Data

Some of the data necessary to specify this model are readily available because of earlier modeling work on the Delaware. The original data (on reach volumes, cross-sectional areas, etc.) were compiled in the Delaware Estuary Comprehensive Study (DECS, 1966) as published by Thomann (1972). The rationale for the choice of 30 reaches between Trenton, New Jersey, and Reedy Island, Delaware, has never been published, and reach lengths vary between ten and twenty thousand feet. In order to save computation time, the possibility of combining the sixteen short reaches into eight long reaches was examined. Although a small amount of detail was lost (according to output of the model described here), it was found that a 22-reach system was adequate (Fig. 5). The physical data on these 22 reaches is presented in Table II.

* The saturation value for oxygen is calculated according to Tennessee Valley Authority (1962).

FIG. 5. Map of area of concern for the Delaware Valley Regional Management Model, showing reach locations for the estuary model. Source: "Ecological Modeling in a Resource Management Framework," © Resources for the Future, Inc.

A. Water Flow

In order to specify the turnover rates of water in each reach, water flow through the reach must be determined. For this, an unpublished model available from the DRBC was used with data collected by the U.S. Geological Survey (USGS, 1970a). In essence, the drainage area of each reach is determined and multiplied by the runoff per square mile (CFSM) for gauged tributaries within the drainage area. The products thus obtained (DELQs) are added cumulatively to the base flow of fresh water at Trenton (the head of the estuary) to obtain estimated flow for each reach. The estimated flow is then corrected for major additions or withdrawals of water to or from sewage treatment plants and other water supply requirements. The results from this model for September 1970 are presented in Table II. The fraction of water lost per day due to fresh-water flow can then be calculated as flow/volume (Table II).

TABLE II

Physical Chacteristics of Estuary

Reach	Upper and lower boundaries (river miles)[a]	Depth[b] (m)	Volume[c] ($\times 10^6$ m^3)	Flow[d] ($\times 10^6$ m^3 day^{-1})	Turnover rate[e] (day^{-1})	Reaeration coefficient[f] (day^{-1})
	133.27					
1		5.68	6.85	10.3	1.50	0.31
	129.2			(4230 cfs)		
2		6.93	10.3	10.6	1.03	0.23
	125.5					
3		7.78	13.0	10.6	0.814	0.18
	121.7					
4		8.12	15.1	10.7	0.709	0.20
	117.8					
5		8.81	18.0	10.8	0.599	0.20
	114.0					
6		8.88	21.4	10.4	0.485	0.25
	110.3					
7		9.74	27.2	10.5	0.386	0.20
	106.6					
8		9.74	31.6	11.1	0.352	0.20
	102.6					
9		9.59	36.4	11.2	0.308	0.14
	98.8					
10		9.28	42.4	11.8	0.279	0.10
	95.0					
11		9.80	52.7	13.2	0.250	0.11
	91.0					
12		8.97	57.5	13.3	0.232	0.13
	87.2					
13		8.36	61.8	13.5	0.219	0.13
	83.5					
14		7.51	67.8	13.7	0.202	0.13
	79.7					
15		6.84	76.2	13.8	0.181	0.13
	75.9					
16		6.44	83.0	13.8	0.166	0.13
	71.6					
17		5.86	87.4	14.4	0.165	0.13
	68.3					
18		5.70	98.8	14.4	0.146	0.18
	64.4					
19		5.43	107	14.4	0.135	0.23
	61.3					
20		5.43	122	18.7	0.153	0.35
	57.7					
21		5.34	139	18.8	0.135	0.25
	53.8					
22		5.25	159	18.8	0.118	0.24
	50.0					

[a] Measured from mouth of Delaware Bay.
[b] Estimated from unpublished data, DRBC.
[c] Converted from Thomann (1972).
[d] Based on DELQ procedure of DRBC (see Appendix II, Table II-3) for September 1970.
[e] Flow per volume.
[f] From Thomann (1972).

B. Inputs of BOD, N, P, and O

There are several exogenous inputs of materials to each reach. These inputs arrive via four sources: (i) municipal sewage treatment plants, (ii) industrial sources, (iii) tributaries (which may include some industrial or municipal sources), and (iv) storm-water overflow.* Four different sources of data (for, unfortunately, different time periods) were used to estimate these inputs. These are discussed in the following paragraphs.

Municipal sewage treatment plant effluents of BOD were taken from DRBC (1970a) for the 1968 calendar year. Inputs of nitrogen and phosphorus were then calculated as 0.15 times BOD for N,† and 0.0625 times BOD for P‡ according to aggregate data given by Rohlich and Uttormark (1972). Oxygen inputs from these sources were assumed to be negligible. Although it would be more accurate to determine the absolute levels of N and P input directly from data on each plant, such data were not readily available, and the level of accuracy of other inputs was not sufficient to warrant the more careful compilation.

Industrial sources of materials were estimated from DRBC unpublished data sheets for each major plant outfall. However, due to the high variability in waste-water flows and many missing or presumably erroneous data points, mean BOD loads calculated for all plants over the last three or four years were compared with previously published BOD loads (DRBC, 1970a; Thomann, 1972). In some cases where total effluent loads were significantly different from previously published data (such as an order of magnitude lower) and were significantly lower than the DRBC allocated load, the discharge estimate was revised upward to previous levels. In no case were observed increases more than 20 percent for any given discharge. When BOD loads were increased, N and P

* It is recognized that a fifth, perhaps significant, source exists, direct land runoff, but without any data available for its estimation it was excluded.

† Only NH_3 and organic nitrogen are considered as inputs to the model (NO_2^- and NO_3^- inputs are ignored).

‡ There are many different forms of phosphorus, only some of which are available for algal growth. Some of the forms not immediately available can, through various chemical and biochemical processes, be converted into forms which are available. The present approach inadequately considers the nature of the aqueous environmental chemistry of phosphorus.

G. FRED LEE
University of Texas at Dallas, Richardson, Texas

When data are available as total phosphate (mg PO_4 l^{-1}) and as no other form (e.g., orthophosphate, organic phosphate, particulate and dissolved forms, etc.), there is no way to make a partition into biologically meaningful components.

ROBERT A. KELLY

loads were increased proportionately. No oxygen was assumed to enter the estuary from industrial sources.

Tributary inputs were calculated from data collected under DRBC sponsorship (stored in EPA's STORET data bank) and USGS water flow records. Concentrations of O, BOD, N, and P at the sampling station closest to the Delaware proper were multiplied by stream flow estimated at that point for the day sampled (see Appendix II, Table II.1). Stream flow was estimated by multiplying CFSM for the date of that sample at the nearest USGS gauging station on the tributary by the total drainage area of the tributary or, where there was no gauging station on a tributary, the CFSM of the closest gauging station to the DRBC sampling point. Because only three years of data are available, and these data were obtained monthly (many stations were sampled only every two or three months), an average input was calculated for the months of July, August, and September for the three-year period, eliminating those data points which had high water flows. Locations and mean estimated water flows for each tributary appear in Table III.

TABLE III

Major Tributaries: Their Locations and Flows

	Tributary	River mile	East or West	Reach	Mean flow[a] September 1970 (cfs)
1	Delaware River	134.5	–	0	4146
2	Assunpink Creek	133.8	E	0	34.4
3	Crosswicks Creek	128.4	E	2	85.3
4	Assiscunk Creek	118.3	E	4	26.6
5	Neshaminy Creek	115.6	W	5	34.6
6	Poquessing Creek	111.7	W	6	3.93
7	Rancocas Creek	111.1	E	6	166
8	Pennypack Creek	109.8	W	7	19.5
9	Pennsauken Creek	105.4	E	8	7.86
10	Frankford Creek	104.6	W	8	12.5
11	Cooper River	101.6	E	9	35.4
12	Newton Creek	96.9	E	10	12.6
13	Timber Creek	95.5	E	10	61.0
14	Schuylkill River	92.5	W	11	324
15	Woodbury Creek	91.6	E	11	10.7
16	Mantua Creek	89.7	E	12	20.8
17	Darby Creek	85.3	W	13	47.2
18	Chester Creek	82.9	W	14	32.7
19	Raccoon Creek	80.7	E	14	20.1
20	Christina River	70.7	W	17	236

[a] Calculated on an areal basis from USGS Surface Water Records. See Table II.1 in Appendix II.

Storm-water overflow inputs of BOD were taken directly from Thomann (1972) with P and N inputs calculated according to data appearing in EPA (1971, p. 180).

The inputs to each reach from each source and the total inputs to each reach are presented in Table IV.

C. Inputs of Heat, Turbidity, and Toxic Materials

Inputs of heat and suspended solids (turbidity) were calculated so that the resulting levels of these exogenous variables in each reach conformed to data observed by the DRBC. Using the STORET data (Appendix II, Table II.2), mean concentrations for the month of September 1970 (five separate sampling dates) were calculated, and the spatial distribution of these variables was plotted on a graph. Linear interpolations between each sampling station yielded a mean concentration of each variable at the midpoint of each reach. The time rate of change equation for each of these variables was set equal to zero, and the input required to attain the observed concentration was solved for.

The time rate of change of turbidity was assumed to be

$$\dot{\sigma}^k = (\omega^{k-1}\sigma^{k-1}\nu^{k-1} + u_\sigma{}^k)/\nu^k - \omega^k\sigma^k - (c_1/d^k)\sigma^k, \qquad (24)$$

where $\dot{\sigma}^k$ is the time rate of change of turbidity, the first term is the input of materials causing turbidity to reach k, the second term $(\omega^k\sigma^k)$ is loss of these materials downstream, and the last term is net loss of these materials to sedimentation.* Solving Eq. (24) at steady state for the input of materials to reach k,

$$\omega^{k-1}\sigma^{k-1}\nu^{k-1} + u_\sigma{}^k = \nu^k[\omega^k\sigma^k + (c_1/d^k)\sigma^k], \qquad (25)$$

or, since upstream inputs have already been determined, the input due to tributaries and other sources is

$$u_\sigma{}^k = \nu^k[\omega^k\sigma^k + (c_1/d^k)\sigma^k] - \omega^{k-1}\sigma^{k-1}\nu^{k-1}. \qquad (26)$$

Levels of turbidity and inputs of turbidity-causing materials are given for each reach in Table V.

In lieu of calculating a complete heat balance for each reach, a heat exchange coefficient for heat loss between water and atmosphere was estimated from data given by Thackston and Parker (1972). According to their formulation, the loss of heat per day per unit surface area of water is equal to an empirical heat exchange coefficient multiplied by the difference between an equilibrium temperature (specific for time of year and location) and the observed

* Since the reaeration coefficient is a function of water flow and depth, it must therefore be related to the level of vertical mixing in a given reach. The settling rate of materials was assumed to be inversely proportional to the mixing level.

TABLE IV Inputs of Organic Material, Nitrogen, and Phosphorus to the Delaware Estuary (Pounds per Day)[a]

Reach	Point sources								
	Industrial			Municipal			Total		
	BOD	N	P	BOD	N	P	BOD	N	P
1	0.	0.	0.	8065.	1210.	504.	8065.	1210.	504.
2	2758.	2940.	807.	1668.	250.	104.	4426.	3190.	911.
3	1111.	3169.	14.	2282.	342.	143.	3393.	3511.	157.
4	1857.	328.	30.	0.	0.	0.	1857.	328.	30.
5	2167.	202.	39.	644.	97.	40.	2811.	299.	79.
6	0.	0.	0.	825.	124.	52.	825.	124.	52.
7	0.	0.	0.	897.	135.	56.	897.	135.	56.
8	5530.	155.	31.	115525.	17329.	7220.	121055.	17484.	7251.
9	9450.	32.	164.	0.	0.	0.	9450.	32.	164.
10	23659.	4380.	552.	141925.	21289.	8870.	165584.	25669.	9422.
11	39973.	20385.	319.	4280.	642.	268.	44253.	21027.	587.
12	25340.	5121.	176.	100564.	15085.	6285.	125904.	20206.	6461.
13	63070.	17539.	455.	7093.	1064.	443.	70163.	18603.	898.
14	16914.	5359.	277.	1803.	270.	113.	18717.	5629.	390.
15	57036.	6800.	1311.	0.	0.	0.	57036.	6800.	1311.
16	0.	0.	0.	673.	101.	42.	673.	101.	42.
17	123807.	38367.	2670.	37279.	5592.	2330.	161086.	43959.	5000.
18	0.	0.	0.	335.	50.	21.	335.	50.	21.
19	12000.	12400.	1087.	0.	0.	0.	12000.	12400.	1087.
20	0.	0.	0.	1365.	205.	85.	1365.	205.	85.
21	0.	0.	0.	0.	0.	0.	0.	0.	0.
22	0.	0.	0.	0.	0.	0.	0.	0.	0.
Total	384672.	117177.	7932.	425223.	63783.	26576.	809895.	180960.	34508.

[a] Organic material is reported in pounds of 5-day biochemical oxygen demand (BOD_5) per day; nitrogen in pounds of nitrogen (N) per day; and phosphorus in pounds of phosphorus (P) per day.

[b] Industrial loads were estimated from unpublished data supplied by the Delaware River Basin Commission.

[c] Based on 1968 sewage treatment plant discharges (DRBC, 1970). Nitrogen and phosphorus loads were estimated from the BOD loads according to data presented in Rohlich and Uttormark (1972).

TABLE IV (*continued*)

Reach	Area sources						Totals, all sources		
	Tributaries[d]			Storm water[e]					
	BOD_5	N	P	BOD_5	N	P	BOD_5	N	P
1	1720.[f]	407.	514.	1360.	367.	103.	11,100	1,980	1,120
2	2090.	931.	483.	0.	0.	0.	6,520	4,120	1,390
3	0.	0.	0.	0.	0.	0.	3,390	3,510	157
4	182.	206.	238.	0.	0.	0.	2,040	534	268
5	2000.	950.	425.	0.	0.	0.	4,810	1,250	504
6	4110.	1810.	724.	0.	0.	0.	4,940	1,930	776
7	211.	159.	150.	1810.	489.	137.	2,920	782	343
8	1880.	564.	439.	12960.	3499.	982.	136,000	21,500	8,670
9	1500.	625.	179.	21260.	5740.	1612.	32,200	6,400	1,950
10	2790.	1785.	431.	9490.	2562.	719.	178,000	30,000	10,600
11	17181.	4455.	5746.	18860.	5092.	1430.	80,300	30,600	7,760
12	1430.	563.	299.	0.	0.	0.	127,000	20,800	6,760
13	2470.	1737.	1000.	1950.	526.	148.	74,600	20,900	2,050
14	879.	1012.	1037.	0.	0.	0.	19,600	6,640	1,430
15	0.	0.	0.	0.	0.	0.	57,000	6,800	1,310
16	0.	0.	0.	0.	0.	0.	673	101	42
17	11900.	3940.	1319.	8320.	2246.	631.	181,000	50,100	6,950
18	0.	0.	0.	0.	0.	0.	335	50	21
19	0.	0.	0.	0.	0.	0.	12,000	12,400	1,090
20	0.	0.	0.	0.	0.	0.	1,370	205	85
21	0.	0.	0.	0.	0.	0.	0.	0	0
22	0.	0.	0.	0.	0.	0.	0	0	0
Total	50343.	19144.	12984.	76010.	20523.	5762.	936,248	220,627	53,254

[d] See Table II.1 in Appendix II.

[e] BOD loads are based on 1964 storm-water overflow in Thomann (1972). Nitrogen and phosphorus loads have been estimated from BOD loads according to data given in EPA (1971).

[f] This quantity does not include inputs from the Delaware River above Trenton, New Jersey.

TABLE V

Concentrations and Inputs of Turbidity, Heat, and Toxic Materials[a]

Reach	A	B	C	D	E
1	7.0	192	24.0	981	2.27
2	7.3	84.2	24.1	39.2	3.13
3	8.5	163	24.2	5.69	3.54
4	8.0	124	24.4	25.9	3.59
5	8.0	160	24.6	23.1	3.40
6	7.8	135	25.3	0	3.13
7	9.0	299	25.1	21.6	2.58
8	8.5	296	25.5	98.9	2.10
9	8.3	472	26.3	76.7	1.79
10	6.8	607	26.6	115	1.48
11	9.0	1030	25.7	131	1.18
12	9.0	880	25.7	33.8	1.05
13	10.0	1080	25.3	50.3	0.96
14	10.7	1250	25.0	44.3	0.87
15	11.3	1480	24.8	48.6	0.78
16	12.0	1710	24.6	33.1	0.71
17	21.0	3410	24.3	86.2	0.66
18	23.8	2970	24.1	26.4	0.60
19	22.3	2240	23.9	22.4	0.55
20	20.8	1740	23.6	396	0.45
21	18.9	2240	23.4	1.42	0.42
22	17.0	2400	23.4	0	0.37

[a] Note: A, turbidity (ϕ), Hellige units; B, input of turbidity-causing material (10^3 lb day^{-1}); C, temperature (θ), degrees C; D, input of heat (10^9 Btu day^{-1}); E, toxic materials concentration (mg l^{-1}).

temperature of a body of water. In terms of temperature (°C), the time rate of change for temperature for each reach can be expressed as

$$\dot{\theta}^k = (\omega^{k-1}\theta^{k-1}\nu^{k-1} + u_\theta{}^k)/\nu^k - \omega^k\theta^k - (0.683/z^k)(\theta^k - \theta^*), \qquad (27)$$

where the first two terms are input and loss of heat due to advective water flow [as in Eq. (24)], 0.683 is the heat exchange coefficient, $1/z^k$ is the surface area of the reach per unit volume (m^{-1}) and θ^* is the equilibrium temperature for September for the Delaware Estuary area. By solving for steady-state and nonestuary heat sources [as in Eq. (26)],

$$u_\theta{}^k = \nu^k(\omega^k\theta^k + (0.683/z^k)\{\theta^k - \theta^*\}) - \omega^{k-1}\theta^{k-1}\nu^{k-1}. \qquad (28)$$

Temperature and heat inputs to each reach are given in Table V.

Due to the lack of data on concentrations of toxic materials (assumed to be phenols for interaction with the regional industrial models)* in the Delaware and any rationale for determining how toxic materials might behave in an estuarine situation, a constant input of 50,000 lb day^{-1} was arbitrarily assigned to each reach. Loss of toxic materials due to decay was assumed to be a constant fraction per day. Thus, behavior of toxic materials in each reach is described by

$$\dot{\phi}^k = (\omega^{k-1}\phi^{k-1}v^{k-1} + u_\phi{}^k)/v^k - \omega^k\phi^k - c\phi^k, \quad (29)$$

where c is a decay rate constant. Concentrations of toxic materials in each reach are given in Table V.

D. Determination of Coefficients for Transfer Functions

The coefficients used in the functions relating relative rates of feeding, respiration, excretion, and death to environmental variables (oxygen concentration, toxics concentrations, etc.) were determined on the basis of known relative sensitivities of the groups to these variables. For example, the effect of oxygen concentration on the respiration rate of fish is much greater than its effect on bacteria. By assuming a sensitivity for one group, the other groups' sensitivities could be estimated. The single exception to this general rule is the effect of temperature on respiration, feeding, and death.

Temperature was assumed to affect given rates equally for each species group: an approximate doubling of the rate for each 10°C rise. Death rates were assumed to increase relatively faster than respiration rates, and both of these are assumed to increase faster than feeding rates (cf. Fig. 3).

The greatest problem arose, however, in attempting to estimate realistic rates for feeding, excretion, and death. The problem is compounded by not having available any data on species composition in the Delaware Estuary, nor any routine estimates of their biomass (chlorophyll *a* concentrations give only a rough estimate of algal densities). Thus, even if it were possible to estimate these

* While there is no doubt that phenols are toxic at high concentrations, their significance in natural water systems is generally related not to toxicity but to the fact that they tend to cause tastes and odors in drinking waters upon chlorination.

G. FRED LEE
University of Texas at Dallas, Richardson, Texas

Phenols are construed as a surrogate for toxic materials in general, and are introduced into the model solely for interaction with the industrial models in the regional modeling effort (Introduction). The level of phenolic equivalents due to all toxic materials entering the Delaware is rapidly approaching or surpassing toxicity levels for at least the higher vertebrates. Phenols in the model have a perceptible influence on the system.

ROBERT A. KELLY

rates per unit biomass, the actual magnitudes of the various rates, as they occur in the Delaware, would be mere guesses.

However, in determining values for the coefficients in the transfer functions (Appendix I, Table I.1), the routine reporting of maximum and minimum oxygen concentrations at five stations in the Delaware maintained by USGS (1970b) served as a benchmark. From the maximum and minimum oxygen concentrations, rough estimates of total oxygen production and consumption could be obtained (see, for example, Whitworth and Lane, 1969). By knowing the approximate algal biomass from chlorophyll *a* data, estimates of oxygen consumption from other organisms could be obtained. With known respiration rates for other groups of organisms (as contained in Desert Biome, 1970), estimates of nonalgal biomass could be obtained. Then, by making assumptions about what happens between the USGS sampling stations, feeding, death, and excretion rates were manipulated to achieve realistic behavior of the model.

It must be made clear at this point that this procedure will not in general yield a unique set of coefficients (i.e., more than one set of coefficients could give equally good "fits" to observed data). Much more research, both in the laboratory and in the field, must be done so that reasonable estimates of these rates can be made. Many integrated modeling and field research programs undertaken through the International Biological Program (IBP) at various sites in the United States and abroad are adding to the knowledge of these rates so that better estimates can be made in the future.*

* A general comment seems in order in the context of data sources and derivations. Water quality is often determined by extreme values, not averages. It is generally related to the effect of certain constituents in the water on one or more beneficial uses. For many beneficial uses, especially those related to a healthy ecosystem, there may be just one or two short-term events each year which determine overall water quality for the entire period. For example, it is not average trace metal concentration which is toxic to benthic organisms, but extremes. A pulse during the course of one evening can completely devastate a population. Under these circumstances there would be no relationship between parameters measured for a trace metal over an annual cycle and the presence or absence of a particular group of organisms. Similarly with respect to eutrophication, it is not generally average numbers of algae to which the public responds. It is the extreme events associated with a bloom, such as piling up of *Cladophora* on the shore, or tastes and odors in drinking water, that register. There seems little likelihood that modeling will in the near future be able to predict reliably either magnitudes or frequencies of dramatic events affecting water quality. As attempts are made to model more subtle effects of chemicals, heat and other inputs to water quality systems, a much better understanding of magnitudes, timing, and variability will need to be incorporated before models can be effective for management purposes.

G. FRED LEE
University of Texas at Dallas, Richardson, Texas

Predicting extreme events is a problem, but stochastic or deterministic introduction of input events such as oil spills, pesticide releases or effluent surges into a model to assess

IV. Some Results and Discussion

Over two hundred runs of the model have been made to date in an attempt to obtain output of the model which is both intuitively realistic and matches, within reason, the known conditions for the Delaware Estuary for September 1970. Of these runs, the "best" fit is presented in Fig. 6. The coefficients for the endogenous functions used in this run are given in Appendix I, Table I.2.

Several points can be made about this run. In general, the pattern (and for the upper reaches of the estuary, the absolute level) of oxygen concentration shows close agreement. The serious disagreement with observed data below reach 16 is probably due to the lack of inclusion of tidal dispersion in the model. As one gets closer to the Bay, as in the lower reaches, tidal dispersion effects should increase [and, according to Thomann (1972), do]. With dispersion included, organic matter entering reach 17 would be spread over two or three reaches, thus eliminating the large drop in DO observed in reach 17.

It is not clear why the BOD predicted by the model behaves so erratically while its closely associated variable, oxygen, behaves so smoothly. The general pattern of BOD is at least realistic, even though the predicted absolute levels do not conform well to observed data.

Predicted phosphorus concentrations agree well with observed data except at the upper end of the estuary. Since input of P to the first reach was calculated from scant data (one observation), a strong case can be made that this input is not representative of conditions prevailing during September 1970 by a factor of two to three.

The observed nitrogen concentration is in terms of total Kjeldahl nitrogen (NH_3 plus organic nitrogen), whereas the predicted nitrogen concentration is in terms of total nitrogen (NH_3 and organic nitrogen plus nitrite and nitrate), since nitrification is not included in the model. O'Connor and Thomann (1971) and Thomas Tuffey (pers. comm.) have indicated that the conversion of NH_3 to NO_2 and NO_3 proceeds at substantial rates in the estuary. If this mechanism were included in the model, the predicted nitrogen concentration of NH_3 plus organic N would be analogous to the measured Kjeldahl nitrogen concentration, and would thus resemble the observed data more closely.

Little can be said about the rest of the output except with regard to algae. The DRBC had stopped taking chlorophyll data by September 1970, so no direct comparisons could be made. However, in examining previous years' data under similar water flow and temperature conditions, the general pattern of

---

effects is one of the more obvious uses of a model. If critical periods in the life of an ecosystem with respect to major perturbing influences can be identified, then perhaps special precautions during these times may minimize dramatic water quality events.

ROBERT A. KELLY

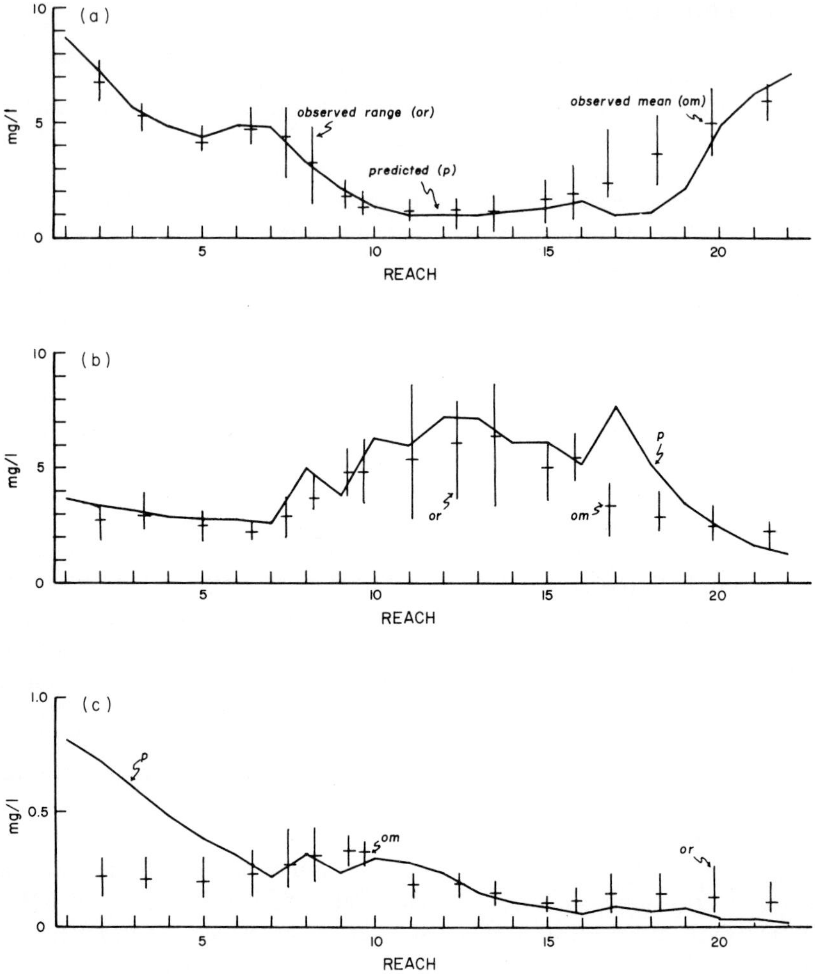

FIG. 6. Predicted and observed values of endogenous variables for September 1970: (a) oxygen concentration; (b) BOD concentration; (c) total phosphate concentration, as P.

increasing algal concentrations in the middle of the estuary with lower concentrations at either end is obvious. With regard to zooplankton, bacteria, and fish, there is no way to assess how close these predictions come to actual estuarine conditions. Bacteria and zooplankton concentrations may be too high by a factor of two or three, and fish concentration is probably too high by an order of magnitude. However, attempts to lower concentrations through increased respiration, excretion, and death rates have so far yielded very

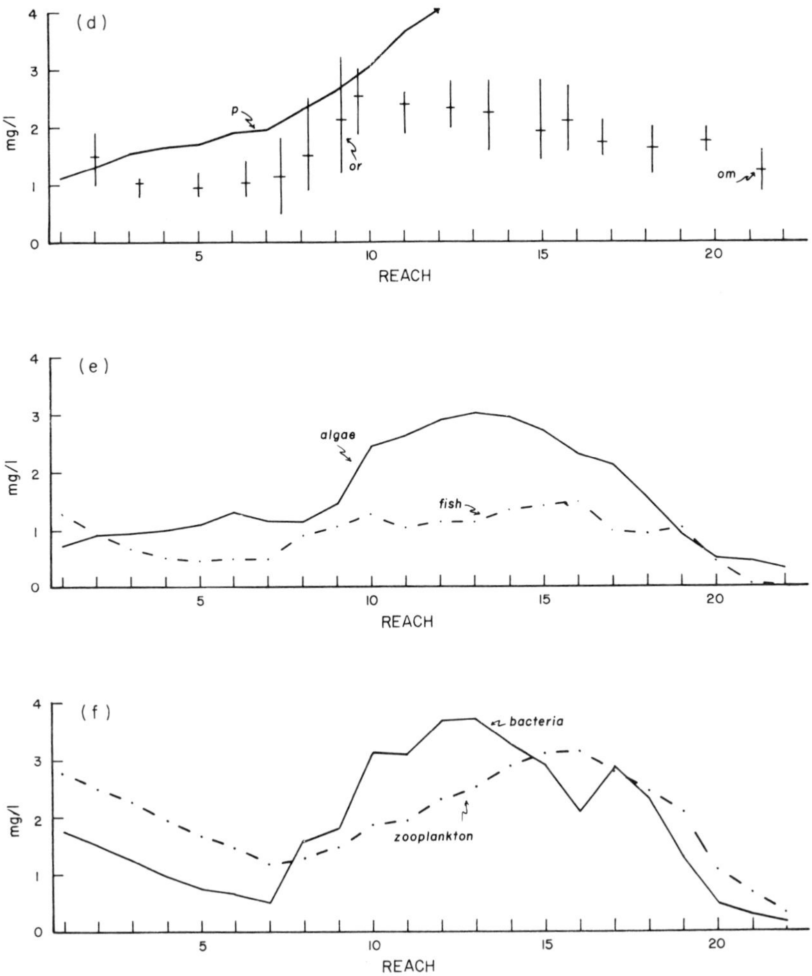

FIG. 6. Predicted and observed values of endogenous variables for September 1970: (d) total Kjeldahl nitrogen concentration; (e) predicted algal and fish concentrations; (f) predicted bacterial and zooplankton concentrations.

unsatisfactory results. The task would be easier if biomass data could be obtained for these species groups.

V. Suggestions for Further Research

There are essentially three areas in which more work must be done before models of this type can be truly predictive:

First, the effects of environmental conditions on feeding, excretion, and death rates of many different species of organisms must be examined. The most important environmental variables to be considered are probably oxygen concentration, temperature, and toxic materials. There is a general lack of information on effects of sublethal concentrations of any material, let alone those which appear important in controlling biological activity in natural environments (pesticides, heavy metals, PCBs, etc.). Consequently, the functional forms which approximate these responses must be guessed. Such guesses are based solely on intuition and thus must be highly subjective.

Second, the distribution of functional groups of organisms in time and space must be better understood. On an aggregate level, biomass determinations of phytoplankton would be much more useful than simple counts. Several other determinations, such as ATP or DNA concentrations, would be useful in estimating metabolic activity or living biomass because of the inability to equate a diatom cell with a blue-green algae cell on a simple numbers basis. Of course, obtaining such data would require a massive amount of work, but if more data of this type were available, accurate ecosystem models could be built which could serve as management tools.

Finally, much more information must be gathered, in detail, on the flows of materials between different components of natural ecosystems. These flows should not be restricted to simple food web determinations, although food web diagrams can serve as useful inputs to ecosystem analysis procedures. Rather, biotic–abiotic interactions as well as chemical–physical interactions must be included. Only by elucidation of the major flows of materials through an ecosystem, and the way that these flows are regulated, can ecosystem models become reliable as management tools.

Appendices

Appendix I. Functional Forms of Equations and Coefficients Used in Model Run Presented

Nutrient Uptake by Algae

In many natural waters, light intensity is observed to decline exponentially with depth. This relationship is commonly expressed in one of two equivalent forms (see examples in Patten, 1968):

$$d\Omega/dz = -k\Omega, \tag{I.1}$$

where Ω is the light intensity in ly $\sec^{-1}$, k is an extinction coefficient, and z is depth in meters, or

$$\Omega_z = \Omega_0 e^{-kz}, \tag{I.2}$$

where Ω_z is the light intensity at depth z, and Ω_0 is the light intensity at the surface.

Although it is well known that the rate of photosynthesis (as measured by oxygen evolution or carbon uptake) is some function of light intensity, the shape of the function is in dispute. Riley (1946) assumed a linear relationship, and Rodhe (1966) presented some evidence to support this view. However, in Rodhe's case, the vertical heterogeneity of phytoplankton was ignored, and thus his results could be misleading.

Theoretical results of Odum (1972b) and empirical observations of Blackmun (cited in Odum, 1971) and Shelef (1968) suggest that the general form

$$Ps = k_1 \Omega/(k_2 + \Omega), \tag{I.3}$$

where k_1 and k_2 are constants, and Ps is the rate of photosynthesis (mg O_2 $l^{-1}hr^{-1}$), is adequate for most purposes. Steele (1966) has suggested a different form for high light intensities, but in the turbid Delaware system this consideration can probably be ignored.

If Eq. (I.3) represents the response of the rate of photosynthesis to light intensity, and Eq. (I.2) the light intensity at any depth, the rate of photosynthesis at a given depth z is

$$Ps_z = k_1 \Omega_0 e^{-kz}/(k_2 + \Omega_0 e^{-kz}). \tag{I.4}$$

By integrating over depth to get the integral rate of photosynthesis and dividing this rate by depth, the mean rate of photosynthesis for all depths is obtained:

$$\bar{P}s = \frac{1}{z_f}\int_0^{z_f} Ps = \frac{1}{z_f}\int_0^{z_f} \frac{k_1 \Omega_0 e^{-kz}}{k_2 + \Omega_0 e^{-kz}}\,dz, \tag{I.5}$$

or

$$\bar{P}s = -\frac{k_1}{z_f k}\ln\left(k_2 + \Omega_0 \exp(-kz_f)\right)\Big|_0^{z_f}, \tag{I.6}$$

or

$$\bar{P}s = \frac{k_1}{z_f k} \ln \frac{k_2 + \Omega_0}{k_2 + \Omega_0 \exp(-kz_f)}. \tag{I.7}$$

By further assuming that the extinction coefficient k can be partitioned among various components—algae, bacteria, and turbidity—the effect of light on photosynthesis can be expressed as

$$\bar{P}s = \frac{k_1}{z_f(0.05 + 0.17x_1 + 0.10x_3 + 0.10\sigma)} \times \ln \frac{430}{30 + 400 \exp[-z_f(0.05 + 0.17x_1 + 0.10x_3 + 0.10\sigma)]}. \tag{I.8}$$

Equation (I.8) is then multiplied by Eq. (5) to obtain the absolute rate of carbon uptake.

TABLE I.1

Functions Used as Rate Reduction Functions (Fig. 4)

Environmental variable	Notation	Function
	In Eq. (4)	
Feeding rate		
Food (or nutrient) concentration	$f_1(x_j)$	$x_j/(c_{i,j} + x_j)$
Oxygen concentration	$f_2(x_5)$	$1 - \exp(-c_i x_5)$[a]
Temperature	$f_3(\theta)$	$c_1 \exp(c_2\theta)$[b]
Toxics concentration	$f_4(\phi)$	$1 - \exp(-c_{1,i}(c_{2,i} - \phi))$[a]
Respiration rate	In Eq. (5)	
Oxygen concentration	$r_1(x_5)$	$1 - \exp(-c_i x_5)$
Temperature	$r_2(\theta)$	$c_1 \exp(c_2\theta)$
Toxics concentration	$r_3(\phi)$	$0.1 + \exp(c_{1,i}(\phi - c_{2,i}))$
Death rate	In Eq. (6)	
Oxygen concentration	$d_1(x_5)$	$0.1 + \exp(-c_{1,i}(x_5 + c_{2,i}))$
Temperature	$d_2(\theta)$	$c_1 \exp(c_2\theta)$
Toxics concentration	$d_3(\phi)$	$0.1 + \exp(c_{1,i}(\phi - c_{2,i}))$

[a] Terms c_i or $c_{1,i}$ and $c_{2,i}$ are species specific coefficients; c_i (as well as $c_{1,i}$ and $c_{2,i}$) in the feeding rate function are different in value from those in the respiration and the death rate functions.

[b] Terms c_1 and c_2 are the same for all species groups, but differ for the feeding. respiration, and death rate.

TABLE I.2

Values of Coefficients Used in Rate Functions

Compartment	Oxygen		Temperature		Toxics		Constant a_i
	c_i[a]		c_1	c_2	$c_{1,i}$	$c_{2,i}$	[Eq. (4)]
Feeding rate							
Algae	3.0		0.17146	0.058779	0.4	10.0	6.0
Zooplankton	0.8		0.17146	0.058779	0.3	10.0	1.5
Bacteria	1.0		0.17146	0.058779	0.9	10.0	3.3
Fish	0.6		0.17146	0.058779	0.25	10.0	1.2
Respiration rate							[Eq. (6)]
Algae	0.8		0.12500	0.069315	0.5	10.2107	8.0
Zooplankton	0.6		0.12500	0.069315	0.4	10.2634	2.3
Bacteria	1.5		0.12500	0.069315	0.8	10.1317	4.5
Fish	0.5		0.12500	0.069315	0.3	10.3512	1.0
Death rate	$c_{1,i}$	$c_{2,i}$					[Eq. (7)]
Algae	4.0	0.02634	0.09391	0.078846	1.0	10.1054	1.0
Zooplankton	0.9	0.11710	0.09391	0.078846	0.6	10.1749	2.0
Bacteria	5.0	0.02107	0.09391	0.078846	3.0	10.0351	3.0
Fish	0.7	0.15050	0.09391	0.078846	0.5	10.2634	0.5

Excretion rate	b_i [Eq. (14)]
Algae	0.04
Zooplankton	0.05
Bacteria	0.09
Fish	0.10

Feeding rate[b] To	From	$c_{i,j}$ (Table I.1)
x_2	x_1	4.0
x_2	x_3	1.25
x_4	x_1	7.0
x_4	x_2	2.0
x_4	x_3	4.5
x_3	x_6	3.0
x_1	x_7	0.151
x_1	x_8	0.015

Boundary inputs[c]	
y_1	4900
y_2	30000
y_3	20600
y_4	0
y_5	100000
y_6	35000
y_7	10700
y_8	9000
y_ϕ	6850
y_σ	0
y_θ	0

[a] Term c_i refers to coefficient so labeled in Table I.1.
[b] For coefficient listed in first row of Table I.1.
[c] Inputs to reach 1 from upstream (kg day^{-1}).

(Phosphorus precipitation constant, k [Eq. (23)] = 0.1; equilibrium temperature, θ^* [Eq. (27)] = 23.3; toxic decay constant, c [Eq. (29)] = 0.4; sedimentation constant, c_1 [Eq. (24)] = 0.2.)

Appendix II. Data on Tributaries Used for Calculating Materials Inputs and Maintstream Water Flows

TABLE II.1

Data Used to Calculate Water Flows and Materials Inputs for Each Tributary

Tributary	Station number		Drainage area (mi^2)			Ratio[c]
	USGS	DRBC	USGS	Used[a]	Added[b]	
1 Delaware at Trenton	1463500	332079	6780	6780		1.0
2 Assunpink Creek	1464000	332081	89.4	90.8		1.02
3 Crosswicks Creek	1464500	332082	83.6	139		1.66
4 Assiscunk Creek	1464500*	332085	83.6	43.3		0.518
5 Neshaminy Creek	1465500	422104	210	233		1.11
6 Poquessing Creek	1465798	422101	21.4	21.4		1.0
7 Rancocas Creek				342[d]		
South Branch	1465850	332089	53.3	154		2.86
North Branch	1467000	332092	111	156	32	1.41
8 Pennypack Creek	1467048	422098	49.8	56.0		1.12
9 Pennsauken Creek				35.1		
South Branch	1467081	332095	9.16	14.3		1.56
North Branch		332097	not used			
10 Frankford Creek	1467088	422092	33.8	36.1		1.07
11 Cooper River	1467150	332100	17.4	40.6		2.33
12 Newton Creek	1467150*	332101	17.4	14.4		0.828
13 Timber Creek				62.8[e]		
Little Timber	1467150*	332107	17.4	3.8		0.218
South Branch	1467150*	332104	17.4	25.4	14.5	1.46
North Branch	1467150*	332105	17.4	19.1		1.10
14 Schuylkill River	1474500	422107	1893	1909		1.01
15 Woodbury Creek	1467150*	332108	17.4	12.3		0.707
16 Mantua Creek	1475000	332110	6.75	50.8		7.53
17 Darby Creek	1475510	422089	37.4	78.6[f]		1.46
Cobbs Creek	1475550		22.0	24.0		1.09
18 Chester Creek	1477000	422095	61.1	66.2		1.08
19 Raccoon Creek	1477120	332113	29.9	45.7		1.53
20 Christina River				565		
Brandywine Creek	1481500	101035	314	329		1.05
Red Clay	1480000		47.0	54[g]		1.15
White Clay	1479000		87.8	108		1.23
Christina River	1478000	101026	20.5	236[h]		3.61

[a] Drainage areas used are from DRBC (1970b) unless otherwise indicated.

[b] Drainage areas added are of the main stem of the tributary not included in areas listed.

[c] Factor by which USGS gauged water flows were multiplied to obtain total flows. Where tributaries branched, total flows for each branch were calculated. These flows were added to obtain total tributary flow. Material fluxes were then summed across branches and converted, on an areal basis, to total watershed inputs to the Estuary.

[d] Areas for North and South Branches estimated from Elson T. Killam Associates (1969).

[e] Calculated from U.S. Army Corps of Engineers (1969).

[f] Area listed is total area of Darby Creek, including Cobbs Creek.

[g] Red Clay is a tributary to White Clay. Areas calculated from Delaware Valley Regional Planning Commission (1969).

[h] Area listed includes Red Clay and White Clay watersheds.

* Indicates USGS station, not on the tributary listed, whose flow was used to calculate flow for that tributary.

TABLE II.2

Sampling Stations on Delaware Estuary Maintained by DRBC

Sampling station	STORET number	River mile	Reach
Trenton Water Filtration Plant	332079		0
Fieldsboro, New Jersey	332061	127.48	2
Florence, New Jersey	332058	122.49	3
Burlington–Bristol Bridge	892080	117.80	4–5
Torresdale, Philadelphia	892077	110.70	6
Palmyra, New Jersey	332055	107.08	7
Northeast STP, Philadelphia	892074	103.96	8
Ben Franklin Bridge	892071	100.15	9
Wharton Street, Philadelphia	892068	98.51	10
Navy Yard, Philadelphia	892065	93.18	11
Paulsboro, New Jersey	332052	87.90	12
Eddystone, Pennsylvania	892062	83.98	13
Marcus Hook, Pennsylvania	332049	78.07	15
Oldmans Point, New Jersey	332046	74.88	16
Cherry Island, Delaware	091011	70.96	17
New Castle, Delaware	091008	65.96	18
Pea Patch Island, Delaware	091005	60.55	20
Reedy Island, Delaware	091002	54.94	21

TABLE II.3

Water Flow Calculation for the Delaware Estuary

An increment in water flow (Δq_i, ft^3 sec^{-1}) for each reach may be calculated by multiplying a drainage area increment ($a_{i,j}$, mi^2) by surface runoff (f_j, ft^3 mi^{-2} sec^{-1}) and adding or subtracting major diversions (b_i, ft^3 sec^{-1}):

$$\Delta q_i = \sum_{j=1}^{18} (a_{i,j} \cdot f_j) + b_i,$$

or, in matrix notation,

$$\Delta Q = A \cdot F + B.$$

Total flow for any given reach is then

$$q_i = q_0 + \sum_{j=1}^{i} \Delta q_j,$$

where q_0 is the base flow at Trenton (ft^3 sec^{-1}).

The A matrix and B vector used in this calculation are presented on the facing page. The numbers used for column headings refer to the *3*rd–*6*th digits in the USGS identification code for each tributary listed in Table II.1, from which surface runoff data is obtained.

TABLE II.3 (*continued*)

Reach	6400	6450	6550	6585	6700	6715	7450	7500	7551	7555	7700	7800	7900	8000	8150	*a*	*b*	*c*	*d*
1	92.5		5.1																31
2		185.9	4.3																
3		7.1	17.4																
4		49.8	14.8																
5		7.8	245.2																
6			28.3	213.9	111.0														−309
7			61.7	14.9		30.7													
8						7.8	36.9												240
9						48.5	8.2												
10						10.0	3.3	62.0											185
11							1913	14.5											205
12								61.9		6.2									
13								19.0	37.4	22.0	81.7								
14											75.1					58.8			
15											14.3					46.6			
16															15.0	3.9			
17												20.5	87.8	47.0	314	4.1		82.1	
18															4.6			6.9	
19																3.3		32.3	
20																18.7		30.5	1700
21																65.2		14.6	
22																	5.5	80.6	

[a] Salem River, station no. 1482500.
[b] Alloways Creek, station no. 1483000.
[c] Mean CFSM of station nos. 7900, 8000, and 8150.
[d] Constant *B* vector, diversions.

Appendix III. Fortran IV Computer Program Used to Solve Model, with Sample Output

```
      BLOCK DATA
      COMMON/ALLSUB/X(11),YU(11),F(11),XSAVE(13,22)
      COMMON/BKDATA/OMEGA(22),DEPTH(22),DIFC(22),VOLUME(22),CTRANS(8),
     1 CL(4),COR(4),COF(4),COD1(4),COD2(4),CTR1(4),CTR2(4),CTF(4),
     2 CTD1(4),CTD2(4),AR(4),AF(4),AD(4),B(4),CONS(4),YINPT(22,6)
C          INITIAL CONDITIONS
      DATA X/0.74,2.79,1.78,1.28,8.73,3.60,1.12,0.81,2.27,6.98,24.00/,
C          INPUTS FROM UPSTREAM
     1 YU/.490E7,.300E8,.206E8,0.0,.100E9,.350E8,.107E8,.900E7,.685E7,
     2    0.0,0.0/,  YINPT/
C          INPUTS FROM EXOGENOUS SOURCES (X1.E3, EXCEPT LAST TWO)
C--BOD
     5 11.1,6.52,3.39,2.04,4.81,4.94,2.92,136.,32.2,178.,80.3,
     6 127.,74.6,19.6,57.0,.673,181.,.335,12.0,1.37,0.00,0.00,
C--TKN
     7 1.98,4.12,3.51,.534,1.25,1.93,.782,21.5,6.40,30.0,30.6,
     8 20.8,20.9,6.64,6.80,.101,50.1,.050,12.4,.205,0.00,0.00,
C--P
     9 1.12,1.39,.157,.268,.504,.776,.343,8.67,1.95,10.6,7.76,
     A 6.76,2.05,1.43,1.31,.042,6.95,.021,1.09,.085,0.00,0.00,
C--TOXICS
     B 22*50.,
C--TURBIDITY  (X1.E5)
     C 1.92,.842,1.63,1.24,1.60,1.35,2.99,2.96,4.72,6.07,10.3,
     D 8.80,10.8,12.5,14.8,17.1,34.1,29.7,22.4,17.4,22.4,24.0,
C--TEMPERATURE (X1.E4)
     E 98.1,3.92,.569,2.59,2.31,0.00,2.16,9.89,7.67,11.5,13.1,
     F 3.38,5.03,4.43,4.86,3.31,8.62,2.64,2.24,39.6,.142,0.00/
C          PHYSICAL CHARACTERISTICS OF WATER BODY
C--VOLUMES OF REACHES (X1.E7)
      DATA VOLUME/ .685,1.03,1.30,1.51,1.80,2.14,2.72,3.16,3.64,4.24,
     1   5.27,5.75,6.18,6.78,7.62,8.30,8.74,9.88,10.7,12.2,13.9,15.9/
C--REAERATION COEFFICIENTS
      DATA DIFC/ .31,.23,.18,.20,.20,.25,.20,.20,.14,.10,.11,
     1           .13,.13,.13,.13,.13,.13,.18,.23,.35,.25,.24/
C--DEPTHS
      DATA DEPTH/5.68,6.93,7.78,8.12,8.81,8.88,9.74,9.74,9.59,9.28,9.80,
     1           8.97,8.36,7.51,6.84,6.44,5.86,5.70,5.43,5.43,5.34,5.25/
C--WATER TURNOVER RATE
      DATA OMEGA/1.50,1.03,.814,.709,.599,.485,.386,.352,.308,.279,.250,
     1           .232,.219,.202,.181,.166,.165,.146,.135,.153,.135,.118/
C          COEFFICIENTS FOR ECOLOGICAL MODEL
C--MICHAELIS CONSTANTS
      DATA CTRANS/ 4.00,2.00,1.25,3.00,7.00,.151,.015,4.50/,
C--OXYGEN EFFECT ON FEED (COF) AND RESPIRATION (COR)
     1 COF / 3.0,0.8,1.0,0.6/,     COR / 0.8,0.6,1.5,0.5/,
C--OXYGEN ON DEATH (COD1) AND TOXICS EFFECT ON FEED (CTF)
     2 COD1/ 4.0,0.9,5.0,0.7/,     CTF / 0.4,0.3,0.9,.25/,
C--TOXICS ON DEATH (CTD1) AND RESPIRATION (CTR1)
     3 CTD1/ 1.0,0.6,3.0,0.5/,     CTR1/ 0.5,0.4,0.8,0.3/,
C--FEEDING RATE CONSTANT (AF) AND RESPIRATION RATE CONSTANT (AR)
     4 AF  / 6.0,1.5,3.3,1.2/,     AR  / 8.0,2.3,4.5,1.0/,
C--DEATH RATE CONSTANT (AD) AND PARTIAL LIGHT EXTINCTION COEFFICIENTS (CL)
     5 AD  / 1.0,2.0,3.0,0.5/,     CL  / .05,.17,0.1,0.1/,
C--EXCRETION RATE CONSTANTS (B) AND MISCELLANEOUS CONSTANTS (CONS)
     6 B   / .04,.05,.09,0.1/,     CONS/ 0.4,0.1,23.3,0.2/,
```

```
C--OTHER BIOLOGICAL CONSTANTS
       7 COD2/   .0263,   .1171,   .0211,   .1505/,
       8 CTR2/ 10.2107,10.2634,10.1317,10.3512/,
       9 CTD2/ 10.1054,10.1749,10.0351,10.2634/
        END
        COMMON/ALLSUB/X(11),YU(11),F(11),XSAVE(13,22)
        COMMON/BKDATA/OMEGA(22),DEPTH(22),DIFC(22),VOLUME(22),BIOL(68),
       1        YINPT(22,6)
        REAL*8 LABEL(10)
        DIMENSION ISAVE(13,22)
        EQUIVALENCE (XSAVE(1),ISAVE(1))
        READ (5,500) LABEL
C
C          CONVERT BLOCK DATA VALUES TO PROPER ORDER OF MAGNITUDE
C
        DO 10 I=1,22
           VOLUME(I)=VOLUME(I)*10000000.
           YINPT(I,6)=YINPT(I,6)*10000.
           YINPT(I,5)=YINPT(I,5)*100000.
     10 CONTINUE
        DO 12 I=1,4
           DO 11 J=1,22
              YINPT(J,I)=YINPT(J,I)*1000.
     11    CONTINUE
     12 CONTINUE
C
C          OVERALL LOOP TO SOLVE FOR STATE OF SYSTEM, REACH BY REACH
C
        DO 400 IR0=1,22
           V=VOLUME(IR0)
C--CALCULATE CONVERSION FACTOR FOR POUNDS OF INPUT TO MG/L OF INPUT
           CONVFP=454./V
C--CALCULATE CONVERSION FACTOR FOR BTU'S TO DEGREES CENTIGRADE EQUIVALENT
           CONVFT=252./V
C--CALCULATE INPUTS TO REACH IR0 IN TERMS OF MG/L
           DO 101 I=1,5
              YINPT(IR0,I)=YINPT(IR0,I)*CONVFP
    101    CONTINUE
C--CALCULATE INPUTS TO REACH IR0 IN TERMS OF DEGREES CENTIGRADE
           YINPT(IR0,6)=YINPT(IR0,6)*CONVFT
C--SUBTRACT FROM N & P INPUTS N & P CONTENT OF BOD INPUT
           YINPT(IR0,2)=YINPT(IR0,2)-.1087*YINPT(IR0,1)
           YINPT(IR0,3)=YINPT(IR0,3)-.02174*YINPT(IR0,1)
C--SUM INPUTS FROM UPSTREAM WITH EXOGENOUS INPUTS
           DO 102 I=6,11
              YU(I)=YU(I)/V+YINPT(IR0,I-5)
    102    CONTINUE
C--CONVERT OTHER UPSTREAM INPUTS TO CONCENTRATION
           DO 103 I=1,5
              YU(I)=YU(I)/V
    103    CONTINUE
           ISAVE(1,IR0)=IR0
C
C          SOLUTION ALGORITHM FOR REACH IR0
C
C--OBTAIN FIRST DERIVATIVE VALUES (F)
           CALL VALUES (IR0)
C--FIND MAXIMUM RATIO OF TIME DERIVATIVES TO VARIABLE
           TEMP=0.
           DO 201 I=1,8
```

```
                  IF (X(I).EQ.0.) GOTO 201
                  T=ABS(F(I)/X(I))
                  IF (T.GT.TEMP) TEMP=T
      201     CONTINUE
C--SET TIME STEP TO CHANGE FASTEST CHANGING VARIABLE BY HALF ITS VALUE
              TSTEP=.5/TEMP
C--UNLESS TSTEP IS GREATER THAN ONE DAT
              IF (TSTEP.GT.1.) TSTEP=1.
C--ITERATE TO STEADY STATE
              DO 204 ITER=1,300
                  CALL VALUES (IR0)
                  DO 202 I=1,11
                     X(I)=X(I)+TSTEP*F(I)
      202         CONTINUE
C--CHECK FOR STEADY STATE
                  DO 203 I=1,11
                     IF (ABS(F(I)).GT.0.0001) GOTO 204
      203         CONTINUE
C--MET STEADY STATE CHECK
                  ISAVE(2,IR0)=ITER
                  GOTO 300
      204     CONTINUE
C--OSCILLATING OR FAR FROM SOLUTION
              GOTO 499
C
C             SOLUTION FOR REACH IR0 OBTAINED
C
      300     DO 301 I=1,11
                 XSAVE(I+2,IR0)=X(I)
      301     CONTINUE
C--CALCULATE INSTREAM INPUTS TO NEXT REACH
              TEMP=OMEGA(IR0)*V
              DO 302 I=1,11
                 YU(I)=X(I)*TEMP
      302     CONTINUE
C
C             END OF OVERALL LOOP
C
      400 CONTINUE
C
C             PRINT RESULTS
C
          WRITE (6,600) LABEL,XSAVE
          WRITE (6,601) (I,(YINPT(I,J),J=1,6),I=1,22)
          STOP
      499 STOP 500
      600 FORMAT ('1',10A8//' STATE OF SYSTEM (MG/L)'//' REACH   ITRTNS',6X,
         .'ALGAE    ZOOP    BACT    FISH     DO      BOD',6X,'N',7X,'P',6X,
         .'TOX      TURB    TEMP'//(I4,I9,5X,11F8.2))
      601 FORMAT (//' INPUTS (MG/L/DAY)'//' REACH',6X,'BOD',6X,'N',7X,'P',
         .6X,'TOX     TURB     TEMP'//(I4,4X,6F8.3))
      500 FORMAT (10A8)
          END
          SUBROUTINE VALUES (IR0)
          COMMON/ALLSUB/X(11),YU(11),F(11),XSAVE(13,22)
          COMMON/BKDATA/OMEGA(22),DEPTH(22),DIFC(22),VOLUME(22),CTRANS(8),
         1 CL(4),COR(4),COF(4),COD1(4),COD2(4),CTR1(4),CTR2(4),CTF(4),
         2 CTD1(4),CTD2(4),AR(4),AF(4),AD(4),B(4),CONS(4),YINPT(22,6)
          REAL*4 OF(4),TF(4),FEE(4),OR(4),TR(4),RES(4),OD(4),TD(4),DTH(4),
         1 EXC(4),PRE(4),YD(11)
C--TEMPERATURE, TOXICS, AND DO
```

```
      TP=X(11)
      TX=X(9)
      OX=X(5)
C--WATER FLOW, DEPTH, AND REAERATION COEFFICIENT
      O=OMEGA(IRO)
      Z=DEPTH(IRO)
      D=DIFC(IRO)
C--OXYGEN SATURATION CALCULATION
      OSAT=14.652+(-.41022+(.007991-.000077774*TP)*TP)*TP
C--TEMPERATURE EFFECT ON FEEDING, RESPIRATION, AND DEATH
      TEMF=.17146*EXP(.058779*TP)
      TEMR=.12500*EXP(.069315*TP)
      TEMD=.09391*EXP(.078846*TP)
C--LIGHT EXTINCTION COEFFICIENT = SUM OF PARTIALS
      DENOM=CL(1)+CL(2)*X(1)+CL(3)*X(3)+CL(4)*X(10)
      XLGT=2.66/(Z*DENOM)
C--FOR LIVING MATERIALS CALCULATE OXYGEN AND TOXIC EFFECTS, AND MULTIPLY
C     TOGETHER TO OBTAIN FEED, RESP, DEATH RATES; CALCULATE EXCRETION
C     RATES AND LOSSES DOWNSTREAM (YD)
      DO 100 I=1,4
         TEMP=X(I)
         OF(I)=1.-EXP(-COF(I)*OX)
         TF(I)=1.-EXP(-CTF(I)*(10.-TX))
         FEE(I)=AF(I)*TEMP*OF(I)*TF(I)*TEMF
         OR(I)=1.-EXP(-COR(I)*OX)
         TR(I)=.1+EXP(CTR1(I)*(TX-CTR2(I)))
         RES(I)=AR(I)*TEMP*OR(I)*TR(I)*TEMR
         OD(I)=.1+EXP(-COD1(I)*(OX+COD2(I)))
         TD(I)=.1+EXP(CTD1(I)*(TX-CTD2(I)))
         DTH(I)=AD(I)*TEMP*OD(I)*TD(I)*TEMD
         EXC(I)=B(I)*TEMP*TEMP
         YD(I)=O*TEMP
  100 CONTINUE
C--CALCULATE MICHAELIS EXPRESSIONS
      TRAN1=X(1)/(CTRANS(1)+X(1))*FEE(2)*0.5
      TRAN2=X(2)/(CTRANS(2)+X(2))*FEE(4)*0.333333
      TRAN3=X(3)/(CTRANS(3)+X(3))*FEE(2)*0.5
      TRAN4=X(1)/(CTRANS(4)+X(1))*FEE(4)*0.333333
      TRAN5=X(6)/(CTRANS(5)+X(6))*FEE(3)
      TRAN6=X(3)/(CTRANS(8)+X(3))*FEE(4)*0.333333
      TRAN7=XLGT*X(7)*X(8)/((CTRANS(6)+X(7))*(CTRANS(7)+X(8)))*FEE(1)
C--CALCULATE FEEDING AND PREDATION RATES FOR LIVING COMPONENTS
      FEE(1)=TRAN7
      FEE(2)=TRAN1+TRAN3
      FEE(3)=TRAN5
      FEE(4)=TRAN2+TRAN4+TRAN6
      PRE(1)=TRAN1+TRAN4
      PRE(2)=TRAN2
      PRE(3)=TRAN3+TRAN6
      PRE(4)=0.
C--CALCULATE FIRST DERIVATIVES
      RESP=0.
      EXCR=0.
      DETH=0.
      DO 101 I=1,4
         F(I)=FEE(I)+YU(I)-RES(I)-PRE(I)-EXC(I)-DTH(I)-YD(I)
         RESP=RESP+RES(I)
         EXCR=EXCR+EXC(I)
         DETH=DETH+DTH(I)
  101 CONTINUE
      F(4)=F(4)-YU(4)+YD(4)
```

```
      TEMP=FEE(1)-RESP
      F( 5)=YU( 5)-O*X( 5)+2.319*TEMP+D*(OSAT-OX)
      F( 6)=YU( 6)-O*X( 6)+EXCR+DETH-FEE(3)
      F( 7)=YU( 7)-.1087*TEMP-O*X(7)
      F( 8)=YU( 8)-.02174*TEMP-(O+CONS(4))*X(8)
      F( 9)=YU( 9)-(O+CONS(1))*X(9)
      F(10)=YU(10)-(O+CONS(2)/D)*X(10)
      F(11)=YU(11)-O*X(11)-(.683/Z)*(X(11)-CONS(3))
      RETURN
      END
```

TEST RUN FOR ECOLOGICAL MODELING SYMPOSIUM.

STATE OF SYSTEM (MG/L)

REACH	ALGAE	ZOOP	BACT	FISH	DO	BOD	N	P	TOX	TURB	TEMP
1	0.74	2.78	1.79	1.28	8.71	3.59	1.12	0.81	2.27	6.98	24.00
2	0.91	2.48	1.50	0.94	7.16	3.34	1.30	0.72	3.13	7.29	24.10
3	0.92	2.21	1.25	0.66	5.59	3.06	1.54	0.60	3.54	8.50	24.20
4	1.00	1.92	0.99	0.51	4.71	2.75	1.62	0.48	3.59	3.01	24.40
5	1.07	1.64	0.78	0.45	4.33	2.62	1.72	0.33	3.40	3.01	24.60
6	1.29	1.44	0.64	0.51	4.37	2.63	1.91	0.31	3.13	7.79	25.25
7	1.19	1.17	0.49	0.46	4.80	2.45	1.99	0.22	2.58	8.99	25.14
8	1.15	1.21	1.62	0.92	3.20	4.80	2.35	0.31	2.10	8.50	25.53
9	1.47	1.42	1.82	1.04	2.06	3.67	2.62	0.24	1.79	8.30	26.35
10	2.47	1.80	3.12	1.25	1.30	6.08	3.04	0.30	1.48	6.80	26.56
11	2.64	1.90	3.10	1.00	1.00	5.81	3.63	0.28	1.13	8.97	25.69
12	2.96	2.23	3.67	1.14	1.00	7.03	4.01	0.24	1.05	8.99	25.34
13	3.05	2.47	3.75	1.15	0.96	7.04	4.59	0.15	0.96	9.99	25.20
14	2.97	2.32	3.26	1.35	1.17	5.87	4.88	0.11	0.87	10.67	24.97
15	2.76	3.09	2.89	1.44	1.30	5.87	5.09	0.08	0.73	11.30	24.84
16	2.32	3.11	2.04	1.46	1.59	5.05	5.33	0.06	0.71	12.01	24.62
17	2.12	2.67	2.82	0.98	0.98	7.34	6.34	0.09	0.66	20.99	24.31
18	1.58	2.40	2.28	0.94	1.14	4.95	6.75	0.07	0.60	23.82	24.11
19	0.94	2.01	1.25	1.02	2.24	3.37	7.53	0.08	0.55	22.32	23.90
20	0.50	1.01	0.47	0.43	5.00	1.95	6.04	0.04	0.45	20.78	23.60
21	0.44	0.60	0.25	0.06	6.33	1.54	6.12	0.03	0.42	13.89	23.40
22	0.34	0.31	0.15	0.00	7.17	1.27	6.21	0.02	0.37	16.99	23.35

INPUTS (MG/L/DAY)

REACH	BOD	N	P	TOX	TURB	TEMP
1	0.736	0.051	0.058	3.314	12.725	36.039
2	0.237	0.150	0.055	2.204	3.711	0.959
3	0.118	0.110	0.003	1.746	5.692	0.110
4	0.061	0.009	0.007	1.503	3.728	0.432
5	0.121	0.018	0.010	1.261	4.036	0.323
6	0.105	0.030	0.014	1.061	2.864	0.0
7	0.049	0.008	0.005	0.835	4.991	0.200
8	1.954	0.097	0.082	0.718	4.253	0.739
9	0.402	0.036	0.016	0.624	5.887	0.531
10	1.906	0.114	0.072	0.535	6.499	0.683
11	0.692	0.133	0.052	0.431	8.873	0.626
12	1.003	0.055	0.032	0.395	6.948	0.148
13	0.548	0.094	0.003	0.367	7.934	0.205
14	0.131	0.030	0.007	0.335	8.370	0.165
15	0.340	0.004	0.000	0.293	8.318	0.161
16	0.004	0.000	0.000	0.273	9.353	0.100
17	0.940	0.158	0.016	0.260	17.713	0.249
18	0.002	0.000	0.000	0.230	13.648	0.067
19	0.051	0.047	0.004	0.212	9.504	0.053
20	0.005	0.000	0.000	0.136	6.475	0.818
21	0.0	0.0	0.0	0.163	7.316	0.003
22	0.0	0.0	0.0	0.143	6.853	0.0

Acknowledgments

The model development was undertaken as an integral part of the research program of the Quality of the Environment Program at Resources for the Future, Inc. I would like to thank Blair T. Bower, Clifford S. Russell, Walter O. Spofford, Jr., all of Resources for the Future, and Robert V. O'Neill of Oak Ridge National Laboratory for comment and criticism on earlier drafts of this manuscript.

REFERENCES

Beers, J. R. (1966). *Limnol. Oceanogr.* **11,** 520.

Brock, T. D., and Brock, M. L. (1966). *Nature (London)* **209,** 733.

Chen, C. W. (1970). *J. San. Eng. Div. ASChE* **96(SA5),** 1085.

Chen, C. W., and Orlob, G. T. (1975). *In* "Systems Analysis and Simulation in Ecology" (B. C. Patten, ed.), Vol. III, pp. 476–588. Academic Press, New York.

DECS (1966). Delaware Estuary Comprehensive Study: Preliminary Results and Findings. Fed. Water Pollut. Contr. Agency Rep., Dept. of the Interior.

Delaware Valley Regional Planning Commission (1969). The Regional Water Supply and Water Pollution Control Plans. DVRPC Plan Rep. 4, Philadelphia, Pennsylvania.

Desert Biome (1970). "Summaries of Recorded Information on Aquatic Biota" (G. W. Minshall, ed.). Idaho State Univ., Pocatello, Idaho.

DiToro, D. M., O'Connor, D. J., and Thomann, R. V. (1971). *Amer. Chem. Soc. Advan. Chem. Ser.* **106,** 131.

DRBC (1970a). Delaware Estuary and Bay Water Quality Sampling and Mathematical Modeling Project. Final Progr. Rep. to Fed. Water Pollut. Contr. Admin.

DRBC (1970b). Biological and Water Quality Investigation of Tributaries to the Delaware Estuary-Bay. Unpubl. Rep.

EPA (1971). "Storm Water Management Model." U.S. Govt. Printing Office, Washington, D.C.

Fuhs, G. W., Demmerle, S. D., Canelli, E., and Chen, M. (1972). Characterization of Phosphorus-Limited Plankton Algae (with reflections on the limiting-nutrient concept), *Amer. Soc. Limnol. Oceanogr. Spec. Symp.* **1,** 113.

Gause, G. F. (1934). "The Struggle for Existence." Reprinted (1964) Hafner, New York.

Hargrave, B. T., and Geen, G. H. (1968). *Limnol. Oceanogr.* **13,** 332.

Holling, C. S. (1966). *Mem. Entomol. Soc. Can.* 1.

IBM (1968). System/360 Continuous System Modeling Program (360A-CX-16X) User's Manual. Int. Business Machines, Inc., White Plains, New York.

Kelly, R. A. (1971). The Effects of Fluctuating Temperature on the Metabolism of Freshwater Microcosms. Doctoral dissertation, Univ. of North Carolina, Chapel Hill, North Carolina.

Killam (Elson T.) Ass. Inc. (1969). A Master Sewerage Plan for Burlington County, New Jersey. Milburn, New Jersey.

Lean, D. R. S. (1973). *Science* **179,** 678.

Lindeman, R. L. (1942). *Ecology* **23,** 399.

Lotka, A. J. (1924). "Elements of Mathematical Biology." Reprinted (1956) Dover, New York.

Mankin, J. B., and Brooks, A. A. (1971). Numerical Methods for Ecosystem Analysis. Oak Ridge Nat. Lab. Publ. ORNL-IBP-71-1.

Morton, S. D., Sernau, R., and Derse, P. M. (1972). *Amer. Soc. Limnol. Oceanogr. Spec. Symp.* **1**, 197.

O'Connor, D. J., and Thomann, R. V. (1971). *In* Tracer Inc., Estuarine Modeling: An Assessment, Rep. to EPA, No. 16070-DZV-02-71, pp. 102–173. NTIS, U.S. Dept. of Commerce, Washington, D.C.

Odum, E. P. (1971). "Fundamentals of Ecology," 3rd ed. Saunders, Philadelphia, Pennsylvania.

Odum, H. T. (1956). *Ecol. Monogr.* **27**, 55.

Odum, H. T. (1967). *In* "Pollution and Marine Ecology" (T. A. Olson and F. J. Burgess, eds.), pp. 99–157. Wiley (Interscience), New York.

Odum, H. T. (1968). Work circuits and system stress, *Proc. Conf. Mineral Cycling,* pp. 81–138. Univ. of Maine Press, Bangor, Maine.

Odum, H. T. (1972a). "Environment, Power, and Society." Wiley (Interscience), New York.

Odum, H. T. (1972b). *In* "Systems Analysis and Simulation in Ecology" (B. C. Patten, ed.), Vol. I, pp. 139–211. Academic Press, New York.

Odum, H. T., Beyers, R. J., and Armstrong, N. E. (1963). *J. Mar. Res.* **21**, 191.

Odum, H. T., Nixon, S. W., and DiSalvo, L. H. (1969). *In* "The Structure and Function of Freshwater Microbial Communities" (J. Cairns, Jr., ed.), pp. 1–30. Virginia Polytechnic Inst. and State Univ., Blacksburg, Virginia.

Patten, B. C. (1968). *Int. Rev. Ges. Hydrobiol.* **53**, 357.

Patten, B. C. (1971). *In* "Systems Analysis and Simulation in Ecology" (B. C. Patten, ed.), Vol. I, pp. 3–122. Academic Press, New York.

Riley, G. A. (1946). *J. Mar. Res.* **6**, 54.

Rodhe, W. (1966). *Mem. Ist. Ital. Idrobiol. Suppl.* **18**, 365.

Rohlich, G. A., and Uttormark, P. D. (1972). *Amer. Soc. Limnol. Oceanogr. Spec. Symp.* **1**, 231.

Russell, C. S., and Spofford, W. O. (1972). *In* "Environmental Quality Analysis: Theory and Method in the Social Sciences" (A. V. Kneese and B. T. Bower, eds.). The Johns Hopkins Press, Baltimore, Maryland.

Shelef, G. (1968). Kinetics of Algal Biomass Production System with Respect to Light Intensity and Nitrogen Concentration. Doctoral dissertation, Univ. of California, Berkeley, California.

Slobodkin, L. B. (1966). "Growth and Regulation of Animal Populations." Holt, New York.

Smith, F. E. (1969). *In* "Eutrophication: Causes, Consequences, Correctives," pp. 631–645. Nat. Acad. of Sci., Washington, D.C.

Spofford, W. O., Russell, C. S., and Kelly, R. A. (1972). Operational Problems in Large-Scale Residuals Management Models. *Conf. Econ. Environm.,* sponsored by Univ.–Nat. Bur. Committee for Econ. Res., New York, and Resources for the Future, Inc.

Steele, J. H. (1968). *Mem. Ist. Ital. Idrobiol. Suppl.* **18**, 383.

Strickland, J. D. H. (1965). *In* "Chemical Oceanography" (J. P. Riley and G. Skivow, eds.), Vol. I, pp. 478–610. Academic Press, New York.

Tennessee Valley Authority (1962). Prediction of Stream Reaeration Rates. TVA, Chattanooga, Tennessee.

Thackston, E. L., and Parker, F. L. (1972). *J. Water Pollut. Contr. Fed.* **44**, 1334.

Thomann, R. V. (1972). Systems Analysis and Water Quality Management. Environmental Res. and Appl. Inc., New York.

Tracor, Inc. (1971). Estuarine Modeling: An Assessment. Rep. to Environmental Protection Agency, No. 16070-DZV-02/71. NTIS, U.S. Dept. of Commerce.

U.S. Army Corps of Engineers (1969). Flood Plain Information, Little Timber Creek, Philadelphia, Pennsylvania.

USGS (1970a). Water Resources for *State*, Part 1, Surface Water Records. Available for Pennsylvania, New Jersey, Maryland, and Delaware.

USGS (1970b). Water Resources Data for *State*, Part 2, Water Quality Records. Available for Pennsylvania, New Jersey, Maryland, and Delaware.

Whitworth, W. R., and Lane, T. H. (1969). *Limnol. Oceanogr.* **14**, 53.

Williams, F. M. (1971). *In* "Systems Analysis and Simulation in Ecology" (B. C. Patten, ed.), Vol. I, pp. 198–267. Academic Press, New York.

Williams, R. B. (1971). *In* "Systems Analysis and Simulation in Ecology" (B. C. Patten, ed.), Vol. I, pp. 543–582. Academic Press, New York.

2

Protein from the Sea: A Comparison of the Simulated Nitrogen and Carbon Productivity of the Peru Upwelling Ecosystem

JOHN J. WALSH* AND STEVEN O. HOWE*

DEPARTMENT OF OCEANOGRAPHY
UNIVERSITY OF WASHINGTON, SEATTLE, WASHINGTON

I. Introduction

Present simulation models of ocean ecosystems are a primitive attempt to describe quantitatively and understand biological production and trophic level interaction within the marine habitat with the eventual promise of utility as a resource management tool (Walsh, 1972). We have concentrated on systems analyses of the quantitative effects of nutrient input to eutrophic and oligotrophic marine systems in our IBP Upwelling Ecosystem study, a small-scale marine counterpart to the large-scale terrestrial Biome programs described in Volume III. Consideration of the future availability of marine protein sources

Contribution No. 889 from the Department of Oceanography, University of Washington.

* Present address: Division of Oceanographic Sciences, Brookhaven National Laboratory, Upton, Long Island, New York.

for human consumption leads to important qualitative as well as quantitative questions, such as the relative carbohydrate to protein composition of organisms in marine food chains. This preliminary simulation analysis explores some possible implications of phytoplankton C/N ratios as an index to the quality of one of the world's richest ecosystems, the Peru upwelling ecosystem.

The most important component of the C/N ratio is nitrogen. Although CO_2 may or may not be limiting in alpine (Gale, 1972), forest (Odum, 1970), lake (Shapiro, 1973), and tide-pool (Byers, 1963) ecosystems, it is unlikely to be limiting in any well-buffered marine ecosystem where the dissolved carbon concentration is at least two orders of magnitude greater than phytoplankton carbon (Wangersky, 1965). In contrast, nitrogen compounds are recycled more slowly than those of phosphorus in oligotrophic (Thomas, 1970) and eutrophic (Ryther and Dunstan, 1971) marine waters, and thus may determine the overall productivity of an area through regulation of protein synthesis. Silicate availability may additionally determine phyletic composition of the phytoplankton. The degree to which the carbohydrate/protein food quality of any individual plant species is altered by nitrogen deficiency and/or carbon storage will affect its suitability as an appropriate diet for either terrestrial (Witkamp, 1966) or marine (Menzel, 1959) herbivores.

Laboratory-induced nitrogen deficiency has generated high phytoplankton C/N atomic ratios of 10–45 (Caperon and Meyer, 1972; Hobson and Pariser, 1970), while in situ phytoplankton populations have lower ratios of ~5–7 in nearshore upwelling areas off Peru (Walsh, 1975), Southwest Africa (Hobson, 1971), California (Holm-Hansen *et al.*, 1966), and Oregon (Small and Ramberg, 1971). The usual problems of detrital contamination of field measurements of particulate matter in the sea are presumably obviated in computing the relative particulate C/N ratio. This carbohydrate and fat to protein index (the usual practice is to multiply phytoplankton N x 6.25 to obtain protein content) of natural populations reflects the coupling of photosynthesis and nutrient uptake. The low particulate C/N ratios of inshore upwelling ecosystems would suggest that nitrogen depletion is not sufficiently severe in these areas to uncouple photosynthesis from nutrient uptake and protein synthesis, with consequent higher carbon storage in carbohydrates and fats (Fogg, 1959; Holm-Hansen *et al.*, 1959).

Comparison of independent ^{14}C and ^{15}N estimates of *net* photosynthetic and nutrient fluxes in these areas, however, do not in general yield a $^{14}C/^{15}N$ uptake ratio of 5–7, but instead a ratio of at least 10–14, or twice that of the particulate material (MacIsaac and Dugdale, 1972; McCarthy, 1972). These isotope incubation experiments may not be estimates of net production. Not all of the nitrogen flux may have been measured, or photosynthetic uncoupling in upwelling ecosystems may be activated by depletion of other nutrients. The role of nutrient limitation, light regulation, and coupling of photosynthesis and nutrient uptake in proscribing natural phytoplankton C/N ratios, and thus perhaps subsequent food chain quality, is examined in two quasi-independent nitrogen and carbon simulation models of the Peru upwelling ecosystem.

II. Methods

Two independent data sets taken at the same time of year at the same place off Punta San Juan, Peru, but three years apart, were used to build and test the nitrogen and carbon models. The drogue cruise track of R/V Anton Bruun in March–April 1966 (validation data) and of the grid and one of the underway temperature maps of the R/V Thomas G. Thompson in March–April 1969 (input data) are shown in Fig. 1. It is not the purpose of this paper to describe these Peru investigations; detailed discussion and appropriate references can be found in Ryther *et al.* (1971) and Walsh *et al.* (1971). Differences between the present two models are in the biological assumptions governing nitrogen and carbon flow through the lower trophic levels of an upwelling ecosystem. Diatomaceous phytoplankton standing crop, spatial extent, upwelled input of nutrients, light availability, and the threshold of nocturnal grazing losses to higher trophic levels are treated the same.

Both models are two-layered, x–y spatial hypotheses directed along the axis of an offshore, 10-m deep Ekman layer with a lower boundary reservoir of high nutrients and low phytoplankton seed. The offshore extent of each model is a 100-km diagonal, 30° to the coast, with a 10-km grid point spacing. The alongshore width is 30 km, with 10-km grid-point spacing to allow for lateral boundary calculations along the axis of the Ekman layer. The spin-up time scale of the models, i.e., steady state, is 10 days, with diel periodicity of nutrient uptake, photosynthesis, grazing, and vertical herbivore migration simulated with periodic functions.

A series of nonlinear, time-varying, partial differential equations are written for the state variables light, nitrate, recycled nitrogen, phosphate, silicate, phytoplankton, zooplankton, and fish in terms of both particulate nitrogen and carbon. An example of the state equations, such as for nitrate in a right-handed Cartesian coordinate system, is

$$\frac{\partial \mathrm{NO_3}}{\partial t} = -\frac{\partial u\mathrm{NO_3}}{\partial x} - \frac{\partial w\mathrm{NO_3}}{\partial z} - \frac{\partial v\mathrm{NO_3}}{\partial y} + K_x\frac{\partial^2 \mathrm{NO_3}}{\partial x^2} + K_z\frac{\partial^2 \mathrm{NO_3}}{\partial z^2} + K_y\frac{\partial^2 \mathrm{NO_3}}{\partial y^2} - \frac{(V_{\max})(\mathrm{NO_3})(\mathrm{PN})}{K_t + \mathrm{NO_3}}.$$

The first three terms represent advection, the second three diffusion, and the last term is phytoplankton uptake of nitrate. Fortran subroutines are written to alter different variables such as $V_{\max}$ to incorporate more biological reality in

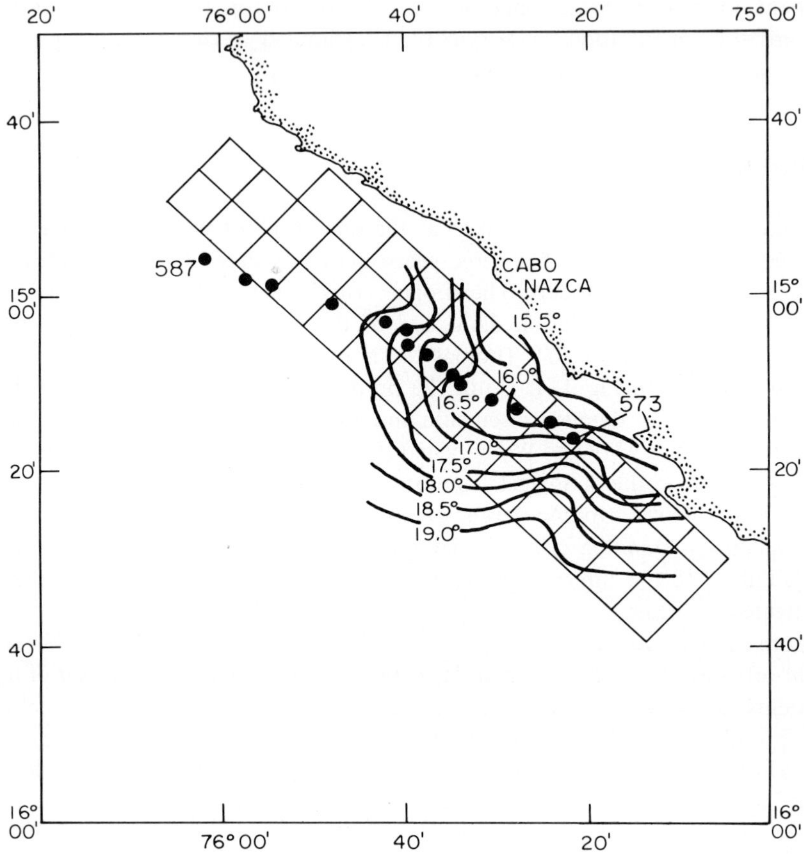

FIG. 1. Peru temperature plume and grid of the Thompson (input data) and drogue track stations (solid circles) of the Anton Bruun (validation data) off Punta San Juan, Peru.

complex submodels at each time step (Walsh and Dugdale, 1972). Further description of parameter evaluation, details of derivation of the particular advection and diffusion fields, and herbivore grazing stress assumptions off Peru in March are described elsewhere (Walsh, 1975). The advection, diffusion, and grazing terms of the carbon and nitrogen phytoplankton equations are the same. Equations are integrated in finite-difference form by an Adams-Bashforth predictor scheme to obtain steady-state solutions after 240 iterations of a one-hour simulated time step. These are not stiff equations (with fast and slow time constants), for herbivore biomass changes are assumed to be zero during the 10-day simulation period. The algorithm is thus numerically stable.

Assumptions of the nitrogen and carbon models are discussed below.

A. Nitrogen Model

The euphotic zone depth (100–1% light level) is below 10 m off Punta San Juan. Unlike the carbon model, nutrient uptake is not a function of light with depth. Nitrogen flux depends on light available during the day, however, with diel periodicity of nutrient uptake expressed as a sinusoid. The nitrogen model allows for ammonia and other recycled nitrogen suppression of nitrate uptake, excretory input to the nutrient pool, and Michaelis–Menten multiple nutrient limitation by nitrogen, phosphorus, or silicate. An average uptake rate of 0.08 h^{-1} over a 12-h day is assumed for each nutrient based on chemostat and shipboard experiments. Respiration is considered a carbon loss, and choice of the limiting nutrient is based on ambient nutrient concentrations and their ratio in healthy unlimited diatoms. Individual diatom species or phyletic changes to dinoflagellates, coccolithophores, or microflagellates are not considered.

B. Carbon Model

Mean net photosynthetic production is calculated as a function of simulated daytime light intensity. Steele's (1962) function for photosynthesis is analytically integrated (Vollenweider, 1965) and then vertically averaged over the light available each hour from 0–10 m at each grid point. Respiration and carbon excretion are included implicitly in the growth expression. Average net photosynthetic carbon uptake rate over 12 h is assumed to be 0.063 h^{-1} based on observations made on the Thompson cruise (Barber *et al.*, 1971), and only silicate is considered as a possible limiting nutrient.

Additional details of the mathematical formulation and definition of state variables are given in the Appendix.

III. Results

Walsh (1975) discusses the relative importance of nonlinear interaction of grazing thresholds, multiple nutrient limitation, and diel periodicity in considering spatial distributions of nutrients and phytoplankton off Punta San Juan, Peru. Figures 2 and 3 from the present 10 x 1 spatial nitrogen model are similar to results obtained in a previous three-dimensional simulation analysis (Walsh, 1975). The circles in the two figures are, respectively, observed nitrate and particulate nitrogen computed from observed particulate phosphorus (PN = 16 x PP) over 0–10 m depth at each of the Anton Bruun drift stations. The triangles connected by line segments are predicted nitrate and particulate nitrogen for each grid point at local time of the closest Bruun stations. The fit of these dynamically computed state variables of the nitrogen model with an

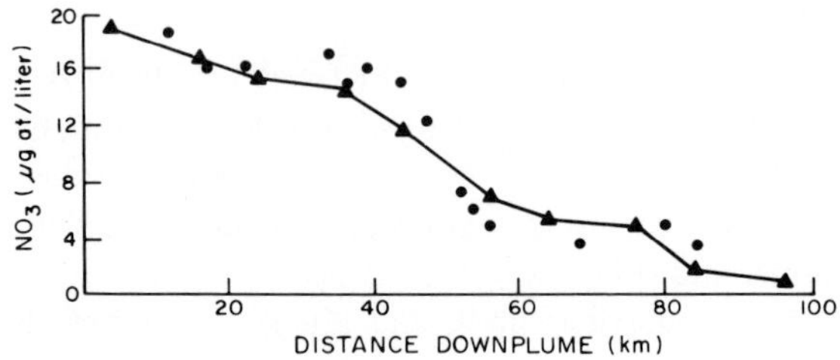

FIG. 2. Simulated (▲) and observed (•) downplume dissolved nitrate of the nitrogen model.

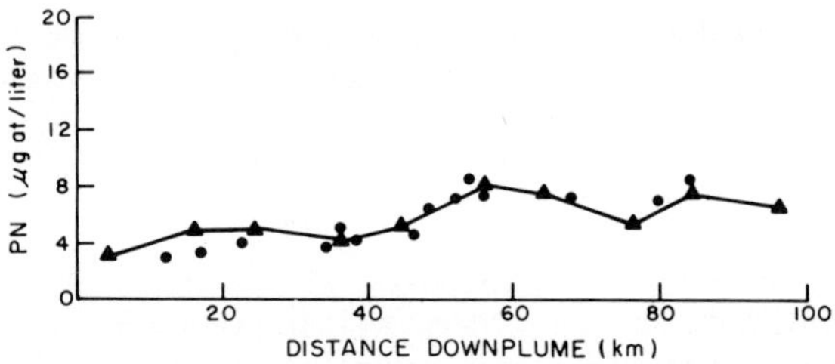

FIG. 3. Simulated (▲) and observed (•) downplume particulate nitrogen of the nitrogen model.

independent set of observed data suggests that the flow field and biological flux assumptions may be reasonable approximations of the spring Peru upwelling ecosystem.

Output of the concurrent carbon model is shown in Figs. 4–6. The first result of the carbon simulation (Fig. 4) is a case in which photosynthesis is a function of light only and not regulated by silicate availability. The same grazing stress, physical driving functions, and diel periodicity of light are applied in the 10 x 1 spatial carbon model as in the nitrogen case, however, and the output of the carbon model is taken at local time of the Bruun stations as well. The predicted carbon agrees fairly well with transformed carbon at most of the inshore stations. Ryther *et al.* (1971) report integrated particulate carbon data over 0–50 m, whereas the euphotic zone appeared to be only 20–30 m deep, and in any case the present model is only concerned with the upper 10-m layer. They state, however, that an empirical 50/1 μg C/μg Chl *a* conversion would properly relate particulate carbon to chlorophyl *a* in the Bruun data and the original 0–10 m mean Anton Bruun Chl *a* data were accordingly transformed to yield the solid circles of Figs. 4 and 5.

The second carbon simulation (Fig. 5) includes multiplicative coupling $(\mathrm{Si}/(K_m + \mathrm{Si}))$ of silicate depression of photosynthesis using a value for K_m of

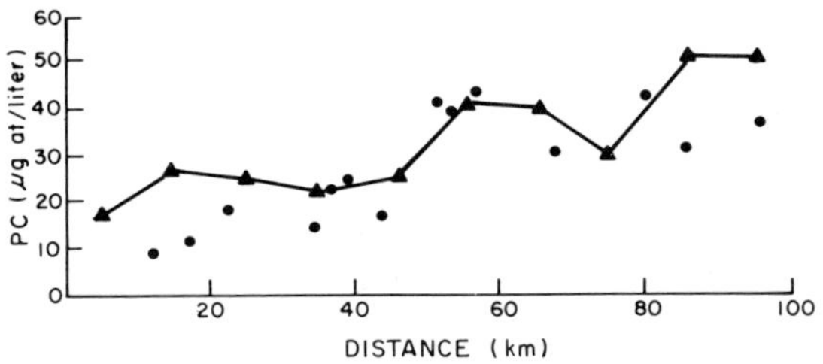

FIG. 4. Simulated (▲) and observed (●) downplume particulate carbon of the carbon model where photosynthesis is uncoupled from silicate regulation.

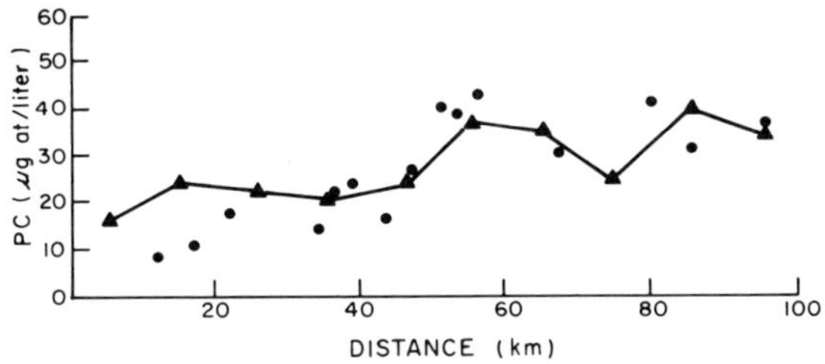

FIG. 5. Simulated (▲) and observed (●) downplume particulate carbon of the carbon model where photosynthesis is coupled to silicate regulation.

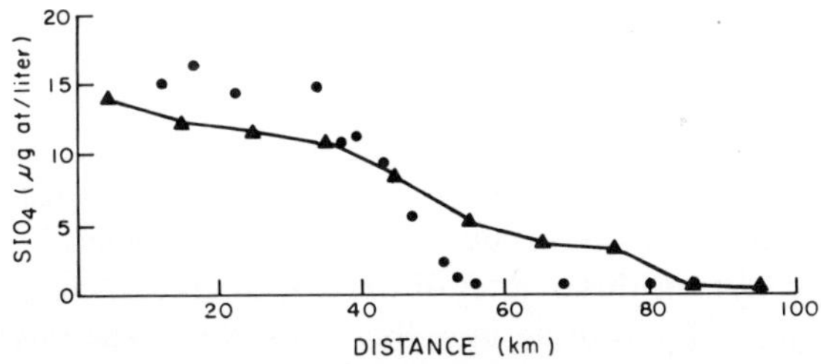

FIG. 6. Simulated (▲) and observed (●) downplume dissolved silicate of the carbon model where photosynthesis is coupled to silicate regulation.

0.5 μg at SiO_4 l^{-1}, which is one-third the value used in the Michaelis–Menten alternative choice expression of silicate limitation in the nitrogen model. The closer fit of predicted and transformed particulate carbon in Fig. 5 suggests that silicate may be regulating carbon as well as nitrogen productivity off Punta San Juan, Peru, in contrast to the conclusion of Ryther *et al.* (1971) that phytoplankton production is not under nutrient limitation in the system. The predicted downplume silicate distribution in this last carbon simulation is shown in Fig. 6 in comparison with the 0–10 m mean silicate concentration observed at each of the Bruun stations.

The carbon and nitrogen models both overestimate their respective observed particulate phytoplankton standing crops very close to shore. The difference between these observed and simulated state variables either may be a function of the present circulation field of the model, or may imply an induction period for recently upwelled phytoplankton in the real world, involving parameters such as chelation availability hypothesized by Barber *et al.* (1971). The respective uptake terms have not yet been modified in these models to consider this additional interaction. The combined results of the two models are shown in Fig. 7 as the predicted downplume C/N ratios at grid points closest to each of the Bruun stations. The mean predicted atomic C/N ratio from the two models is 4.8, compared to an observed 5.6 mean of 85 samples collected on the 1969 Thompson cruise, a 4.0 mean of five Bruun samples (Dugdale and Goering, 1970), and a 6.1 mean observed over offshore and onshore stations on a 1966 cruise of the R/V Unanue to an area just south of Punta San Juan (Strickland, 1970).

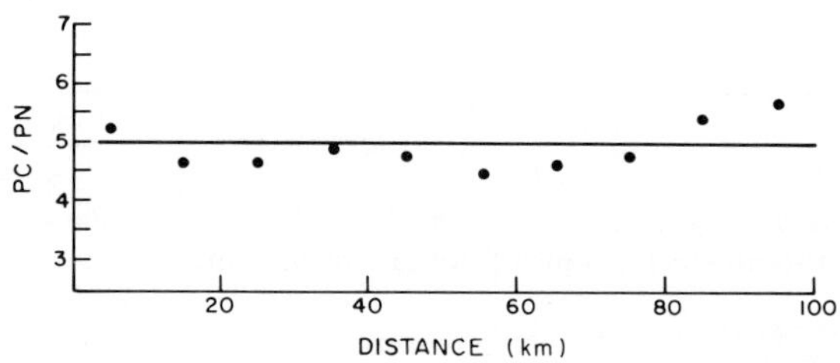

FIG. 7. Predicted C/N ratios at each of the Bruun stations.

The reasonable match of predicted and observed C/N ratios suggests that the Peru system within 40 km of the coast may be regulated by both silicate availability and the intrinsic rate of phytoplankton increase evolved under typical energy input levels of an upwelling ecosystem. The phytoplankton may be up on the plateau of the respective hyperbolas of growth versus ambient nitrogen, phosphorus, and perhaps light in terms of both maximal carbon and nitrogen productivity in the upper 10 m of the inshore regions. Such a hypothesis suggests that a C/N ratio of ~5/1 may be about as low as can be expected for in situ diatoms (the C/N ratio of pure protein is ~3.6), and thus may reflect the maximum natural protein content of individual phytoplankton in upwelling areas. As a phytoplankton population is advected offshore, away from the source of upwelled nutrients, and crosses the ecotone to an ambient oligotrophic system where nitrogen may limit the total yield of algal biomass, the algae begin to operate on the nonlinear part of the hyperbolic nitrogen uptake function and C/N ratios increase. Additional protein can perhaps be induced inshore, of course, by either growing more organisms over a larger upwelling area or providing those present with an energetically cheaper source of reduced nitrogen.

IV. Conclusions

The similar, but independently derived, daytime uptake rate constants of 0.08 and 0.063 h^{-1} for, respectively, nitrogen- and carbon-based growth (V_{max} and P_{max}) and the agreement of predicted C/N ratio of ~5/1 from these two models suggest that both the C/N uptake and phytoplankton protoplasmic ratios are about the same, at least off Peru. For phytoplankton, losses would be the same for each elemental pool; i.e., herbivores eat organisms, not carbon or nitrogen. Dugdale and Goering (1967) have observed lower uptake ratios of 6/1 for $^{14}C/^{15}N$ studies in the Arabian Sea, but particulate data were not presented. It is not clear why the more recent studies (MacIsaac and Dugdale, 1972; McCarthy, 1972) report higher uptake ratios, although it is possible that all nitrogen sources in the dissolved organic nitrogen pool are not being examined (Walsh, 1975) and that our present ^{15}N mass spectrometer technique may be underestimating the actual flux of nitrogen (Pavlou, 1972). A series of field experiments have been initiated to examine more closely the implications of ^{14}C and ^{15}N experiments and their significance to the coupling of photosynthesis and nutrient uptake in marine phytoplankton.

The silicate limitation observed in both models does not appear to uncouple photosynthesis and nitrogen uptake and thus agrees with recent chemostat experiments (Harrison *et al.*, in prep.), which indicate that C/N ratios of three different diatom species under silicate limitation are approximately the same as in unlimited cells, while nitrogen limitation of the same species doubles C/N ratios in comparison with the unlimited case. The requirement of silicate limitation in both models to match the observed downplume particulate nitrogen and carbon does, however, raise the question of phyletic shifts in the phytoplankton and why diatoms are not immediately replaced by nonsilicate requiring dinoflagellates, resulting in a higher overall phytoplankton productivity offshore.

Dinoflagellates such as *Gonyaulax polyedra* have been observed to take up nutrients and presumably grow at one-half to one-fourth the rate of diatoms (Eppley *et al.*, 1969) and have C/N protoplasmic ratios of 8–9 (Strickland *et al.*, 1969), or almost twice that of unlimited diatoms. Dinoflagellates are found at the edges of the Peru plumes (Blasco, 1971), but their slow growth rate may induce a time lag in downplume species succession. Their higher C/N ratios, presumably the result of their cellulose polysaccharide (Fritsch, 1961) rather than silica skeletal structure, may not make them as desirable a protein source as diatoms, although rats have been successfully fed on a diet of dried *Gonyaulax polyedra* whose protein composition is similar to casein (Patton *et al.*, 1967). The role of red tides in upwelling and other coastal areas is still an open question, but the presently inexplicable dominance of *Gonyaulax polyedra* at the onset of upwelling off Baja California in a silicate-rich nutrient regime (Walsh *et al.*, 1974) highlights our lack of knowledge about qualitative aspects of marine ecosystems.

The C/N ratio of phytoplankton in pelagic ecosystems must act as a protein gate for input to the system, but subsequent transfer and storage of protein within the higher trophic levels has been little studied on a synoptic basis. Curl's (1962) pioneer study of producers, herbivores, and consumers off the New York coast suggests a seasonal C/N ratio of 5–7 for all three trophic levels, implying a predominantly diatom-based system and that the phytoplankton protein content may be a reasonable index of the protein content of higher trophic levels.

MacIsaac and Dugdale (1972) have suggested that lower and less variable phytoplankton C/N uptake ratios may occur in more stable oligotrophic or eutrophic regions where presumably the food supply is more constant and the organisms have adjusted their protein composition to available nutrients. In this respect, a seasonal study of boreal and tropical zooplankton C/N content (Omori, 1969) demonstrated a range of 5–10 in the environmentally unstable higher latitudes in contrast to an almost constant seasonal C/N ratio of ~4 near the Marshall Islands in a presumably more stable or monotonous environment. If such tenuous hypotheses are correct, one might expect to find similar 5/1 C/N ratios for the higher trophic levels of the Peru upwelling system as in the case of a 5.17 C/N mean recently obtained for 10 homogenized specimens of *Engraulis ringens* (Whitledge, pers. comm.). A question might then be asked as to how one could further increase the protein content and yield from either individual or all trophic levels of an upwelling ecosystem.

V. Discussion

Consideration of ways to manipulate the Peru upwelling or other marine ecosystems to increase protein yield to man is a highly speculative and controversial endeavor (Ryther, 1969; Alverson *et al.*, 1970; Ryther, 1970), considering our present dearth of knowledge about the basic mechanisms of these ecosystems. Although lower C/N values of 2–3 for particulate matter in the sea usually imply selective leaching of carbohydrate from detrital material in aphotic regions (Degens, 1970), it may be theoretically possible to increase the C/N content of individual phytoplankton by growing them on alternative reduced nitrogen sources such as sewage effluent, thereby increasing their protein content *if* no other deleterious tradeoffs result from this type of manipulated nutrient addition, and *if* sufficient silicate is available to yield the presumably more protein-rich diatoms. It is not clear how a large-scale nutrient input of this sort could effectively be introduced into an already rich upwelling ecosystem. There are some pilot studies underway, however, to induce artificial upwelling in the Caribbean (under the direction of Dr. Oswald Roels of the Lamont-Doherty Observatory), and to feed diluted sewage to phytoplankton to increase oyster production (under the direction of Dr. John Ryther at Woods Hole Oceanographic Institution).

Increased trophic level efficiency of protein transfer within a clupeid–phytoplankton-based upwelling food chain is another possibility for increasing protein yield besides increasing the areal extent of upwelling or the protein content of individual food, although *Engraulis ringens* may already by partially phytophagous in the Peru system. In any case, there are some additional market psychology problems. Humans presently prefer to use anchoveta as fish meal input to poultry industries rather than consume them directly; in 1969 a ton of anchoveta was worth ~$10 in comparison to a ton of salmon worth ~$800 (Paulik, 1970). It is clear, however, that if protein scarcity becomes an increasing reality in the future, present fisheries models which operate in an environmental vacuum (Gulland, 1972) will be of little additional use as a total resource management tool until they are more realistically coupled to simulation analyses of the lower trophic levels and their physical and chemical driving forces. The present simulation of phytoplankton carbon and nitrogen production over a 10-day period of the March–April Peru upwelling ecosystem suggests that we may be able to begin to both quantitatively and *qualitatively* represent biological and physical inputs to more long-term considerations of protein yield from clupeid or other food chains of marine ecosystems.

Acknowledgments

We would like to thank Drs. T. Whitledge, S. Pavlou, R. Holmes, R. Dugdale, H. Conway, C. Davis, and P. Harrison for discussion and use of presently unpublished data. This research was supported by National Science Foundation Grants GA-31093, GB-35880X, and GX-33502 and is a contribution from our Chemostat Laboratory, the U.S. IBP Upwelling Study, and the Coastal Upwelling Ecosystems Analysis (CUEA) program of the International Decade of Ocean Exploration (IDOE).

Appendix. State Equations and Algebraic Subroutines

The equations for the eleven state variables are as follows:

1. *Light (I)*:

The subroutine involving the algebraic equations for the light state variable is discussed below in relation to calculation of particulate carbon.

2. *Nitrate (NO_3):*

$$\frac{\partial NO_3}{\partial t} = -\frac{\partial uNO_3}{\partial x} - \frac{\partial wNO_3}{\partial z} + K_y \frac{\partial^2 NO_3}{\partial y^2} - \frac{V_{max}(NO_3)(PN)}{(K_t + NO_3)},$$

where u and w are, respectively, the downstream and vertical velocities specified at each grid point, K_y is a constant lateral eddy coefficient of $\sim 10^6$ cm^2 sec^{-1}, K_t is the half saturation constant of 1.5 μg at NO_3 l^{-1} for nitrate uptake, and V_{max} is the maximum uptake rate of NO_3 as a function of ammonia and daylight expressed as $V_{max} = (0.11 - 0.02\ NH_3)(\sin 0.2618t)$.

3. *Recycled nitrogen represented by ammonia (NH_3):*

$$\frac{\partial NH_3}{\partial t} = -\frac{\partial uNH_3}{\partial x} - \frac{\partial wNH_3}{\partial z} + K_y \frac{\partial^2 NH_3}{\partial y^2} - \frac{V(NH_3)(PN)}{(K_r + NH_3)}$$

$$+ 0.67 \left[\frac{G_{max}(PN - P_z)(ZPN)}{K_p + (PN - P_z)}\right] + 0.67 \left[\frac{G_{max}(PN - P_0)(FPN)}{K_p + (PN - P_0)}\right],$$

where V is the maximum uptake of 0.08 h^{-1} (0.11 h^{-1} adjusted to maximum of a sinusoid $V = 0.11 \sin 0.2618t$) for all other nutrients, K_r is the 1.5 μg at NH_3 l^{-1} half saturation constant for ammonia uptake, G_{max} is the maximum specific grazing rate of herbivores expressed as a cosine function to simulate nocturnal grazing loss, i.e., $G_{max} = 0.008 \cos\ (0.2618t + 1.571)$, P_z is the zooplankton grazing threshold of 0.5 μg at PN l^{-1} and 2.5 μg at PC l^{-1}, P_0 is the anchoveta grazing theshold of 3.0 μg at PN l^{-1} and 15.0 μg at PC l^{-1} necessary to parameterize phytoplankton patchiness, and K_p is the half saturation grazing constant of 1.5 μg at PN l^{-1} and 7.5 μg at PC l^{-1}.

4. *Phosphate (PO_4):*

$$\frac{\partial PO_4}{\partial t} = -\frac{\partial uPO_4}{\partial x} - \frac{\partial wPO_4}{\partial z} + K_y \frac{\partial^2 PO_4}{\partial y^2} - \frac{V(PO_4)(PP)}{(K_s + PO_4)}$$

$$+ 0.13 \frac{[G_{max}(PN - P_z)(ZPN)]}{K_p + (PN - P_z)} + 0.13 \frac{[G_{max}(PN - P_0)(FPN)]}{K_p + (PN - P_0)},$$

where PP is phytoplankton particulate phosphorus and K_s is the 1.5 μg at PO_4 l^{-1} half saturation constant of phosphate uptake. The excretion input of phosphorus in relation to nitrogen recycling is based on analysis of anchoveta impact on nutrient regeneration (Walsh, 1975).

5. *Silicate (SiO_4):*

$$\frac{\partial SiO_4}{\partial t} = -\frac{\partial uSiO_4}{\partial x} - \frac{\partial wSiO_4}{\partial z} + K_y \frac{\partial^2 SiO_4}{\partial y^2} - V\frac{(SiO_4)(PSi)}{(K_n + SiO_4)},$$

where PSi is particulate silicate and K_n is the 1.5 μg at SiO_4 l^{-1} half saturation constant for silicate uptake. No silicate is assumed to be regenerated through herbivore excretion.

6. *Phytoplankton particulate nitrogen (PN)*:

$$\frac{\partial PN}{\partial t} = -\frac{\partial uPN}{\partial x} - \frac{\partial wPN}{\partial z} + K_y \frac{\partial^2 PN}{\partial y^2} - \frac{[G_{max}(PN-P_z)(ZPN)]}{K_p + (PN-P_z)}$$

$$-\frac{[G_{max}(PN-P_0)(FPN)]}{K_p + (PN-P_0)} + \min\left\{\left[\frac{V_{max}(NO_3)(PN)}{(K_t + NO_3)}\right.\right.$$

$$\left.\left.+\frac{V(NH_3)(PN)}{(K_r + NH_3)}\right], \frac{[V(PO_4)(PP)]}{(K_s + PO_4)}, \frac{[V(SiO_4)(PSi)]}{(K_n + SiO_4)}\right\},$$

where the lowest of the three uptake expressions is selected for input to phytoplankton at each time step of the model.

7. *Phytoplankton particulate carbon (PC)*:

$$\frac{\partial PC}{\partial t} = -\frac{\partial uPC}{\partial x} - \frac{\partial wPC}{\partial z} + K_y \frac{\partial^2 PC}{\partial y^2} - \frac{[G_{max}(PC-P_z)(ZPC)]}{K_p + (PC-P_z)}$$

$$-\frac{G_{max}(PC-P_0)(FPC)}{(K_p + (PC-P_0)} + \frac{(PC)(P_{max})[\exp(1-I_z/I_{max}) - \exp(1-I_t/I_{max})]}{10K},$$

where all terms in the equation, except for the last one, are similar to those used in the phytoplankton PN equation. The last term represents mean net photosynthetic increase in phytoplankton particulate carbon as given by the analytical integration of Steele's (1962) function for photosynthesis as a function of light from the surface to 10 m depth. P_{max}, the maximum net specific photosynthetic uptake rate of 0.10 h^{-1} at the peak of the sinusoid, is based on analysis of growth kinetics experiments done by Barber *et al.* (1971). I_{max}, the light intensity at which P_{max} occurs, was assumed to be approximately 33% of maximum surface insolation during March–April off Punta San Juan, Peru. If the surface insolation (I_t) during an hour t is found by assuming that I_{total}, the total daily insolation (ly day^{-1}), is distributed as a sine function of time over the light day beginning at 0600 and ending at 1800 hours, then

$$I_t = (I_{total})(0.1309 \sin 0.2618\, t),$$

while the light at 10 m depth is found by applying Lambert's Law and assuming a time and spatial varying extinction coefficient for this 0–10-m layer,

$$I_z = I_t\, e^{-10K},$$

where K is recalculated each iteration as a function of phytoplankton carbon by

$$K = 0.04 + 0.0021\, PC + 0.021(PC)^{0.67}.$$

8. *Zooplankton particulate nitrogen (ZPN)*:

$$\frac{\partial \text{ZPN}}{\partial t} = [1.0-0.2-0.8]\left[G_{\max}\frac{(\text{PN}-P_z)(\text{ZPN})}{K_p+(\text{PN}-P_z)}\right],$$

where the first part of the first term is grazing input to zooplankton, the second part is assimilation loss, and the third part is excretory return to the nutrient pools, i.e., $\partial\text{ZPN}/\partial t = 0$.

9. *Zooplankton particulate carbon (ZPC)*:

$$\frac{\partial \text{ZPC}}{\partial t} = [1.0-0.2-0.8]\,\frac{[G_{\max}(\text{PC}-P_z)(\text{ZPC})]}{K_p+(\text{PC}-P_z)},$$

with analogous logic to the zooplankton equation above.

10. *Fish particulate nitrogen (FPN)*:

$$\frac{\partial \text{FPN}}{\partial t} = [1.0-0.2-0.8]\,\frac{[G_{\max}(\text{PN}-P_0)(\text{FPN})]}{K_p+(\text{PN}-P_0)},$$

with the parts of the first term analogous to the zooplankton equations, i.e., $\partial\text{FPN}/\partial t = 0$, and similarly for the fish carbon budget.

11. *Fish particulate carbon (FPC)*:

$$\frac{\partial \text{FPC}}{\partial t} = [1.0-0.2-0.8]\,\frac{[G_{\max}(\text{PC}-P_0)(\text{FPC})]}{K_p+(\text{PC}-P_0)}.$$

These partial differential equations are then transformed to finite-difference form and their simultaneous steady-state solutions are obtained numerically as outlined in the text.

REFERENCES

Alverson, D. L., Longhurst, A. R., and Gulland, J. A. (1970). *Science* **168,** 503.

Barber, R. T., Dugdale, R. C., MacIsaac, J. J., and Smith, R. L. (1971). *Inv. Pesq.* **35,** 171.

Blasco, D. (1971). *Inv. Pesq.* **35,** 61.

Byers, R. J. (1963). *Ecol. Monogr.* **33,** 281.

Caperon, J., and Meyer, J. (1972). *Deep-Sea Res.* **19,** 601.

Curl, H. (1962). *Rapp. Cons. Explor. Mer.* **153,** 183.

Degens, E. T. (1970). *In* "Organic Matter in Natural Waters" (D. W. Hood, ed.), Publ. I, pp. 77–106. Univ. of Alaska, Fairbanks, Alaska.

Dugdale, R. C., and Goering, J. J. (1967). *Limnol. Ocenogr.* **12,** 196.

Dugdale, R. C., and Goering, J. J. (1970). *Anton Bruun Rep.* **5,** 5.2–5.8.

Eppley, R. W., Rogers, J. M., and McCarthy, J. J. (1969). *Limnol. Oceanogr.* **14,** 912.

Fogg, G. E. (1959). *Symp. Soc. Exp. Biol.* **13,** 106.

Fritsch, F. E. (1961). *In* "The Structure and Reproduction of the Algae" (F. E. Fritsch, ed.), Vol. I, pp. 664–720. Cambridge Univ. Press, London and New York.

Gale, J. (1972). *Ecology* **53,** 494.
Gulland, J. A. (1972). Population Dynamics of World Fisheries. Washington Sea Grant Publ. WSG 72-1, Seattle, Washington.
Harrison, P. J., Conway, H. L., Holmes, R. W., and Davis, C. O. (in preparation).
Hobson, L. A. (1971). *Inv. Pesq.* **35,** 195.
Hobson, L. A., and Pariser, R. J. (1970). *J. Exp. Mar. Biol. Ecol.* **6,** 69.
Holm-Hansen, O., Nishida, K., Moses, V., and Calvin, M. (1959). *J. Exp. Bot.* **10,** 109.
Holm-Hansen, O., Strickland, J. D. H., and Williams, P. M. (1966). *Limnol. Oceanogr.* **11,** 548.
MacIsaac, J. J., and Dugdale, R. C. (1972). *Deep-Sea Res.* **19,** 209.
McCarthy, J. J. (1972). *Limnol. Oceanogr.* **17,** 738.
Menzel, D. W. (1959). *J. du Conseil* **23,** 308.
Odum, H. T. (1970). *In* "A Tropical Rain Forest" (H. T. Odum and R. F. Pigeon, eds.), pp. I-191–I-281. U.S. AEC, Oak Ridge, Tennessee.
Omori, M. (1969). *Mar. Biol.* **3,** 4.
Patton, S., Chandler, P. T., Kalan, E. B., Loeblich, A. R., Fuller, G., and Benson, A. A. (1967). *Science* **158,** 789.
Paulik, G. J. (1970). The Anchoveta Fishery of Peru. Paper No. 13, Center for Quantitative Sci. in Forestry, Fisheries and Wildlife, Univ. of Washington, Seattle, Washington.
Pavlou, S. P. (1972). Phytoplankton Growth Dynamics. Tech. Ser. 1, Chemostat Methodology and Chemical Analyses. Spec. Rep. No. 52, Dept. of Oceanogr., Univ. of Washington, Seattle, Washington.
Ryther, J. H. (1969). *Science* **166,** 72.
Ryther, J. H. (1970). *Science* **168,** 505.
Ryther, J. H., and Dunstan, W. M. (1971). *Science* **171,** 1008.
Ryther, J. H., Menzel, D. W., Hulburt, E. M., Lorenzen, C. J., and Corwin, N. (1971). *Inv. Pesq.* **35,** 43.
Shapiro, J. (1973). *Science* **179,** 382.
Small, L. F., and Ramberg, D. A. (1971). *In* "Fertility of the Sea" (J. Costlow, ed.), Vol. II, pp. 475–492. Gordon and Breach, New York.
Steele, J. H. (1962). *Limnol. Oceanogr.* **7,** 137.
Strickland, J. D. H. (1970). Research on the Marine Food Chain. Progr. Rep., July 1969. Part IV, IMR 70-5, Univ. of California, San Diego, California.
Strickland, J. D. H., Holm-Hansen, O., Eppley, R. W., and Linn, R. J. (1969). *Limnol. Oceanogr.* **14,** 23.
Thomas, W. H. (1970). *Limnol. Oceanogr.* **15,** 380.
Vollenweider, R. A. (1965). *In* "Primary Productivity in Aquatic Environments" (C. R. Goldman, ed.), pp. 425–457. Univ. of California Press, Berkeley, California.
Walsh, J. J. (1972). *Science* **176,** 969.
Walsh, J. J. (1975). *Deep-Sea Res.* **22,** 201.
Walsh, J. J., and Dugdale, R. C. (1972). *In* "Nutrients in Natural Waters" (H. E. Allen and J. R. Kramer, eds.), pp. 171–191. Wiley (Interscience), New York.
Walsh, J. J., Kelley, J. C., Dugdale, R. C., and Frost, B. W. (1971). *Inv. Pesq.* **35,** 25.
Walsh, J. J., Kelley, J. C., Whitledge, T. E., MacIsaac, J. J., and Huntsman, S. A. (1974). *Limnol. Oceanogr.* **19,** 553.
Wangersky, P. J. (1965). *Amer. Sci.* **53,** 358.
Witkamp, M. (1966). *Ecology* **47,** 194.

3

A Simulation of the Mesoscale Distribution of the Lower Marine Trophic Levels off West Florida

JAMES J. O'BRIEN*

DEPARTMENTS OF METEOROLOGY AND OCEANOGRAPHY
FLORIDA STATE UNIVERSITY, TALLAHASSEE, FLORIDA

and

JOSEPH S. WROBLEWSKI†

DEPARTMENT OF OCEANOGRAPHY
FLORIDA STATE UNIVERSITY, TALLAHASSEE, FLORIDA

Another version of this work has appeared in *Investigación Pesquera* 37, (2), 193 (1973).

* Present address: Office of Naval Research, Arlington, Virginia.

† Present address: Department of Oceanography, Dalhousie University, Halifax, Nova Scotia.

I. Introduction

A. Overview

A simulation of the flow of the biologically limiting nutrient through the trophic levels (phytoplankton, zooplankton, pelagic fish, detritus, and limiting nutrient dissolved in the water column) of a marine ecosystem over a continental shelf is presented. Interrelated processes of this time-dependent, spatial, nonlinear, physical–chemical–biological model include advection, diffusion, several biotic and abiotic environmental conditions, and numerous biological processes. The importance of advection in determining the spatial distribution of the biotic components in an upwelling region is demonstrated.

The maximum rate of nutrient uptake by phytoplankton, V_m, is found to be a fundamental time scale to which both physical and biological processes can be related. A nondimensional parameter S, evolved from formulation of the model, scales the effects of advection and diffusion relative to the rate of biological turnover in determining spatial solutions. The magnitude of S is dependent on the value of V_m.

The spatial distributions of biotic components are calculated for both phosphate- and nitrate-limiting situations. Localities of strong upwelling of nutrient-rich waters into the euphotic zone show greater phytoplankton and zooplankton standing stocks.

The rates of the system are explored. Sensitivity analyses conducted on the model formulation determine the most important controlling factors in the system dynamics to be herbivore grazing and excretion rates.

B. BACKGROUND

The building of mathematical models to understand the processes of ecological systems began with the classic prey–predator equations of Lotka (1925) and Volterra (1926). Since then marine scientists have attempted to describe processes of biological production in the ocean by deducing equations which represent the interaction of biological elements with their environment. Raymont (1966) and Patten (1968) have prepared excellent review articles on mathematical models of plankton production.

Investigations of the dynamics of marine ecosystems have followed two approaches. Energy flow studies (e.g., Odum, 1957; Slobodkin, 1962; and Macfadyen, 1964) have provided significant contributions to the understanding of aquatic ecosystem energetics. A second approach concerns studies of the cycling of material through the biotic and abiotic elements of the system (e.g., Curl, 1962; Gerking, 1962; and Dugdale and Goering, 1967). The interconvertibility of these approaches was demonstrated by Odum (1962). Smayda (1966) made the first comprehensive quantitative analysis of phytoplankton dynamics in an upwelling region. Walsh and Dugdale (1971) have pioneered the construction of spatial models of upwelling ecosystems.

II. Purpose of the Study

The objective of this chapter is to describe a numerical model of the mesoscale spatial distributions of the lower marine trophic levels in waters over a continental shelf. The study area is the Gulf of Mexico off west Florida during winter and spring months of the year. Dugdale's (1967) concept of biomass estimation in terms of content of the biologically limiting nutrient is the basis of this simulation. Specifically, the relative proportion of the limiting nutrient in each of the biotic components (phytoplankton, zooplankton, pelagic fish, organic detritus, and dissolved nutrients) is calculated.

Conservation of mass within the marine ecosystem is a basic assumption. The total amount of limiting nutrient in the system is accounted for by either mass transport into or out of the area, or uptake and release by biological components within the system.

Present biological and physical oceanographic conceptions about the area are incorporated into this study. The mechanism of current-induced upwelling is used to generate a circulation pattern over the shelf such that nutrient-rich bottom waters are brought into the euphotic zone. This addition of limiting

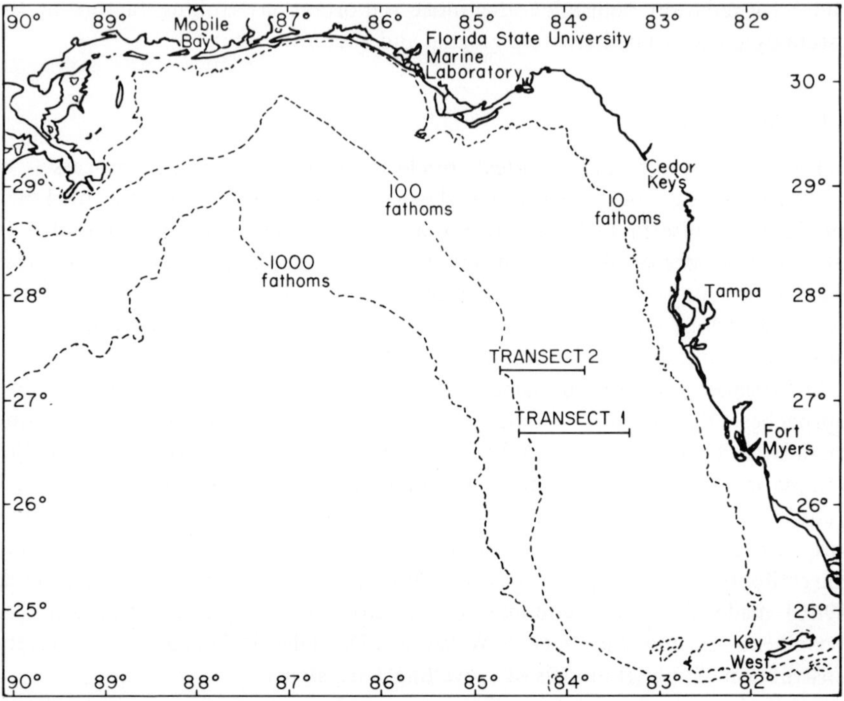

FIG. 1. Model area: the continental shelf off west Florida. Transects 1 and 2 are locations of observational data collected for input into the simulation model.

nutrient enables greater biological production in otherwise nutrient-poor waters.

Figure 1 is a map of the Gulf of Mexico waters under consideration. Bottom topography consists of a gently sloping bottom extending to approximately 200 km seaward where a sharp continental shelf break is encountered. Here the bottom plunges to depths of greater than 1000 m. During winter and spring, storms and winds mix the shelf waters to the bottom. The Florida Loop Current, the major current occurring in the eastern Gulf of Mexico, is found running south close to the shelf break during this period.

Field studies of the area by Bogdanov *et al.* (1968) and Austin (1971) indicated that upwelling occurs offshore during the cold period of the year, accompanied by increases in plankton biomass and concentration of main commercial species of fish on the southwestern part of the Florida Shelf.

The problem consists of integrating the physical and biological dynamics concurrently. The approach followed here is to construct a composite of the biological, chemical, and physical interrelations into a single system. After formulation of the biological dynamics, physical flow patterns, environmental conditions, and necessary initial conditions, calculations of time and space changes of the system components are made using numerical techniques.

An intent of this study is to define objectives for subsequent field research programs. Insufficient knowledge of physiological processes and food chain dynamics limits understanding of the marine ecosystem. It is hoped this research will identify the important biological processes which govern the continental shelf ecosystem. Future feedback between field observations and model development is intended.

III. System Representation and Formulation

The idealized model area is shown in cross section in Fig. 2. The oceanic section is generalized as a two-dimensional box with the top a free surface and the bottom bounded by a flat continental shelf. A straight vertical coast constitutes the right-hand boundary and the seaward boundary of integration is at the continental shelf break. The x direction is taken as positive toward the coast and the depth z as positive upward. An assumption necessary for computational economy is that there are no longshore variations; that is, $\partial \cdot / \partial y = 0$.

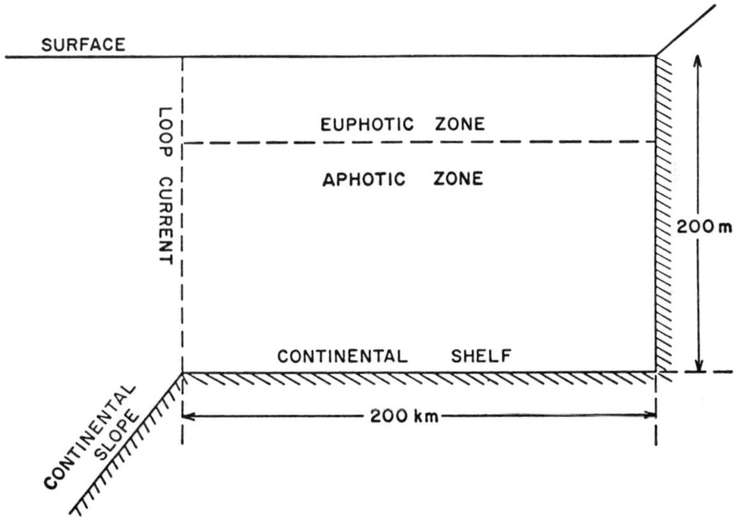

FIG. 2. The idealized oceanic area in cross section with the top a free surface, the bottom a flat continental shelf, the right-hand boundary the vertical coast, and the seaward boundary at the continental shelf break. The water column is divided into a euphotic and an aphotic zone.

The cross section is divided into a grid of 41 by 82 rectangles in the x and z directions, respectively. The dimensions of each grid box are approximately 2.5 m in depth by 5 km in width. A theoretical flow field is specified at each grid point. The onshore water velocity u is taken as positive along x toward the coast, the vertical velocity w is positive upward, and the longshore horizontal velocity v is negative in a southerly direction along the coast.

We assume the phytoplankton dynamics $P(x, z, t)$ determining biomass concentration of the phytoplankton component P at an arbitrary point on the spatial grid are governed by

$$\frac{dP}{dt} = \frac{\partial P}{\partial t} + u\frac{\partial P}{\partial x} + w\frac{\partial P}{\partial z} - \frac{\partial}{\partial x}\left[\nu_{\mathrm{h}}\frac{\partial P}{\partial x}\right] - \frac{\partial}{\partial z}\left[\nu_{\mathrm{v}}\frac{\partial P}{\partial z}\right] = \text{biological terms.} \qquad (1)$$

The first three terms on the left-hand side represent collectively the change of P following a water parcel moving within the ocean. The first term is local change and the other terms are advective changes. The last two terms on the left-hand side represent turbulent diffusion, where ν_{h} and ν_{v} are horizontal and vertical eddy diffusivities, respectively. These diffusion terms represent the diffusion of P by chaotic motion and motion whose time and space scales are smaller than those which can be resolved by the grid size used in the model. These five terms define the operator on the extreme left.

The biological terms include any biotic processes which might be included in a marine food chain model. Formulations for phytoplankton growth, processes that diminish the phytoplankton stock (predation by higher trophic levels, sinking, mortality, etc.), and any losses or additions to the phytoplankton component which are biological in nature are included here. Equations similar to Eq. (1) can be written for zooplankton, detritus, and dissolved nutrients. The biological processes of the model are indicated in Eqs. (2)–(6):

$$\frac{dP}{dt} = \underset{\mathrm{I}}{\text{growth of } P \text{ from } N \text{ uptake}} - \underset{\mathrm{II}}{\text{extracellular release}} - \underset{\mathrm{III}}{\text{grazing by } Z} - \underset{\mathrm{IV}}{\text{predation by } F}, \qquad (2)$$

$$\frac{dZ}{dt} = \underset{\mathrm{III}}{\text{growth of } Z \text{ by grazing on } P} \text{ and } \underset{\mathrm{V}}{D} - \underset{\mathrm{VI}}{\text{excretion}} - \underset{\mathrm{VII}}{\text{natural death}} - \underset{\mathrm{VIII}}{\text{predation by } F}, \qquad (3)$$

$$\frac{dD}{dt} = \underset{\mathrm{VII}}{\text{input from } Z \text{ natural death}} - \underset{\mathrm{V}}{\text{grazing by } Z} - \underset{\mathrm{IX}}{\text{bacterial regeneration}} - \underset{\mathrm{XI}}{\text{sinking}}, \qquad (4)$$

$$\frac{dF}{dt} = \underset{\text{IV}}{\text{growth of } F \text{ by predation on } P \text{ and } Z} - \underset{\text{X}}{\text{excretion}}, \quad \text{VIII} \tag{5}$$

$$\frac{dN}{dt} = \underset{\text{I}}{\text{loss from uptake by } P} + \underset{\text{II}}{\text{extracellular release by } P} + \underset{\text{VI} \quad \text{X}}{\text{excretion by } Z \text{ and } F} + \underset{\text{IX}}{\text{input of decomposed } D}, \tag{6}$$

where P is phytoplankton, Z is zooplankton, D is zooplankton detritus, F is pelagic fish, and N is limiting nutrient dissolved in the water column. The specific formulation of Roman numeralled terms I–X will be discussed in the next section, while the sinking term XI is formulated later.

Figure 3 summarizes the system dynamics allowed within any single spatial block. The biological processes are shown as transfers (arrows) between components. The advective and diffusive fluxes between spatial blocks are represented by heavy arrows.

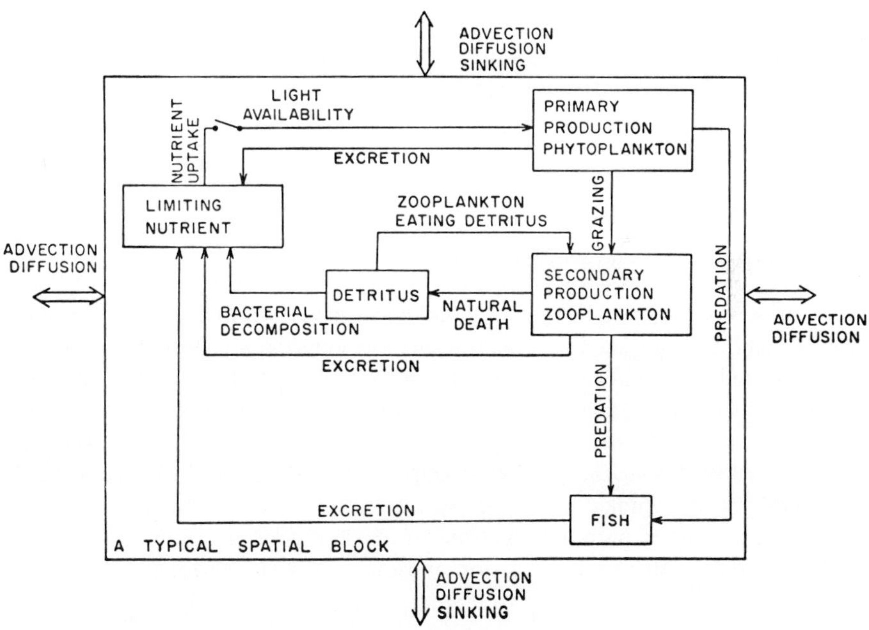

FIG. 3. Biological processes within a spatial block. The model's biotic components include phytoplankton, zooplankton, pelagic fish, detritus, and the biologically limiting nutrient dissolved in the water column. Light arrows denote flow pathways of the limiting nutrient between trophic levels and the dissolved nutrient. Heavy arrows indicate fluxes between spatial blocks.

IV. Mathematical Formulation of the Biological Dynamics

Marine species which have feeding habits in common and can be assigned to one trophic level are treated collectively. Therefore, species composition is not considered. A simple food chain of three trophic levels is conceived. The dynamics (nutrient uptake and turnover or feeding rates) of each biotic component, nutrient exchanges between trophic levels, and effects of changes in resource availability are investigated.

The five biotic components are expressed in concentration of the biologically limiting nutrient. The combined amount in the phytoplankton, zooplankton, detritus, fish, and nutrient dissolved in the water column is defined as N_t, the spatially averaged amount of limiting nutrient on the continental shelf area amongst all components. Since N_t is assumed constant,

$$N_t = N' + P' + Z' + D' + F', \tag{7}$$

where primes denote dimensional quantities. In this case, the units are concentration, e.g., μg at NO_3–N per liter. Subsequently, all quantities will be nondimensionalized. These will be denoted by the absence of primes.

The amount of the limiting nutrient in *P, Z, D,* and *F* is often expressed as microgram atom nutrient per gram biomass. One can obtain the units used in this study upon multiplying by grams biomass per liter to give microgram atom nutrient per liter.

A. The Phytoplankton Equation

As this is a material flow study, biological processes that deal with uptake, release, and transfer of limiting nutrient are considered. Dugdale (1967) has demonstrated the applicability of Monod enzyme kinetics to the uptake of limiting nutrients by marine phytoplankton. Subsequent laboratory and field work (Eppley and Coatsworth, 1968; Thomas and Dodson, 1968; MacIsaac and Dugdale, 1969; Eppley *et al.,* 1969; Dugdale and MacIsaac, 1971) support the hypothesis that uptake rates of nitrate, ammonium, and phosphate by phytoplankton are hyperbolic functions of nutrient concentration.

The Michaelis–Menten formulation of Monod (1942) kinetics is

$$V = V_m N'/(K + N'),$$

where V is the specific uptake rate (h^{-1}) of nutrient N' (concentration), V_m is the maximum uptake rate, and K is the Michaelis constant or half-saturation constant. Parameter K is the concentration supporting half the maximum uptake rate.

The dynamic equation for phytoplankton growth is

$$\frac{dP'}{dt'} = \underset{\text{I}}{V_m N'P'/(K+N')} - \underset{\text{II}}{BP'} - \underset{\text{III}}{E_z[1-\exp(-d_p P')]Z'} - \underset{\text{IV}}{\Phi F'\,[\theta P'/(\theta P'+Z')]}, \quad (8)$$

where the numbered terms correspond to those of the descriptive equations (2)–(6). The first term on the right represents uptake of the limiting nutrient N' by phytoplankton P' according to Michaelis–Menten kinetics.

A fundamental assumption in this research is that growth of phytoplankton by nutrient uptake is governed by Michaelis–Menten kinetics. One advantage of not considering the nutrient's molecular form within the cell is that the cell "nutrient pool" phenomenon can be ignored. Once the cell has taken up the nutrient, it can be considered as belonging to the phytoplankton component, whether it is immediately fixed into biomass or stored for later anabolism.

Term II represents the collective losses of limiting nutrient from the phytoplankton component. It encompasses the processes of extracellular release by phytoplankton and the autolysis of senescent cells. Phytoplankton excretion is usually expressed as percentage of carbon fixed by photosynthesis (Watt, 1966). We consider phytoplankton excretion as the extracellular release of amino acids containing the limiting nutrients of the model, namely phosphate and nitrate.

The third term (III) represents zooplankton predation; E_z is the maximum herbivore grazing ration with units of h^{-1}. McAllister (1970) documents the utility of this nonlinear phytoplankton-dependent formulation of grazing by zooplankton. The fourth term of Eq. (8) describes fish predation upon phytoplankton which will be discussed in context of the fish dynamics.

B. The Zooplankton Equation

Cushing (1969) expressed the need for a model giving an estimate of zooplankton grazing capacity over a wide range of algal densities. Laboratory studies have indicated zooplankton grazing rates to be a linear function of phytoplankton density for low concentrations of phytoplankton. When phytoplankton are abundant there exists a maximum rate of grazing attainable by the zooplankton (McAllister, 1970).

Parsons *et al.* (1967) have modified the expression of Ivlev (1945),

$$I = E_z\{1-\exp[-d_p(P'-P^*)]\},$$

where I is the rate of ingestion per unit concentration of grazer at phytoplankton concentration P', E_z is the maximum rate of ingestion attainable by the zooplankton, d_p is the Ivlev constant which modifies the rate of change of ingestion with food concentration, and P^* is the threshold concentration of phytoplankton at which feeding begins. At concentrations below P^*, the zooplankton starve.

In the euphotic zone of our model, P' never approaches concentration values which would dictate the inclusion of P^*. The constant $1/d_p$ is the concentration of P at which the specific ingestion rate I is approximately two thirds of the maximum rate E_z.

The zooplankton dynamics in the model are governed by

$$dZ'/dt' = \underset{\text{III}}{E_z(1-\exp(-d_p P'))Z'} + \underset{\text{V}}{E_D D'Z'} - \underset{\text{VI}}{\Gamma[E_z(1-\exp(-d_p P'))Z']Z'} - \underset{\text{VII}}{d_D Z'}$$

$$- \underset{\text{VIII}}{\Phi F'[Z'/(\theta P' + Z')]}, \tag{9}$$

where D' (term V) is the detritus component which serves as a partial food source for zooplankton. The parameter d_D (term VII) is the natural death rate for zooplankton. The term VII is the flux of zooplankton biomass being transferred to the detritus component. Coefficient E_D (term V) is the grazing coefficient ($\text{conc}^{-1}\,\text{h}^{-1}$) for zooplankton on detritus. This grazing rate is $E_D D$.

Field and laboratory studies (Cushing, 1969; Parsons *et al.*, 1967) have indicated that zooplankton excrete more of ingested material when phytoplankton are abundant. A direct relationship between zooplankton grazing on phytoplankton and excretion is therefore assumed. Term VI is linearly related to term III by the excretion coefficient Γ. When ingestion of phytoplankton is small, zooplankton excretion is small.

C. The Detritus Equation

Detritus consists of dead zooplankton biomass only and is grazed upon by living zooplankton, a simplification of the marine food chain:

$$dD'/dt' = \underset{\text{VII}}{d_D Z'} - \underset{\text{V}}{E_D Z'D'} - \underset{\text{IX}}{lD'} - \underset{\text{XI}}{w_s\,\partial D'/\partial z'}. \tag{10}$$

Term VII is the source of detritus; term V is the loss due to zooplankton grazing. Term IX is a simple linear parameterization of the regeneration of detritus by bacteria into dissolved limiting nutrient. A sensitivity analysis (Section X) demonstrates that the model dynamics are not crucial to the value of l. Therefore a more sophisticated formulation does not seem appropriate. The e-folding time $1/l$ may be interpreted as the time for two thirds of the detritus standing concentration to be regenerated in the absence of all other gain and loss mechanisms. Term XI parameterizes gravitational sinking of detritus, to be discussed later.

D. The Fish Equation

As the time scale of concern here is, at most, several weeks, we assume there will be no net increase in fish biomass F over the period of model integration. Hence, nekton grazing on P and Z must balance nekton excretion. Thus the derivative

$$dF'/dt' = \Phi F' - HF' = 0,$$

where Φ is the nekton grazing rate and H is the nekton excretion rate (h^{-1}). The fish are not affected by water movement or turbulent diffusion as are the other biotic components. Also, fish mobility in this preliminary ecosystem model is ignored.

Species composition of the nekton trophic level is not considered, but the pelagic fish are regarded as planktivores, that is, omnivorous filter feeders. Fish grazing is a function of gill raker efficiency. It is assumed that the fraction theta of the phytoplankton and all the zooplankton are within sizes retainable by the filtering mechanism.

The fish dynamics equation is

$$dF'/dt' = \underset{\text{IV}}{\Phi F' \theta P'/(\theta P' + Z')} + \underset{\text{VIII}}{\Phi F' Z'/(\theta P' + Z')} - \underset{\text{X}}{HF'}, \tag{11}$$

or

$$dF'/dt' = \Phi F' - HF',$$

and, as indicated above, $\Phi = H$ in the model. Term IV is the rate of ingestion of P by F. Term VIII is the rate of ingestion of Z by F and term X is the subsequent, instantaneous, and total excretion of the grazed ration into dissolved nutrients.

E. The Nutrient Equation

The local change in dissolved limiting nutrient N' is the sum of all nutrient excretion processes (term II, VI, and X), plus nutrient regenerated from detritus (term IX), minus that taken up by phytoplankton (term I):

$$dN'/dt' = -\underset{\text{I}}{V_m N'P'/(K+N')} + \underset{\text{II}}{BP'} + \underset{\text{VI}}{\Gamma E_z [1-\exp(-d_p P')] Z'^2} + \underset{\text{X}}{HF'} + \underset{\text{IX}}{lD'}. \tag{12}$$

If spatial effects were not of concern, this would be the complete formulation of a one-box model, as discussed by Verhoff (1971). Formulation of the physical processes will be discussed in Section XI.

V. Scaling of the Biological Dynamics

In the physical sciences, researchers find it rewarding to nondimensionalize their equations. This approach is sometimes an enigma to those unfamiliar with the procedure. We have chosen to solve Eqs. (8)–(12) after scaling by the phytoplankton growth parameter V_m and the total amount of limiting nutrient N_t.

The biological formulation, Eqs. (8)–(12), contains explicitly the parameters V_m, K, B, d_D, Γ, l, E_z, E_D, d_p, Φ, H, and θ, and implicitly the initial concentrations of N', P', Z', D', and F'. We anticipate that by nondimensionalizing the equations the number of parameters will be reduced. Solving one nondimensional case is then equivalent to solving many dimensional cases. To transform back to dimensional units one multiplies the nondimensional solutions by the scaling parameters.

We are fundamentally interested in the role of Michaelis–Menten kinetics within the phytoplankton trophic level. The flow of nutrient material through the primary producers is initially a function of the uptake rate of nutrient by phytoplankton from the water column. The maximum rate of nutrient uptake by phytoplankton, V_m, may be measured in concentration of the limiting nutrient or as the doubling rate of phytoplankton per 24 hours. It is a fundamental rate to which all other biological rates may be related. Thus, we let time be scaled by V_m, $t' = t/V_m$, where t' has units of hours, t is nondimensional, and V_m has units of h^{-1}. Let P', N', D', F', and Z' be scaled by N_t, the amount of limiting nutrient in all biological components averaged over the entire continental shelf region.

Using these scaling relationships, Eqs. (8)–(12) become

$$dP/dt = NP/(\alpha + N) - \beta P - \epsilon_z(1 - \exp(-\delta_p P))Z - \phi F\theta P/(\theta P + Z), \quad (13)$$

$$dZ/dt = \epsilon_z(1 - \exp(-\delta_p P))Z + \epsilon_D ZD - \gamma\epsilon_z(1 - \exp(-\delta_p P))Z^2 - \delta_D Z - \phi FZ/(\theta P + Z), \quad (14)$$

$$dD/dt = \delta_D Z - \lambda D - \epsilon_D ZD, \quad (15)$$

$$dF/dt = \phi F\theta P/(\theta P + Z) + \phi FZ/(\theta P + Z) - \eta F, \quad (16)$$

$$dN/dt = -NP/(\alpha + N) + \beta P + \gamma\epsilon_z(1 - \exp(-\delta_p P))Z^2 + \eta F + \lambda D, \quad (17)$$

where $N = N'/N_t$, $P = P'/N_t$, $Z = Z'/N_t$, $D = D'/N_t$, $F = F'/N_t$, $\alpha = K/N_t$, $\beta = B/V_m$, $\epsilon_z = E_z/V_m$, $\delta_p = d_p N_t$, $\phi = \Phi/V_m$, $\delta_D = d_D/V_m$, $\lambda = l/V_m$, $\epsilon_D = E_D N_t/V_m$, $\gamma = \Gamma N_t$, $\eta = H/V_m$, $t = t'V_m$, and $\theta = \theta$.

All quantities in Eqs. (13)–(17) are nondimensional. P, Z, N, F, and D are all fractions; if multiplied by 100 they are percent of N_t in the system at any time, i.e., a standing stock. One time unit equals V_m^{-1} in hours. The reader must bear in mind these scaling relationships when comparing our results with real oceanographic process rates.

VI. Environmental Considerations

The model is designed to apply to shelf waters during late fall to early spring, when winds create a well-mixed water column.

A euphotic zone, defined here as the part of the water column within which photosynthesis can take place, is assumed to extend to a maximum depth of

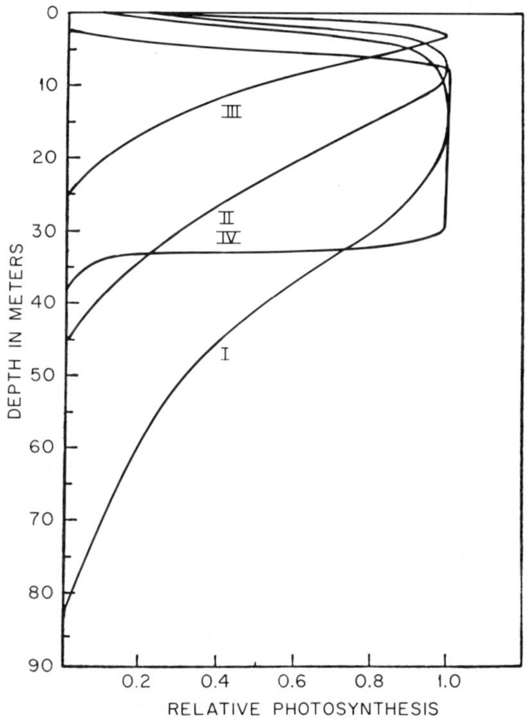

FIG. 4. Relative photosynthesis as a function of depth for a homogeneously distributed phytoplankton population under bright sunlight in the open ocean (after Yentsch, 1963): I, maximum transparency; II, average transparency; III, minimum transparency; IV, model curve defines the euphotic zone and the relative rate of photosynthesis in the spatial model. (Reprinted in slightly modified form with permission from George Allen and Unwin Ltd., London.)

40 m. As photosynthesis is a function of light intensity and light attenuates with depth, the rate of photosynthesis with depth in the euphotic zone is not constant. In the spatial model the growth rate of phytoplankton is subjected to a photosynthesis-depth curve (Fig. 4).

We define a photosynthesis-depth function, $E(z)$, which approximates Curve II (Yentsch, 1963) of Fig. 4:

$$E(z) = 0.5[1 + \tanh(10[Z - Z_0])],$$

where z_0 is approximately 30 m. To account for light inhibition of photosynthesis in the upper 5 m of the water column, we redefine

$$E(0-2.5 \text{ m}) = 0 \qquad \text{and} \qquad E(2.5-5 \text{ m}) = 0.5.$$

Note $E(z) < 0.01$ below 35 m. There is considerable room for improvement in the specification of this important environmental parameter. Phytoplankton cell self-shading effects are not considered.

The Michaelis–Menten uptake term in Eqs. (13) and (17) is multiplied by this dimensionless function thereby restricting increases in phytoplankton to the defined euphotic zone. Phytoplankton advected or diffused below 40 m experience only losses due to excretion and predation, since $E(z) = 0$. Zooplankton grazing is allowed to take place wherever phytoplankton occur.

VII. Values for the Variable Biological Parameters

The model is integrated using specific values of the parameters α, β, ϵ_z, ϵ_D, λ, δ_D, γ, ϕ, η, δ_p, and θ which are based on oceanographic measurements. In some cases food chain theory is resorted to as a guide to the appropriate parameter value. The rationale for selecting the specific values of the parameters used in this study follows.

A. Michaelis Constant ALPHA

Dugdale (1967), Eppley *et al.* (1969), and others document the Michaelis constant K as a small concentration of the limiting nutrient for which the specific uptake rate is one half the maximum uptake rate. MacIsaac and Dugdale (1969) present euphotic zone values of the order of 10^{-1} μmoles l^{-1} in nitrogen-poor regions which experience seasonal upwelling of water with concentrations of 10–20 μmoles l^{-1} of nitrate. Thomas (1970) has found a Michaelis constant of 0.75 μmole NO_3–N l^{-1} for natural tropical Pacific phytoplankton populations limited by nitrogen.

If the dissolved nutrient concentration at 2000 m off the West Florida continental shelf is N_t, then for a nitrate-limited system $N_t = 30\ \mu\text{g at NO}_3\text{-N l}^{-1}$ (Collier, 1958) and

$$\alpha = \frac{K}{N_t} = \frac{0.75\ \mu\text{g at NO}_3\text{-N l}^{-1}}{30\ \mu\text{g at NO}_3\text{-N l}^{-1}} = 0.025.$$

Thomas and Dodson (1968) have reported a K value of $0.12\ \mu\text{g at PO}_4\text{-P l}^{-1}$ for tropical waters where phosphate is limiting. Assuming the total amount of biologically available phosphate to be $4.0\ \mu\text{g at PO}_4\text{-P l}^{-1}$ (Collier, 1958), then

$$\alpha = \frac{K}{N_t} = \frac{0.12\ \mu\text{g at PO}_4\text{-P l}^{-1}}{4\ \mu\text{g at PO}_4\text{-P l}^{-1}} = 0.03.$$

A range of α values (10^{-1} to 10^{-2}) has been investigated. Lower values of α correspond to phytoplankton utilizing extremely low concentrations of limiting nutrient, permitting a higher percent of limiting nutrient in the phytoplanton component. A larger value of α results in less efficient uptake.

B. Phytoplankton Excretion Coefficient BETA

The coefficient β parameterizes the combined losses of the limiting nutrient by P not accounted for by zooplankton grazing and fish predation. A major loss process is extracellular release of photosynthetic products. The magnitudes published for excretion are controversial due to experimental difficulties, but a significant portion of the compounds excreted may be amino acids. Fogg (1966) has found the proportion of photosynthetic production lost through extra-cellular release to range from 5 to 34% in natural phytoplankton populations. Watt (1966), Fogg (1966) and Thomas (1970) found that nutrient deficiencies, as nitrogen starvation, cause an increase in the release of dissolved organic matter from algae.

Consider the loss of phytoplankton biomass due to the excretion process alone. The time $T = 1/\beta$ is the e-folding time taken for the process $-\beta P$ to reduce P to one third of any arbitrary initial value. A typical model value for β is 0.025. If V_m is of the order $10^{-1}\ \text{h}^{-1}$ for nutrient-limited systems (Eppley *et al.*, 1969), then the real time scale for loss of absorbed nutrient is

$$T' = T/V_m = (\beta V_m)^{-1} \approx 40\ \text{h}.$$

The excretion time scale β^{-1} must be longer than the growth time scale or else the phytoplankton population will collapse.

C. Zooplankton Parameters EPSILON-Z, EPSILON-D, and DELTA-P

The determination of zooplankton grazing rates in the sea is very difficult. Laboratory measurements by Parsons *et al.* (1967) of the grazing ration expressed as a ratio of the weight of phytoplankton carbon ingested to the weight of the animals varied from 10 to 60% for varying concentrations of phytoplankton. It is not obvious what value the nondimensional grazing coefficient ϵ_z should attain. We choose the value of parameters ϵ_z and ϵ_D to give a reasonable ecological efficiency by zooplankton.

The expression for phytoplankton-dependent zooplankton grazing in the model $\epsilon_z\,[1-\exp(-\delta_p P)]\,Z$ yields a curve whose slope initially depends on δ_p and asymptotically reaches the maximum ϵ_z. The quantity δ_p^{-1} is the concentration at which the zooplankton grazing rate is approximately two thirds of the maximum allowable rate ϵ_z. In the model δ_p is taken to be 1.2, giving a grazing rate of $(2/3)\epsilon_z$ at a nondimensional phytoplankton concentration of 0.83. Since the model solutions almost always yield $P < 0.8$, the model utilizes the nonlinear part of the curve.

D. Zooplankton Excretion and Death Coefficients GAMMA and DELTA-D

A correct formulation of zooplankton excretion processes is essential for marine ecosystem models, as excretion is an important mechanism by which nutrients are recycled back into the water column (Cushing, 1969). Much of the grazed phytoplankton passes through the zooplankton gut undigested, especially in regions of high phytoplankton concentration.

Parameter γ of the term $\gamma\epsilon_z\,[1-\exp(-\delta_p P)]\,Z^2$ linearly relates zooplankton excretion to zooplankton grazing. In the model, the amount grazed minus the amount excreted equals the nutrient assimilated by zooplankton. Neglecting the relatively small detritus food source for illustrative purposes, the limiting nutrient assimilated by the zooplankton is

$$dZ/dt = \epsilon_z(1-\exp(-\delta_p P))Z - \gamma\epsilon_z(1-\exp(-\delta_p P))Z^2.$$

If both sides of the equation are divided by the quantity grazed, we have

$$[\epsilon_z(1-\exp(-\delta_p P))Z]^{-1}\, dZ/dt = 1-\gamma Z.$$

For growth of zooplankton, γZ must be less than 1. As zooplankton concentrations are typically less than $0.25N_t$, γ may be estimated to be approximately 4.0.

Upon death the zooplankton becomes part of the detritus component. The limiting nutrient within the detritus component may be directly utilized by zooplankton, or regenerated to dissolved nutrient by bacterial decomposition. The value of δ_D is taken as 0.10. Thus two thirds of the Z population dies in time $t = 1/\delta_D$ or $10/V_m$ hours. If V_m is of the order 10^{-1} to $10^{-2}\,\mathrm{h}^{-1}$ (Eppley *et al.*, 1969), then the e-folding time is on the order of $\simeq$ 4 to 40 days, if its food source is absent.

E. Nutrient Regeneration Constant LAMBDA

The coefficient of bacterial decomposition of detritus into limiting nutrient is λ. In Eq. (15), λ is nondimensionalized by V_m^{-1}. A value of $\lambda = 0.5$ would be equivalent to a regeneration time lag of the order of one or two days. This seems an appropriate regeneration time for the limiting nutrients (phosphate and nitrate) considered in the model (Ketchum and Corwin, 1965).

F. Fish Grazing, Excretion, and Gill Raker Efficiency Coefficients PHI, ETA, and THETA

The coefficient ϕ is taken to be 0.25, based on ecological efficiency considerations. As fish biomass is constant in this model, the excretion coefficient η is identical to the grazing rate ϕ.

The gill raker efficiency coefficient θ is arbitrarily set at 0.10. This value means that 90% of the phytoplankton is passed by the gill rakers.

VIII. Steady-State Values of the Biological Components

We have formulated a complex system describing the dynamics of the biological components. In order to interpret the spatial effects, it is desirable first to understand the system within the classical one-box framework, where most physical processes can be ignored. In this respect, Eqs. (13)–(17) describe the dynamics of a nutrient-limited biological system. The properties of the system are determined by investigating the time-dependent solutions of the biotic components as governed by these equations. An analytical solution is not possible due to nonlinearity of the relationships. Equations (13)–(17) are consequently integrated numerically for the range of parameter values discussed above.

Figure 5 shows one time-dependent solution of these equations. The abscissa is nondimensional time. One time unit is V_m^{-1} hours, and is of the order of one day (25 h for $V_m = 0.04\ \text{h}^{-1}$). The curves express the standing stock of each biotic component as a fraction of N_t. We assume at $t = 0$, that the onset of intense upwelling or deep mixing by a storm injects an excess amount of the limiting nutrient, $N = 0.25$, into the closed system. We arbitrarily choose $P = 0.33$, $Z = 0.33$, $F = 0.09$, and $D = 0.0$ as the other initial concentrations.

Figure 5 shows the rapid growth of phytoplankton by uptake of the limiting nutrient. The dissolved nutrient decreases accordingly. The zooplankton curve asymptotes to a steady-state standing stock value. Detritus increases from zero at time $t = 0$ to a small fraction (4%) of N_t. Fish biomass remains constant. Note that $P + N + Z + D + F = 1$ for all time.

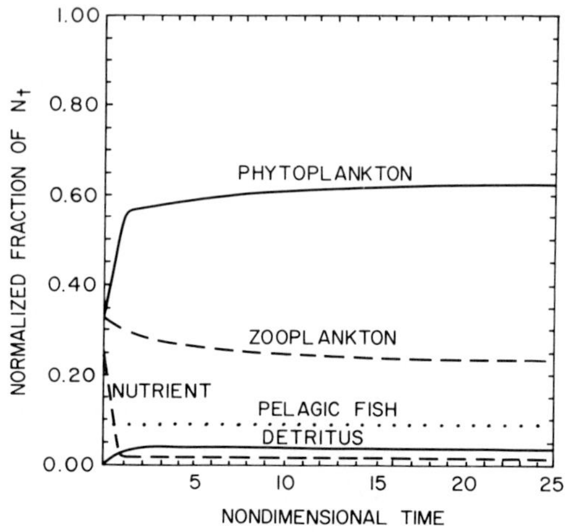

FIG. 5. Time-dependent standing stock concentrations of the biotic components *P, Z, N, D,* and *F.* The abscissa is nondimensionalized time ($t = t'V_m$). The ordinate is the concentration of the biotic component as a fraction of the total amount of limiting nutrient in the system ($N_t = 1.0$). The parameter values for this solution are: $\alpha = 0.02$, $\beta = 0.25$, $\epsilon_z = 0.63$, $\epsilon_D = 0.60$, $\delta_p = 1.20$, $\gamma = 2.30$, $\lambda = 0.50$, $\delta_D = 0.10$, $\theta = 0.10$, $\phi = \eta = 0.25$.

The overall model behavior is *not* critically dependent on the initial conditions. The biotic components approach the same steady-state values when initial concentrations are changed. This result is not shown for brevity.

The response of the one-box system to fluctuations in environmental conditions is of interest. Since in nature the rate of photosynthesis varies with light intensity, we simulate the effect of diurnal variation of sunlight upon the growth kinetics of phytoplankton. The phytoplankton nutrient uptake term is multiplied by a periodic function such that the rate of phytoplankton growth becomes a function of time of day. Optimal light conditions initiate an increased photosynthetic rate by plants, but this peak lasts for only several hours of the day. Sufficient nutrient uptake may be accomplished within this short period to sustain the phytoplankton over 24 h (Glooschenko *et al.*, 1972).

We choose a periodic function which gives the same time-averaged amount of nutrient uptake as would a constant rate over 24 h:

$$f(t) = \pi(2\tau)^{-1} \sin(\pi t/\tau),$$

where t is nondimensional time equal to $t'V_m$ and τ represents the day length fraction of one nondimensional day. When $f(t) \leqslant 0$, $f(t) = 0$ is set to simulate night conditions. Note that

$$\int_0^\tau f(t) = 1,$$

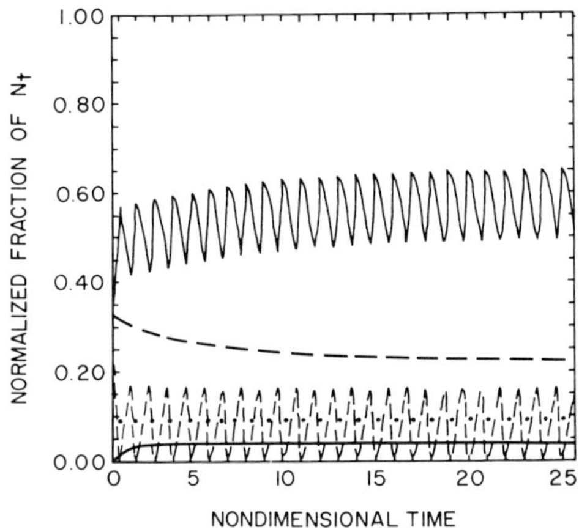

FIG. 6. Effect of diurnal photosynthetic rate variation upon the standing stock concentration of phytoplankton, zooplankton, dissolved nutrient, detritus, and pelagic fish. The solution includes the diurnal photosynthetic rate function with parameter $\tau = 0.50$. Line representations and all parameter values are the same as in Fig. 5.

which states that over one nondimensional day the amount of nutrient uptake is exactly equal to uptake with or without diurnal variation, if N is a constant over the entire period.

A one-box solution including this diurnal variation of nutrient uptake rate by phytoplankton is presented in Fig. 6. The value of τ used here is 0.50. Each time unit on the abscissa of Fig. 6 is one V_m^{-1}. All parameter values are the same as in Fig. 5.

The daily fluctuations in phytoplankton standing stock result in a negatively correlated oscillation of the dissolved nutrient concentration. The phytoplankton oscillations are damped out by numerous feedbacks of the food chain system, such that the zooplankton and detritus components show no daily fluctuations.

McAllister (1970) has shown that nocturnal zooplankton grazing behavior places a greater stress upon the phytoplankton standing crop than does a constant grazing or grazing only during the day. A nocturnal grazing stress upon the system has yet to be investigated with the model.

IX. Standing Stocks and Ecological Efficiencies of the Food Chain Model

In 1962, Curl attempted to measure the standing crops of carbon, nitrogen, and phosphorus and their transfer between marine trophic levels in the continental shelf waters south of New York. Sampling errors resulted in poor measurements of phytoplankton, herbivore, and carnivore standing stocks, and subsequent estimation of trophic level production became difficult. In view of the complexity of food web structure and the natural fluctuations in standing crops both in time and space, it is doubtful that a total understanding of the biological dynamics of an area can be attained by sampling alone. An approximate model of trophic level dynamics such as presented here may provide the basis for investigating factors which cause changes in standing stocks.

Major fluctuations in phytoplankton standing stock reflect the availability of nutrient in the water column. Zooplankton and fish production depend on the food supply. This model permits a study of the effect of nutrient limitation on standing crops of phytoplankton, zooplankton, and detritus.

A standing stock is defined as the concentration of limiting nutrient in a trophic level at a certain time. In this model trophic level biomass or standing stock is a function of assimilated rather than merely ingested food. Thus,

$$\text{standing stock} = \text{nutrient gains} - \text{nutrient losses}.$$

The amount of food intake by a trophic level is proportional to its biomass.

Due to the confusion of terminology of various authors on the trophic level concept, we follow Kozlovsky's (1968) lead in explicitly defining *net productivity* as the energy, or here limiting nutrient accumulation, which occurs in a trophic level. *Production* is explicitly defined as *that portion of the productivity of a trophic level which is available to the next trophic level.* In other words, production is less than productivity since some of the net productivity is lost through processes such as death or excretion. It should be remembered that a closed system is assumed, with conservation of mass. Slobodkin's (1960) definition of ecological efficiency originally formulated for energy transfers can be directly applied to nutrient transfer systems. We define

$$\text{ecological efficiency} = \frac{\text{production}}{\text{food intake}} = \frac{\text{energy passed to next higher trophic level}}{\text{ingestion at the trophic level}}.$$

For example, the ecological efficiency between phytoplankton and zooplankton in the model using steady-state values from Fig. 5 and Eqs. (13) and (14) is

$$\frac{\text{production of } Z}{P \text{ and } D \text{ available to } Z} = \frac{\epsilon_z[1-\exp(-\delta_p P)]\,Z + \epsilon_D ZD - \gamma\epsilon_z[1-\exp(-\delta_p P)]\,Z^2 - \delta_p Z}{[NP/(\alpha+N)] - \beta P + \delta_D Z - \lambda D}$$

$$= 0.2071.$$

This means that 4.8 times as much phytoplankton is being produced than zooplankton in terms of biomass. This value is within the limits established in present food chain theory. A simplified 10% ecological efficiency between trophic levels of an aquatic food chain has been postulated by Lindeman (1942). Slobodkin (1962) suggests higher ecological efficiencies of 20–25% in general for herbivores, and Ryther (1969) postulates 20% efficiency for zooplankton in upwelling areas.

The ecological efficiency for fish production using model equations (13), (14), and (16) and steady-state values from Fig. 5 is 0.2295. This fish production efficiency is quite acceptable, especially in light of work by Gulland (1970) and Lasker (1970). It is the steady-state standing stocks which determine the efficiencies, and the standing stocks are not known until the model is run. Parameter values have been varied to study the effect of food supply on the steady-state standing stocks and resulting ecological efficiencies. In development of the model, the tendency has been for more realistic ecological efficiencies to occur following specification of more realistic model formulations.

We conclude reasonable efficiencies can be obtained with the existing model, Eqs. (13)–(17), within the range of parameter values deduced.

X. Sensitivity Analysis

To determine the relative importance of biological processes occurring within the system, the technique of sensitivity analysis (Tomovic, 1963) has been employed. This formal mathematical analysis elegantly replaces the less economical, empirical procedure of adjusting the model's parameters and observing the resulting solutions.

Model sensitivity is defined as the displacement from equilibrium a model experiences due to quantitative variation in an individual parameter. If a model component or the whole system changes substantially with small variation in a particular parameter, then the value of that parameter is important and must be estimated with precision (Smith, 1970). Any continuing research on the ecosystem should focus on the study of mechanisms found highly influential in the simulation model. On the other hand, sensitivity analysis is no panacea. It can only sort out important processes already included in the model. It cannot anticipate which additional processes should be included.

Model sensitivity is investigated both empirically and analytically. First, the steady-state values of the model components are empirically computed in a simulation run. One parameter is increased by a fixed percent, and the model solution recalculated. The process is repeated for each parameter. The results give estimates of quantities such as $\partial P/\partial\beta$, $\partial N/\partial\alpha$, and so forth.

The method for analytical sensitivity analysis involves derivation of partial differential equations describing the rate of change of biological components with respect to change in the parameters. These simultaneous equations are then solved for values of the partial derivatives $\partial P/\partial\beta$, $\partial N/\partial\alpha$, and so forth. The two methods of analysis may be compared. To within roundoff error they were found identical.

A. Analytical Sensitivity Analysis

Following Tomovic (1963) the steady-state forms of Eqs. (13)–(17) are differentiated with respect to each parameter, the independent variables for this analysis. The results are represented by a square coefficient matrix A. Matrix A is always the same upon differentiation with respect to any parameter. We then have the matrix equation $AX = B$, where vector B is dependent on the parameter chosen. The solution to the set of simultaneous equations for the rate of change of the components with respect to a change in the individual parameters is

$$X = A^{-1}B,$$

where A^{-1} is the matrix inverse. As the sum of the steady-state equations, Eqs. (13)–(17), is zero, the simple closure relationship

$$N + P + Z + D + F = 1$$

must be used in the sensitivity analysis.

We arbitrarily neglect the nutrient equation. The same solutions would result if the phytoplankton, zooplankton, or detritus equation were chosen to be neglected. Since fish biomass is constant, the fish equation is disregarded as trivial.

Differentiating the remaining equations with respect to α, for example, gives the square coefficient matrix A,

$$\begin{bmatrix} A_{11} & A_{12} & A_{13} & A_{14} \\ A_{21} & A_{22} & A_{23} & A_{24} \\ A_{31} & A_{32} & A_{33} & A_{34} \\ A_{41} & A_{42} & A_{43} & A_{44} \end{bmatrix} \begin{bmatrix} \dfrac{\partial P}{\partial\alpha} \\ \dfrac{\partial Z}{\partial\alpha} \\ \dfrac{\partial D}{\partial\alpha} \\ \dfrac{\partial N}{\partial\alpha} \end{bmatrix} = \begin{bmatrix} \dfrac{N}{(\alpha + N)^2} \\ 0 \\ 0 \\ 0 \end{bmatrix},$$

where

$$A_{11} = N/(\alpha + N) - \beta - \delta_p \epsilon_z Z \exp(-\delta_p P) + \theta^2 \phi FP/(\theta P + Z)^2 - \theta\phi FP/(\theta P + Z),$$

$$A_{12} = -\epsilon_z [1 - \exp(-\delta_p P)] + \theta\phi FP/(\theta P + Z)^2,$$

$$A_{14} = \alpha/(\alpha + N)^2,$$

$$A_{21} = [\delta_p \epsilon_z Z \exp(-\delta_p P)](1-\gamma Z) + \theta\phi FZ/(\theta P + Z)^2,$$

$$A_{22} = \epsilon_z [1-\exp(-\delta_p P)](1-2\gamma Z) + \epsilon_D D + \phi FZ/(\theta P + Z)^2 - \delta_D - \phi F/(\theta P + Z),$$

$$A_{23} = \epsilon_D Z,$$

$$A_{32} = -\epsilon_D D + \delta_D,$$

$$A_{33} = -\epsilon_D Z - \lambda,$$

$$A_{13} = A_{24} = A_{31} = A_{34} = 0,$$

$$A_{41} = A_{42} = A_{43} = A_{44} = 1.$$

A standard Gaussian elimination subroutine is used to solve for A^{-1}. The column vector B for α is shown above.

Table I is the result of the matrix solution for each parameter column vector. The values in Table I are the evaluated partial derivatives of P, Z, N, and D with respect to the parameters α, β, ϵ_z, λ, γ, δ_D, θ, ϕ, ϵ_D, and δ_p and normalized by the ratio of the parameter to the model component. That is, the first value in Table I is $(\alpha/P)(\partial P/\partial\alpha)$. The steady-state values for P, Z, N, and D and the parameter values used in the calculations are specified in Fig. 5. Component F is not included as it has a constant concentration of limiting nutrient in the model. As fish excretion coefficient η is equivalent to the fish grazing coefficient ϕ, only ϕ is considered in the analysis. For definition purposes, the quantitative change in a component is the component's *sensitivity* to that particular parameter. The sensitivity of greatest magnitude (−116, 90) is that by Z for γ. The smallest sensitivity (−1.66) indicates minimum fluctuations in detritus concentration for changes in β.

TABLE I

Sensitivity Analysis: Partial Derivative Values of the Biotic Components with Respect to the Biological Parameters

	P	Z	N	D
α	−2.60	−3.23	99.29	−2.53
β	−1.70	−2.12	65.18	−1.66
ϵ_z	−45.67	102.80	68.79	80.52
λ	7.88	−7.63	−3.20	−84.32
γ	50.50	−116.90	−41.40	−91.62
δ_D	18.81	−57.11	−19.57	55.26
θ	−4.75	10.85	5.44	8.50
ϕ	22.17	−51.53	−16.21	−40.37
ϵ_D	−4.07	12.37	4.24	−11.97
δ_P	−30.79	69.28	46.37	54.28

Relative sensitivity is obtained by setting the largest normalized change equal to one, and expressing all other sensitivities as a percent of this largest change (Smith, 1970). Upon observation of which normalized value is the greatest, we divide all other normalized values by this quantity. Table II results from this procedure. All normalized values have been divided by the quantity $[(\gamma/Z)(\partial Z/\partial\gamma) \times 100]$. We denote the total operations upon the original partial derivative thus far as the partial derivative with an asterisk, e.g., $(\partial P/\partial\alpha)^*$.

TABLE II

Relative Sensitivity Analysis: Normalized Table I Values Divided by $|(\gamma/z)(\partial z/\partial\gamma) \times 100|$

	P	Z	N	D
α	−0.022	−0.028	0.849	−0.022
β	−0.014	−0.018	0.557	−0.014
ϵ_Z	−0.390	0.879	0.588	0.688
λ	0.067	−0.065	−0.027	−0.721
γ	0.432	−1.000	−0.354	−0.783
δ_D	0.161	−0.488	−0.167	0.473
θ	−0.041	0.093	0.046	0.073
ϕ	0.190	−0.441	−0.139	−0.345
ϵ_D	−0.035	0.106	0.036	−0.102
δ_P	−0.263	0.592	0.397	0.464

The largest relative change of all components is that of zooplankton as affected by its excretion coefficient γ. The negative sign (−1.0) indicates a decrease in Z if γ is increased. Higher values of γ mean less assimilation. The next largest relative sensitivity (0.879) is $(\partial Z/\partial\epsilon_z)^*$, the positive effect of zooplankton grazing for the zooplankton standing crop. A larger maximum grazing coefficient results in more zooplankton biomass, as one would expect. The third greatest sensitivity (0.849) is the dependence of the dissolved limiting nutrient N upon the nondimensional Michaelis–Menten constant α. A higher α indicates a less efficient uptake of the nutrient by phytoplankton, a greater concentration of N remaining biologically unavailable in the water column.

The magnitude of the $(\partial D/\partial\gamma)^*$ relationship between detritus and the zooplankton excretion coefficient γ is (−0.783). The effect is indirect, but understandable. The more zooplankton excrete, the less zooplankton biomass is available eventually to become detritus. The effect (−0.721) of the bacterial regeneration rate coefficient λ upon the detritus component is to lower the amount of D in the system. A higher λ shortens the time for regeneration of detritus into available nutrient, as the *e*-folding time scale for detritus is λ^{-1}. The dependence of detritus concentration upon the zooplankton grazing coefficient ϵ_z (0.688) is again related to zooplankton standing crop.

The seventh largest sensitivity is that by zooplankton biomass for the Ivlev constant. The $(\partial Z/\partial \delta_p)^*$ relationship (0.592) indicates the importance of the nonlinear zooplankton grazing formulation. The indirect dependence of N upon ϵ_z (0.588) is obscure, except in light of the coupling of system processes. Greater zooplankton grazing results in fewer phytoplankton and more nutrient in the water column.

The amount of dissolved limiting nutrient would expectedly increase (0.557) with greater phytoplankton extracellular release, and zooplankton biomass would decrease (−0.488) with a greater death rate.

The above are the ten largest normalized sensitivities. Also of interest are parameters with the least influence upon the biotic component value. It is enlightening to discover that the effect of parameters α and β, the phytoplankton nutrient uptake and extracellular release coefficients, have the smallest effect upon the standing stocks of phytoplankton, zooplankton, and detritus.

There are 40 relationships expressed in each of Tables I and II, all of which can be interpreted in terms of food chain relationships or physiological processes. The relative sensitivities of Table II express the importance of system processes without regard to magnitude of the coefficients or components involved. Using the normalized sensitivities, the overall importance of the parameters can be estimated by assigning each scaled partial derivative an order from one to forty based on its absolute value in Table II.

Table III shows the order assignment of these forty partial derivatives. For example, the order of $(\partial Z/\partial \gamma)^*$ is read under the γ parameter column along the Z component row. Next we sum the columns. Using high relative sensitivities for low-order assignment, the lowest column totals represent overall importance of parameters. "Overall" means the summed influence of the parameter on standing stocks of the system's biotic components.

TABLE III

Overall Importance of Parameters: Ordering of the Forty Model Sensitivity Values of Table II. High Relative Sensitivities Are Given Low-Order Assignments

	ϵ_Z	γ	δ_P	δ_D	ϕ	λ	α	θ	ϵ_D	β
P	16	14	19	22	20	28	36	31	33	39
Z	2	1	7	10	13	29	34	26	24	38
N	8	17	15	21	23	35	3	30	32	9
D	6	4	12	11	18	5	37	27	25	40
Σ	32	36	53	64	74	97	110	114	114	126

From Table III, the zooplankton grazing coefficient ϵ_z is the most important parameter in the biological dynamics. Closely following is the zooplankton excretion coefficient γ. The Ivlev constant δ_p and the zooplankton death coefficient δ_D are also highly influential on the system's steady state. Clearly, the system is very sensitive to zooplankton dynamics.

The fish grazing coefficient ϕ is fifth in importance. The regeneration rate parameter λ has the next overall importance on the system, followed by the nondimensionalized Michaelis–Menten constant α. The fish gill raker efficiency constant θ is followed by ϵ_D and finally β. The parameter with the lowest sensitivity in Table III, namely β, is indicated to have the least overall effect upon the system.

B. Interpretation of Sensitivity Analysis

Sensitivity analysis gives an indication of which of the included processes greatly affect the system's steady state. To achieve a better understanding of the system, the processes characterized as important should then be studied more closely in the laboratory and field. In our formulated food chain model, the herbivore grazing dynamics are paramount. The importance of zooplankton grazing in marine plankton systems has been documented by Cushing (1969), Riley (1946), and Steele (1959). Observations of herbivore excretion enhancing the growth of phytoplankton have been recorded (Cushing, 1969; Walsh and Dugdale, 1971). The sensitivity analysis implies that the parameters needed to be most accurately measured for this system are those of zooplankton grazing, excretion and death, and fish predation. Current research is concerned with these estimations.

The fundamental importance of V_m, the growth rate of the phytoplankton, should not be forgotten. Parameter V_m is implicit in the sensitivity analysis via the scaling parameters. For example, a change in ϵ_z implies a change in either V_m or the maximum grazing ration E_z. As $t = t'(V_m)$, we are in essence investigating all other processes relative to the growth time scale of the phytoplankton.

XI. Water Circulation on the Shelf

A known physical circulation is essential to any spatial description of biological production. The west Florida coastal water is influenced by the presence of a strong (50–150 cm sec^{-1}) southerly flow, the Loop Current, located off the continental shelf break. It occurs approximately 200–250 km off the west coast of Florida during the winter season (Leipper, 1970). Where the Loop Current meanders against the continental shelf, bottom friction may result in shoreward transport of water along the continental shelf bottom. Conservation of mass would require displaced bottom water to be forced upward and offshore. This process is termed "current-induced upwelling" (Hsueh and O'Brien, 1971). Observational evidence in the form of oxygen isopleths and isolines of σ_t indicates the seasonal occurrence of upwelling (Bogdanov *et al.*, 1968; Austin, 1971).

A. Formulation of the Flow Field

A simple linear model of a homogeneous ocean with a straight vertical coast and a flat continental shelf (Fig. 2) is employed. Boundary constraints placed upon the flow are:

1. flow at the coast is zero; i.e., $u' = v' = 0$;
2. flow at the bottom is zero; i.e., $u' = v' = 0$;
3. there is no wind stress at the free surface; i.e., $\partial u'/\partial z' = \partial v'/\partial z' = 0$.

At the seaward edge of the shelf region, the horizontal velocity of the offshore current is prescribed as a function of depth. The magnitude of the v' velocity at the shelf break boundary is described by a cosine function,

$$v' = -C \cos(3\pi z/2d),$$

simulating a southerly (negative) flow in the upper layers and a countercurrent northerly (positive) flow above the bottom. The characteristic speed is C and the depth of the water column is d. Upon neglecting the nonlinear acceleration terms, assuming constant density and hydrostatic equilibrium, and taking $\partial \cdot /\partial y = 0$, we can write the steady-state u' momentum equation as

$$fv' - g\,\partial\zeta'/\partial x' + A_h\,\partial^2 u'/\partial x'^2 + A_v\,\partial^2 u'/\partial z'^2 = 0, \tag{18}$$

where primes denote dimensional terms, $\partial\zeta'/\partial x'$ is sea slope in the x direction, f is the Coriolis parameter, g is the acceleration of gravity, and A_v and A_h are assumed constant vertical and horizontal eddy viscosities.

The v' momentum equation at steady state can be written

$$-fu' + A_h\,\partial^2 v'/\partial x'^2 + A_v\,\partial^2 v'/\partial z'^2 = 0. \tag{19}$$

The parameter space can be reduced by nondimensionalizing these equations. Let

$$u' = Cu, \qquad v' = Cv,$$
$$x' = (A_h/f)^{1/2}x, \qquad w' = (A_v/A_h)^{1/2}Cw,$$
$$z' = (A_v/f)^{1/2}z, \qquad \zeta' = -[(fA_h)^{1/2}C/g]\zeta,$$

where C is the typical horizontal speed of the Loop Current. The nondimensionalized equations of motion for the cross section of the ocean shown in

Fig. 2 is

$$-v = -\partial\zeta/\partial x - \partial^2 u/\partial x^2 - \partial^2 u/\partial z^2, \tag{20}$$

and

$$u = \partial^2 v/\partial x^2 + \partial^2 v/\partial z^2. \tag{21}$$

Since $\partial v/\partial y$ is taken to be zero, the continuity equation

$$\partial u/\partial x + \partial w/\partial z = 0, \tag{22}$$

may be used to calculate w.

At the surface, the Coriolis force fv balances the pressure gradient force $\partial\zeta/\partial x$. However, within the bottom Ekman layer, friction along the bottom decreases the v velocity. The geostrophic balance is upset and water is transported onshore. The water must flow seaward at the top to provide mass balance.

B. The Simulated Circulation Pattern

Figure 7 is a graphical representation of the physical circulation pattern over the shelf used in this model. The vertical velocities are shown in Fig. 7a. Note that the strong positive velocities (upwelling) occur over most of the shelf while negative (downwelling) velocities are found inshore and at the shelf break boundary.

Figure 7b shows the meridional circulation. The stream function is defined as

$$\partial\psi/\partial z = -u \qquad \text{and} \qquad \partial\psi/\partial x = w,$$

since $\partial u/\partial x + \partial w/\partial z = 0$ in this model. This picture conceptually represents flow of the ocean in the x, z-plane. The magnitude of the flow is proportional to spacing of the contours, and the sign of the stream function specifies direction. The flow is always parallel to the stream function contours.

Two gyres over the shelf are found. One gyre weakly spinning in a clockwise direction is located over the shoreward part of the shelf. The second gyre located approximately 130 km offshore spins strongly in a counterclockwise direction. This motion results in strong upwelling of bottom waters from 120 to 180 km offshore and downwelling near the coast and seaward boundary. There is flow onto the shelf near the bottom and off the shelf at the surface. The longshore flow v is not shown as the biological distributions are homogeneous in this direction.

C. Combination of Biological and Physical Formulations

The biological dynamics can now be placed into the simulated physical system. Boundary constraints placed on the advection and diffusion of biotic

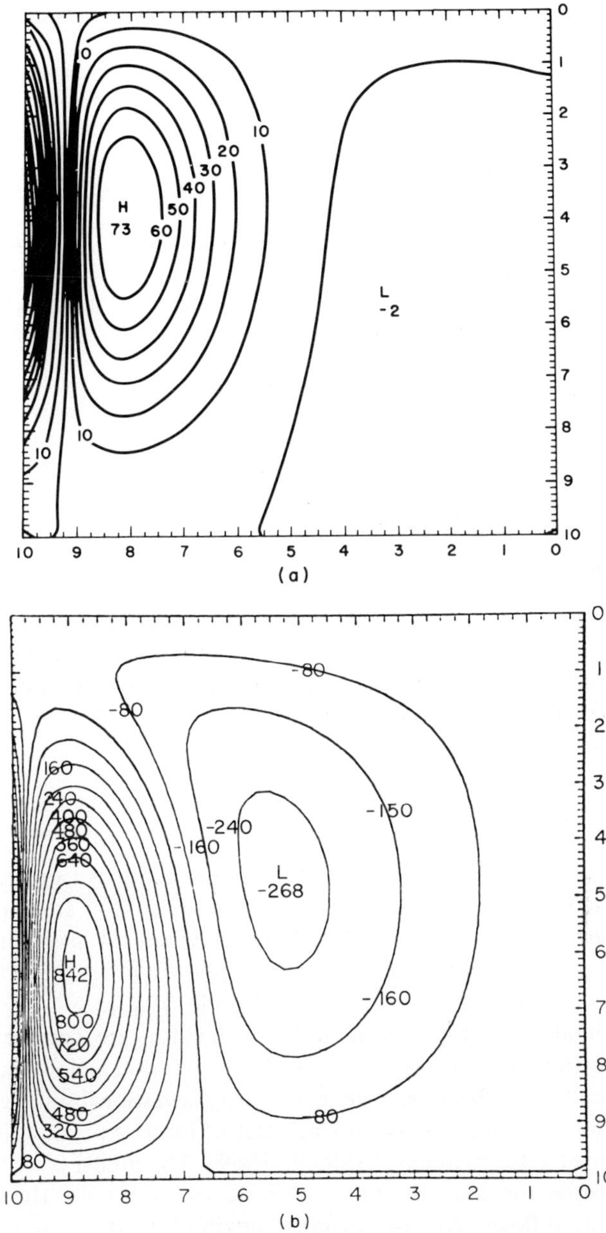

FIG. 7. Contours of the vertical w velocities (a) and the streamfunction ψ (b) over the oceanic section. Positive w indicates an upward velocity; negative w represents a downward velocity. Negative ψ values represent a clockwise gyre. Positive ψ values denote a counterclockwise gyre. Strong upwelling is indicated from approximately 100–180 km offshore, with downwelling occurring at the coast and seaward boundary. Each enumerated vertical unit is 20 m and each horizontal unit represents 20 km. Figure 7a contours from -2.30 to 0.73 with a contour interval of 0.1. Figure 7b contours from -0.56 to 0.80 by 0.08.

components are:

1. no transport across the free surface,
2. no transport through the bottom or coastal boundary,
3. transport across the seaward boundary is determined by the flow field.

We define a nondimensional ratio $S = C/[V_m(A_h/f)^{1/2}]$ (O'Brien and Wroblewski, 1973), and the Rossby number $R_0 = C(A_h f)^{-1/2}$. The diffusion coefficients are nondimensionalized as $\nu_h = \nu_h'[V_m(A_h/f)]^{-1}$ and $v_v = v_v'[V_m(A_v/f)]^{-1}$. The final nondimensional form of Eq. (1) is

$$\frac{\partial P}{\partial t} + S\left(u\frac{\partial P}{\partial x} + w\frac{\partial P}{\partial z}\right) - \nu_h\frac{\partial^2 P}{\partial x^2} - \nu_v\frac{\partial^2 P}{\partial z^2}$$

$$= PN/(\alpha + N) - \beta P - \epsilon_z[1 - \exp(-\delta_p P)]Z - \phi F[\theta P/(\theta P + Z)]. \tag{23}$$

The total derivatives (left-hand side) for the Z and N components are similar. No spatial distribution of pelagic fish is considered.

Notice that the advection processes are scaled by the parameter S. This nondimensional ratio is related to the maximum advection speed in the horizontal, the maximum rate of uptake of limiting nutrient, and the turbulent part of the flow field. The value of S is discussed later, but we note that the rates of the physical–chemical–biological model are all scaled by the biological turnover rate V_m^{-1}.

D. Sinking of Detritus

The subject of sinking of plankton components is controversial. Field observations of the vertical distribution of phytoplankton are confused by the combined effects of gravitational sinking of senescent phytoplankton cells, neutral buoyancy maintained by actively photosynthesizing cells, and the vertical advection of cells by water motion (Steele, 1959).

In this model, phytoplankton and zooplankton are considered neutrally buoyant and thus subject to effects of vertical advection and diffusion. Only the dead zooplankton, or detritus, is subjected to a gravitational sinking velocity.

Laboratory measurements of sinking rates of inactive planktonic material are on the order of 1 to 10 m day^{-1} (Steele, 1959). The model is investigated with sinking velocities for detritus within this order of magnitude. The sinking rate is formulated as follows. Average assumed sinking velocity is $w_s = 7$ m day^{-1} or 0.0081 cm sec^{-1}. Detritus sinking is scaled as

$$w_s' = (A_v/A_h)^{1/2} C w_s, \tag{24}$$

where w_s is the nondimensional sinking rate, C is maximum horizontal velocity, and A_v/A_h is the ratio of the vertical and horizontal eddy viscosities. Note the sinking velocity w_s' is scaled in the same manner as the vertical velocity w'.

The total derivative for the detritus component may then be expressed as

$$\frac{\partial D}{\partial t} + S\left(u\frac{\partial D}{\partial x} + w\frac{\partial D}{\partial z} + w_s\frac{\partial D}{\partial z}\right) - \nu_h\frac{\partial^2 D}{\partial x^2} - \nu_v\frac{\partial^2 D}{\partial z^2} = \delta_D Z - \lambda D - \epsilon_D ZD. \tag{25}$$

The term for gravitational sinking of detritus is $w_s\ \partial D/\partial z$.

These final biological equations are expressed in finite differences and solved for each spatial grid point using the u and w velocities of the simulated circulation pattern. Following is a description of the general procedure.

E. Finite-Difference Schemes

A grid mesh is created over the cross section of the study area (Fig. 2) with 41 by 82 divisions in the x and z directions, respectively. The vertical grid lines in the x direction are indexed by the letter j and the horizontal lines in the z direction by the letter k. Nondimensional time is denoted by the index m. Thus $P_{j,k}^m = P(m\Delta t, j\Delta x, k\Delta z)$ and similarly for the other variables. The steady-state values of the u and v velocities are found using the over-relaxation techniques (Isaacson and Keller, 1966).

In the spatial, time-dependent, physical–chemical–biological model, the concentration of a biotic component such as phytoplankton is calculated by considering the amount present in the grid box, the concentration in adjacent boxes, transport into and out of the box, and the biological dynamics occurring within the box. Let us consider a simplified version of Eq. (23) describing the concentration of phytoplankton, $P(x, z, t)$, where all the biological terms are included in the symbol B,

$$\partial P/\partial t = -S(\partial uP/\partial x + \partial wP/\partial z) + \nu_h\ \partial^2 P/\partial x^2 + \nu_v\ \partial^2 P/\partial z^2 + B.$$

The advection terms are written in flux form by using the continuity equation (22). The change of phytoplankton in the center of the box is dependent on the P concentration at its nearest four grid neighbors. We refer to the stencil shown in Fig. 8. Notice that the values of the u velocities are stored at grid corners, the w velocities at the center of the grid walls, and the calculated average concentrations of P, Z, D, and N at the center of the grid box.

The time-difference formulation of the finite-difference scheme, where the concentration of the biotic component is calculated at the center of the box, is

$$P_{j,k}^{m+1} = P_{j,k}^{m-1} - \frac{\Delta t}{2\Delta x}\,[\bar{u}^z\bar{P}^x]_{j,k}^m - \frac{\Delta t}{\Delta z}\,[w\bar{P}^z]_{j,k}^m$$

$$+\frac{2\Delta t\nu_h}{(\Delta x)^2}\,[P_{j+1,k}^m + P_{j-1,k}^m - 2P_{j,k}^{m-1}]$$

$$+\frac{2\Delta t\nu_{\mathrm{v}}}{(\Delta z)^2}\,[P_{j,k+1}^{m}+P_{j,k-1}^{m}-2P_{j,k}^{m-1}]$$

$$+\,2\Delta t B_{j,k}^{m-1},$$

where the advective terms are defined as

$$[\bar{u}^z\bar{P}^x]_{j,k}=[(u_{j,k}+u_{j,k+1})(P_{j+1,k}+P_{j,k})-(u_{j-1,k}+u_{j-1,k+1})\times(P_{j-1,k}+P_{j,k})],$$

and

$$[w\bar{P}^z]_{j,k}=[w_{j-1,k}(P_{j,k}+P_{j,k-1})-w_{j-1,k+1}(P_{j,k}+P_{j,k+1})].$$

The advective scheme is a quadratic conservative, finite-difference method recommended by Grammeltvedt (1969).

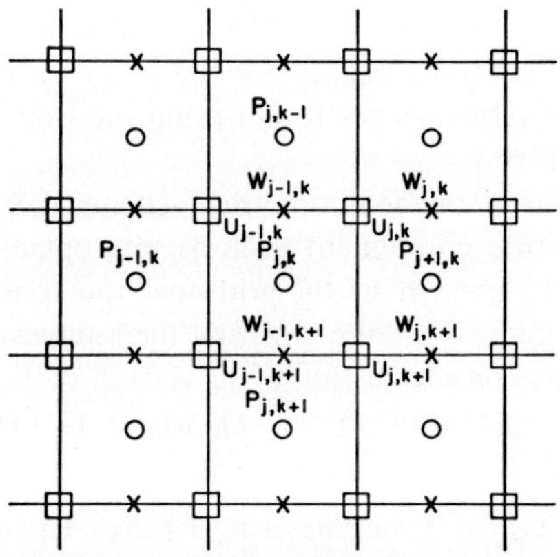

FIG. 8. Stencil showing the grid point location of the u (squares) and w (crosses) velocities and the biotic component P (circles) concentration used in integrating the spatial, physical–chemical–biological model. The vertical grid lines in the x direction are indexed by the letter j and the horizontal lines in the z direction by the letter k.

The time-dependent, spatial phytoplankton dynamics represented by the symbol B may be defined as

$$B_{j,k}^{m-1}=(P_{j,k}^{m-1}N_{j,k}^{m-1})/(\alpha+N_{j,k}^{m-1})-\beta P_{j,k}^{m-1}-\epsilon_z[1-\exp(-\delta_p P_{j,k}^{m-1})]Z_{j,k}^{m-1}-\phi F[\theta P_{j,k}^{m-1}/(\theta P_{j,k}^{m-1}+Z_{j,k}^{m-1})].$$

When advection and diffusion are neglected this reduces to the common Euler method.

The maximum time step allowed by linear computational stability is difficult to determine exactly for this complicated model. However, necessary conditions for stability are

$$\Delta t < \Delta x/\sqrt{2}u_{\max}, \qquad u_{\max} = \max_{j,\,k,\,m} |Su|$$

$$\Delta t < \Delta z/\sqrt{2}w_{\max}, \qquad w_{\max} = \max_{j,\,k,\,m} |Sw|$$

$$\Delta t < (\Delta x)^2/8\nu_{\mathrm{h}}, \qquad \Delta t < (\Delta z)^2/8\nu_{\mathrm{v}}.$$

In this model the advective criteria are the more stringent.

The calculated values of *P, Z, N,* and *D* are the time-dependent, standing stocks of the biotic components within the spatial blocks.

XII. Parameter Values Specific to the Florida Shelf

The model has been formulated and processes scaled such that a minimum of characteristic values need to be specified. By identifying the region for investigation, a number of physical parameter values are determined.

Model area geometry specifies the depth d and width b of the basin. The depth of the water column is taken to be a constant 200 m over a 200-km-wide shelf. The Coriolis parameter f is approximately $5 \times 10^{-5}\ \mathrm{sec}^{-1}$ for the 27°N lat. Observational data provide a characteristic maximum horizontal velocity C of 50 cm sec^{-1} for the Loop Current. The relationship between the horizontal and vertical diffusivities is taken as proportional to the ratio of the chosen horizontal and vertical Austauch coefficients,

$$\nu_{\mathrm{v}}'/\nu_{\mathrm{h}}' = A_{\mathrm{v}}/A_{\mathrm{h}}.$$

Specifying A_{h} equal to $2 \times 10^8\ \mathrm{cm}^2\ \mathrm{sec}^{-1}$ and A_{v} equal to 200 $\mathrm{cm}^2\ \mathrm{sec}^{-1}$, and taking ν_{v}' as 1 $\mathrm{cm}^2\ \mathrm{sec}^{-1}$, then ν_{h}' is $10^6\ \mathrm{cm}^2\ \mathrm{sec}^{-1}$.

Using these values of C, f, and A_{h}, the Rossby number as defined has a value of 0.5. This low Rossby number validates the use of the Hsueh and O'Brien (1971) theory for a circulation over the shelf driven by offshore currents.

The value of the biological turnover rate V_{m}, characteristic of the model area waters, is required in calculating the influence of advective processes upon biological spatial distributions. Since $S = C/[V_{\mathrm{m}}(A_{\mathrm{h}}/f)^{1/2}]$ and the value of V_{m} is dependent on the biological and chemical characteristics of an area, it is readily seen how different spatial solutions may arise. The standing stock of phytoplankton in the euphotic zone depends not only on the cell division rate V_{m}, but also on the magnitude of cell advection below the euphotic zone, and on the resupplying of nutrients to the euphotic zone by upwelling water.

The limiting nutrient characteristics of the model area also specify the appropriate value of the Michaelis–Menten constant K to be used in the biological dynamics. As K is nondimensionalized by N_t, an approximate value of the total amount of limiting nutrient available in the system is needed. The value of N_t has been estimated by identifying the probable limiting nutrient in Florida shelf waters and measuring its concentration in biologically inactive waters being supplied to the system. The biological parameters characteristic of the area which need be supplied are V_m, K, and N_t.

Field tests (Wroblewski, 1972) as well as published data (Bogdanov *et al.*, 1968) indicate that both phosphorus and nitrate concentrations in model area waters may be growth-limiting. Water-reactive phosphate was found to be almost undetectable in these waters, indicating that phosphate may be the limiting nutrient. Values of V_m and K for a phosphate-limiting sea were determined by Thomas and Dodson (1968), who experimented with the diatom *Chaetoceros gracilis,* isolated from the Costa Rica Dome, an upwelling area in the northeastern tropical Pacific. *Chaetoceros gracilis* occurs in abundance at times in the model area.

The V_m value found by Thomas and Dodson was 0.22 μg at PO_4–P l^{-1} which gave a growth rate of 2.49 cell divisions per 24 hours (0.10 h^{-1}). The K value, the concentration of limiting nutrient at which the phytoplankton growth rate is $V_m/2$, was found to be 0.12 μg at PO_4–P l^{-1}. The corresponding non-dimensionalized value of α used in the model is 0.03, estimating N_t as 4.0 μg at PO_4–P l^{-1} for the model area waters (Collier, 1958).

Phosphate is quickly recycled in coastal waters (Ketchum and Corwin, 1965). Regeneration rates of utilizable phosphorus from zooplankton excretion products and phytoplankton cell autolysis may be on the order of hours (Steele, 1959). Also, the low concentration of phosphate in the model waters may not be indicative of phosphorus limitation since phytoplankton require relatively small quantities of the nutrient for unlimited growth.

From observational data, nitrate is a possible limiting nutrient. Thomas (1970) reports a V_m value of 7 μg at NO_3–N l^{-1}, which gave 1.22 doublings per 24 h (0.05 h^{-1}) for a mixed phytoplankton population from the nitrate-limited eastern tropical Pacific. Observations show concentrations of less than 1 μg at NO_3–N l^{-1} in euphotic zone waters of the model region. The K value for nitrate is reported by Thomas (1970) as 0.75 μg at NO_3–N l^{-1}. This would give an α value of 0.025, taking N_t equal to 30 μg at NO_3–N l^{-1} (Collier, 1958).

Recent research (O'Brien, 1972) has indicated that, while several nutrients may be in low enough concentrations to affect the rate of phytoplankton growth, at steady state generally only one nutrient is found to be limiting. Which nutrient becomes most important is determined primarily by rate of supply versus rate of withdrawal of the various nutrients. Use of one specific V_m value in the model may be justified instead of considering a synergistic combination of several V_m values.

The importance of advection in the spatial model is determined by the value of the S parameter, which is inversely dependent on the value of V_m. If V_m is high, biological turnover rate is rapid and the importance of advection on the concentration of the biotic component is diminished. Strong turbulence also lowers the effect of advection.

The value of the S parameter used in the model based on the V_m value for a PO_4-limiting sea was 0.9. The value calculated for a hypothesized NO_3-limiting sea is twice that, namely 1.8. We find that advection effects are twice as important in a nitrate-limiting sea than in phosphate-limited waters. The high V_m for phosphate results in a higher biological turnover rate. The spatial distribution of plankton in the phosphate-limiting case is then determined more by phytoplankton growth than by advective processes. The usefulness of the S parameter lies in its simple formulation and general applicability to any nutrient-limited aquatic ecosystem.

XIII. Spatial Distributions of Biotic Components in Absence of Advection

It is instructive to run the spatial model considering time dependency and diffusion, but without advection. The u, v, and w velocities are taken to be zero. The biological components are initialized at nondimensional concentrations of $P = 0.33$, $Z = 0.33$, $N = 0.25$, $F = 0.09$, and $D = 0.00$, at time $t = 0$ over all space. The biological parameter values used are the same as in Fig. 5.

The spatial solution for an elapsed model time $t = 8$ (equivalent to three to six days in nature) for the phytoplankton standing stock concentration appears in Fig. 9. The euphotic and aphotic zones are well-defined. The absence of contours except at the upper and lower boundary of the euphotic zone illustrates homogeneous distributions within the euphotic and aphotic zones. The gradient between the zones arises from the diffusion effect which acts to make the transition between zones smooth rather than abrupt. Photoinhibition at the surface is indicated.

The euphotic zone concentrations of dissolved limiting nutrient, phytoplankton, zooplankton, and detritus components are comparable with solutions for the one-box model (Fig. 5). As environmental conditions permit only dissolved nutrient and sinking detritus to exist below the euphotic zone, the aphotic zone is devoid of phytoplankton and zooplankton. Phytoplankton (Fig. 9) are maintained only in the area where the depth–photosynthesis curve is positive. Phytoplankton initially existing below the euphotic zone are grazed to extinction since no threshold is included in the grazing formulation. Zooplankton (not shown) are confined to the area where their phytoplankton food source exists. The zooplankton in the aphotic zone starve to death after consuming all the initial phytoplankton there.

A low concentration of dissolved limiting nutrient (not shown) occurs in the euphotic zone as a consequence of uptake by phytoplankton. Highest concentrations of N occur in the aphotic zone, where sinking detritus is decomposed and there are no nutrient losses.

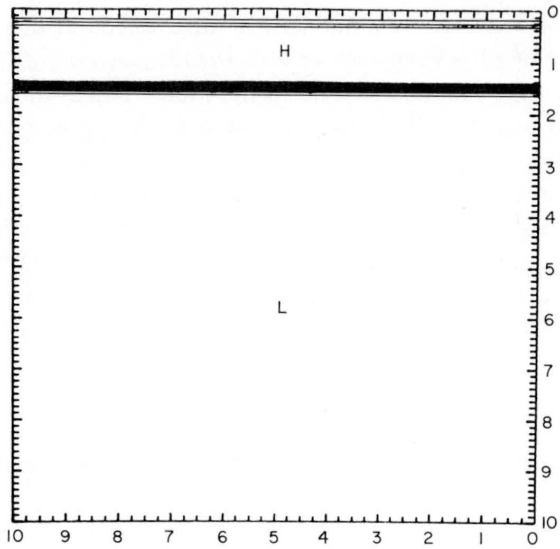

FIG. 9. Spatial distribution of phytoplankton standing stock concentration at time $t = 8.0$ in the absence of advection. Contours are from 0.0 to 0.63, contour interval of 0.05.

The purpose of this exercise has been to demonstrate the transition from a one-box model to one with 3200 boxes. We see that without advection the zones of biological activity as defined by environmental effects (light, nutrients) are well-defined, homogeneous, and generally uninteresting. The use of models with only vertically stacked boxes (e.g. Riley *et al.*, 1949) is thus justified in situations where horizontal advection is negligible.

XIV. The Spatial Solutions for an Advective, Phosphate-Limiting Sea

We next consider effects of advection on spatial distributions of biotic components. The u, v, and w velocities of the final biological equations are those predicted from the steady-state solution of Eqs. (20)–(22). These specify the magnitude and direction of water velocity at each point on the spatial grid. For the following solutions the model waters are considered to be phosphate limited, and appropriate values of V_m, K, N_t, and S are used.

Calculations are made of changes in the spatial distribution of phytoplankton, zooplankton, dissolved nutrient, and detritus as they are advected and diffused from their initial homogeneous concentrations of $P = 0.33$, $Z = 0.33$, $N = 0.25$, $F = 0.09$, and $D = 0.0$. The time-dependent solutions which include advection of P, Z, N, and D below the euphotic zone and upwelling of nutrient-rich bottom waters are shown in Figs. 10–13.

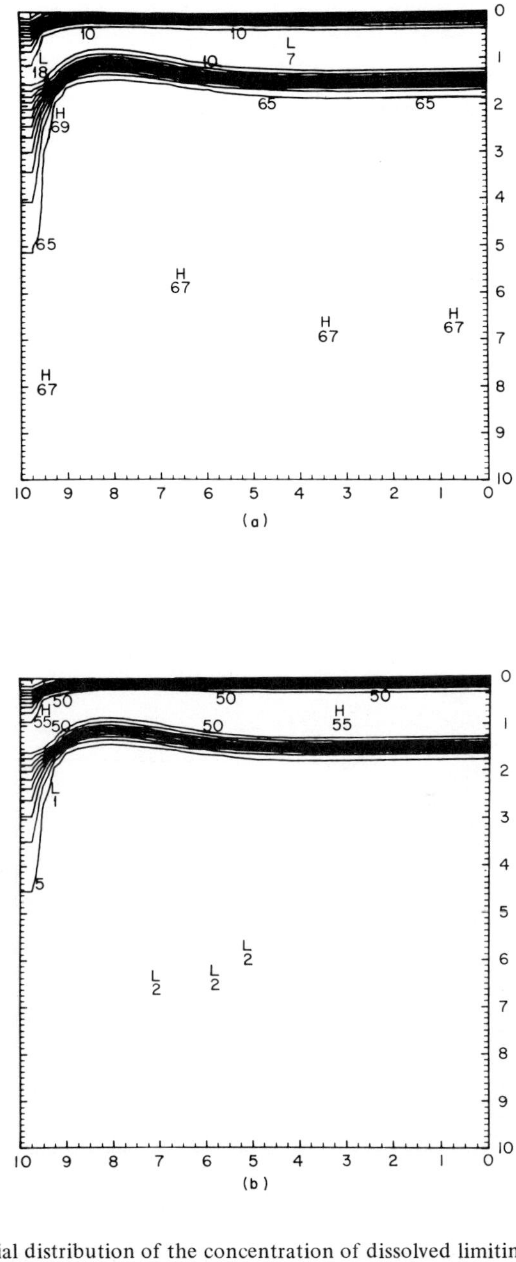

FIG. 10. Spatial distribution of the concentration of dissolved limiting nutrient (a) and phytoplankton standing stock (b) at time $t = 4.0$ for an advected, PO_4-limiting sea. Figure 10a contours from 0.06 to 0.69 by 0.05. Figure 10b contours from 0.01 to 0.55, contour interval of 0.05.

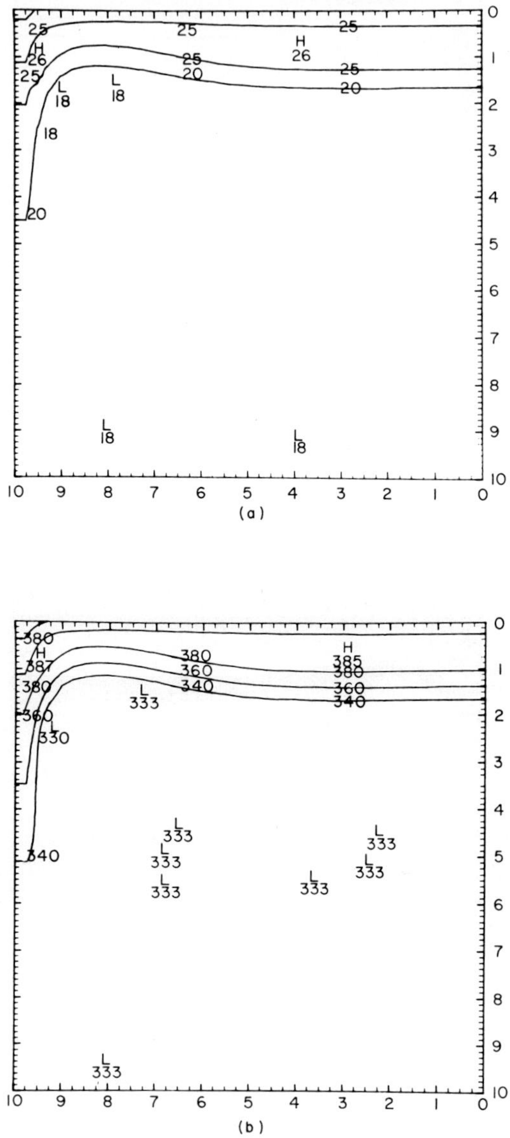

FIG. 11. Spatial distribution of the concentration of zooplankton (a) and detritus (b) at time $t = 4.0$ for an advected, PO_4-limiting sea. Figure 11a contours from 0.18 to 0.26, contour interval of 0.05. Figure 11b contours from 0.033 to 0.039, contour interval of 0.002.

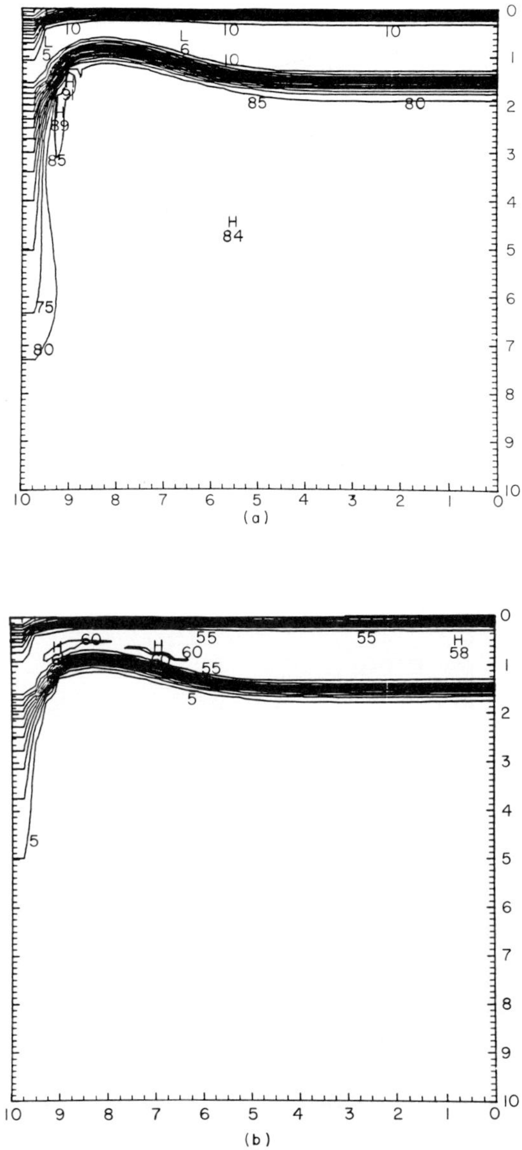

FIG. 12. Spatial distribution of the concentration of dissolved limiting nutrient (a) and phytoplankton standing stock (b) at time $t = 8.0$ for an advected PO_4-limiting sea. Figure 12a contours from 0.05 to 0.91, contour interval of 0.05. Figure 12b contours from 0.0 to 0.62, contour interval of 0.05.

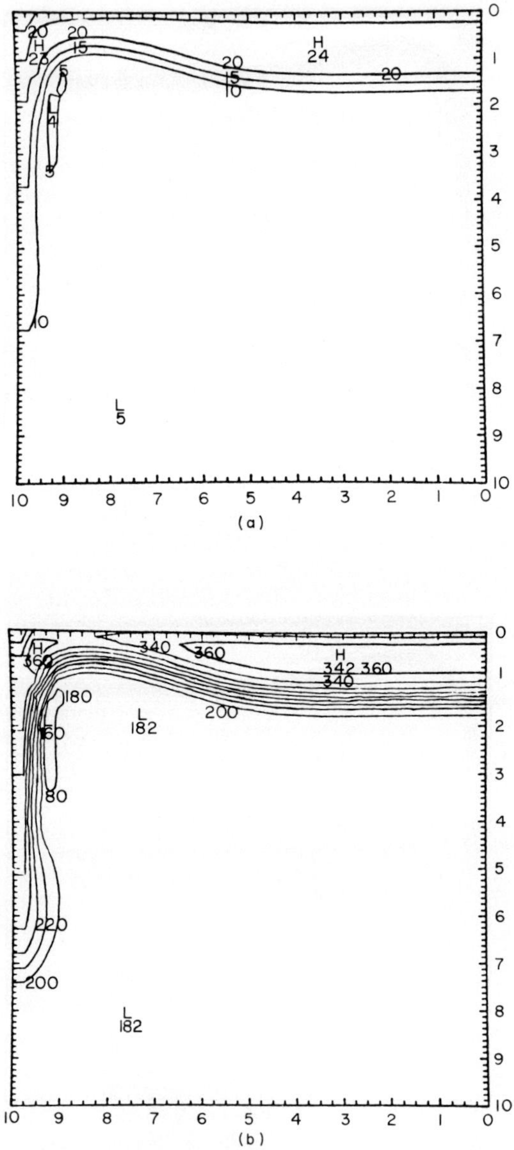

FIG. 13. Spatial distribution of the concentration of zooplankton (a) and detritus (b) at time $t = 8.0$ for an advected, PO_4-limiting sea. Figure 13a contours from 0.04 to 0.23, contour interval of 0.05. Figure 13b contours from 0.016 to 0.037, contour interval of 0.002.

At time $t = 4$ (40 h), definite zonations in the distribution and concentration of the N, P, Z, and D components are observed. Limiting nutrient has significantly decreased in the euphotic zone. Upwelling of nutrient-rich bottom water is evident in Fig. 10a where contour lines indicate an upward advection of water into the euphotic zone. Figure 10b shows the phytoplankton standing stock at 40 h. Almost no phytoplankton exists below the euphotic zone, except in regions of strong downwelling at the outer boundary of the shelf.

Figure 11a depicts the highest concentrations of zooplankton in areas where phytoplankton are abundant. Zooplankton continues to decrease in the aphotic zone as the phytoplankton there are grazed to extinction. Highest concentrations of detritus (Fig. 11b) appear in the euphotic zone. The weakening of the gradient below the euphotic zone is due to gravitational sinking of detritus.

The steady-state P, Z, N, and D standing stock values for phosphate-limiting conditions as given by the one-box biological model are

$$\begin{aligned} P &= 0.6190 \\ Z &= 0.2303 \\ N &= 0.0246 \\ D &= 0.0361 \\ F &= 0.0900 \\ \hline N_t &= 1.0000 \end{aligned}$$

These values are being approached by time $t = 8$ (80 h) in the euphotic zone (Figs. 12 and 13).

The spatial model differs from the one-box system in that biotic components become more concentrated in certain boxes than in others. Each box approaches its own steady state. The concentration of dissolved limiting nutrients (Fig. 12a), for example, is greatest in aphotic zone boxes where uptake is small and regeneration of detritus takes place. Transport of this high concentration into adjacent boxes depends on the direction of the circulation pattern. The final steady state within each box depends on the biological dynamics occurring within the box, transport into and out of the box, and the gradient existing between adjacent boxes.

XV. The Spatial Solutions for an Advective, Nitrate-Limiting Sea

The spatial model's response to variation in V_m, N_t, K, and the S parameter was investigated by considering the shelf region to be nitrate limited rather than phosphate limited. The results are shown in Figs. 14 and 15. We begin with the

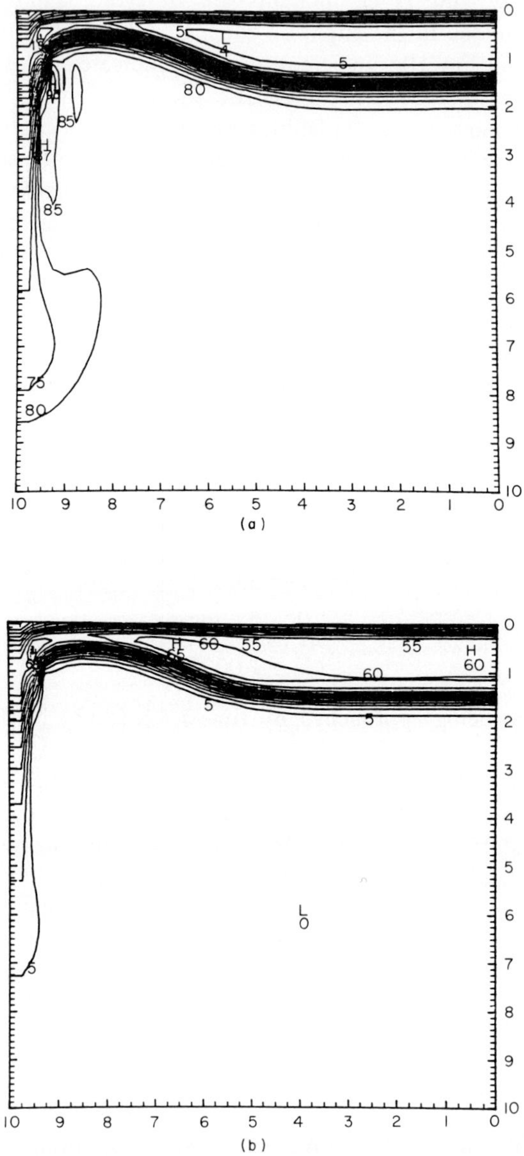

FIG. 14. Spatial distribution of the concentration of dissolved limiting nutrient (a) and phytoplankton standing stock (b) at time $t = 8.0$ for an advected, NO_3-limiting sea. Figure 14a contours from 0.04 to 0.93, contour interval of 0.05. Figure 14b contours from 0.0 to 0.68, contour interval of 0.05.

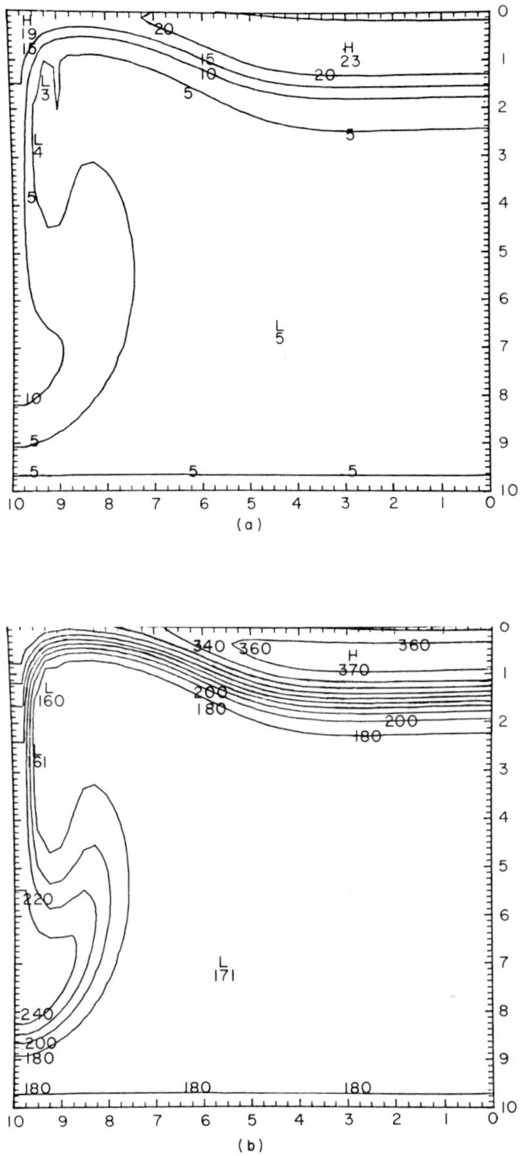

FIG. 15. Spatial distribution of the concentration of zooplankton (a) and detritus (b) at time $t = 8.0$ for an advected NO_3-limiting sea. Figure 15a contours from 0.03 to 0.23, contour interval of 0.05. Figure 15b contours from 0.017 to 0.037, contour interval of 0.002.

same spatially homogeneous initial values of the biological components as in the phosphate-limiting case. Major differences in the time-dependent solutions are observed.

As in the phosphate-limiting case, the model approaches steady-state concentrations for P, Z, N, and D (Figs. 14 and 15) by $t = 8.0$ (160 h, since V_m for the NO_3-limiting situation is 0.05 h^{-1}). The steady-state values given by the nonspatial biological model using the values of K and N_t for NO_3 are

$$
\begin{aligned}
P &= 0.6271 \\
Z &= 0.2340 \\
N &= 0.0124 \\
D &= 0.0365 \\
F &= 0.0900 \\
\hline
N_t &= 1.0000
\end{aligned}
$$

One may at first be surprised by the similarity of the euphotic zone P values in the PO_4- and NO_3-limiting cases. However, we must return to dimensional space before the values can be interpreted. Upon multiplying by N_t, the steady-state phytoplankton values are (0.619)(4.0 μg at PO_4 l^{-1}) = 2.47 μg at PO_4 l^{-1} in the phosphate-limiting case and (0.627)(30.0 μg at NO_3 l^{-1}) = 18.8 μg at NO_3 l^{-1} in the nitrate-limiting case. Relatively small additions of phosphate and larger amounts of nitrate would be expected to increase the phytoplankton stock proportionately. This scaling has been hidden by normalizing by N_t.

Comparing Figs. 14b and 12b, the spatial distributions of phytoplankton at $t = 8.0$ for the two different nutrient-limited cases, the phytoplankton euphotic zone concentration is seen to be higher in the nitrate-limiting case (Fig. 14b) than in the phosphate-limiting situation (Fig. 12b). A high of 68% of N_t is found in the region of maximum upwelling of nitrate-rich waters as compared to a high level of 62% of N_t in this region in the phosphate case. The higher phytoplankton standing stock results from (i) the slightly lower value of α, enabling phytoplankton to utilize smaller concentrations of dissolved nutrient, and (ii) the greater value of S, enabling increased transport of nutrient-rich aphotic zone water into the euphotic zone. The increased width of the gradient band results from an increased diffusion rate. Extension of the depth to which phytoplankton are advected below the euphotic zone at the shelf break boundary reflects the increased effect of advection.

A comparison of Figs. 14a and 12a, the two solutions for the dissolved limiting nutrient concentration and distribution at $t = 8.0$, shows lower N concentrations in the euphotic zone for the nitrate-limiting case. This decrease in dissolved nutrient results from increased uptake by phytoplankton. Zooplankton (compare Figs. 15a and 13a) shows a higher biomass standing stock in a nitrate-limited euphotic zone, a result of increased phytoplankton standing stock. The greater effects of advection and diffusion in the nitrate-limiting sea are made quite evident by comparison of the spatial distributions of the detritus component at $t = 8.0$ (Figs. 15b and 13b). The increased detritus sinking rate is reflected in the lower depth of the euphotic–aphotic zone gradient band in Fig. 15b.

Response of the spatial model to variation in the parameters V_m, K, N_t, and S is complicated. Different spatial distributions of the biotic components result as advection, scaled by S, and diffusion, scaled by V_m, become important. Localities of upwelling of nutrient-rich bottom waters show greater phytoplankton and zooplankton production. The steady-state standing stock of phytoplankton is dependent on the efficiency of the nutrient uptake parameter K. And finally, the biological turnover rate V_m specifies the time scale of biological processes such as growth, death, excretion, and regeneration. These variations collectively account for differences in the phosphate and nitrate spatial solutions after the same elapsed model time.

XVI. Critique

We have constructed a time-dependent, physical–chemical–biological spatial model of the transfer of biologically limiting nutrient between the lower marine trophic levels of a continental shelf ecosystem. A theoretical circulation pattern is employed to advect and diffuse phytoplankton, zooplankton, detritus, and limiting nutrient dissolved in the water column.

Evaluation of the model as an approximation of the real world is in order. The major strength of the model is formulation of numerous physical and biological processes into a workable framework. A weakness is the knowledge of correct expressions and parameter values which describe the real world ecosystem. Sensitivity analysis indicates which of the included processes strongly influence the system, but we have no measure other than realism of the solutions of the expressions' correctness. The spatial solutions of the model are quite smooth in the horizontal direction, except near the locale of strong upwelling. Small-scale patchiness observed in continuous sampling across any oceanic area is not resolved here, due to the 5 km grid spacing.

Yet, the model is encouraging. Reasonable biological production and ecological efficiencies can be simulated using the formulations presented.

The limitation of the model is in simulating the complexity of the actual ecosystem. Species composition of the trophic levels is not specified. The food chain is confined to primary and secondary producers with an omnivorous tertiary level. Mobility of fish biomass is neglected. Temperature effects on biological rates is ignored. Model solutions are toward a steady state, while the real system is in constant flux. More realistic circulation patterns are required by this type of model, as spatial distributions of the biological components are very dependent on advection, especially in nutrient-limited seas experiencing upwelling.

The predictive ability of this type of model in locating concentrations of marine food resources is a fundamental goal. A feedback between model theory and field observations is essential. This feedback is a major objective of the Coastal Upwelling Ecosystem Analysis (CUEA) program of the National Science Foundation-International Decade for Ocean Exploration (NSF-IDOE).

XVII. Summary

The flow of biologically limiting nutrient through the lower trophic levels of a marine ecosystem has been investigated. A time-dependent, spatial model of the distribution of biological components is constructed which incorporates a theoretical circulation pattern for the West Florida continental shelf region.

The formulation of this model has evolved a parameter S, whose value is dependent on the biological kinetics and organized water motion of a specific area. This parameter scales the effects of advection and diffusion relative to the rate of biological turnover in determining spatial solutions. The phytoplankton specific growth rate V_m is shown to be a fundamental time scale to which both physical and biological processes can be related.

Acknowledgments

This research has been supported primarily by the Office of Naval Research and the Oceanography Section of the National Science Foundation under Grant No. GA-29734. Part of the research was conducted at the National Center for Atmospheric Research (NCAR) in Boulder, Colorado. Partial support has been derived from the CUEA project of IDOE.

The Computer Facility at NCAR provided CDC 7600 and 6600 time and the Florida State University Computing Center provided CDC 6400 time.

Our appreciation is extended to Professor Albert Collier, Dr. Robert C. Harriss, Dr. Richard Iverson, Dr. R. Winston Menzel, and Dr. John Winchester for their encouragement. We wish to thank Murice O. Rinkel, Assistant Director, State University System of Florida Institute of Oceanography for making possible field observations necessary for this study.

This research reflects the interdisciplinary interests, ideas, and teamwork of the Mesoscale Atmospheric-Oceanic Interaction Research Group at Florida State University. Our gratitude is extended to the members of this team for their aid in solving this multidisciplinary oceanographic problem.

REFERENCES

Austin, H. M. (1971). The characteristics and relationships between the calculated geostrophic current component and selected indicator organisms in the Gulf of Mexico Loop Current System. Doctoral dissertation, Florida State Univ., Tallahassee, Florida.

Bogdanov, D. W., Sokolov, V. A., and Khromov, N. S. (1968). *Oceanology* **8**, 371.

Collier, A. (1958). U.S. Dept. of Interior, Fish, and Wildlife Serv., Spec. Sci. Rep.: Fisheries 249.

Curl, H. (1962). *Rapp. Proc. Verb. Cons. Perm. Int. Explor. Mer.* **153**, 183.

Cushing, D. H. (1969). Morning Review Lectures of the *Int. Oceanogr. Congr., 2nd, Paris, UNESCO* pp. 103–115.

Dugdale, R. C. (1967). *Limnol. Oceanogr.* **12**, 685.

Dugdale, R. C., and Goering, J. J. (1967). *Limnol. Oceanogr.* **12**, 196.

Dugdale, R. C., and MacIsaac, J. J. (1971). *Inv. Pesq.* **31**, 299.

Eppley, R. W., and Coatsworth, J. L. (1968). *J. Phycol.* **4**, 151.

Eppley, R. W., Rogers, J. N., and McCarthy, J. J. (1969). *Limnol. Oceanogr.,* **14**, 912.

Fogg, G. E. (1966). *In* "The Extracellular Products of Algae" (*Oceanogr. Mar. Biol. Ann. Rev.* **4**) (H. Barnes, ed.), pp. 195–212. George Allen and Unwin Ltd., London.

Gerking, S. D. (1962). *Ecol. Monogr.* **32**, 31.

Glooschenko, W. A., Curl, H., and Small, L. F. (1972). *J. Fish. Res. Bd. Canada* **29**, 1253.

Grammeltvedt, A. (1969). *Mon. Weather Rev.* **97**, 384.

Gulland, J. A. (1970). *In* "Marine Food Chains" (J. H. Steele, ed.), pp. 296–315. Univ. of California Press, Berkeley, California.

Hsueh, Y., and O'Brien J. J. (1971). *J. Phys. Oceanogr.* **1**, 180.

Isaccson, E., and Keller, H. B. (1966). "Analysis of Numerical Methods." Wiley, New York.

Ivlev, V. S. (1945). *Usp. Sovrem. Biol.* **19**, 88.

Ketchum, B. H., and Corwin N. (1965). *Limnol. Oceanogr. Suppl.* **10**, 148.

Kozlovsky, D. G. (1968). *Ecology* **49**, 48.

Lasker, R. (1970). *In* "Marine Food Chains" (J. H. Steele, ed.), pp. 265–283. Univ. of California Press, Berkeley, California.

Leipper, D. F. (1970). *J. Geophys. Res.* **75**, 637.

Lindeman, R. L. (1942). *Ecology* **23**, 399.

Lotka, A. J. (1925). "Elements of Physical Biology." Williams and Wilkins, Baltimore, Maryland.

Macfadyen, A. (1964). *In* "Grazing in Terrestrial and Marine Environments" (D. Crisp, ed.). Blackwell, Oxford.

MacIsaac, J. J., and Dugdale, R. C. (1969). *Deep-Sea Res.* **16**, 415.

McAllister, C. D. (1970). *In* "Marine Food Chains" (J. H. Steele, ed.), pp. 419–457. Univ. of California Press, Berkeley, California.

Monod, J. (1942). "Recherches sur la croissance de cultures bacteriennes." Hermann, Paris.

O'Brien, J. J., and Wroblewski, J. S. (1973). *J. Theor. Biol.* **38**, 197.

O'Brien, W. J. (1972). Presented at 35th Ann. Meeting, Amer. Soc. Limnol. and Oceanogr., Tallahassee, Florida.

Odum, E. P. (1962). *Jp. J. Ecol.* **12**, 108.

Odum, H. T. (1957). *Ecol. Monogr.* **27**, 55.

Parsons, T. R., LeBrasseur, R. J., and Fulton, J. D. (1967). *J. Oceanogr. Soc. Japan* **23**, 10–17.

Patten, B. C. (1968). *Int. Rev. Ges. Hydrobiol.* **53**, 357.

Raymont, J. E. G. (1966). *Advan. Ecol. Res.* **3**, 117.

Riley, G. A. (1946). *J. Mar. Res.* **6**, 54.

Riley, G. A., Stommel, H., and Bumpus, D. F. (1949). *Bull. Bingham Oceanogr. Coll.* **12**, 1.
Ryther, J. H. (1969). *Science* **166**, 72.
Slobodkin, L. B. (1960). *Amer. Nat.* **94**, 213.
Slobodkin, L. B. (1962). *Advan. Ecol. Res.* **1**, 69.
Smayda, T. J. (1966). Inter-American Tropical Tuna Comm., Bull. 11, pp. 355–612.
Smith, F. E. (1970). *In* "Ecological Studies–Analysis and Synthesis" (D. E. Reichle, ed.), Vol. 1, pp. 7–18. Springer-Verlag, Berlin and New York.
Steele, J. H. (1959). *Biol. Rev.* **34**, 129.
Thomas, W. H. (1970). *Limnol. Oceanogr.* **15**, 386.
Thomas, W. H., and Dodson, A. N. (1968). *Biol. Bull.* **134**, 199.
Tomovic, R. (1963). "Sensitivity Analysis of Dynamic Systems." McGraw-Hill, New York.
Verhoff, F. H. (1971). *J. Theor. Biol.* **33**, 131.
Volterra, V. (1926). *J. Cons. Explor. Mer.* **3**, 1.
Walsh, J. J., and Dugdale, R. C. (1971). *Inv. Pesq.* **35**, 309.
Watt, W. D. (1966). *Proc. Roy. Soc. London B* **164**, 521.
Wroblewski, J. S. (1972). An ecological model of the lower marine trophic levels on the continental shelf off West Florida. M.S. thesis, Florida State Univ., Tallahassee, Florida.
Yentsch, C. S. (1963). *Oceanogr. Mar. Biol. Ann. Rev.* **1**, 157.

PART II

MODELS OF TERRESTRIAL ECOSYSTEMS

This section contains three chapters devoted to terrestrial ecosystems.

Chapter 4 by Drs. Lugo, Sell, and Snedaker bridges the transition from sea (Part I) to land in a treatment of mangrove ecosystems in south Florida. A preliminary effort to simulate mangrove ecosystems based on information assembled during a regional study, the South Florida Ecological Study coordinated by the U.S. Department of the Interior, is described. Mangroves serve to build land and protect coastal areas during high tides and storms. They export organic matter to adjacent waters and function as nursery grounds for commercially important fish and shrimp species. They occur in desirable locations for commercial development, which creates social and economic conflicts. Means for assessing tradeoffs in determining optimal use are required, and simulation modeling is explored as one available tool. Objectives of the present simulation were to evaluate effects of terrestrial runoff and tidal flushing on nutrient cycling, productivity, and detritus export and accumulation; to assess mangrove contributions to water quality; and to gauge the utility of modeling in research planning. A three-compartment (mangrove biomass, detritus, and nutrients) nonlinear differential equation model, calibrated with field data from several sites, is simulated by analog computer to assess: effects of mangrove development under present conditions, tide effects on organic production and export, effects of mangrove succession on nutrient content of adjacent bays, and nutrient runoff effects on mangroves. It is concluded that mangrove forests attain biomass steady states in phase with the frequency of tropical hurricanes, that detritus storage and export are functions of tidal amplitude, photosynthesis is sensitive to terrestrial nutrient imputs, mangrove zonation and vigor may be associated with nutrient availability rather than solely salinity as previously held, and mangroves control nutrients in adjacent waters during succession. The study serves as an excellent example of proper interplay between small-scale models and empirical research in a balanced approach to ecosystem investigations.

Chapter 5 presents selected pertubation responses of a well-known forest simulation model, JABOWA, developed by Dr. Botkin. Such models are considered to provide theoretical and conceptual guidance in problems involving ecosystem perturbations where complex and elaborate experiments cannot be conveniently performed. JABOWA was originally designed to simulate stand growth and responses to both tree and site manipulations in the Hubbard Brook Ecosystem Study, New Hampshire. Thirteen forest tree species are included in the version of the model discussed. Three principal subroutines provide the annual growth increment of each tree, new saplings, and mortality of individual trees. A fourth principal subroutine computes measures of environmental conditions for each plot. These major subroutines are described, noting that both tree species and abiotic environment are defined by a few simple characteristics, competition is expressed only as competition for light, and survivorship is determined by differential addition of new saplings and differential species mortality. Two basic strategies are recognized in respect to light gradients and successional stage: shade-intolerant (r-selected) and shade-tolerant (K-selected) species capable of, respectively, rapid growth in a clearing versus durability and persistence under deep shade. In the computer trials, perturbations involve altering annual diameter increments of each tree to simulate fertilization or stress effects implicitly. Basal area is the prime measure of forest response. When perturbations do not alter favored strategies under normal environmental conditions, ecosystem responses are readily predicted by the model. Otherwise, behavior is complex and paradoxical predictions may be generated. Because JABOWA considers individual forest stand species, it tends to demonstrate the importance of species interactions in producing whole ecosystem dynamic characteristics.

Chapter 6 by P. Sollins, W. F. Harris, and N. T. Edwards describes simulation of the physiology of a second-growth mesophytic deciduous forest in eastern Tennessee. A forest ecosystem is persistent despite the relatively rapid flow through it of organic matter and water and, to a lesser extent, inorganic nutrients. Nutrients may be taken to control water and organic matter accumulation and transfer, and the latter are carriers of nutrients. The model was constructed as part of the International Biological Program Eastern Deciduous Forest Biome study (see Volume III, Part I of this series) to mimic seasonal accumulations and transfers of organic matter. The chapter describes the model, an organic matter budget from which it is derived, and simulated forest behavior under perturbation conditions. The modeling approach was to develop a physiological conceptualization of the entire forest ecosystem, including both primary production and decomposition processes, from the budget analysis. Mathematical functions describe organic matter transfers between compartments in response to abiotic variables such as temperature, moisture, and solar radiation. Differential equations formed by summing functions for incoming and outgoing transfers are solved numerically by fixed-step Runge-Kutta integration. Simulations include the seasonal cycle of forest CO_2 exchange and perturbations due to increased bud kill, stand thinning, and defoliation. In the first, the forest is suggested to "inhale" CO_2 during the growing season and "exhale" it again during spring and fall decomposition periods. Bud kill, e.g., due to ionizing radiation or insect outbreaks, decreased production by the affected species (*Liriodendron tulipifera*) and increased production of other vegetation components. Selective overstory thinning decreased leaf biomass of both overstory (more) and understory (less), and temporarily increased herbaceous biomass; forest floor respiration and dry matter accumulation both increased. Defoliation due to simulated increased consumption by insects reduced active tissue biomass significantly and woody tissue increment, net primary production, and leaf production only slightly. In evaluating the model, the authors indicate the main weakness to be an inadequate data base, particularly for root processes; unrealistic formulations in some cases, reflecting lack of understanding of mechanisms; and insufficient linkage to real-life problems. The basic approach of modeling forest physiology versus tree growth is seen as reasonable, however. Further developments should emphasize development and coupling of nutrient, water, and energy models to the organic matter model, and then devising resource management strategies to be tested with the models and put into action.

4

Mangrove Ecosystem Analysis

ARIEL E. LUGO

DEPARTMENT OF BOTANY

MAURICE SELL

DEPARTMENT OF ENVIRONMENTAL ENGINEERING SCIENCES

and

SAMUEL C. SNEDAKER*

RESOURCE MANAGEMENT SYSTEMS PROGRAM
SCHOOL OF FOREST RESOURCES AND CONSERVATION
UNIVERSITY OF FLORIDA, GAINESVILLE, FLORIDA

* Present address: Rosenstiel School of Marine and Atmospheric Sciences, University of Miami, Miami, Florida.

I. Introduction

The extensive land development activities in south Florida have been accompanied by progressive loss of natural resources and deterioration of environmental quality. Governmental agencies, in response to public concern, have initiated studies of regional ecosystems whose objectives are to describe, quantitatively, the patterns and mechanisms of man's impact on the structure and function of the landscape units. One of the regional studies is the South Florida Ecological Study coordinated by the United States Department of the Interior. An objective of this study is to survey knowledge of the ecology of south Florida, identify major gaps, and initiate research necessary to bridge these gaps. Models of energy flow were used to identify research needs in the study, and are reported elsewhere (Lugo *et al.*, 1971). The concept was to use qualitative ecosystem models as planning tools for research such that data to be collected would be quantitative and in a form amenable to computer simulations. The original planning model for the mangrove research is depicted in Fig. 1. This model served to identify major compartments, flows, and forcing functions believed to be important in the local and regional functions of the mangrove ecosystem. Research was then designed to obtain baseline data expressed per unit area for the storages, and per unit area per unit time for the flows and forcing functions. Emphasis, in general, was placed on those data that were not available for south Florida mangrove ecosystems.

This paper reports a first attempt to simulate some of the information collected prior to and during the first phase of the study. As research progresses, results from these simulations will be validated and the model will be refined. Therefore, this report should be considered preliminary.

II. Regional Role of Mangrove Ecosystems

Since the early studies of Davis (1940), mangroves have been recognized for the value they contribute to regions in which they grow. Davis, for example, discussed their role as land builders and protectors of coastal areas during periods of high tides and strong winds caused by hurricanes. Golley *et al.* (1962) described the role of mangroves as exporters of labile organic matter to adjacent bays. More recently, Heald (1969) and Odum (1969) measured and traced this organic output to the food webs supporting fisheries of the southern coastline of Florida. Sastrakusumah (1971) described the role of mangrove estuaries as nursery grounds for the pink shrimp (*Penaeus duorarum*), and Clark (1971) did the same for juvenile fish species in Everglades National Park.

The regional importance of mangroves, and the fact that they occur in desirable locations for development, creates a dilemma within certain sectors of

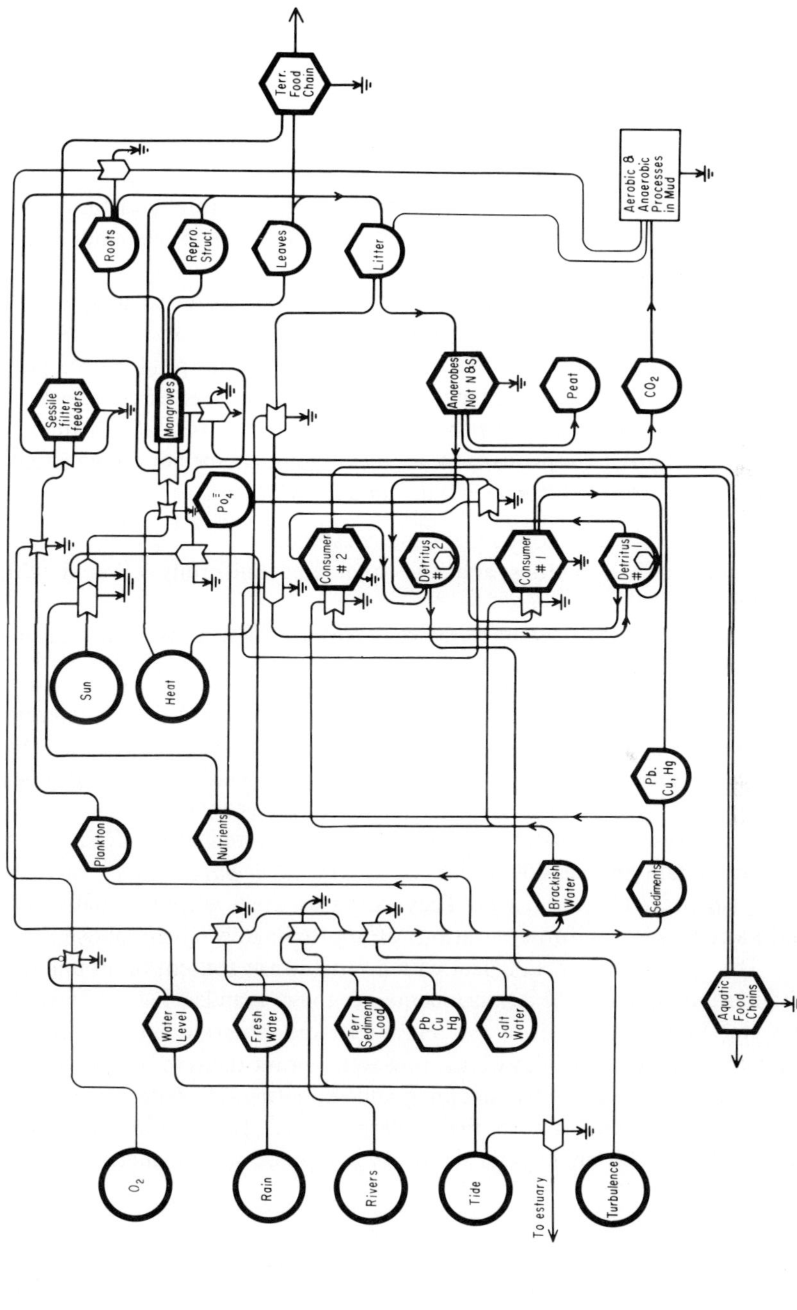

FIG. 1. Planning model for mangrove research in south Florida. Symbols are those of the energy language described by Odum (1972). The model is from Lugo *et al.* (1971), and it depicts the major energy flows and interactions believed to be operating in mangrove forests in south Florida.

society. Criteria for decisions concerning the development or protection of these areas are needed in order to optimize their value to man. Questions that need to be answered to manage mangroves for maximum regional service are related to the sensitivity of mangrove ecosystems to environmental factors which control their structure and function. In the planning model (Fig. 1), these factors are illustrated as forcing functions and include oxygen, rainfall, terrestrial runoff, tides, turbulence, sun, and heat. These were considered to be the major energy sources to which mangrove ecosystems respond with structural and functional adaptations. Many of these factors interact with each other and Fig. 1 summarizes the pathways that they may control.

In the early phases of this study, a field survey was made to describe the extant mangrove types in the Ten Thousand Island area of south Florida (Fig. 2). It was determined from this survey that the earlier phytosociological classifications (e.g., Davis, 1940), based on zonation and species composition, were not applicable to the mangrove area of interest.* Instead, distinctive physiognomic patterns (irrespective, to a large extent, of species composition) were observed to be closely associated with the following:

1. location with respect to the mainland coastline;
2. topographic irregularities and the total area uniformly covered with mangroves; and,
3. integrating these two, characteristics of surface water movement resulting from tides and seasonally high water.

Based on these observations, four mangrove forest types (basin, fringe, riverine, and overwash) were described and will be reported on in detail elsewhere. This paper focuses specifically on the mangrove overwash type, although generalizations are extended to encompass mangroves in general.

Mangrove overwash forests are dominated by the red mangrove (*Rhizophora mangle*) and contain an occasional black mangrove (*Avicennia nitida*)† or white mangrove (*Laguncularia racemosa*). They occur as small islands, usually very long and narrow, with uniform elevations of approximately 30 cm above mean sea level. The substrate is organic peat with a variable ash content of around 6%. The distribution and spatial arrangements of these islands (Fig. 2) serve to restrict water flow in and out of the relatively large estuarine bays contiguous with the mainland. The low elevations, however, permit them to be overwashed at high tide. Surface flushing by incoming and retreating tide water carries with it loose debris and detritus into the estuarine bays. This results in a much reduced standing stock of organic debris as compared with mangrove islands

* The Ten Thousand Island area is characterized by a low, essentially level, topographic relief, and thus the "zones" are so broad as to be unrecognizable.

† Synonymous with *A. germinans*.

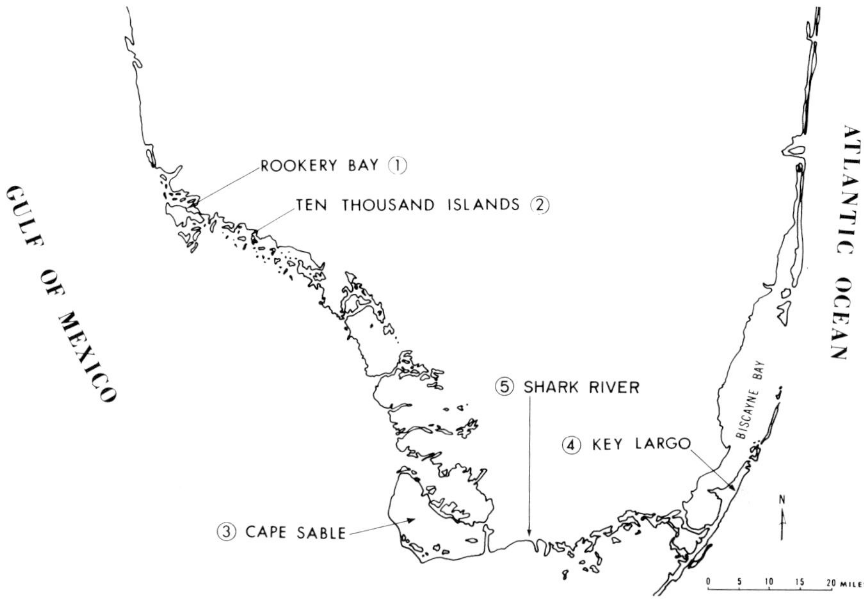

FIG. 2. Map of south Florida depicting areas where mangrove research has been conducted. Our research sites are Rookery Bay, Faka Union Bay, and Fakahatchee Bay. Also shown are the research sites of Heald (1969) and Odum (1969) in Shark River, those of Miller (1972) on Key Largo, and Cape Sable where Davis (1940) described the largest stands of mangroves in Florida. (From Lugo *et al.,* 1975.)

(fringe forests) which are inundated but not overwashed at high tide (1565 ± 245 g m^{-2} versus 5165 ± 295 g m^{-2}, $\bar{x} \pm 1$ s.d.).

To initiate interpretation of the available data, and for purposes of this simulation, we chose to study the relative importance of terrestrial runoff and tidal flushing in nutrient cycling and organic production in the overwash forest type. The simulation model is shown in Fig. 3 and described below. The objectives of the simulation were:

1. to study the relative effects of terrestrial runoff and tides on nutrient cycling and productivity in mangrove forests;
2. to study the effect of tidal flushing on accumulation and export of detritus in and from mangrove forests;
3. to study the role of mangroves on water quality; and
4. to assess the usefulness of modeling for research planning.

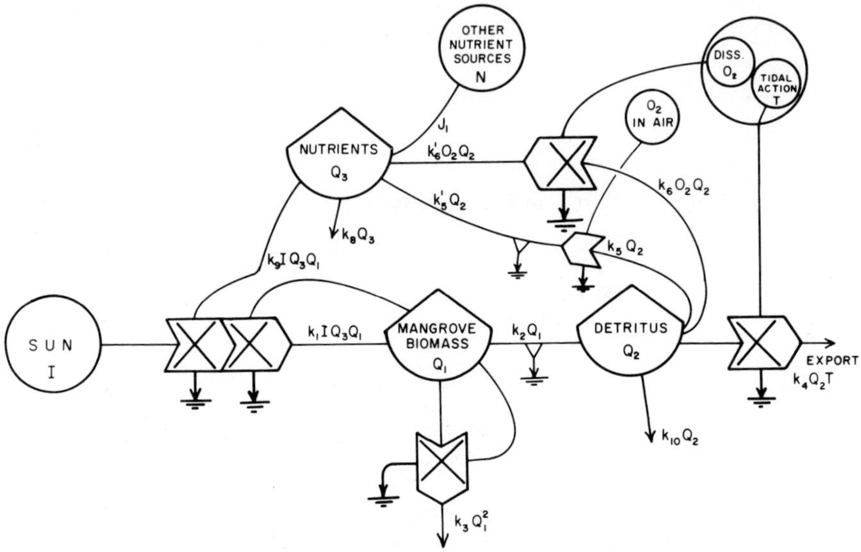

FIG. 3. Simulation model showing expressions that describe each pathway, and changes in state level of each compartment in the mangrove forest model. $\dot{Q}_1 = k_1 IQ_3 Q_1 - k_3 Q_1^2 - k_2 Q_1$, $\dot{Q}_2 = k_2 Q_1 - k_4 Q_2 T - k_5 Q_2 - k_6 O_2 Q_2 - k_{10} Q_2$, $\dot{Q}_3 = k_5' Q_2 + k_6' O_2 Q_2 + J_1 - k_8 Q_3 - k_9 IQ_3 Q_1$. Symbols are described by Odum (1972).

III. Description of the Model

In this model, the radiant energy of sunlight I interacts with nutrients Q_3 and mangrove biomass Q_1 and is converted through the process of gross photosynthesis into organic matter. Some of this gross production is respired by the forest, some is stored as a net increase in forest biomass, and some is deposited in the forest floor as detritus. The detritus Q_2 may be exported from the forest floor to the estuary by tidal action. Some of it is grazed in situ by mangrove consumers, and some is decomposed or accumulated as peat. Decomposition may occur under influence of the oxygen-saturated waters of incoming tides, or by atmospheric oxygen when the forest floor is exposed to the air. Decomposition of detritus within the mangrove system represents a source of nutrients for photosynthesis. Other nutrient sources are from terrestrial runoff, tidal waters, rainfall, and sediment storage. Of these, terrestrial runoff is the most significant. In the model, they are all grouped as a single source. Some nutrients Q_3 are not used and are lost from the system; the rest are sequestered through plant photosynthesis, thus completing the cyclic loop in the model.

A. Model Structure and Computer Program

Each flow or pathway in Fig. 3 can be described by an equation that indicates the flow to be a linear or nonlinear relationship with a storage or forcing function in the model. An example of a linear relationship is the flow from mangrove biomass to detritus, k_2Q_1. Gross photosynthesis flow into biomass is an example of a nonlinear relationship, $k_1IQ_3Q_1$. Nonlinear interactions in this model were assumed to be multiplicative and are identified by the multiplier symbol. The expressions for each pathway are shown in Fig. 3. Also shown in the figure are the differential equations describing rates of change in mangrove biomass, detritus, and nutrient levels. The rate coefficients (k's) were assumed to remain constant throughout the simulation. Table I contains all the rate coefficients used in the various simulations. Tables II–IV (pp. 124–126)

TABLE I

Rate Coefficients Used for Flows in the Mangrove Model[a]

Coefficient	Value[b]	
k_1		
High metabolism, low nutrients	2.55×10^{-9}	m^4/kcal g nutrients
High metabolism, high nutrients	4.25×10^{-11}	m^4/kcal g nutrients
Mean metabolism, low nutrients	1.32×10^{-9}	m^4/kcal g nutrients
Mean metabolism, high nutrients	2.20×10^{-11}	m^4/kcal g nutrients
k_2	8.4×10^{-2}	y^{-1}
k_3		
High metabolism	1.68×10^{-5}	m^2/g C y
Mean metabolism	1.25×10^{-5}	m^2/g C y
k_4	5.12	$m^{-1}\ y^{-1}$
k_5	1.8×10^{-2}	m^2/g C y
k_5'	1.44×10^{-3}	m^2/g C y
k_6	1.02×10^{-2}	m^3/g O_2 y
k_6'	8.2×10^{-4}	m^3/g O_2 y
k_8		
High metabolism, low nutrients	3.5×10^{-1}	y^{-1}
High metabolism, high nutrients	5.8×10^{-3}	y^{-1}
Mean metabolism, low nutrients	1.88	y^{-1}
Mean metabolism, high nutrients	3.13×10^{-2}	y^{-1}
k_9		
High metabolism, low nutrients	2.05×10^{-10}	m^4/kcal g nutrients
High metabolism, high nutrients	3.4×10^{-12}	m^4/kcal g nutrients
Mean metabolism, low nutrients	1.05×10^{-10}	m^4/kcal g nutrients
Mean metabolism, high nutrients	1.75×10^{-12}	m^4/kcal g nutrients
k_{10}	3.68×10^{-1}	y^{-1}

[a] Figure 3 identifies the pathways in the model where these constants are used.
[b] Note that y = year.

summarize the data that were used. The following equations, Eqs. (1)–(3), demonstrate a sample calculation:

$$\text{gross photosynthesis} = k_1 I Q_3 Q_1, \tag{1}$$

where

$$I = 1.46 \times 10^6 \text{ kcal m}^{-2} \text{ y}^{-1},$$

$$Q_3 = 100 \text{ g nutrients m}^{-2},$$

$$Q_1 = 10{,}500 \text{ g C m}^{-2}.$$

Using a high estimate of gross photosynthesis (3920 g C m^{-2} y^{-1}),

$$k_1 = \frac{3920 \text{ g C m}^{-2} \text{ y}^{-1}}{(1.46 \times 10^6 \text{ kcal m}^{-2} \text{ y}^{-1})(100 \text{ g nutrients m}^{-2})(10{,}500 \text{ g C m}^{-2})}, \tag{2}$$

or

$$k_1 = 2.55 \times 10^{-9} \text{ m}^4 \text{ kcal}^{-1} \text{ g nutrients}^{-1}. \tag{3}$$

Equations (4)–(6) following demonstrate the procedure for analog computer scaling, as detailed by Patten (1971). The scaled equations for mangrove biomass, detritus, and nutrients are shown in Eqs. (6)–(8), respectively.

For mangrove biomass Q_1,

$$dQ_1/dt = k_1 I Q_3 Q_1 - k_2 Q_1 - k_3 Q_1^2. \tag{4}$$

Inserting numerical values of the rate coefficients (Table I) and dividing and multiplying each variable on the right by its assigned maximum value gives

$$\frac{dQ_1}{dt} = 2.55 \times 10^{-9}(3.65 \times 10^6)\left[\frac{I}{3.65 \times 10^6}\right](800)\left[\frac{Q_3}{800}\right](3 \times 10^4)\left[\frac{Q_1}{3 \times 10^4}\right]$$
$$- 0.084(3 \times 10^4)\left[\frac{Q_1}{3 \times 10^4}\right] - 1.68 \times 10^{-5}(3 \times 10^4)^2\left[\frac{Q_1}{3 \times 10^4}\right]^2. \tag{5}$$

Dividing both sides by the assigned maximum value for mangrove biomass gives the scaled equation

$$\frac{dQ_1/dt}{3 \times 10^4} = 7.45\left[\frac{I}{3.65 \times 10^6}\right]\left[\frac{Q_3}{800}\right]\left[\frac{Q_1}{3 \times 10^4}\right] - 0.084\left[\frac{Q_1}{3 \times 10^4}\right]$$
$$- 0.504\left[\frac{Q_1}{3 \times 10^4}\right]. \tag{6}$$

Similar equations for detritus and nutrients are shown in Eqs. (7) and (8), respectively:

$$\frac{dQ_2/dt}{10^4} = 0.252\left[\frac{Q_1}{10^4}\right] - 10.24\left[\frac{Q_2}{10^4}\right]\left[\frac{T}{2}\right] - 0.018\left[\frac{Q_2}{10^4}\right] - 0.0816\left[\frac{Q_2}{10^4}\right]\left[\frac{O_2}{8}\right] - 0.368\left[\frac{Q_2}{10^4}\right], \tag{7}$$

$$\frac{dQ_3/dt}{800} = 0.018\left[\frac{Q_2}{10^4}\right] + 0.082\left[\frac{Q_2}{10^4}\right]\left[\frac{O_2}{8}\right] + 0.455 - 0.35\left[\frac{Q_3}{800}\right] - 22.4\left[\frac{I}{3.65 \times 10^6}\right]\left[\frac{Q_3}{800}\right]\left[\frac{Q_1}{3 \times 10^4}\right]. \tag{8}$$

The model with the data is presented in Fig. 4, and Fig. 5 is the analog computer program.

B. Field Data

Field studies were conducted in the Ten Thousand Island area in southwestern Florida from August 1971 to February 1973. Metabolic data were collected in a zoned mangrove forest located at Rookery Bay, near Naples.

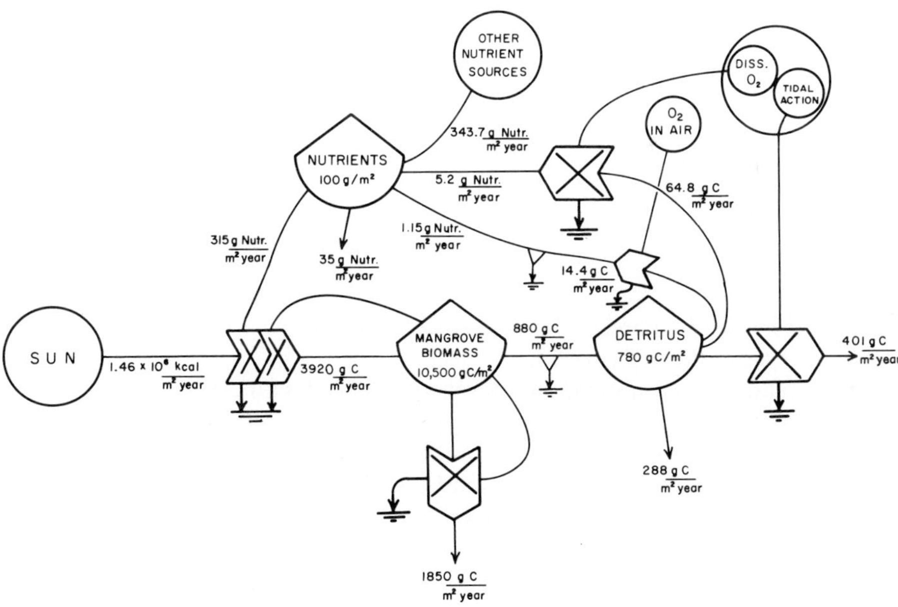

FIG. 4. Simulation model with data values utilized for scaling the analog computer. Symbols are described by Odum (1972).

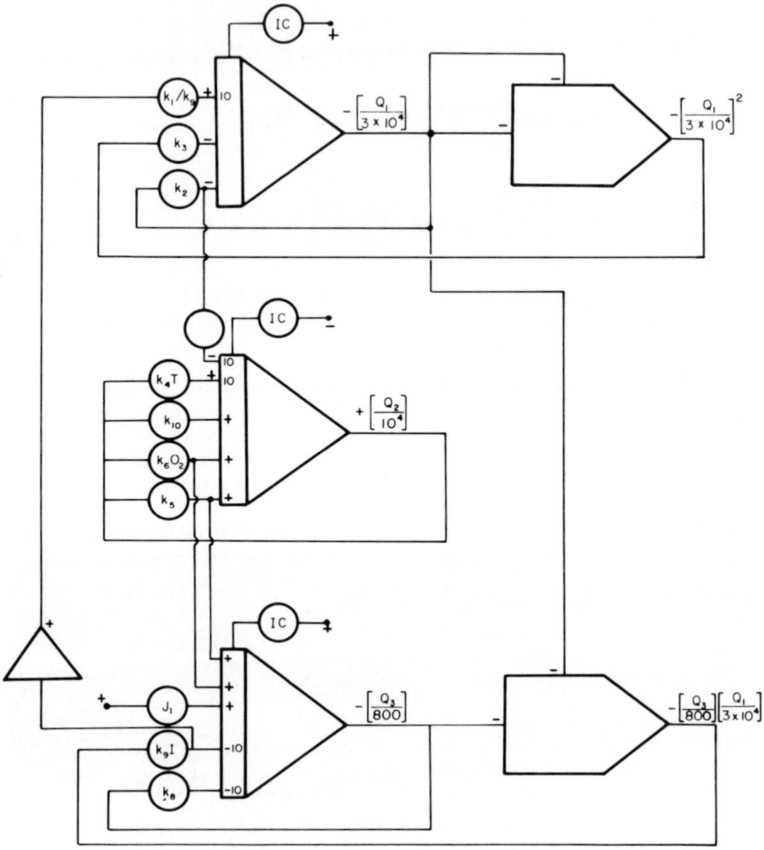

FIG. 5. Analog computer diagram of the simulation model.

Biomass data were obtained in a mangrove overwash forest in Faka-Union Bay some 20 miles to the east (Fig. 2).

Photosynthesis and respiration rates were obtained with a Beckman infrared carbon dioxide gas analyzer, powered by gasoline and diesel generators, and located in a tent inside the mangrove forest. These metabolic rates were measured on a diurnal basis as described by Odum *et al.* (1970). The apparatus was assembled with solenoid valves, timers, blowers, pumps, and so forth, so that four forest compartments could be monitored in sequence at 15-min intervals. Over 50 diurnal records of photosynthesis and respiration (sun and shade leaves, tree stems, prop roots, pneumatophores, and seedlings of each mangrove species) were obtained during the period of study. A detailed account of the procedures and results of these measurements will be reported elsewhere.

Biomass determinations were based on replicated 5 x 5 m clearcut harvests in a mangrove overwash forest. The particular study site was considered to be representative of the overwash type of mangrove stands in the Ten Thousand Island area. For each species in each harvest area, leaf, stem (by size class), detritus, prop root, and fruit and flower biomass levels were determined. In addition, leaf, prop root, and trunk area indices were calculated by species for each site.

Leaf fall was determined from 20 litterfall baskets located at Rookery Bay and emptied periodically during the study. Water quality of incoming and outgoing tides, river runoff, swamp waters, precipitation throughfall, and of adjacent canals was determined seasonally at Rookery Bay. In addition, concurrent studies by the Environmental Protection Agency and the University of Miami's Rookery Bay Project provided monthly data on water quality, water flows, tidal characteristics of the larger area outside Rookery Bay, and also provided data on periphyton productivity on mangrove prop roots. Rates for detritus export into the bays and its decomposition in the forest were obtained from the literature. This included work of Golley *et al.* (1962) in Puerto Rico, and Heald (1969) in the Shark River estuary, southeast of the present study site (Fig. 2). Nutrient uptake associated with photosynthesis was calculated from photosynthetic rates, assuming a 3% ash content plus a 1% nitrogen content in mangrove biomass. Table II summarizes the metabolic data that were obtained, Table III characterizes the structure and biomass data for the overwash study site, and Table IV depicts all values used in scaling equations for the analog computer simulation. The simulation was done on an Electronic Associates, Inc., Miniac analog computer.

IV. Results

Simulation results are presented in Figs. 6–17. For convenience, the results are presented under four headings: mangrove development under current (initial) conditions, effects of tides on mangrove production and export, effects of mangrove succession on nutrient content of adjacent waters, and effects of nutrient runoff on mangroves.

A. Mangrove Development under Current (Initial) Conditions

Three simulations were done utilizing the measured standing crop, and high and mean estimates of metabolic rates (gross photosynthesis and total respiration for a 24-h period) as initial conditions. For each run, the temporal

TABLE II

Photosynthesis and Respiration of the Experimental Mangrove Forest in South Florida[a]

	Red mangrove					White mangrove						Periphyton		Total	
		Leaves					Leaves				R_{24} Trunk				
Site 3[b]	LAI	P_{net}	R_{night}	P_{gross}	R_{24}	LAI	P_{net}	R_{night}	P_{gross}	R_{24}	and root	P_{net}	R_{24}	P_{gross}	R_{24}
High estimates															
Plot 7	2.05	5.88	0.88	6.76	1.76	1.80	2.52	0.54	3.06	1.08	1.33	0.9	0.9	10.72	5.07
Plot 8	3.65	10.47	1.57	12.04	3.14	–	–	–	–	–	1.33	0.9	0.9	12.94	5.37
Mean estimates															
Plot 7	2.05	2.82	0.48	3.30	0.96	1.80	1.04	0.30	1.34	0.60	1.33	0.9	0.9	5.54	3.79
Plot 8	3.65	5.03	0.85	5.88	1.70	–	–	–	–	–	1.33	0.9	0.9	6.78	3.93

[a] Values in g C m^{-2} day^{-1}. High estimates are based on an optimal day. Mean estimates represent the mean of all determinations made for that ecosystem component. Net daytime photosynthesis is P_{net}; gross photosynthesis is P_{gross}; nighttime respiration is R_{night}; and 24-h respiration is R_{24}. Leaf area index is LAI.

[b] Overwash mangrove forest.

TABLE III

Structural Indices and Biomass Characteristics of a Mangrove Overwash Forest in South Florida

Parameter	Mean	Std. error of mean
Area indices		
Basal ($m^2\ ha^{-1}$)	14.68	0.33
Leaf ($m^2\ m^{-2}$)	3.74	0.12
Trunk ($m^2\ m^{-2}$)	0.15	–
Prop root ($m^2\ m^{-2}$)	1.31	–
Biomass[a]		
Leaves	710.0	22.0
Fruit	12.8	15.3
Wood	7043.0	7.0
Prop roots	4695.0	711.0
Detritus	1565.0	234.5

[a] Grams per dry weight organic matter per square meter.

patterns of gross photosynthesis, mangrove and detritus standing crop, and detritus export were studied. These results are depicted in Figs. 6–8.

Figure 6a suggests that the overwash mangroves in the Ten Thousand Islands are about 10 y from reaching steady state with respect to standing crop, gross photosynthesis, and detritus export. The steady-state standing crop is dependent on initial metabolism rates, reaching a maximum of 16,200 g C m^{-2} and 13,100 g C m^{-2} at high and mean estimates, respectively. The rate of detritus accumulation in the forest also depends on metabolism of the stand, with steady-state values of 1230 and 1000 g C m^{-2}, high and mean estimates, respectively. These detritus values, however, do not appear to be as sensitive to changes in metabolism as is the total standing crop of the system. Figures 7a and b compare the ratios of gross photosynthesis to detritus export with high and mean initial rates of metabolism. At high rates of metabolism this ratio reaches a steady-state value of 8.7 compared to a value of 5.3 for the mean metabolic rate.

If one sets the initial conditions of standing crop to a value similar to those measured for successional plots in the study area, one observes the rate of mangrove succession at mean and high metabolism rates. These simulations are depicted in Figs. 6b and 8. Figure 6b shows the changes in mangrove standing crop and detritus storage in the forest at mean and high metabolism rates. It is clear that steady-state conditions are retarded by the lower metabolic rate. At the initial mean metabolism rates the system takes about 23 y to reach steady-state conditions compared to 12 y at high metabolic rates. The mean

TABLE IV

Values Used in Simulations of the Mangrove Model for Initial Conditions of Forcing Functions, State Variables, and Flows

	Value	
	Initial	Maximum
Forcing functions		
Sunlight, I	4000 kcal m^{-2} day^{-1}	10,000 kcal m^{-2} day
Tide, T	10 cm	2 m
Dissolved oxygen, O_2	4 g m^{-3}	8 g m^{-3}
State variables		
Mangrove biomass, Q_1	10,500 gC m^{-2}	30,000 gC m^{-2} (current values)
	300 gC m^{-2}	30,000 gC m^{-2} (successional conditions)
Detritus, Q_2	780 gC m^{-2}	10,000 gC m^{-2}
Nutrients, Q_3	100 g m^{-2}	800 g m^2 (low nutrients)
	6000 g m^{-2}	8000 g m^{-2} (high nutrients)
Flows		
Gross photosynthesis, kIQ_3Q_1	10.72 gC m^{-2} day^{-1} (high value)	
	5.54 gC m^{-2} day^{-1} (mean value)	
Respiration of mangroves, $k_3Q_1^2$	5.07 gC m^{-2} day^{-1} (high value)	
	3.79 gC m^{-2} day^{-1} (mean value)	
Litter fall, k_2Q_1	2.41 gC m^{-2} day^{-1}	
Export of detritus by tidal flushing, k_4Q_2T	1.1 gC m^{-2} day^{-1}	
Decomposition of detritus when mangroves are dry, k_5Q_2	0.16 gC m^{-2} day^{-1} (only 3 months of year)	
Decomposition of detritus when mangrove forest floor is water-covered, $k_6Q_2Q_2$	0.12 gC m^{-2} day^{-1} (during 3 months of dry season)	
	0.30 gC m^{-2} day^{-1} (during 6 months of wet season)	
Grazing on detritus and other losses, $k_{10}Q_2$	0.786 gC m^{-2} day^{-1}	
Nutrients derived from decay of		
detritus, k_5Q_2	0.0128 gC m^{-2} day^{-1} (3 months of dry season)	
detritus $k_6O_2Q_2$	0.0336 gC m^{-2} day^{-1} (3 months of dry season + 6 months of wet season)	
Nutrients from other sources, J_1	0.94 g nutrients m^{-2} day^{-1}	
Nutrient uptake by mangroves, $k_9IQ_3Q_1$	0.846 g nutrients m^{-2} day^{-1}	
Nutrients not used by mangroves, k_8Q_3	0.094 g nutrients m^{-2} day^{-1}	

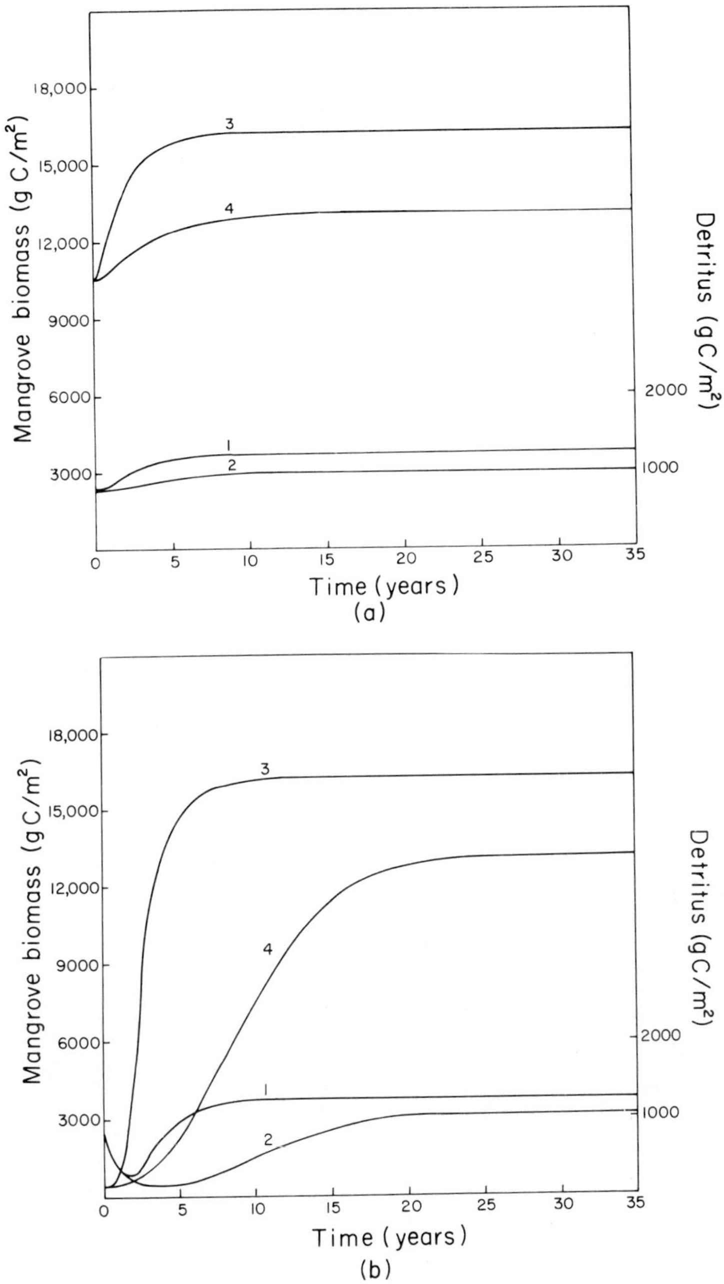

FIG. 6. Standing crop of biomass and detritus storage in mangrove forests with (a) starting conditions at present levels of mangrove biomass, and (b) early successional levels of mangrove biomass. Each simulation was run at high and mean rates of metabolism. Initial gross photosynthesis as follows (in g C m^{-2} day^{-1}): (1) detritus, 10.72; (2) detritus, 5.54; (3) mangrove biomass, 10.72; (4) mangrove biomass, 5.54. Table IV contains all values used for initial conditions.

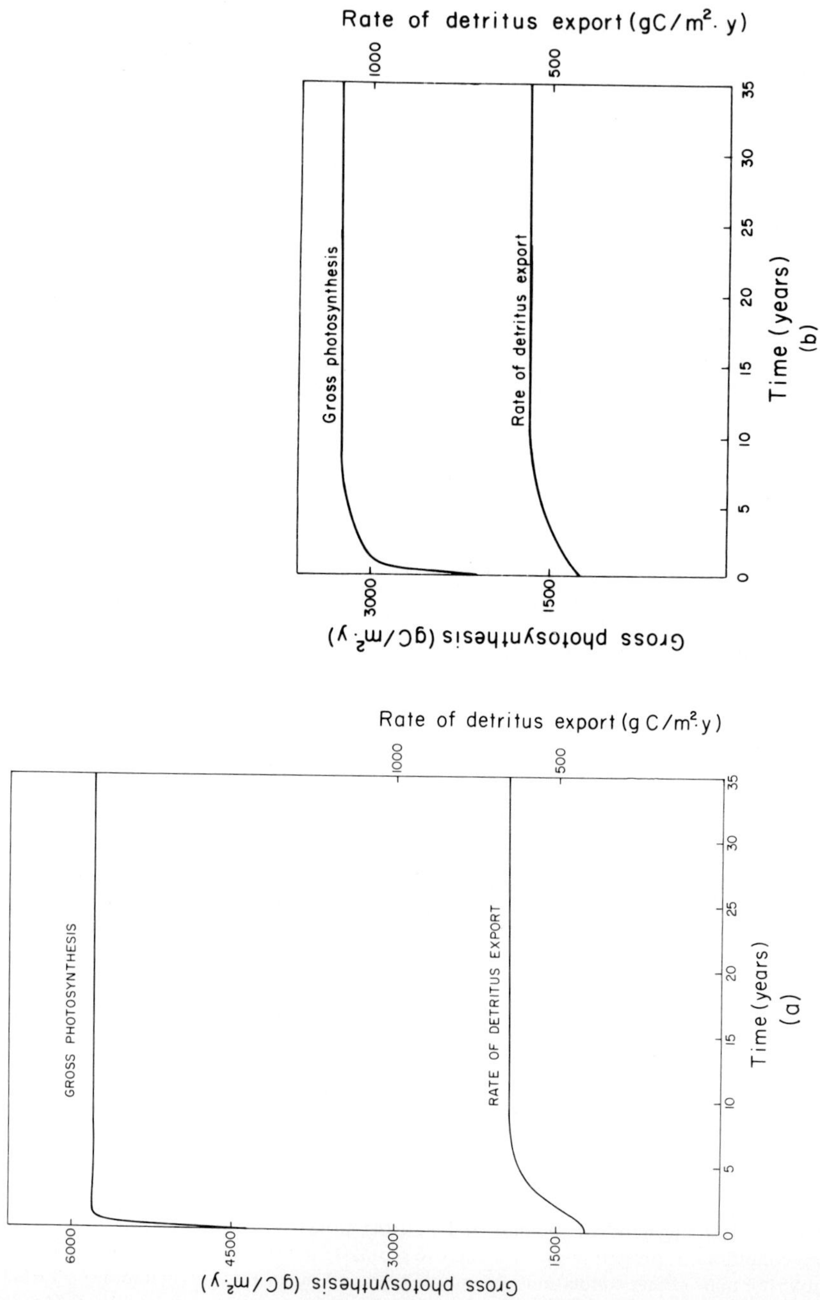

FIG. 7. Rates of gross photosynthesis and detritus export for mangroves undergoing succession, with initial conditions of current biomass levels and (a) high and (b) mean metabolism rates. Tidal depth 10 cm. Table IV contains all values used for initial conditions.

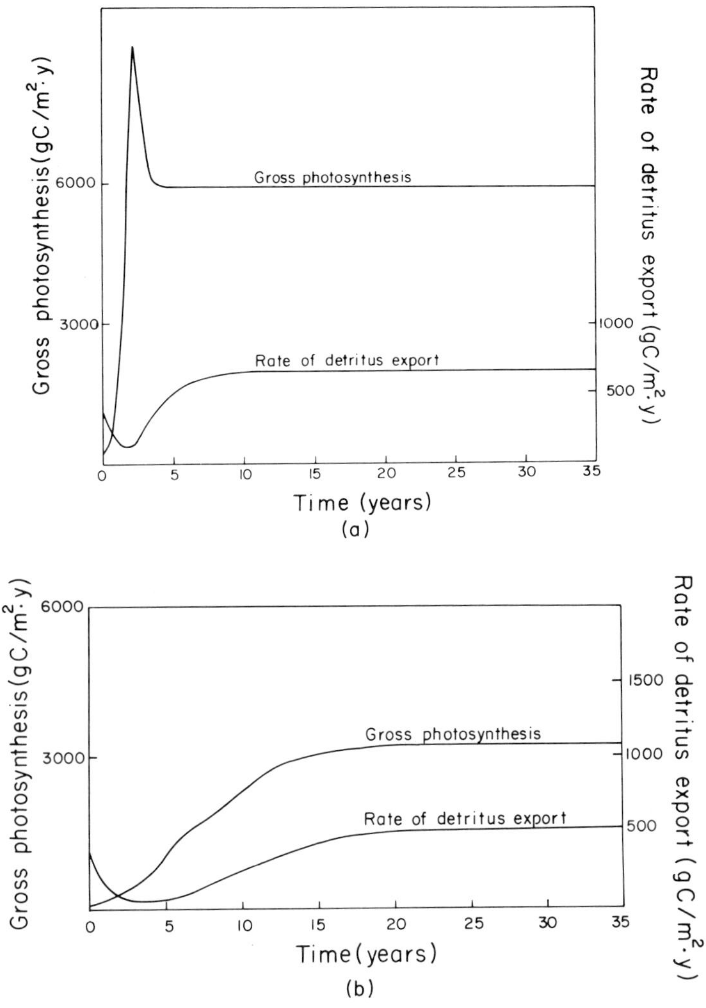

FIG. 8. Rates of gross photosynthesis and detritus export for mangroves undergoing succession with initial conditions of 300 g C m^{-2} of mangrove biomass and (a) high and (b) mean rates of metabolism. Tidal depth 10 cm. Table IV contains all values used for initial conditions.

rates of gross photosynthesis and respiration seem to depict a more realistic picture of the true development of mangrove systems. This is not surprising since mangroves are not expected to metabolize at optimal rates every day of the year throughout their development. The ratio of gross photosynthesis to detritus export is depicted in Fig. 8 for both high (8a) and mean (8b) metabolic rates.

High rates of metabolism caused an overshoot in the gross photosynthesis curve. This overshoot was not observed when initial mean rates were utilized but was observed in several runs with high rates of metabolism. It appears that overshooting is a property of a fast-growing system exhibiting surges of high production during short time periods. The initial decrease in detritus export shown in this simulation (Fig. 8) is due to the time lag in establishment of a system with sufficient production for export. Table V summarizes the levels and steady-state values obtained from this first set of simulations.

B. Effects of Tides on Mangrove Production and Export

Figure 9 depicts the effect of different tidal levels on the rate of detritus export and gross photosynthesis in a mangrove forest with initial conditions of high metabolism and measured standing crop. It is clear that tides have less effect on gross photosynthesis as compared to their effects on the rate of

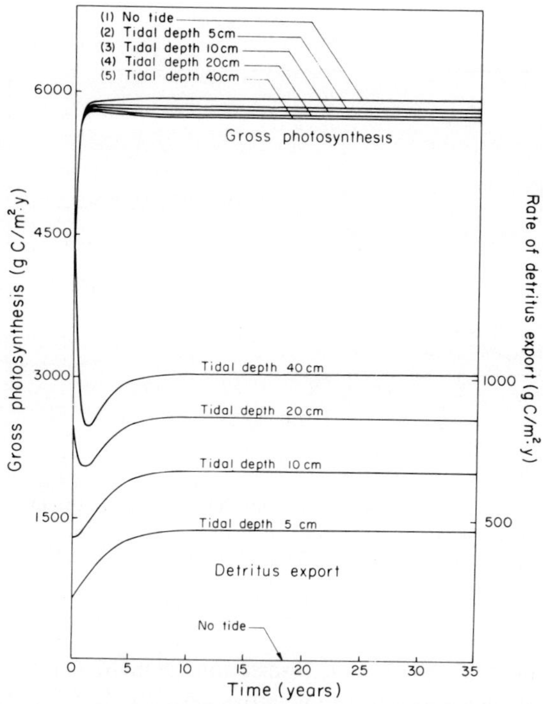

FIG. 9. Rates of gross photosynthesis and detritus export in a mangrove forest submitted to several tidal amplitudes and high rates of metabolism. Table IV contains all values used for initial conditions.

TABLE V

Steady-State Levels of Biomass, Detritus, and Nutrients; Ratio of Gross Photosynthesis to Detritus Export and Time Required to Reach Steady State in Four Simulations with Different Starting Conditions of Biomass and Metabolism Rates

		Level at steady state (g m^{-2})			Time to steady state (years)		
Initial conditions[a]	Ratio of P_{gross} to Detritus Export	Biomass	Detritus	Nutrients	Biomass	Detritus	Nutrients
1	8.7	16,200	1230	90	$7\frac{1}{2}$	6	5
2	5.8	16,200	1230		10	8	
3	8.7	13,100	1000		$12\frac{1}{2}$	8	
4	6.0	13,100	1000		23	19	

[a]

	Mangrove biomass (g C m^{-2})	P_{gross} (g C m^{-2} day^{-1})	Respiration (g C m^{-2} day^{-1})
1	10,500	10.72	5.07
2	10,500	5.54	3.79
3	300	10.72	5.07
4	300	5.54	3.79

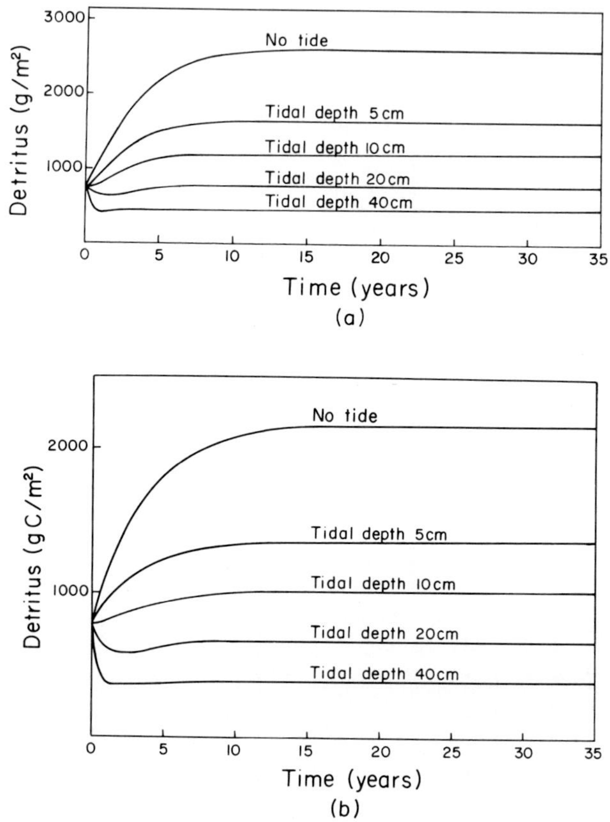

FIG. 10. Level of detritus storage in a mangrove forest submitted to several tidal amplitudes and (a) high and (b) mean rates of metabolism. Table IV contains all values used for initial conditions.

detritus export. High tidal levels slightly decrease the rate of gross photosynthesis, but significantly increase the rate of mangrove detritus export. Storage of detritus at the above initial conditions is shown in Fig. 10a. As tide levels increase, the amount of detritus accumulating in the forest decreases. Maximum detritus accumulation in the forest is observed in absence of a tide when output to the bay is zero. This is consistent with observations of large peat accumulations in areas devoid of extensive flushing action. As observed earlier, the levels of gross photosynthesis, export and storage of detritus, and time required to reach steady-state conditions are lower for those simulations based on mean rates of metabolism (Figs. 10b and 11). All the values obtained in this set of simulations are summarized in Table VI.

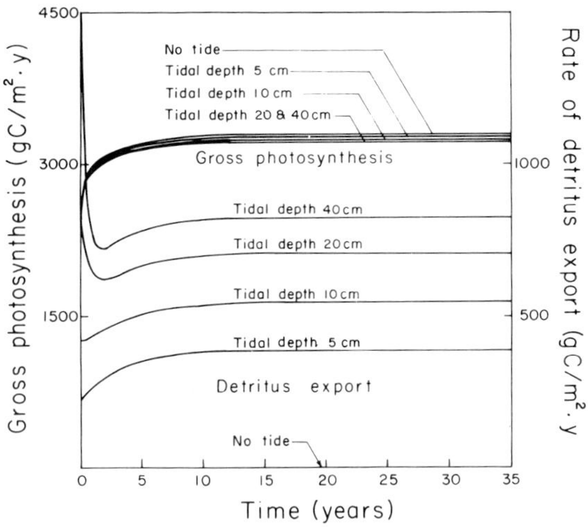

FIG. 11. Rates of gross photosynthesis and detritus export at several tidal depths from a mangrove forest at initial conditions of mean metabolism. Table IV contains all values used for initial conditions.

C. Effects of Mangrove Succession on Nutrient Content of Adjacent Waters

Rates of gross photosynthesis and nutrient storage as a function of time with various initial conditions of standing crop are plotted in Figs. 12 and 13. Figure 12 represents the high metabolic rate and Fig. 13 the mean rate. Overshooting of gross photosynthesis is again observed at high metabolic rates and steady state is reached in 6–8 y. The simulations demonstrate that, depending upon the initial conditions of mangrove biomass, nutrients more or less, are going to be found in waters adjacent to the mangroves. At low mangrove biomass, nutrients in the water are higher and remain in the transient state longer than at higher standing crops of mangrove forest. Results are summarized in Table VII.

D. Effects of Nutrient Runoff on Mangroves

Since the high metabolic rate estimate yielded higher and faster steady-state values than those observed in Florida (see below), these last simulations were run with initial conditions set at the mean metabolic values. In Figs. 14–16 the amount of nutrients in runoff was varied, and in Fig. 17 it was held constant, but the amount of nutrients lost from the system was altered. The initial conditions of tides and biomass were the same as those found in the south

TABLE VI

Steady-State Levels and Time Required to Reach Steady State of Gross Photosynthesis, Detritus Export, and Standing Crop in Nine Simulations with Different Starting Conditions for Tidal Levels and Mangrove Biomass

	P_{gross} (g C m^{-2} day^{-1})		Rate of Detritus export (g C m^{-2} day^{-1})		Detritus (g C m^{-2})	
Initial conditions tide level (cm)	Time to steady state (years)	Level at steady state	Time to steady state (years)	Level at steady state	Time to steady state (years)	Level at steady state
A[a]						
0	5	16.3	0	0	10	2600
5	$2\frac{1}{2}$	16.1	8	1.26	8	1650
10	5	15.9	$7\frac{1}{2}$	1.84	$7\frac{1}{2}$	1200
20	5	15.8	$7\frac{1}{2}$	2.36	5	800
40	5	15.7	$7\frac{1}{2}$	2.77	3	450
B[b]						
0					15	2170
5			8	1.0	12	1350
10	8	8.9	10	1.53	10	1000
20			8	1.89	8	650
40			8	2.19	5	400

[a] Mangrove biomass, 10,500 g C m^{-2}; P_{gross}, 10.72 g C m^{-2} day^{-1}; respiration, 5.07 g C m^{-2} day^{-1}.
[b] Mangrove biomass, 10,500 g C m^{-2}; P_{gross}, 5.54 g C m^{-2} day^{-1}; respiration, 3.79 g C m^{-2} day^{-1}.

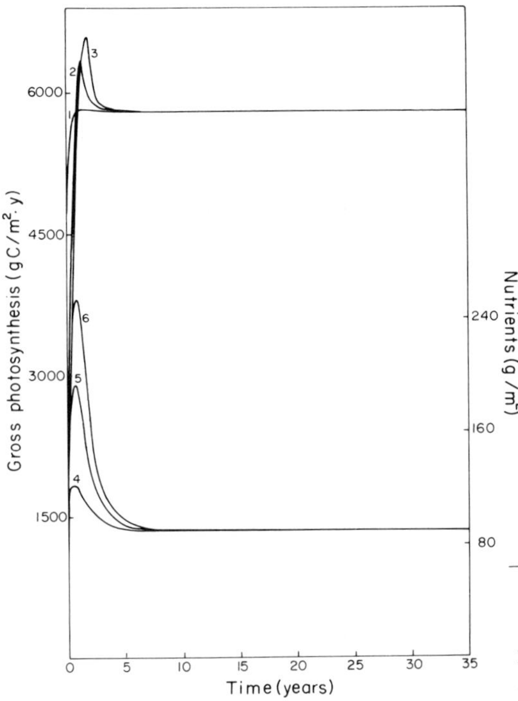

Initial biomass (g C m^{-2})

Gross photosynthesis		Nutrients	
1	10,500	4	10,500
2	5250	5	5250
3	2625	6	2625

FIG. 12. Effects of mangrove biomass level on nutrient level in mangrove forests with high metabolism. Table IV contains all values used for initial conditions.

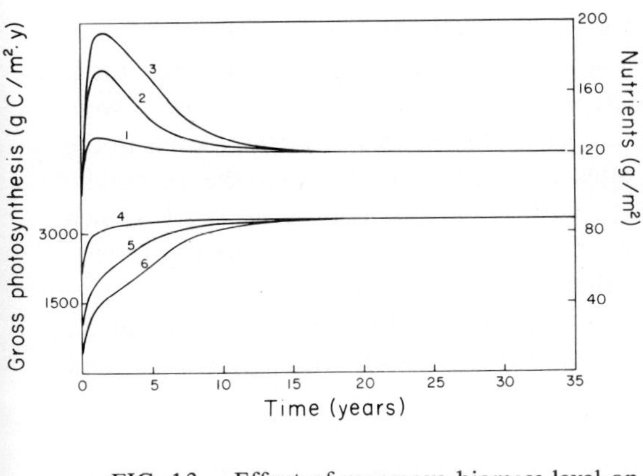

Initial biomass (g C m^{-2})

Nutrients		Gross photosynthesis	
1	10,500	4	10,500
2	5250	5	5250
3	2625	6	2625

FIG. 13. Effect of mangrove biomass level on nutrient level in mangrove forests with mean rates of metabolism. Table IV contains all values used for initial conditions.

TABLE VII

Steady-State Levels of Gross Photosynthesis and Nutrients in Water in Eight Simulations with Different Initial Conditions for Mangrove Biomass and Metabolism Rates

	P_{gross}, g C m^{-2} day^{-1}		Nutrients, g m^{-2}	
Initial condition	Time to steady state (years)	Level at steady state	Time to steady state (years)	Level at steady state
A[a]				
Mangrove biomass				
10,500 g C m^{-2}	$2\frac{1}{2}$	16.2	6	90
5250 g C m^{-2}	5	16.2	$7\frac{1}{2}$	90
2625 g C m^{-2}	5	16.2	$8\frac{1}{2}$	90
300 g C m^{-2}	5	16.2	–	–
B[b]				
Mangrove biomass				
10,500 g C m^{-2}	8	8.9	$7\frac{1}{2}$	120
5250 g C m^{-2}	15	8.9	15	120
2625 g C m^{-2}	18	8.9	18	120
300 g C m^{-2}	20	8.9	–	–

[a] P_{gross} of 10.72 g C m^{-2} day^{-1}; respiration of 5.07 g C m^{-2} day^{-1}.
[b] P_{gross} of 5.54 g C m^{-2} day^{-1}; respiration of 3.79 g C m^{-2} day^{-1}.

Florida study area. In the simulations in Figs. 14 and 16, the initial conditions of nutrient levels were raised from 100 g m^{-2} to 6000 g m^{-2}.

The model was sensitive to nutrient availability. Figures 15 and 16 show that, when nutrient runoff was set at zero, mangrove biomass decreased steadily. The steady-state level depends on the fertility of the site as shown in Fig. 16, where the initial nutrient levels were set at 6000 g m^{-2}. At higher rates of nutrient runoff, however, the steady-state value of biomass was independent of the initial nutrient conditions although the transient patterns of biomass growth were different for each initial setting. Figure 14 shows changes in nutrient levels in the forest associated with different rates of nutrient runoff and the metabolic behavior of the forest. In Fig. 17, the pattern of gross photosynthesis is shown to be sensitive to the efficiency of nutrient uptake from surrounding waters. This suggests a role for a mechanism that would concentrate nutrients.

These simulations demonstrate the dependency of mangrove ecosystems on nutrient inputs derived from the land. Decomposition within the forest and inputs from the sea do not seem to be enough to maintain the observed rates of metabolism in these systems. Table VIII summarizes the relative importance of nutrient sources to the mangrove forests, and Table IX summarizes the results of this set of simulations.

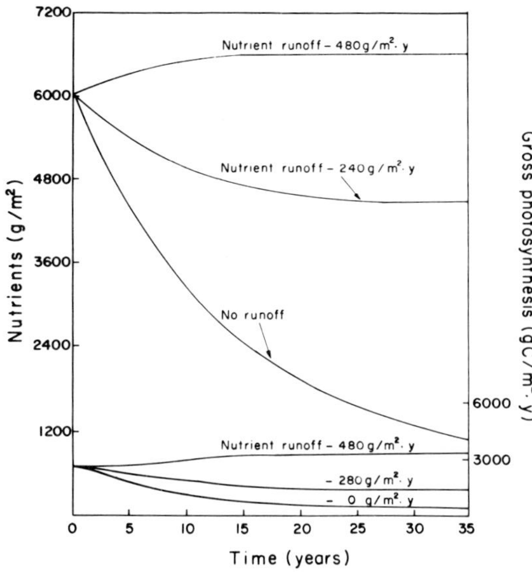

FIG. 14. Rates of gross photosynthesis and level of nutrients in mangrove forests with initial conditions of high nutrient level (6000 g m^{-2}), mean rates of metabolism, and three rates of nutrient runoff. Gross photosynthesis, 5.54 g m^{-2} day^{-1}. Table IV contains all values used for initial conditions.

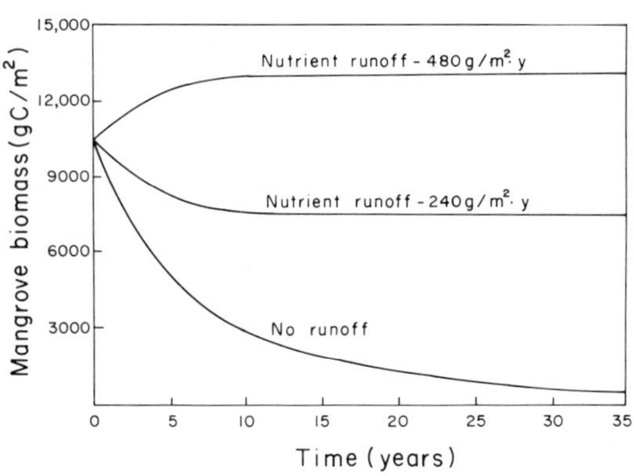

FIG. 15. Mangrove biomass at current initial conditions, mean rate of metabolism, and three levels of nutrient runoff. Initial gross photosynthesis, 5.54 g C m^{-2} day^{-1}; initial nutrients, 100 g m^{-2}. Table IV contains all values used for initial conditions.

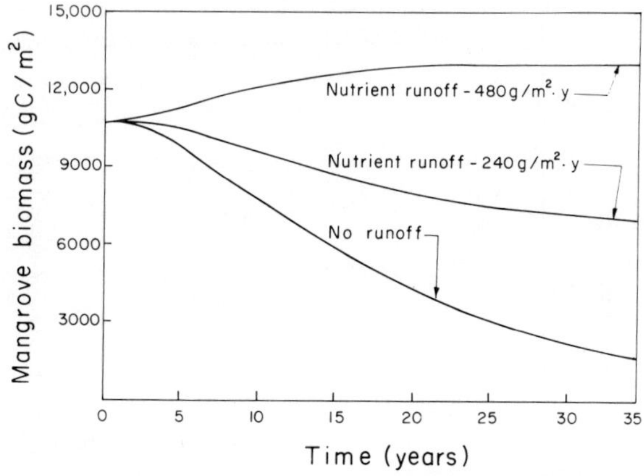

FIG. 16. Mangrove biomass with high initial conditions of nutrients, mean rate of metabolism, and three levels of nutrient runoff. Initial nutrients, 6000 g m^{-2}; gross photosynthesis, 5.54 g m^{-2} day^{-1}. Table IV contains all values used for initial conditions.

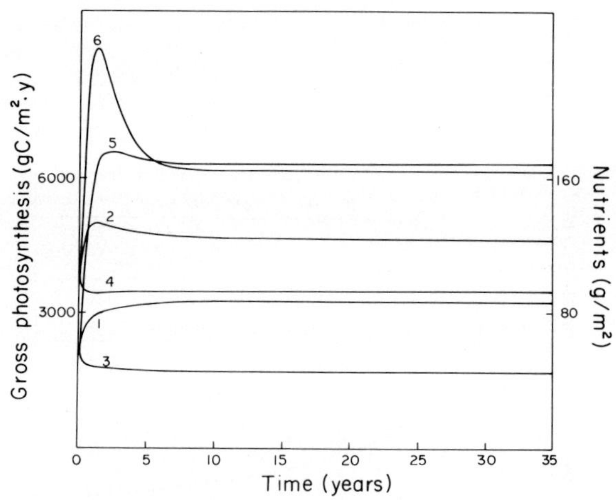

FIG. 17. Gross photosynthesis and nutrient levels as affected by variations in efficiency of nutrient uptake at mean rates of metabolism. Gross photosynthesis (nutrients unused): (1) 188 g m^{-2} y^{-1}, (3) 376 g m^{-2} y^{-1}, (5) all nutrients used. Nutrients (nutrients unused): (2) 188 g m^{-2} y^{-1}, (4) 376 g m^{-2} y^{-1}, (6) all nutrients used. Table IV contains all values used for initial conditions.

TABLE VIII

Phosphorus and Nitrogen Inputs into Rookery Bay, Naples, Florida[a]

Source	Wet season[b]	Dry season[b]
River runoff	6.9	52.8
Tidal flow	0.3	4.9
Rainfall	0.0001	–

[a] Values were obtained by multiplying the concentration of nutrients by water flow in a wet and a dry month (November and March, respectively).

[b] kg day^{-1}.

TABLE IX

Steady-State Levels and Time Required to Reach Them, for Mangrove Biomass, Gross Photosynthesis, and Nutrients in Simulations with Different Initial Conditions for Nutrient Runoff, Nutrient Levels, and Efficiency of Nutrient Use

Initial conditions	P_{gross}, g C m^{-2} day^{-1}		Nutrients, g m^{-2}	
	Time to steady state (years)	Level at steady state	Time to steady state (years)	Level at steady state
A[a]				
Nutrient input of 240 g m^{-2} $year^{-1}$	2	7.67	2	74
480 g m^{-2} $year^{-1}$	2	16.0	5	92
720 g m^{-2} $year^{-1}$	2	24.1	5	116
Nutrient losses, 0 g m^{-2} $year^{-1}$	2	17.0	5	94
35 g m^{-2} $year^{-1}$	2	16.0	5	92
70 g m^{-2} $year^{-1}$	2	14.8	5	89
B[b]				
Nutrient input of 240 g m^{-2} $year^{-1}$	5	3.84	5	78
480 g m^{-2} $year^{-1}$	8	8.9	10	126
720 g m^{-2} $year^{-1}$	13	15.5	12	160
Nutrient losses, 0 g m^{-2} $year^{-1}$	$7\frac{1}{2}$	17.3	8	164
188 g m^{-2} $year^{-1}$	8	8.9	10	126
376 g m^{-2} $year^{-1}$	5	4.66	5	93

[a] P_{gross} of 10.72 g C m^{-2} day^{-1}; respiration of 5.07 g C m^{-2} day^{-1}; initial nutrients, 100 g m^{-2}

[b] P_{gross} of 5.54 g C m^{-2} day^{-1}; respiration of 3.79 g C m^{-2} day^{-1}; initial nutrients, 100 g m^{-2}

TABLE X

Comparison of the Mangrove Model's Prediction of Steady-State Levels of Mangrove Biomass, Detritus Standing Crop, Gross Photosynthesis, and Detritus Export with Independent Field Determinations of the Same Parameters

Ecosystem parameter	Model prediction: High metabolism rate	Model prediction: Mean metabolism rate	Observed values
State variable			
Mangrove biomass	16,200 g C m^{-2}	13,100 g C m^{-2}	8700 g C m^{-2} on riverine forest in Florida (Snedaker *et al.*, unpublished) 14,000 g C m^{-2} in Panama (Golley, 1968)
Detritus	1230 g C m^{-2}	1000 g C m^{-2}	782 ± 123 to 2582 ± 147 g C m^{-2} (Snedaker *et al.*, unpublished)
Flow (process variable)			
Gross photosynthesis	15.9 g C m^{-2} day^{-1}	8.9 g C m^{-2} day^{-1}	8.7 g C m^{-2} day^{-1} (Lugo *et al.*, unpublished); Mean for the study sites
Detritus export	1.84 g C m^{-2} day^{-1}	1.5 g C m^{-2} day^{-1}	1.1 g C m^{-2} day^{-1} (Golley *et al.*, 1962) 0.67 g C m^{-2} day^{-1} (Heald, 1969)

V. Discussion

A. Steady-State Biomass and Metabolic Rates

The validity of the model was tested by comparing some of the simulation results with values reported in the literature for other stands. In Fig. 6 the model predicted a maximum biomass at steady-state of 16,200 to 13,100 g C m^{-2}, depending upon the initial rate of metabolism. These values are higher than our own biomass estimates in the other plots that were harvested (Table X). However, all the experimental plots were under the influence of recent hurricanes and, thus, presumably below steady-state conditions. The descriptions of Davis (1940) and Craighead (1971) of mature stands in Cape Sable suggest higher possible biomass levels for Florida. The biomass predictions based on mean metabolic rates are closer to measured values than those based on the high metabolism estimate. Of interest is the length of time required to reach a steady-state condition with respect to biomass. When metabolism rates are set at the high value, the forest reaches steady state in $7\frac{1}{2}$ y. However, when the mean values are used, steady state takes 10 y from the present and 23 y from an early stage of succession. Literature reports on the age of mangrove stands seem to indicate that most mangroves reach maturity at about 20–25 y (Davis, 1940; Craighead, 1971). These estimates agree with the model predictions based on a mean rate of metabolism. Wadsworth (1972) reports a growth rate for mangroves of 3.0 m in 18 months and indicates that stands are usually 25 y old with a few large individuals exceeding 50 y. These represent the survivors from previous hurricanes. Examination of hurricane frequencies in Puerto Rico and our study area reveal a probable mean frequency of 24 and 20 y, respectively (Table XI). Thus, it appears that mangrove ecosystems are adapted to rapid growth following hurricane disturbance and are thus capable of regaining steady-state conditions during the time period between hurricanes. The model prediction of a steady state in 10 y with current initial conditions also agrees with the 20-y hurricane cycle as the last hurricane in the area was Hurricane Betsy in 1965.

TABLE XI

Comparison of Hurricane Frequencies in Two Mangrove Areas, and the Predictions of Two Simulations for the Time Required by Mangroves to Reach Steady State[a]

Model predictions	High metabolism	Mean metabolism
Time to reach steady-state biomass level		
Current initial conditions	7	10
Initial conditions of low biomass		
(300 g C m^{-2})	10	23

[a] Literature reports related to steady-state mangroves. Frequency of hurricanes: in the study area, 20 y (Brunn *et al.*, 1962); in southern Puerto Rico, 24 y (Anon., 1972).

In Figs. 7 and 8 the model predicted steady-state values for gross photosynthesis of 8.9–15.9 g C m^{-2} day^{-1}, depending on initial conditions. These values are within the range of our metabolic studies in south Florida mangroves (Table X). This is also true for the detritus storage and export values (Table X). The conclusion at this point is that the simulations using mean rates of metabolism as initial conditions provide a closer estimate of observed processes than those obtained with high rates of metabolism. This indicates that mangroves operate at a maximum metabolic rate for short time periods and that our metabolism estimates are in line with real world values.

B. Metabolic Basis of Mangrove Vigor and Zonation

For a long time the zonation and vigor of mangrove stands has been related to gradients of salinity (Davis, 1940). A good reason for this is the impressive array of adaptations to a saline environment. However, no one has suggested nutrients as another important causal factor. The simulations in Figs. 14–16 suggest that nutrients seem to be an important factor in determining the vigor of mangrove ecosystems. Figures 9 and 11 suggested that the gross photosynthetic rate of a stand was not affected much by tidal fluctuation, but that the amount and rate of detritus export was a function of tidal action. In Figs. 14–17 it was shown that the amount of nutrients available to mangroves and the efficiency of their use were the main determinants of succession rates and biomass levels that could be developed at steady state. Table VIII demonstrates that most of the nutrients available to mangroves are of terrestrial origin. In addition, descriptions of mangrove stands by Davis (1940), Craighead (1971), Macnae (1967), and Walsh (1967) indicate that the largest and most vigorous mangrove stands are always associated with riverine conditions where detritus accumulation is low. These observations have been confirmed by the writers reporting on mangroves in several locations in Puerto Rico, Florida, Costa Rica, and Venezuela.

Thus, it appears that nutrient availability for photosynthesis is an important determinant of the vigor of a mangrove stand and that salinity adaptations may represent the energetic cost of tapping these nutrient sources where competition is controlled (i.e., reduced) by other factors. The overall gain in nutrients is greater than nutrient losses (in detritus) to the estuarine bays. This is demonstrated in our simultaneous measurements of photosynthesis and respiration for a given day.

Red mangroves exposed to riverine inputs exhibit maximum rates of photosynthesis which are twice as high as white mangroves growing more inland, away from the riverine input. Similarly, these red mangroves had a higher metabolic rate than black mangroves growing inside the swamp. The black mangroves had a higher metabolism rate than a red mangrove growing in the same habitat. Night respiration rates in the leaves were the same in all trees.

C. Implications for Nutrient Recycling

Figures 12 and 13 implied that successional development in mangroves was associated with their control of regional water quality. The role of mangrove forests as nutrient sinks has been described by Hesse (1961) and by Harriss (pers. comm.) for heavy metals. In the present study area a proliferation of molluscs and other filter feeders, and periphyton on prop roots and pneumatophores was observed. The metabolic work of these organisms represents mechanisms for nutrient recycling and uptake and seems to be coupled to proliferation of fine absorptive mangrove roots growing in close proximity. These adaptations would increase the efficiency of nutrient uptake from tidal or riverine waters passing over the swamp. As suggested in Fig. 17, increasing the efficiency of nutrient uptake would enhance the capability of the swamps to maximize productivity and biomass.

D. Implications for Mangrove Development

In spite of biological adaptations for the maximization of nutrient uptake and recycling from flowing waters, the simulations in Figs. 15 and 16 demonstrate that in situ decomposition in mangrove forests is not enough to maintain observed high rates of gross photosynthesis and biomass storage. This is partially due to detritus export to the bay which drains part of the nutrient currency of the system. Thus, mangroves require a steady input of terrestrial nutrients in order to maintain their characteristic rates of growth. Many of the land developments in south Florida include channelization of water flows and diversion of terrestrial runoff to coastal areas. These waters are usually diverted for human consumption or discharged through canals directly into the sea without exchange with mangrove stands in the estuaries. This strategy is not ecologically sound since the analog simulations suggest that under these conditions mangroves would have to utilize the sediment storage of nutrients, which, over a period of time, is not adequate to maintain the original growth rates. The result is loss of mangrove vigor and selection for smaller stands. Mangrove management requires maintenance of terrestrial runoff patterns, and perhaps these forests could be used for tertiary sewage treatment. This strategy would require periodic stresses similar in intensity to hurricanes in order to create growing conditions leading to surges in nutrient uptake.

VI. Summary and Conclusions

It appears that the use of models for planning research is a valuable tool for programs involving one or many researchers attempting to understand the regional relationships among ecosystem types. A preliminary model of a mangrove forest in south Florida has yielded the following information that was not apparent from examination of data prior to the present study:

1. Mangrove forests appear to reach a steady state, with respect to their biomass, in phase with the frequency of tropical hurricanes in regions where they occur.
2. The storage of organic detritus in the forest and its export to the bays is a function of tidal amplitude, but tides do not seem to affect gross photosynthetic rates as much as they do detritus accumulation versus export.
3. Gross photosynthesis appears to be sensitive to terrestrial input of nutrients, and the development of mangrove biomass is dependent on the quantity of nutrients and the efficiency of nutrient uptake.

4. Mangrove zonation and vigor may thus be a function of nutrient availability rather than solely salinity as previously thought.

5. During succession, mangroves exert significant control over the amount of nutrients in adjacent waters but, if terrestrial runoff is reduced, they do not have the capacity to maintain themselves at the same level of production. This is due to loss of nutrients in detritus export to the bays and suggests that there must be selective pressure for mechanisms of recycling within the mangrove forest.

Acknowledgments

Many groups have supported the overall mangrove research which resulted in this preliminary report. Financial support was provided by the University of Florida's Division of Sponsored Research and by the USDI Bureau of Sport Fisheries and Wildlife (Contract No. 14-16-008-606). The Environmental Protection Agency provided essential on-site logistic support and also made available some of the data used in the simulations. Additional data were obtained from Mr. Bernard Yokel, Director of the University of Miami's Rookery Bay Project (OWRR Grant No. 14-31-0001-3170). The authors also wish to acknowledge assistance of members of the University of Florida's Departments of Botany and Environmental Engineering Sciences, and the Center for Aquatic Sciences.

REFERENCES

Anonymous (1972). Aguirre Power Plant Complex Environmental Rep. Puerto Rico Water Res. Authority, San Juan, Puerto Rico.

Brunn, P., Chiu, T. Y., Gerritsen, F., and Morgan, W. H. (1962). Storm Tides in Florida as Related to Coastal Topography. Eng. Progr. at Univ. Florida Eng. Ind. Exp. Station, Gainesville, Florida.

Clark, S. H. (1971). Factors Affecting the Distribution of Fishes in Whitewater Bay, Everglades National Park, Florida. *Sea Grant Tech. Bull.* **8,** Univ. Miami, Coral Gables, Florida.

Craighead, F. C. (1971). "The Trees of South Florida." Univ. of Miami Press, Coral Gables, Florida.

Davis, J. H. (1940). The ecology of geologic role of mangroves in Florida. *Papers Tortugas Lab.* **32,** 305.

Golley, F. B. (ed.) (1968). Mineral Cycling in Tropical Forest Ecosystems. Prepared by a research team of the Univ. of Georgia, Inst. of Ecol., Athens, Georgia (manuscript).

Golley, F. B., Odum, H. T., and Wilson, R. F. (1962). *Ecology* **43,** 9.

Heald, E. J. (1969). The Production of Organic Detritus in a South Florida Estuary. Doctoral dissertation, Univ. of Miami, Coral Gables, Florida.

Hesse, P. R. (1961). *Plant and Soil* **14,** 335.

Lugo, A. E., Snedaker, S. C., Bayley, S., and Odum, H. T. (1971). "Models for Planning and Research for the South Florida Environmental Study." Univ. of Florida, Gainesville, Florida.

Lugo, A. E., Evink, G., Brinson, M. M., Broce, A., and Snedaker, S. C. (1975). *In* "Tropical Ecological Systems" (Golley and Medina, eds.), p. 335. Springer-Verlag, Berlin and New York.

Macnae, W. (1967). *In* "Estuaries" (G. H. Lauff, ed.), pp. 432–441. AAAS Publ. 83, U.S. Govt. Printing Office, Washington, D.C.

Miller, P. C. (1972). *Ecology* **53**, 22.

Odum, H. T. (1972). *In* "Systems Analysis and Simulation in Ecology" (B. C. Patten, ed.), Vol. II, pp. 139–211. Academic Press, New York.

Odum, H. T., Lugo, A., Cintron, G., and Jordan, C. F. (1970). *In* "A Tropical Rain Forest" (H. T. Odum and R. F. Pigeon eds.), pp. I-103–I-164. AEC Div. Tech. Informat. and Ed., Oak Ridge, Tennessee.

Odum, W. E. (1969). The Structure of Detritus-Based Food Chains in a South Florida Mangrove System. Doctoral dissertation, Univ. of Miami, Coral Gables, Florida.

Patten, B. C. (1971). *In* "Systems Analysis and Simulation in Ecology" (B. C. Patten, ed.), Vol. I, pp. 3–121. Academic Press, New York.

Sastrakusumah, S. (1971). A Study of the Food of Juvenile Migrating Pink Shrimp, *Penaeus duorarum* Burkenroad. *Sea Grant Tech. Bull.* **9**, Univ. of Miami, Coral Gables, Florida.

Wadsworth, F. (1972). An Evaluation of the Mangroves West of Jobos Bay, Puerto Rico. *In* Aguirre Power Plant Complex Environmental Rep., Appendix C. Puerto Rico Water Res. Authority, San Juan, Puerto Rico.

Walsh, G. E. (1967). *In* "Estuaries" (G. H. Lauff, ed.), pp. 420–431. AAAS Publ. 83, U.S. Govt. Printing Office, Washington, D.C.

5

The Role of Species Interactions in the Response of a Forest Ecosystem to Environmental Perturbation

DANIEL B. BOTKIN

ECOSYSTEMS CENTER, MARINE BIOLOGICAL LABORATORY
WOODS HOLE, MASSACHUSETTS

I. Introduction

Twentieth-century civilization has brought about so many changes in the environment that one wonders about the ability of natural communities to recover from perturbations. We know so little about the general attributes of complex systems that it is difficult at this time to make a priori judgments about the dynamics of ecosystems. Without some theoretical or conceptual guidance, a strictly empirical approach must include a multitude of time-consuming and elaborate experiments. For communities with long-lived species, such as forests, it is questionable that experiments could ever be of sufficient duration to expose subtle but perhaps crucial changes. Although current knowledge makes prediction difficult, even if knowledge were plentiful, our minds would have difficulty assessing the probable outcome of any perturbation of a system as complex as a biological community.

Confronted with these problems, we have created a computer model of a forest ecosystem which we believe is an appropriate tool to study perturbation effects on such ecosystems. The characteristics of the simulated forest community are the result of dynamics and interactions of the individual species, not of a priori constraints concerning the whole system. Thus, one's faith in the model depends on one's judgment of the accuracy of these assumptions as well as the realism with which the model reproduces species dynamics and the observed behavior of the entire forest.

The model is constructed with a biological orientation, reflecting currently available ecological information and assumptions. It is dynamic in the sense that changes in the state of the simulated forest are the result of its current state plus random components. In such a model, the cumulative effects of perturbations are not necessarily obvious beforehand nor analytically predictable from the initial conditions.

No claim is made that the model is a definitive treatment of a forest ecosystem, only that it is a reasonable approximation of current understanding of the population dynamics of forest tree species, and that it successfully reproduces general dynamic characteristics of a forest (Botkin *et al.*, 1972a).

In this study the model is operated to investigate the response of a forest ecosystem to one kind of environmental disturbance. Because so little is known about the action of a perturbing agent in a real forest, we have chosen a simplified case, in which perturbation of the environment is considered to affect directly a single factor, the annual diameter growth of all tree species, in a uniform way. Four cases are considered: one concerning secondary succession in a forest with permanent alteration of site conditions, and three concerning the response to perturbation of forests at three different stages in development: equilibrium, approaching equilibrium, and far from equilibrium. In traditional ecological terms, these represent climax, and late and early successional stages, respectively.

To understand and evaluate the simulated experiments, one must understand the basis of the model, a detailed description of which follows.

II. Description of the Model*

The model simulates growth of individual trees on small plots within a forest. It was originally designed for use in the Hubbard Brook Ecosystem Study to

* This section is reprinted from Botkin *et al.* (1972a,b) (*IBM J. Res. Develop.*, copyright 1972 by International Business Machines Corporation, and *J. Ecol.*) and is included for the sake of completeness. Material appears with permission of the publishers.

capitalize on the original vegetation survey of that study (Bormann *et al.*, 1970), to "grow" the trees, and to allow for manipulation of both stand and site characteristics. The version of the program used here includes the thirteen species* found in the Hubbard Brook forest. However, the model's underlying concepts are general and each species is defined by a few general characteristics derived from its entire geographic range (see Table I, p. 153). In theory, the model could be readily modified to include any nonhydrophytic species whose relevant characteristics are known, and a modified version is currently being developed for the forests of Isle Royale National Park, Michigan.

To understand the model, the reader should have in mind the general pattern of succession in northern New England. Studies of succession in this area indicate that, in the year following a large clearcutting, pin cherry germinates in great abundance to form dense, evenly aged stands. Subsequent germination of this species does not occur (Marks, 1971). Under some conditions, white or yellow birch may germinate with pin cherry and continue to enter the stand in subsequent years until light levels become low. It appears that the birches germinate and survive under some shading and can germinate in openings or partial clearings that would not provide sufficient light for the cherry species (Marquis, 1969). In general, under heavy shade, species such as sugar maple and beech germinate or sprout on better sites, red spruce and balsam fir on colder or poorer sites. Analyses of original plot data from the Hubbard Brook vegetation survey do not yield strong correlations for heavy shade conditions between the number of trees, saplings, and seedlings of a given species that are present on a plot. At higher elevations of the Hubbard Brook watershed, the soils tend to be shallower and rockier, and nonboreal species grow poorly and are subjected to a differentially severe mortality. Observations of seedling and sapling growth of the major shade-tolerant species indicate that under shaded conditions a number on the order of 0 to 2 individual saplings may be added in any one year to a 10 x 10-m plot (Forcier, pers. comm.). The occurrence of saplings of the tolerant species is not clearly related to light intensity at the forest floor, while seed germination, growth, and survival of young stems undergo very great yearly variations suggesting cycles of 3–5-y duration.

In creating the model, we have attempted to introduce a minimal number of assumptions and to find the simplest mathematical expression for each factor that was consistent with observation. The model was built step by step, beginning with optimum growth of single trees, then adding the effect of less

* Sugar maple (*Acer saccharum*†), beech (*Fagus grandifolia*), yellow birch (*Betula alleghaniensis*), white ash (*Fraxinus americana*), mountain maple (*Acer spicatum*), striped maple (*A. pensylvanicum*), pin cherry (*Prunus pensylvanica*), chokecherry (*P. virginiana*), balsam fir (*Abies balsamea*), red spruce (*Picea rubens*), white birch (*Betula papyrifera*), mountain ash (*Sorbus americana*), and red maple (*Acer rubrum*).

† Nomenclature is according to Gleason (1963).

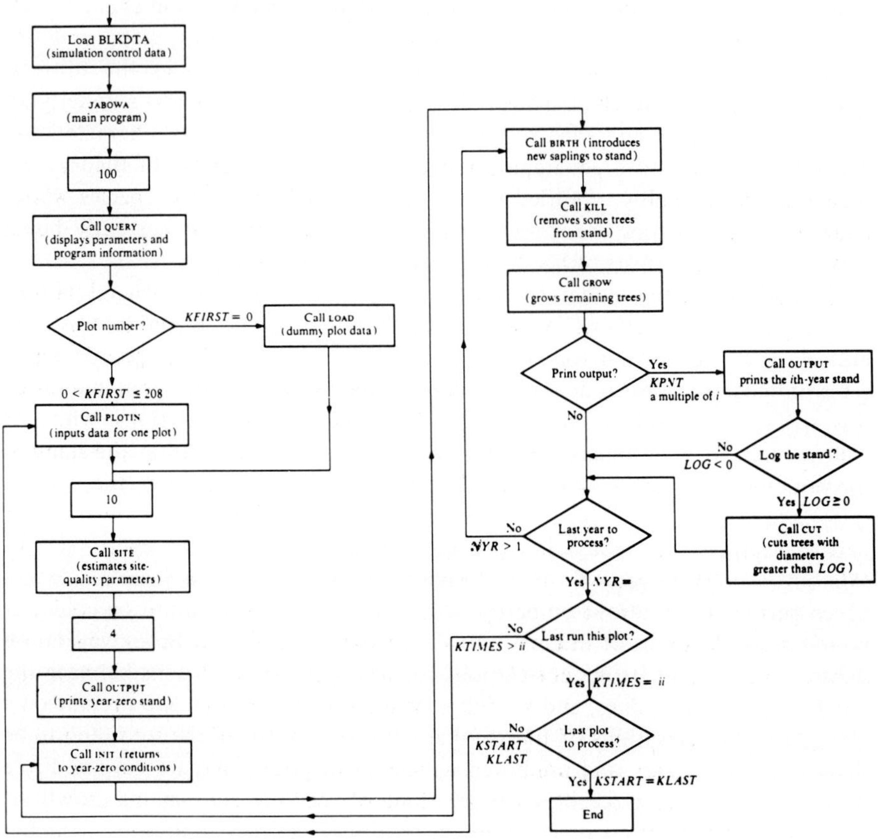

FIG. 1. Flow chart of JABOWA, Version 1. From Botkin *et al.*, 1972a. (Reprinted with permission from the *IBM Journal of Research and Development*. Copyright 1972 by International Business Machines Corporation.)

than optimum light and moisture levels on growth, and allocation of the growth resources among competing trees. New factors have been introduced only when it has become clear that model results are not consistent with observation.

The original version of the program, called JABOWA, is written entirely in Fortran IV, and uses only standard library routines and a good random number generator. It is designed for interactive use with a time-sharing system remote terminal and has been operated successfully at a number of computer centers. Other versions of the program include one in APL (Smart, 1972) and one designed for use with interactive television graphics. The experiments discussed here were conducted with a version designed for batch operation to carry out large simulations involving many plots.

A flow chart of the program (Fig. 1) shows that the innermost loop of JABOWA contains calls for each plot year to three major subroutines: GROW, which deterministically provides the annual growth increment for each tree; BIRTH, which stochastically adds new saplings; and KILL, which stochastically decides which trees die. A fourth subroutine, SITE, calculates measures of environmental conditions for each plot. The other subroutines are subservient to these and are discussed elsewhere (Botkin *et al.,* 1972a).

In the following description of the major subroutines, it should be clear that each tree species and the abiotic environment are defined in the model by a few simple characteristics. Direct competition among individuals is restricted to competition for light (taller trees shade smaller ones, and species with more leaves for a given diameter shade smaller competitors more than other species; under shaded conditions photosynthesis is higher for shade-tolerant species than intolerant ones, and vice versa). Species strategy is also invoked by species specific survival probabilities and by differential addition of new saplings in relation to light at the forest floor. Because the annual probability of an individual's survival is related to the maximum known lifetime of its species, individuals of long-lived species have a better chance of survival in any one year than those with short maximum lives.

In regard to the gradients of light and successional stage, two basic kinds of strategy are defined in the model: those of shade-intolerant or *r*-selected species, and shade-tolerant or *K*-selected species. The former strategy capitalizes on catastrophe, so that many young individuals germinate when light intensity is high, soon after a clearing occurs in a forest. The shade-tolerant strategy involves ability to grow at all ages in comparatively deep shade, and to sacrifice rapid growth for durability and persistence. There are intermediate strategies in the real and model forests, such as that of yellow birch, but the dichotomy is drawn strongly here for clarity.

A. Subroutine GROW

This subroutine uses a tree growth model to augment the dbh's* of all trees on a 10 x 10-m plot by an amount representing one year's growth. The model consists of a basic growth rate equation for each species that may be taken to represent the rate of growth of a tree with optimum site quality and no competition from other trees. For each plot-year, this growth rate is decreased by factors taking into account shading and shade tolerance, soil quality, and average climate as measured by the number of growing degree-days.

Many growth curves tend to be sigmoid and our final growth equation exhibits this overall property. We realize that some readers may feel the

* dbh is the forestry measure of diameter at breast height.

equations which follow are occasionally based upon rather arbitrary assumptions, but we hope they will concur that there is no unique solution to forest growth simulation and that many equations based upon different assumptions could yield quite similar results.

A tree growing in the open collects an amount of radiant energy roughly proportional to its leaf area. The JABOWA growth rate equation for a tree growing under optimum conditions has the form

$$\delta(D^2H) = R \cdot LA \cdot [1-(DH/D_{max}H_{max})], \tag{1}$$

where D is the dbh of the tree, H its height, D_{max} and H_{max} maximum values of these quantities for a given species, LA leaf area, and R a constant. The equation states that the change in volume (D^2H) of a tree over a period of one year is proportional to the amount of sunlight the tree receives, decreased by a factor $(1 - DH/D_{max}H_{max})$, which takes some account of the energy required to maintain the living tissue. The right-hand side of Eq. (1) is later multiplied by additional factors to take shading, climate, etc., into account. Values used in JABOWA, Version 1, for D_{max}, H_{max}, and other parameters are given in Table I.

The height H (cm) of a tree with dbh D (cm) is assumed to be given by the following expression (Ker and Smith, 1955):

$$H = 137 + b_2D - b_3D^2. \tag{2}$$

The number 137 represents breast height (cm), and the constants b_2 and b_3 are chosen for each species so that $H = H_{max}$ and $dH/dD = 0$ when $D = D_{max}$. One finds

$$b_2 = 2(H_{max} - 137)/D_{max}, \qquad b_3 = (H_{max} - 137)/D^2_{max}. \tag{3}$$

Rate of change of height decreases with increasing diameter [Eq. (2)], and actual change in height becomes negligible for large diameters.

From the above discussion, one can readily see that JABOWA currently makes no adjustment for the forester's concept of "form factor." If data were available, it would be comparatively simple to modify Eq. (2) of subroutine GROW to reflect differences in site quality if we knew how the constants H_{max} and D_{max} varied with measures of growing degree-days, and the other requirements for autotrophic plant growth. Field checks would be necessary to quantify such relationships reliably; therefore, this was not attempted.

The leaf weight for a tree of species i is taken to be

$$\text{WEIGHT} = C_iD^2, \tag{4}$$

where C_i is a constant. This equation states that the ratio of leaf weight to stem basal area is a constant from sapling age to death. Data from Hubbard Brook (Whittaker, 1970) and elsewhere (Perry *et al.*, 1969; Kittredge, 1948; Baskerville, 1965) indicate that the actual exponent ranges from 1.5 to about 3.

TABLE I[k]. Basic Parameters[a]

	G^f	C^g	*AGEMX*	Type[j]	D_{max}/H_{max}	b_2	b_3	$DEGD_{min}$[e]	$DEGD_{max}$[e]	*WMIN*	*WMAX*
Sugar maple	170	1.57[b]	200[d]	2	152.5[d]/4011[d]	50.9	0.167	2000	6300	300	
Beech	150	2.20[b]	300[d]	2	122[d]/3660[d]	57.8	0.237	2100	6000[h]	300	
Yellow birch	100	0.486[b]	300[d]	1	122[d]/3050[d]	47.8	0.196	2000	5300	250	
White ash	130	1.75	100	2	50/2160[d]	80.2	0.802	2100	10700	320	
Mt. maple	150	1.13[b]	25	2	13.5/500	53.8	2.00	2000	6300	320	
Striped maple	150	1.75	30	2	22.5/1000	76.6	1.70	2000	6300	320	
Pin cherry	200	2.45[c]	30	1	28[c]/1126[c]	70.6	1.26	1100	8000	190	
Chokecherry	150	2.45	20	1	10/500	72.6	3.63	600[d]	10000	155	
Balsam fir	200	2.5	80[d]	2	50/1830[d]	67.9	0.679	1100	3700	190	
Spruce	50	2.5	350[d]	2	50/1830[d]	67.9	0.679	600	3700	190	
White birch	140	0.486	80	1	46/1830[d]	73.6	0.800	1100	3700	190	600[i]
Mt. ash	150	1.75	30	2	10/500	72.6	3.63	2000	4000	300	
Red maple	240	1.75	150[d]	2	152.5[d]/3660[d]	46.3	0.152	2000	12400	300	

G, growth constant; *C*, leaf area constant; *AGEMX*, maximum age (y); D_{max}, maximum known diameter (cm); H_{max}, maximum known height (cm); b_2 and b_3, constants in the equation $H = 137 + b_2 D - b_3 D^2$ relating height to diameters; $DEGD_{min}$ and $DEGD_{max}$, minimum and maximum degree-days; *WMIN* and *WMAX*, minimum and maximum values for the index of evapotranspiration (mm of water per year available for evapotranspiration).

[a] Values not otherwise referenced were developed during the course of the study.

[b] R. H. Whittaker (1970, personal communication).

[c] Marks (1971).

[d] Harlow and Harrar (1941).

[e] Climatological ranges in growing degree-days, obtained by matching northern and southern limits of range maps (Fowells, 1965) to January and July mean world isotherms.

[f] Growth constants adjusted for reasonable growth of individual tree in full sun with climate and soil factors equal to 1 (values of *G* will give ~2/3 of maximum diameter at 1/2 maximum age starting from an 0.5-cm stem).

[g] Actual leaf area in square meters is $\sim CD^2/15$ for *D* in centimeters.

[h] Northern strain.

[i] Calculated for New York City.

[j] Type 1 is shade-intolerant; type 2 is shade-tolerant.

[k] From Botkin *et al.* (1972b). Reproduced with permission from the *Journal of Ecology*.

The error associated with these estimates is unknown. Increasing the exponent in the leaf weight–diameter relationship [Eq. (4)] has the effect of steepening this curve for intermediate-aged trees. The exponent could be as small as 1 or as large as 3 without drastically altering the overall shape of the final growth curve.

Assuming that leaf area is proportional to leaf weight, and defining $G = RC_i$ we can write Eq. (1) in the form

$$\delta D = GD[1 - DH/D_{max}H_{max}]/[274 + 3b_2D - 4b_3D^2]. \tag{5}$$

A curve of diameter versus time resulting from this equation corresponds to an unusually large tree, and reflects the fact that the simulator should be capable of producing trees of any species as large as have ever been observed. Because of the way trees are killed by the simulator, however, the presence of such large trees would be an extremely rare event, and the usual dominant trees produced by the simulator are considerably smaller than the maximum values given in Table I.

The constant G in Eq. (5) sets the initial growth rate of young trees of species i [the solution $D(t)$ of the equation is asymptotic to D_{max} as $t \to \infty$]. Given a maximum observed age for each species, AGEMX, the constant G was arbitrarily chosen so that D/D_{max} was $\frac{2}{3}$ for a tree of half the maximum age; this choice of G gives reasonable growth rates for most species. An alternate method to derive G is given elsewhere (Botkin *et al.*, 1972b).

The growth rate equation actually used in subroutine GROW is obtained by multiplying the right-hand side of Eq. (5) by additional factors: r(AL), representing effects of shading, shade tolerance, and actual site insolation; T(DEGD), representing climatological effects; and S(BAR), taking some account of soil quality:

$$\delta D = \frac{GD[1 - DH/D_{max}H_{max}]}{[274 + 3b_2D - 4b_3D^2]} \cdot r(\mathrm{AL}) \cdot T(\mathrm{DEGD}) \cdot S(\mathrm{BAR}). \tag{6}$$

These three additional factors are now discussed in detail.

Assuming that a layer of leaves uniformly absorbs light, it can be shown that light intensity at a height h is related to light intensity Q_0 above the top of the canopy by (Kasanaga and Monsi, 1954; Loomis *et al.*, 1967; Perry *et al.*, 1969):

$$Q(h) = Q_0 \exp(-k \int_h^\infty \mathrm{LA}(h')\, dh'),$$

where LA(h) is the distribution with height of leaf area per unit plot area, and k is a constant. In subroutine GROW, this equation is replaced by

$$\mathrm{AL} = \mathrm{PHI}\ e^{-k \cdot \mathrm{SLA}}, \tag{7}$$

where AL is available light for a given tree, SLA is the "shading leaf area," defined as the sum of leaf areas [obtained from Eq. (4)] of all higher trees on the plot [with heights obtained from Eq. (2)], and PHI is the annual insolation in appropriate units. Currently, JABOWA uses a default value of 1 for PHI; a desirable improvement would consist of a subroutine for generating a value of PHI based on latitude and aspect. The constant k in Eq. (7) is adjusted for reasonable shading beneath a dense canopy, and it has been found that $k = 1/6000$ [in units reciprocal to those of Eq. (4)] gives good results for 10 x 10-m plots.

Version I of JABOWA recognizes two types of trees: shade tolerant and shade intolerant. For these two degrees of tolerance, the quantity $r(\mathrm{AL})$ appearing in Eq. (6) is

$$r(\mathrm{AL}) = \begin{cases} 1-\exp[-4.64(\mathrm{AL}-0.05)] & \text{(shade tolerant)} \\ 2.24[1-\exp(-1.136(\mathrm{AL}-0.08))] & \text{(shade intolerant).} \end{cases} \quad (8)$$

In each case, AL is to be obtained from Eq. (7). The function r, which may be thought of as a representation of photosynthetic rates, for the two degrees of shade tolerance, contains constants chosen to give reasonable fits to measured photosynthesis curves (Kramer and Kozlowski, 1960). Note that the annual insolation PHI can be expressed in any units if appropriate changes are made in the constants appearing in Eq. (8).

The function $T(\mathrm{DEGD})$ in Eq. (6) represents an attempt to take account of the temperature effect on photosynthetic rates. It is assumed that each species will have an optimum temperature, and photosynthesis will decrease symmetrically above and below this optimum. A rough index of these thermal effects is obtained from the number of growing degree-days per year (40° F base) for the site. This quantity is defined as the sum of $(T - 40)$ over all days of the year for which the average temperature T exceeds 40° F. Inasmuch as such detailed temperature profiles do not exist for most forest sites, we use an approximation involving only January and July average temperatures. If one assumes that the annual temperature profile is sinusoidal, it is easy to compute the number of degree-days using the average annual temperature as a base. If this average is not too far from 40° F, a correction to the 40° F base can be made by approximating the temperature curve by straight lines near the average annual temperature. In this way, the following approximate expression for the number of growing degree-days is obtained:

$$\mathrm{DEGD} = \frac{365}{2\pi}(T_{\mathrm{Jul}} - T_{\mathrm{Jan}}) - \frac{365}{2}\left(40 - \frac{T_{\mathrm{Jul}} + T_{\mathrm{Jan}}}{2}\right) + \frac{365}{\pi}\frac{[40 - (T_{\mathrm{Jul}} + T_{\mathrm{Jan}})/2]^2}{T_{\mathrm{Jul}} - T_{\mathrm{Jan}}}, \quad (9)$$

in which all temperatures are in degrees Fahrenheit.

For each species, we now set

$$T(\text{DEGD}) = \frac{4(\text{DEGD} - \text{DEGD}_{\text{min}})(\text{DEGD}_{\text{max}} - \text{DEGD})}{(\text{DEGD}_{\text{max}} - \text{DEGD}_{\text{min}})^2}. \tag{10}$$

This function is a parabola (see Hellmers, 1962) having the value zero at minimum and maximum values of DEGD, and a value between zero and one for any value of DEGD between the extremes. Values of DEGD_{min} and DEGD_{max}, representing the extremes for which each species will grow, can be obtained reasonably accurately by comparing species range maps (Fowells, 1965) to lines of constant DEGD estimated from maps of the January and July world isotherms (Trewartha, 1968; U.S. Department of Commerce, 1968). There are admittedly many microenvironmental effects, such as exposure to wind and available nutrients, that are completely neglected in this approximation. However, the number of growing degree-days proves to be a useful measure of gross thermal effects upon plant growth.

Equation (6) contains the factor $S(\text{BAR})$, which is simply

$$S(\text{BAR}) = 1 - (\text{BAR}/\text{SOILQ}), \tag{11}$$

where BAR is total basal area on the plot, SOILQ is the maximum basal area on the plot to which trees will grow, and the function $S(\text{BAR})$ is a crude expression of competition for soil moisture and nutrients on the plot.

B. Subroutine BIRTH

In each year, new saplings of each species enter a plot on the basis of their relative tolerance to shade and whether the degree-day and soil moisture conditions allow growth of that species. The model is designed in this regard to mimic the specific behavior of each species and the process of succession described previously. It assumes a seed source available for each species, but only those species that can grow are added to the stand. The available growing degree-days, DEGD, at the site is compared with the species vectors of minimum and maximum values of growing degree-days, DMIN(i) and DMAX(i), and a similar comparison of growing season evapotranspiration, SOILM, is made against the vector of species requirements WMIN(i) to produce a list of allowable species. For all but the birches and cherries, a random choice is made from the allowable species list and a random choice of zero, one, or two new trees of the selected species is added. The diameter of the added trees depends upon the parameter SIZE (default value 0.5 cm) and a small random addition.

The idea of randomly selecting which shade-tolerant tree species to add is not the desperate expedient that it appears to be at first thought. The Hubbard Brook plots show many inexplicable differences in species composition between plots within the elevational bands discussed by Bormann and co-workers (1970).

In addition, upon the better sites, with the single exception of beech, the existing plot data do not yield strong correlations between number of trees, saplings, and seedlings of a given species that are present on a plot. At higher elevations of the Hubbard Brook watershed, the soils tend to be shallower and rockier, and the nonboreal species grow poorly and are subjected to differentially severe killing (see discussion of subroutines GROW and KILL).

A range of 0 to 2 individual saplings is a reasonable rate of annual introduction for the shade-tolerant species on a 10 x 10-m plot (Forcier, 1969). Seed germination, growth, and survival of young stems undergo great yearly variations, suggesting cycles of 3–5-y duration, and the occurrence of saplings is not clearly related to light intensity at the forest floor. Therefore, random introduction approximates our current knowledge well. In practice, JABOWA, Version I, produces stands that are similar to those observed at Hubbard Brook, although it is possible that the above described procedure may produce stands that underemphasize the major species (beech, in particular) while overemphasizing the importance of minor stand components (e.g., red maple). We believe that biasing the probabilities for species' entry should await extensive testing of the present random choice algorithm. Further field study is also clearly necessary to determine conditions that promote sapling survival of the shade-tolerant species.

The four intolerant successional species are handled quite differently. If the total leaf area on a plot (variable WEIGHT) is below a first threshold (variable CHERRY), between 60 and 75 new cherry trees are added to the stand. The number of cherries added by the simulator is far below the dense thickets observed in the field. Cherries are short-lived trees, and we felt that more realistic modeling would result in an excess of largely unnecessary computation without markedly improving the overall simulation of stand dynamics. In future versions of the program, when the simulation will be used to calculate turnover of water, minerals, and energy, the initial number of cherry stems will be made to agree with observations. As the program now functions, it starts with too few cherry stems and gives them a higher probability of survival than is observed, but the number of cherries 10 y following a clearcut is realistic. If leaf area is greater than the cherry species' cutoff but less than a second threshold value (variable BIRCH), then the two birch species can enter the plot. Between zero and thirteen trees are added as a random choice weighted by stand density, so that the shadier the plot the fewer the trees that are added.

The two cherry and two birch species both have overlapping ranges. Chokecherry has a wider geographical range than pin cherry (Harlow and Harrar, 1941), but the latter species is more prevalent near its optimum than choke cherry. Although the two birch species have overlapping ranges, yellow birch occurs much farther south than white birch, while white birch survives farther north and on much shallower and more rocky sites. In our simulator the choice between cherry or birch species is weighted by site and species constants to reflect the observed ranges.

C. SUBROUTINE KILL

Little data are available regarding survival rates of trees. We have therefore made simple assumptions regarding tree death. We assume that from sapling age to maturity there are some causes of tree mortality that are age independent. For actively growing trees, we assume that no more than 2% of the saplings of a species should reach the maximum age for that species. This gives a probability that a tree will die in any year of

$$p = 1-(1-\epsilon)^n.$$

If $p = 0.98$ when n = AGEMX, the maximum age of the species, and ϵ is the death probability, then

$$(1-\epsilon)^{\text{AGEMX}} = 2 \times 10^{-2},$$

which gives approximately

$$\epsilon = 4.0/\text{AGEMX}.$$

Trees whose annual increment is below a certain value (0.01 cm in the present version of the program) are subjected to a second death mechanism which assures that such a tree would have only a 1% chance of surviving 10 y in the forest with its annual increment remaining below the minimum (the probability that such a tree will die in any one year is 0.368).

The basic rationale is that a tree which cannot maintain a certain minimum growth rate cannot survive for long in the forest, but that in addition there are numerous other events, such as severe wind, lightning, parasitism, and defoliation, which may result in the death of any tree at any time. While in reality some of these events may be truly random, others are not. The assumption of this model is that even the nonrandom events have only a certain probability of affecting any one tree and that the sum of effects of all such events approaches a random probability of killing any tree in any year. That such biotic and abiotic factors affect each species differently is represented by use of the maximum known age of each species to determine the survival probability for any year.

D. SUBROUTINE SITE

This subroutine produces indices of the quality of the site for growing trees. It is customary for forest site indices to be based upon the observed height–age relationship of dominant trees of certain key species that hopefully appear on the plot. It is time consuming to accumulate such data and often quite difficult to obtain it for disturbed stands. For JABOWA, Version I, we were guided by a different philosophy: we wished to index site quality not by what was currently growing on a plot but by estimates of exogenous influences, growing degree-days, and an index of actual evapotranspiration. We believe this approach also raises interesting questions about the minimum amount of light and moisture necessary as well as the optimum levels needed for tree and stand growth.

The growing degree-days parameter, DEGD, has already been described in the discussion of subroutine GROW. To estimate the value of DEGD for an individual site, we need long-term estimates of mean January and July temperatures. It is axiomatic that this information is not available for most sites where we may be interested in using JABOWA, and we shall usually have to resort to subterfuge. Conversion of mean monthly temperatures, BASET, from a nearby U.S. Weather Bureau first-order weather station, is done by subroutine SITE using an average lapse rate for the elevation difference between plot height, IELEV, and the base station, BASEH. For the growing season months the lapse rate used is 3.6° F/1000 ft, while the January minimum uses 2.2° F/1000 ft (Sellers, 1965). Note that subroutine SITE accepts elevation in feet, rainfall in inches per month, and temperatures in degrees Fahrenheit; conversion to metric units is carried out in the subroutine. No attempt was made to account for differences in monthly temperatures which result from differing aspects. All the Hubbard Brook plots have more or less southerly exposure, and differences in mean monthly temperature associated with aspect differences were assumed small; for other areas this may not be a reasonable approximation.

Latitudinal as well as coastal influences on climate are pronounced in New England, and it is hoped that, by using the closest long-term weather station and adjusting for elevation difference, reasonable microclimatic indices can be developed for nearby sites. Elevational transference should be kept less than about 1000 m, as the growing degree-day calculation is sensitive to large elevational changes and is not completely linear.

For deep well-drained forest soils in New England, it is assumed that soil moisture stress is generally not sufficient to restrict stand growth. However, for shallow, rocky soils it appears that many species may have difficulty becoming established and, accordingly, we have incorporated an index of actual evapotranspiration, SOILM. The index is developed as a modified Thornthwaite water balance calculation (Sellers, 1965). The maximum available moisture storage, STRMAX, is the minimum of soil depth, TILL, or 10 m, multiplied by the moisture storage per unit depth of the fine soil fraction, TEXT (Thornthwaite and Mather, 1957), discounted by the percentage rock in the soil mantle. The percentage rock in the soil mantle was not estimated for the Hubbard Brook plots, although the percentage of surface area covered by boulders or rock outcrops, IROCK, was observed, and this was used as a substitute.

Monthly precipitation values are needed for the water balance calculation and, lacking other information, these are assumed to be the same as for the base station, BASEP. The only other parameter needed for the calculation of SOILM is an estimate of the proportion of the current month's precipitation to be added to SOILM if potential evapotranspiration exceeds current storage. The default value for this parameter, EXCESS, is 0.25. No provision was made to adjust EXCESS as a function of STRMAX, although such adjustment may later be necessary.

The initial assumption of no moisture restriction on stand growth for New England forest stands is corroborated by the values of SOILM that we obtained. Using the 30-y Woodstock, New Hampshire, weather station data, assuming IROCK = 0, and IELEV ≃ 300 m, then TILL has to fall below 0.45 m before SOILM starts to contract. At higher elevations there is even less heat and hence an even smaller requirement for soil moisture storage, so that we find that at 1500 m TILL can fall to 0.33 before SOILM starts to contract.

We suspect that something similar to a "wind chill factor" accelerates tree mortality as timberline is approached, but the difficulty of quantifying the wind chill concept with the available sparse data prevented any direct assessment of the chill factor within subroutine SITE. An indirect assessment of the chill factor has probably been built into the simulator in that only those species which can be grown can be added to the stand (tests of DMIN(i) versus DEGD, and WMIN(i) versus SOILM are made by subroutine BIRTH). In particular, for shallow, rocky, high-elevation sites the value of SOILM falls precipitously. It is probable that such localities form a set that largely intersects the set of wind chill sites. Later efforts should be directed toward determining whether or not this indirect assessment of the chill factor is indeed adequate.

The parameter SOILQ measures the maximum basal area of a stand of trees under optimum growing conditions for a 10 x 10-m plot. If rock outcrops reduce the available plot area, the SOILQ should be reduced accordingly. The parameter IROCK is used by subroutine SITE in this capacity; later efforts may show that this approximation needs to be strengthened.

III. Simulation Experiments

It is reasonable to assume that most perturbations of a real ecosystem would have a combination of simultaneous effects. For example, a real change in the environment of a forest might affect such characteristics of trees as seed germination, survival rate of seedlings, saplings, and mature trees, average size and shape of individuals, as well as rates of various physiological processes. However, there are very few cases at this time for which we can specify all ecological effects of a real environmental disturbance. If details were available, many kinds of real perturbations could be simulated readily by the JABOWA model. Lacking such detail, we can gain insight into the behavior of a forest ecosystem by modifying single characteristics of tree species in separate experiments.

Because of the importance of growth rates, we have conducted simulated experiments which study the response of the model to changes in the annual diameter growth increment of individual trees. In these experiments, a normal plot is one in which the annual diameter increment of each tree is calculated according to standard algorithms of the model. A perturbed plot is one in which the annual diameter increment of each tree is increased or decreased over what it would be normally by a fixed percentage common to all species and all years in which the perturbation is simulated. Thus, in a perturbed plot, ΔD, the annual diameter increment of any tree, is

$$\Delta D = \Delta D' \cdot k, \tag{12}$$

where $\Delta D'$ is the increment calculated normally, and k is the fractional increase or decrease.

Some readers may find it helpful to consider these growth changes analogous to fertilization or stress. While real mechanisms of fertilization or stress would very likely be species specific, for simplicity we have chosen to assume a uniform effect as a starting point. Experiments that assume a species specific change in growth have been conducted and will be reported elsewhere (Botkin and Brittain, 1976).

Four cases are considered. Three concern perturbations of forest stands at different stages: stands at equilibrium, approaching equilibrium, and following a catastrophic clearing. The fourth case concerns development of a forest under a permanent alteration of site conditions.

In considering the results of these experiments, we become involved in the question of stability of ecosystems. Although there have been many discussions of ecological stability, the concept suffers from a variety of interpretations. In the physical sciences, absolute stability generally refers to a system at equilibrium that returns to equilibrium following perturbation. Since we rarely know whether an ecosystem has been at equilibrium prior to disturbance, the concept is not usually applicable in this strict sense. Here we will call an ecosystem stable if, regardless of whether it begins from equilibrium, it eventually becomes indistinguishable from another identical one which did not undergo perturbation.

Relative stability refers to the relative response to a uniform perturbation of one system in comparison to another. One ecosystem is said to have greater relative stability than a second if, in a graph of a measure of the ecosystem versus time, the area under the curve from the end of the perturbation to the time that equilibrium is reached is smaller for the first than for the second.

The term "stability" is used here in a strict sense, which is sufficient for the present discussion. A fuller discussion of the ecological use of this term is presented elsewhere (Botkin and Sobel, 1975).

In each experiment, a treatment consists of 100 plots with identical initial conditions. In all cases, each plot begins as a clearcut just starting secondary succession. Three of the cases involve simulated growth alterations of 50% for a duration of 50 y. These are drastic changes, larger than one would reasonably expect for many real events, but they bring out clearly the dynamic response of the model to perturbation. Perturbations of smaller percentages have results similar in kind but smaller in amount.

Basal area (the circular cross-sectional area of tree stems calculated from diameters measured at 1.37 m above the ground) is used as a measure of the forest. The model also provides, as output, density (number of individuals per unit area) and a variety of biomass values. In general, density varies inversely and biomass directly with basal area.

IV. Results

A. Case I. Perturbation of a Forest at Equilibrium

Figure 2 shows the effect of an annual growth reduction of one half with a duration of 50 y on an equilibrium forest. Previous simulations have shown that the model predicts that a northern hardwood forest at elevations between 610

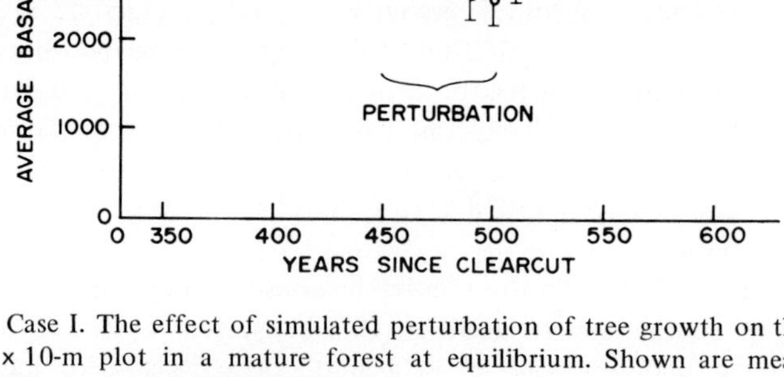

FIG. 2. Case I. The effect of simulated perturbation of tree growth on the total basal area per 10 x 10-m plot in a mature forest at equilibrium. Shown are means and 95% confidence intervals for 100 replicates in each treatment. K is the constant multiple of the annual diameter increment of each tree. In this case, the annual growth of each tree is multiplied by 0.5 each year for 50 y during the perturbation. Data: all species, 610 m elevation; •, normal; ○, $K = 0.5$ (from year 450 to 499).

and 762 m in northern New Hampshire reaches equilibrium in approximately 400 y. This result is consistent with analysis of data from real forests in Wisconsin (Loucks, 1970). That the forest is at equilibrium is shown by the graph of basal area of the normal forest, which does not change significantly from year 350–600. The response of the forest in this case is straightforward. The model is stable and recovery is symmetric in the sense that basal area of the perturbed stand becomes statistically equivalent to the normal in 50 y.

Simulation for stands with a growth increase of 1.5 times normal also was carried out. The results were comparable to those illustrated; the basal area increased during the perturbation and then decreased to normal within 50 y.

B. CASE II. PERTURBATION OF A FOREST APPROACHING EQUILIBRIUM

Figure 3 shows the effect of an annual growth reduction of one half with a duration of 50 y on a forest approaching equilibrium. The basal area of the forest reaches a peak between years 150 and 200 following clearcutting and decreases to equilibrium, which is reached by year 350. The results again appear straightforward and are not surprising. The forest is stable; once the perturbation is removed, the stand returns to normal. Again, recovery is generally symmetric

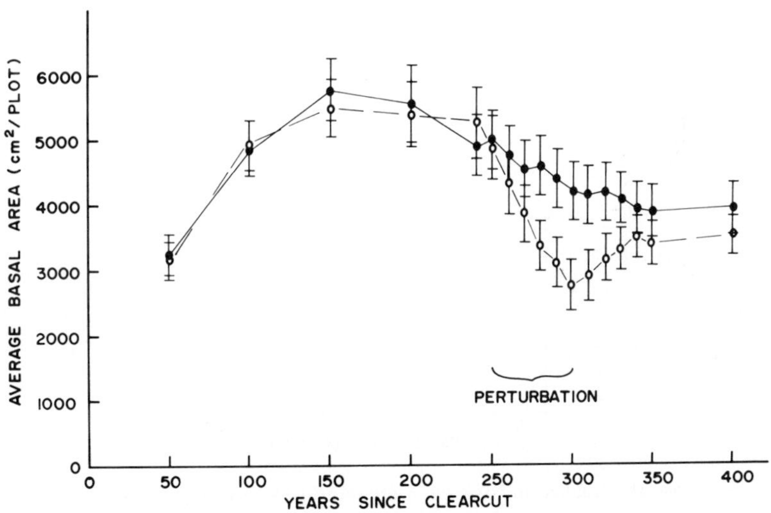

FIG. 3. Case II. The effect of a simulated perturbation of tree growth on the total basal area per 10 x 10-m plot in a forest approaching equilibrium. Shown are the means and 95% confidence intervals for 100 replicates in each treatment. For an explanation of K see Fig. 2. Data: all species, 610 m elevation; •, normal; ○, $K = 0.5$ (from year 250 to 299).

in that the time to recovery (40 y) is approximately the same as the duration of the perturbation.

C. CASE III. PERTURBATION OF A FOREST BEGINNING SECONDARY SUCCESSION

Figure 4 compares secondary succession in a normal forest with one that underwent an annual growth reduction of one half during the first 50 y following clearcutting. The basal areas of the two stands become statistically equivalent for a single decade, the first following removal of the perturbation. Then, the stand in which growth was decreased shows a greater basal area than the normal one. This difference remains significant for 100 y; the stands become statistically equivalent at year 180. Although the forest is stable in the sense that eventually the perturbed stand becomes indistinguishable from normal, it is relatively less stable than the previous cases, and shows what appears to be a peculiar and paradoxical response: reduction in growth results in subsequent greater productivity and standing biomass. What accounts for this response? The explanation lies in interactions among species. These are revealed most clearly through examination of the fourth case. We will examine this case now, and later return to Case III.

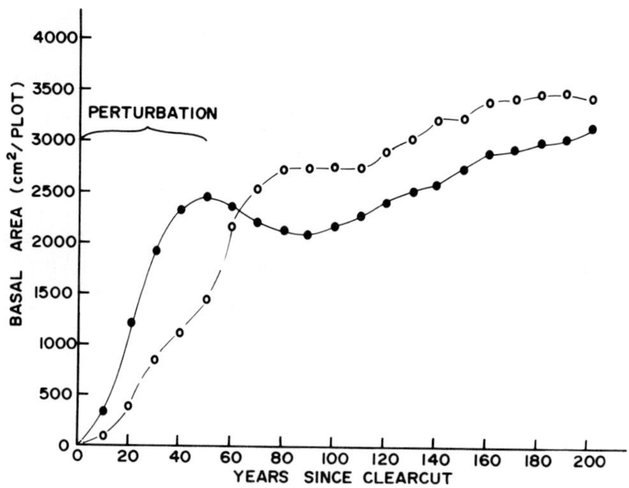

FIG. 4. Case III. Change in total basal area during secondary succession for a normal stand and one in which annual diameter increment of each tree is reduced by one half during the first 50 y. Shown are mean values for 100 replicates in each treatment. On the basis of 95% confidence intervals, the plots are significantly different for all years between 0 and 180 except year 60. The stands remain statistically equivalent after year 180. Data: all species, 762 m elevation; •, normal; ○, $K = 0.5$ (from year 0 to 49).

D. Case IV. Permanent Change in Site Conditions

Figure 5 shows the changes in basal area during the first century of secondary succession for three stands: normal, annual diameter growth increased, and decreased. In these experiments, growth changes are 10% of normal. If such growth changes resulted from fertilization or stress of a real forest, one would think them severe, but the model predicts that at no time during the first century of secondary succession are the perturbed stands significantly different from normal. Simulated experiments of greater severity, involving annual changes of 50 and 100%, have greater, but similar, effects. For example, when k of Eq. (12) was set at 1.5, the treated and normal stands were statistically equivalent at the third, fourth, and fifth decade. When k is set equal to 2.0, the long-term average total basal area is significantly different from the normal forest, but the difference is in a ratio of approximately 1.5 to 1.0 for the means. In some way the forest community seems to be buffered against the disturbance.

The explanation lies in the response of individual species with different strategies. Species characterized by the shade-tolerant strategy, such as red spruce (Fig. 6a) or sugar maple (not illustrated), have a basal area 20–30% greater than normal when annual growth is increased 10%, and a similar reduction when growth is decreased 10%.

The response of species characterized by the shade-intolerant strategy, such as white birch (Fig. 6b), is quite different. When annual growth is increased, white birch grows slightly faster than normal during the first two decades, but it has a basal area lower than normal in all subsequent decades. Similarly, in stands

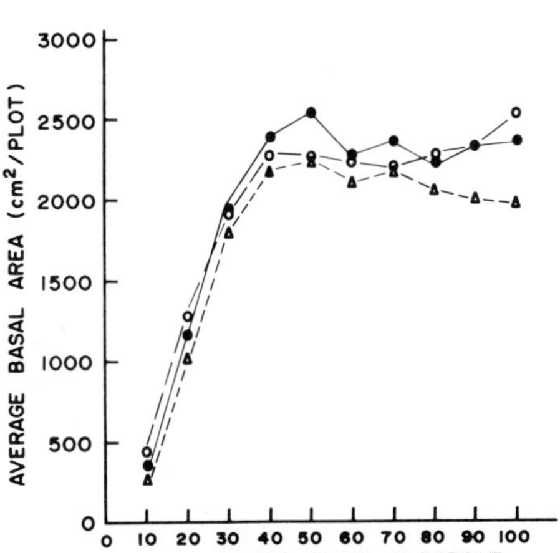

FIG. 5. Simulated change in average total basal area per 10 x 10-m plot during the first 100 y of secondary succession following clearcutting for forest stands whose trees had normal, impeded, and augmented growth. Shown are means for 100 replicates in each treatment. Data: all species, 762 m elevation; •, normal; ○, $K = 1.1$; △, $K = 0.9$.

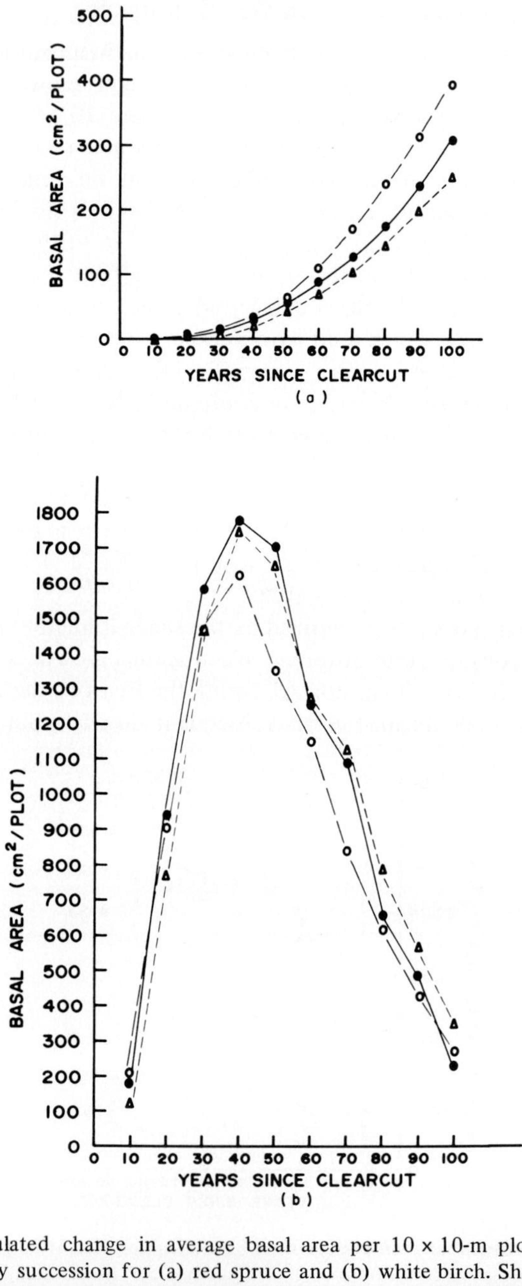

FIG. 6. Simulated change in average basal area per 10 x 10-m plot during the first 100 y of secondary succession for (a) red spruce and (b) white birch. Shown are means for 100 replicates in each treatment. Treatments are the same as those given in Fig. 5. Data: 762 m elevation; ●, normal; ○, K = 1.1; △, K = 0.9.

where annual growth is reduced, white birch has a basal area equal to or greater than the normal after year 50.

Why does this happen? Under the growth increase, shade-tolerant species do better and become stronger competitors against white birch. Red spruce and sugar maple grow taller and develop more leaves faster than normal, shading increases faster, and the resultant ability of shade-tolerant species to suppress white birch counters the direct effect of growth enhancement on white birch.

When growth of all species is decreased uniformly, light intensity near the ground decreases slower than normal. Thus, conditions of uniform growth suppression of all species tend to favor the intolerant species strategy, while conditions of growth enhancement tend to favor the shade-tolerant strategy. Since shade-intolerant species, typical of early stages of succession, are generally faster growing and more highly productive, this result may seem surprising, but it is clearly a direct consequence of the assumptions of the model. The general implication is that interactions among species lead to indirect effects which tend to buffer the ecosystem against changes, and to create responses that would not occur in a carefully managed monoculture. Another interpretation is that growth suppression slows the rate of development of the forest, that is, it expands the time scale, while growth enhancement speeds the development or compresses the time scale.

Ironically, an observer who knew nothing about the treatments, who used the productivity of white birch as an index of fertility, and who made measurements at year 50, would conclude that the untreated stand was most fertile. If his measurements were made after year 70, he would conclude that the highest fertility for white birch belonged to the stand that was in fact least fertile.

With this insight in mind, we can understand the predictions of the model for Case III, although the results for this case are more complex than in Case IV. Figure 7a shows the basal area of yellow birch in a normal stand and one in which annual growth was decreased to one half normal during the first 50 y of secondary succession. This shade-intolerant species shows only a slight decrease in basal area from normal during the perturbation, but subsequently its basal area increases more rapidly than normal and remains significantly different from normal from year 60 to year 180. The uniform suppression of growth of all species during the first 50 y of secondary succession leaves yellow birch in a better competitive position than normal. Part of the reason for this, as seen before, is that conditions of uniform growth suppression tend to favor the shade-intolerant strategy. Because suppression occurs at the start of secondary succession, the normal roles of other short-lived species are upset, with the result that the competitive position of the relatively longer-lived yellow birch is further improved.

For example, balsam fir, the shortest lived of the shade-tolerant species, shows a relatively great suppression of growth during the perturbation (Fig. 7b).

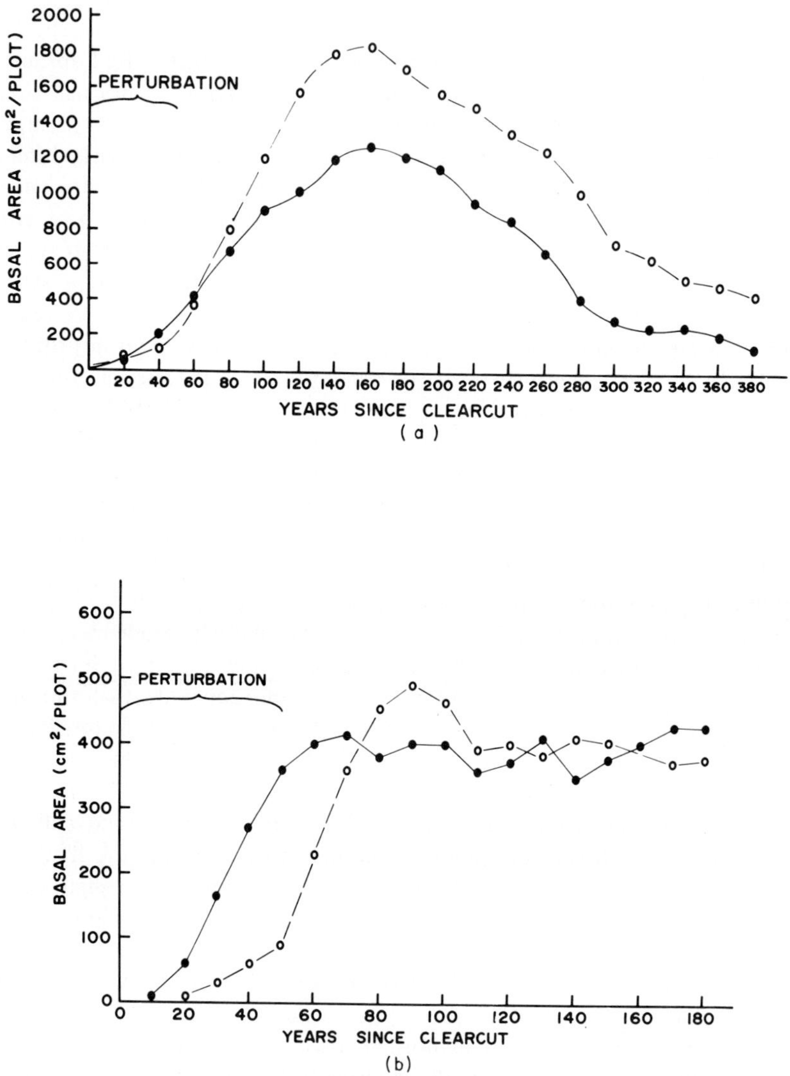

FIG. 7. Simulated change in average basal area during secondary succession for (a) yellow birch and (b) balsam fir in a normal forest stand and one in which the annual diameter increment of each tree was reduced by one half during the first 50 y. Shown are means for 100 replicates in each treatment. Based on 95% confidence intervals, the means in (a) are significantly different for all years until year 200 except 60 and 80. In (b) means are significantly different for years 10–60. Data: 762 m elevation; •, normal; ○, $K = 0.5$ (from year 0 to 49).

However, it also shows a brief overshoot following the release from the perturbation. Both yellow birch and balsam fir benefit from the effect of the perturbation on pin cherry, the most intolerant and shortest lived of the major species. The duration of growth suppression is longer than the expected lifetime of pin cherry, which normally dominates the first few decades of secondary succession. The peak basal area contributed by pin cherry is reduced to one fourth of normal, and essentially no individuals of this species are left at the end of the perturbation.

V. Discussion

These simulation experiments suggest some generalizations regarding the response of forests to perturbation in the growth of individual trees. When the perturbations do not alter the kinds of species strategy that are favored under normal environmental conditions, then the response of the ecosystem is easily predicted and resembles the response of a linear system. When the perturbations occur at early stages in succession, the growth perturbations change the kinds of species strategy that are favored under the environmental conditions. When this occurs, the behavior of the ecosystem is not simple, and a knowledge of the dynamics of individual species and the interactions among species is needed.

The model makes some seemingly paradoxical predictions: uniform enhancement of the growth of individuals of all species does not affect all species uniformly; shade-intolerant species do better if their growth is restricted, as long as their competitors suffer equal restrictions; a uniform but temporary reduction in growth of all individuals leads subsequently to greater community productivity. These paradoxes are resolved when one realizes that, in the experiments, growth of individuals is changed as a uniform fraction of what it would have been for each individual under the competitive and environmental conditions existing in that year for that individual. This, however, is consistent with our concept of the dynamic behavior of living organisms in a competitive, mixed species and mixed age–class community.

Perturbations that reduce tree growth tend to favor shade-intolerant species, while perturbations that increase growth tend to favor shade-tolerant species. Temporary alteration of the growth of all individuals at early stages in secondary succession affect short-lived early successional species more than longer-lived species characteristic of later successional stages. The combination of these factors can improve the competitive position of some species in a way that leads to greater producitivity of those species and even greater total community productivity.

An observer who had only a short time to make observations during a perturbation could easily be led astray. For example, when growth is restricted during the first 50 y of succession an observer comparing disturbed and normal stands at year 10 or at year 20 might conclude that the ultimate fate of the disturbed stand would be lower biomass and lower productivity. If he knew that suppression of growth tended to expand the time scale, and understood the behavior of individual species and species interactions, then he might be able to improve his prediction.

Any real fertilization or stress in a forest would very likely have complex effects that were species specific. It is possible that a real perturbation could change what we have considered to be fundamental species constants. We have seen that in some cases interactions among species tend to buffer the ecosystem against change. It is conceivable that, if appropriate constants were properly scaled, the sole effect of a perturbation might be to compress or expand the time scale and in no other way affect the forest.

It is clear that the model would make different predictions for a carefully managed monoculture. For example, if a monoculture were managed so that poorly growing individuals were removed and vigorous ones protected, it is possible that fertilization would result in growth that would resemble that of red spruce in Case IV. Put another way, the model suggests that measurement of the response of a carefully nurtured individual or monoculture to fertilization or stress would not provide an accurate estimate of the response of an unmanaged, mixed species, mixed age–class forest. The model points out the importance of species interactions, and suggests that the behavior of a mixed species ecosystem is different in kind from the behavior of a single species monoculture.

Acknowledgments

I wish to acknowledge the role of J. F. Janak and J. R. Wallis of the IBM Thomas J. Watson Research Center, Yorktown Heights, New York, in development of the JABOWA model. I thank the other investigators of the Hubbard Brook Ecosystem Study for pertinent data, and D. Gazis for initiating the cooperative research between the Hubbard Brook Study and IBM research that led to the model's development. This is contribution No. 64 of the Hubbard Brook Ecosystem Study. The work was partially supported by NSF Grants, Nos. GB 1144, GB 4169, GB 742, GB 757, GB 14325, GB 14289, and GB 30968X.

REFERENCES

Baskerville, G. L. (1965). *Forest Sci. Monogr.* **9**.

Bormann, F. H., Siccama T. G., Likens, G. E., and Whittaker, R. H. (1970). *Ecol. Monogr.* **40**, 373.

Botkin, D. B., and Brittain, E. G. (1976). Simulated tree species with C4 photosynthetic pathway: The effect on forest composition predicted by a computer model (in preparation).

Botkin, D. B., and Sobel, M. J. (1975). *Amer. Nat.* **109**, 625.

Botkin, D. B., Janak, J. F., and Wallis, J. R. (1972a). *IBM J. Res. Develop.* **16,** 101.
Botkin, D. B., Janak, J. F., and Wallis, J. R. (1972b). *J. Ecol.* **60,** 849.
Forcier, L. (1969). pers. comm.
Fowells, H. A. (1965). "Silvics of Forest Trees of the United States," USDA Handbook 271. U.S. Govt. Printing Office, Washington, D.C.
Gleason, H. A. (1963). "The New Brittain and Brown Illustrated Flora of the Northeastern United States and Adjacent Canada." Hafner, New York.
Harlow, W. M., and Harrar, E. S. (1941). "Textbook of Dendrology." McGraw-Hill, New York.
Hellmers, H. (1962). *In* "Tree Growth" (T. T. Kozlowski, ed.), p. 275. Ronald Press, New York.
Kasanaga, H., and Monsi, M. (1954). *Jap. J. Bot.* **14,** 304.
Kittredge, J. (1948). "Forest Influences." McGraw-Hill, New York.
Kramer, P. J., and Kozlowski, T. T. (1960). "Physiology of Trees." McGraw-Hill, New York.
Loomis, R. S., Williams, W. A., and Duncan, W. G. (1967). *In* "Harvesting the Sun" (A. San Pietro, F. A. Greer, and T. J. Army, eds.), p. 291. Academic Press, New York.
Loucks, O. L. (1970). *Amer. Zool.* **10,** 17.
Marks, P. (1971). The Role of *Prunus pensylvanica L.* in the Rapid Revegetation of Disturbed Sites. Doctoral dissertation, Yale Univ., New Haven, Connecticut.
Marquis, D. A. (1969). *In Birch Symp. Proc.* p. 40. Northeastern Forest Exp. Station, Upper Darby, Pennsylvania.
Perry, T. O., Sellers, H. E., and Blanchard, C. O. (1969). *Ecology* **50,** 39.
Sellers, W. D. (1965). "Physical Climatology." University of Chicago Press, Chicago.
Smart, J. S. (1972). An APL Version of JABOWA, the Northeastern Forest Growth Simulator. IBM Res. Rep. RC 3985.
Thornthwaite, C. W., and Mather, J. R. (1957). *Publ. Climatol.* **10,** 181.
Trewartha, G. T. (1968). "An Introduction to Climate." McGraw-Hill, New York.
U.S. Department of Commerce (1968). "Climatic Atlas of the United States." U.S. Govt. Printing Office Washington D.C.
Whittaker, R. H. (1970). Unpublished data.

6

Simulating the Physiology of a Temperate Deciduous Forest

PHILLIP SOLLINS,* W. FRANKLIN HARRIS
AND NELSON T. EDWARDS

ENVIRONMENTAL SCIENCES DIVISION
OAK RIDGE NATIONAL LABORATORY, OAK RIDGE, TENNESSEE

* Present address: School of Forestry, Oregon State University, Corvallis, Oregon.

I. Introduction

A forest, although not an organism of well-defined genetic and behavioral characteristics, is a system and may be studied as such. Its most conspicuous feature is that it grows, but at a slow rate compared to the flux of material through it.

Organic matter and water comprise the vast majority of this flux while inorganic materials (mainly dissolved ionic forms) account for a very small portion. Sometime around 1967 the idea appeared (although seemingly never in print) that in an analysis and simulation of a forest system we should concentrate first on organic matter and water transfers and consider these as carriers determining or at least affecting transfer rates of the so-called nutrient elements. The nutrient elements should in turn be considered mainly as factors controlling the rates of transfer and accumulation of organic matter and water. Economically, to man, this is eminently reasonable since organic matter (wood and fiber) and water are the two forest products of greatest interest. Nutrient content of the water leaving forested areas is important, but only because the nutrients are factors controlling organic matter production in streams and lakes. Fate of contaminants in a forest ecosystem is an important exception, but, particularly in this case, organic matter and water models seem prerequisite.

As part of the Eastern Deciduous Forest Biome Project (Oak Ridge site) we have, based almost entirely on data collected prior to IBP (International Biological Program) here and elsewhere in the southeastern United States (e.g., Peters *et al.*, 1969; Sollins and Anderson, 1971), constructed a budget of organic matter storage, accumulation, and transfer in a site designated as one of the intensive study areas of the IBP. From this budget we have constructed a simulation model which mimics what in our best estimate is the seasonal physiology of the forest—the mechanisms behind the accumulations and transfers of organic matter. We have used this model to guide research plans and also to provide preliminary answers to questions about behavior of the system under different conditions of perturbation—at least those perturbations for which we might ignore problems of water and nutrient relations in even a preliminary study.

The detailed data behind the budget calculations have been described elsewhere (Sollins *et al.*, 1973; Edwards and Sollins, 1973; Reichle *et al.*, 1973a), as have aspects of the complete budget (Harris *et al.*, 1975). Some of this information has been repeated here in the hope of presenting a reasonably self-contained document. Our state variable, material balance modeling philosophy has also been presented elsewhere (Reichle *et al.*, 1973b), and in this paper we present primarily the organic matter model itself—its structure and behavior—and results of simulations of forest behavior under perturbation conditions.

II. Site Description

The site is located in an area of second-growth mesophytic forest on the U.S. AEC Oak Ridge Reservation (35° 55′ N lat, 84° 17′ W long) at an elevation of 290 m within the ridge and valley physiographic province of eastern Tennessee. The study area is a 500-m^2 plot that was the site of an earlier experiment involving tagging with ^{137}Cs (Auerbach *et al.*, 1964; Olson, 1965), which produced much of the baseline data used in this study.

Vegetation consists of second-growth mesophytic hardwood forest dominated by yellow poplar, with several other species, including various oaks (*Quercus velutina, Q. alba, Q. coccinea, Q. rubra,* and *Q. prinus*), short-leaf pine (*Pinus echinata*), and hickories (mainly *Carya tomentosa*), attaining canopy height. The subordinate vegetation is diverse but consists mainly of redbud (*Cercis canadensis*), flowering dogwood (*Cornus florida*), and ash (*Fraxinus pennsylvanica*). Christmas fern (*Polystichum acrostichoides*) and woody hydrangea (*Hydrangea arborescens*) account for nearly 90% of the shrub and groundcover biomass, although many other species are present. Species composition of the overstory and understory vegetation is summarized in Table I.

TABLE I

Species Composition in 1970 of Trees on the 500 m^2 Study Plot[a]

Species	Overstory trees (%)	Understory trees (%)
Cercis canadensis	19.3	36.4
Nyssa sylvatica	19.3	0.0
Oxydendron arboreum	18.1	3.7
Pinus echinata	17.5	0.0
Sassafras albidum	8.8	0.0
Juniperus virginiana	6.3	0.0
Quercus spp.	6.1	7.4
Acer rubrum	4.6	6.9
Cornus florida	0.0	29.5
Fraxinus spp.	0.0	13.8
Carya spp.	0.0	2.3
Total	100%	100%

[a] *Liriodendron tulipifera* accounted for 76.0% (10.12 kg DW m^{-2}), other overstory trees for 17.7% (2.36 kg DW m^{-2}), and understory trees 6.4% (0.85 kg DW m^{-2}) of the total aboveground biomass (13.30 kg DW m^{-2}).

The history of the study area is unknown but aerial photographs taken in 1924 and 1939 show evidence of earlier clearing, perhaps for pasture. Increment cores taken from trees near the 500-m^2 plot during 1970 indicate a mean stand age of 48 y for dominant and codominant trees. In 1962 the basal area of the 500-m^2 plot (based on stems $\geqslant$2.54 cm dbh) was 19.2 m^2 ha^{-1}; by 1970 it had increased to 22.1 m^2 ha^{-1}. Height of the largest tree in this plot was about 30 m in 1971 (42 cm dbh).

The entire study area is located in a region of karst topography underlain by Knox dolomite. Soils in the sinkhole bottoms, in which most of the research is conducted, are of the Emory series of cherty silt loam and are alluvial in origin. Analysis of soil profile characteristics near the 500-m^2 plot (Sollins *et al.*, 1973) indicates the existence of at least one erosion cycle. A zone of high organic matter content and low bulk density extends downward about 10 cm. Beneath is a layer of alluvium 30 cm thick probably deposited while the land was cleared. Below this the organic matter content again increases with depth, suggesting existence of at least one other profile developed before the erosion cycle.

The Oak Ridge climate is designated as humid–mesothermal in the classification system of Thornthwaite (1948). Winters are generally mild with an average of only 10.7 cm of precipitation falling as snow. Winter weather patterns are frontal in nature while thunderstorm activity dominates the summer weather, with a distinct period of rainfall activity occurring usually in July, although occasionally in June or August. The growing season for *Liriodendron* extends from approximately April 7 to October 15, 180–200 days. Monthly climatic data for the Oak Ridge area have been summarized by Sollins *et al.* (1973).

III. Organic Matter Budget

In our analysis of forest organic matter processes we found it convenient to consider vegetation and consumers separately from the litter and soil organic matter, and chose problems for study which we felt required analysis of the former components in greater detail than the latter. This in turn was because our estimates of standing crop, annual increment, and turnover are, for reasons discussed below, much better for the vegetation than for litter and soil.

A. Primary Production Budget

Primary producers were divided into four major groups (Table I): dominant overstory yellow-poplars, other overstory trees, understory trees, and the groundcover flora (including cryptogams, shrubs, and tree seedlings). Overstory trees (subsystems A and B) were divided into foliage, branches, bole, and large roots. However, because branch and bole typically were not separated during harvest of understory trees and saplings, these compartments were combined in the understory budget (subsystem C). Groundcover flora was considered a single compartment because of the lack of any significant woody component. Fine roots of the four groups were combined into a single compartment because of the lack of any experimental method for separation by species. Standing

TABLE II

Standing Crop of Vegetation on the 500-m^2 Study Plot near Oak Ridge, Tennessee[a]

Plant component	Yellow-poplar trees	Other overstory trees[b]	Understory trees[b]	Groundcover	Standing deadwood	Total alive
Leaf						
1962	204	90	48	–[c]	–	342
1970	247	76	51	–	–	374
Reproductive parts						
1962	n.d.[d]	n.d.	n.d.	–	–	20
1970	n.d.	n.d.	n.d.	–	–	25
Branch						
1962	1674	582	198	–	n.d.	2454
1970	2174	537	213	–	202	2924
Bole						
1962	5774	1849	542	–	n.d.	8165
1970	7694	1746	589	–	629	10029
Stump and central roots						
1962	1155	370	162	–	n.d.	1687
1970	1539	349	176	–	n.d.	2064
Large lateral roots (>0.5 cm)						
1962	954	306	134	–	n.d.	1394
1970	1271	288	145	–	n.d.	1704
Fine lateral roots (⩽0.5 cm)						
1962	251	111	60	20	n.d.	442
1970	335	103	82	20	n.d.	540
Total aboveground						
1962	7652	2521	788	50	n.d.	11031
1970	10115	2359	853	50	831	13352
Total						
1962	10012	3308	1144	70	n.d.	14554
1970	13260	3099	1256	70	–	17710

[a] All values are g m^{-2} DW of organic matter.
[b] Composition of overstory and understory vegetation is shown in Table I.
[c] –, Item not applicable or not separated.
[d] No data available.

TABLE III

Nonrespiratory Components of Net Primary Production during 1966[a]

Component	Yellow-poplar	Other overstory trees	Understory trees	Flora and seedlings	Total
Litterfall (leaf)	226	82	50	–[b]	359
Insect consumption	6	2	2	0[c]	10
Reproductive parts	14	6	3	–	23
Litterfall (small branches and twigs)	26	7	4	–	37
Branch death	0	13	–	–	13
Bole death	0	39	52[d]	–	91
Branch increment	63	–5.6	1.9	–	59
Bole increment	240	–12.9	5.9	–	233
Groundcover turnover	–	–	–	30	30
Stump and central root increment	48	–2.6	1.8	–	47
Stump, central root and large lateral root death	0	7	6	–	13
Large lateral root increment	40	–2.3	1.4	–	39
Fine lateral root increment[e]	10.5	–1	2.8	0[c]	12.3
Fine lateral root death[e]	113	42	25	20	200
Total aboveground	575	132	119	30	856
Total belowground	212	43	37	20	342
Total	787	175	156	50	1168

[a] All values in grams dry weight organic matter m^{-2} y^{-1}.
[b] Item not separated or not applicable.
[c] Item not measured but assumed to be negligible.
[d] Bole and branch death combined for understory trees.
[e] Estimate probably low (see text).

deadwood derived from the three tree groups was combined for the same reason.

The procedure was first to calculate standing crop of organic matter in each compartment (peak value in the case of fluctuating compartments such as foliage and fine roots, Table II), next to compute the annual increment (difference between peak values during successive years, Table III), and finally, when possible, to calculate the amount transferred annually through pathways connecting compartments (Tables III and IV).

The assumptions and data behind these calculations are presented fully elsewhere (Sollins *et al.*, 1973). In general, we relied on locally measured data wherever possible, although in many cases we were forced to use data for

TABLE IV

Estimated Respiratory Losses from Vegetation Compartments during 1966[a]

Vegetation compartment	Yellow-poplar trees	Miscellaneous canopy trees	Understory trees	Groundcover flora and seedlings	Total
Leaves (dark x 2)	0.300	0.109	0.066	0.061[b]	0.536
Branches	0.177	0.051	0.071[c]	–	0.299
Bole	0.123	0.033	–	–	0.156
Central roots and stump	0.024	0.007	0.003	–	0.034
Large lateral roots (>0.5 cm)	0.102	0.028	0.013	–	0.143
Fine roots (<0.5 cm)	0.252	0.092	0.061	0.016	0.420
Total	0.978	0.320	0.214	0.077	1.588

[a] Data are in kg DW organic matter per m^2 of ground area per year, calculated from CO_2 values by assuming organic matter consists of $C_6H_{10}O_5$.

[b] Based on respiration rate of leaves (2.0 kg CO_2 kg^{-1} y^{-1}).

[c] Branch and bole respiration combined for understory trees and calculated based on branch respiration rate (0.15 kg CO_2 kg^{-1} y^{-1}).

European and Japanese forests of quite different character. When transfers were not measured, mass conservation was assumed in calculating amounts transferred annually along a pathway. As an example, consider transfer to yellow-poplar bole (Q_3). Conservation of mass means that $\Delta Q = \Sigma$ Incomes $- \Sigma$ Losses or Σ Incomes $= \Delta Q + \Sigma$ Losses. Thus, because bole respiration was 0.123 kg m^{-2} (Table IV), mortality was 0.0 (Table III) and growth (annual increment) 0.240 kg m^{-2} (Table III), transfer amounted to a net 0.363 kg m^{-2} during the year.

Photosynthetic inputs (gross primary production) were estimated in this way. (Measured values were used only for comparison, as discussed below.) After all transfers were estimated we constructed our annual budget (Fig. 1).

1. *Comparison of Gas Exchange and Harvest Estimates of Gross Primary Production*

Results of preliminary gas exchange experiments (Dinger *et al.*, 1972) gave a value for net daytime photosynthesis plus nighttime respiration of 9.31 g CO_2 fixed per gram dry weight of leaves per growing season, or 3.33 kg CO_2 m^{-2} fixed by all primary producers assuming, until more data are available, that all species have the same photosynthetic rate.

Based on budget calculations (Fig. 1), organic matter flowing into primary producers was estimated at 2.76 kg m^{-2} y^{-1} dry weight. This represents CO_2

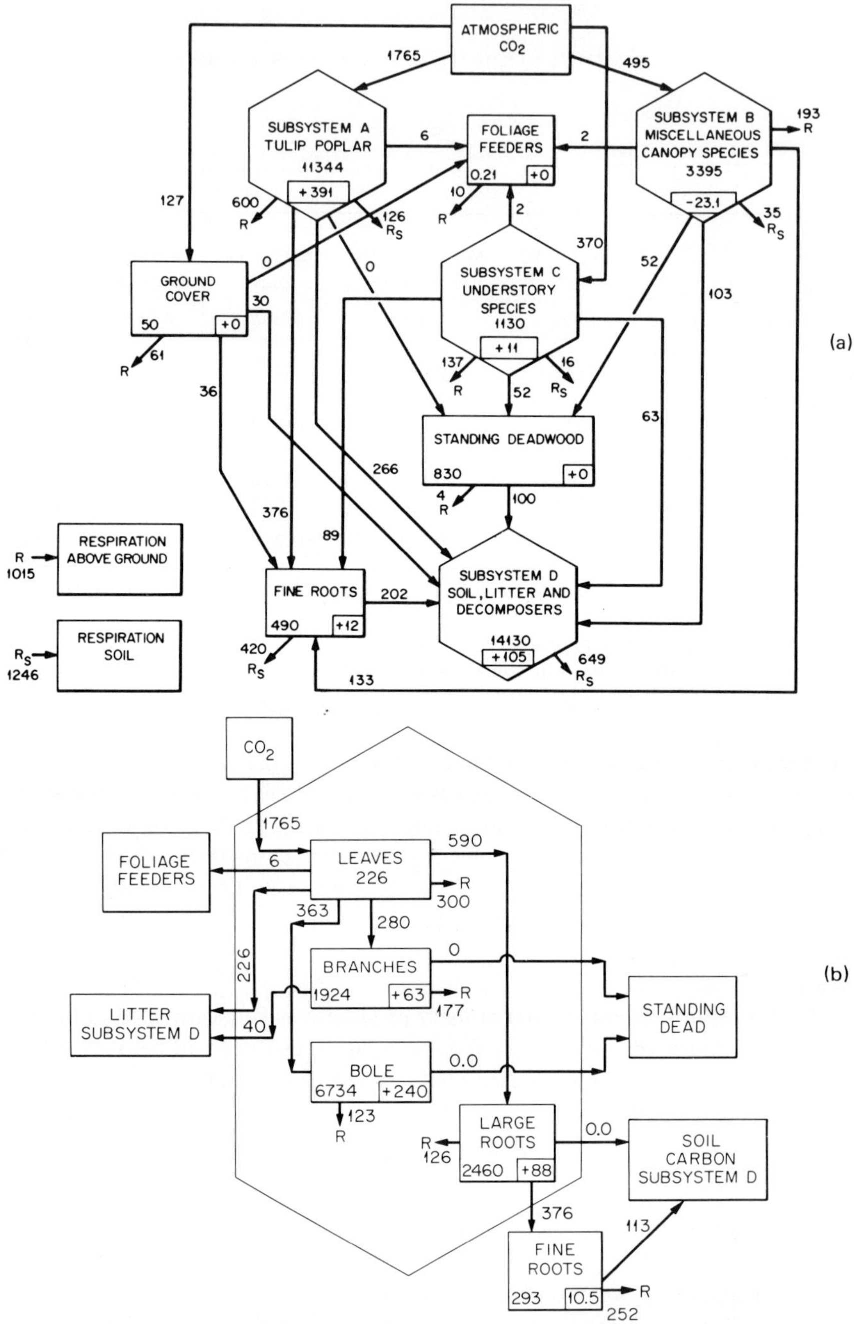

FIG. 1. Organic matter budget for a *Liriodendron*-dominated mesic forest ecosystem at Oak Ridge, Tennessee. All compartment values are grams dry weight per square meter; all transfer and increment values are grams per square meter per year. Compartment increment values are shown in boxes within the compartment. Values for all compartments except leaves are for the 1965–1966 dormant season. Leaf values are peak biomass. Transfers are annual totals based on an average of the 1965 and 1966 data. R and R_s refer to aboveground and belowground respiration, respectively.

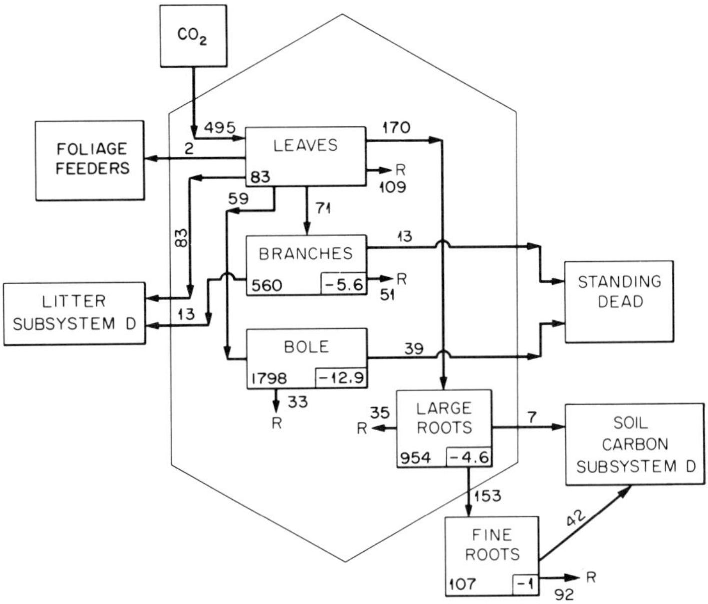

(c)

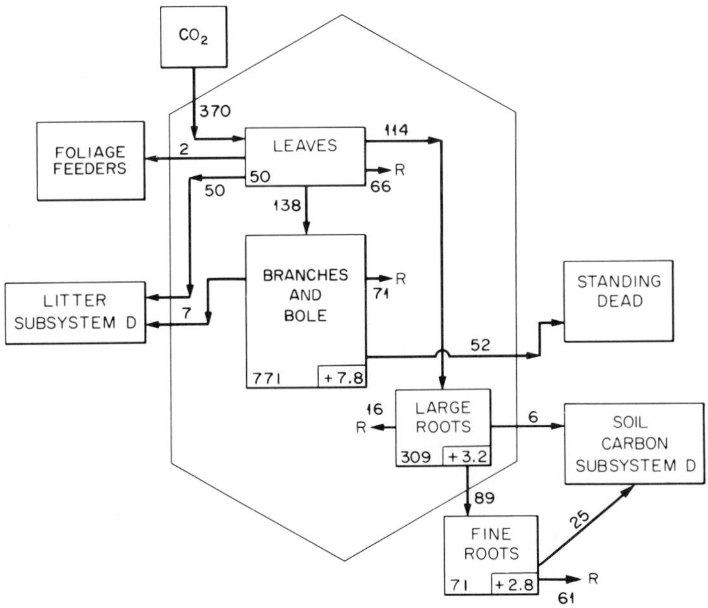

(d)

(a) Overview of the entire forest ecosystem showing linkages between the four major subsystems and miscellaneous compartments.
(b) Detailed budget analysis of subsystem A (yellow-poplar trees). Transfer, increment, and standing crop values for fine roots refer to the subsystem A portion only.
(c) Detailed budget analysis of subsystem B (other overstory trees). Transfer, increment, and standing crop values for fine roots refer to the subsystem B portion only.
(d) Detailed budget analysis of subsystem C (understory trees). Transfer, increment, and standing crop values for fine roots refer to the subsystem C portion only.

fixation plus nutrients absorbed by root systems and distributed throughout the trees. For comparison with gas exchange estimates of gross photosynthesis, mineral uptake must be subtracted from gross production. However, nutrient uptake is typically less than 6% of net aboveground production (Henderson, unpubl. data), and would be an even smaller fraction of gross production.

The factor of 0.614 for conversion of CO_2 to $C_6H_{10}O_5$ was used to convert organic matter production to CO_2 fixation. With this factor, gross photosynthesis based on budget calculations was 4.46 kg CO_2 m^{-2} y^{-1}.

Considering that the gas exchange value was extrapolated from limited measurements and that many budget calculations were based on literature estimates or educated guesses, the 34% difference is not surprising. Botkin *et al.* (1970) attempted the same comparison for the pine-oak forest at Brookhaven National Laboratory, and obtained similar agreement using more extensive gas exchange data. Their gross production values were 3.38 kg CO_2 m^{-2} y^{-1} based on gas exchange measurements and 2.95 based on harvest data.

2. *Comparison of Standing Crop and Metabolic Parameters among Deciduous Forest Stands*

The overall standing crop of vegetation on plot LT1 in 1966 (the midpoint of the study) was 16.1 kg m^{-2} dry weight of organic matter, of which 12.2 constituted aboveground components. Art and Marks (1971) list seven temperate deciduous stands between 40 and 50 y age with a mean aboveground standing crop of 12.4 kg m^{-2}. This value compares well with the estimate for our stand.

Belowground standing crop of plot LT1 was estimated at 3.9 kg m^{-2} resulting in a root : shoot ratio of 0.32. This value is larger than previous values of 0.13 for a Belgian oak-ash forest (Duvigneaud and Denayer DeSmet, 1970) and 0.18 for a mixed deciduous forest in Sweden (Andersson, 1970). However, the difference is probably due to improved harvest methods, allowing more complete root harvest (Harris *et al.*, 1972), rather than to any real differences between the stands.

Net primary production (NPP = gross production − all autotrophic respiratory losses) for the study plot was 1.2 kg m^{-2} y^{-1} (0.9 kg m^{-2} y^{-1} aboveground NPP). Art and Marks (1971) located data for only three stands of comparable age and standing crop. These averaged 0.7 kg m^{-2} y^{-1} aboveground NPP. Based on budget results, annual production of organic detritus in the *Liriodendron* forest amounted to 0.76 kg m^{-2} y^{-1} of organic matter. Adding root respiration estimated at 0.6 kg m^{-2} y^{-1} gives a total of 1.36 kg m^{-2} y^{-1}, which could potentially appear as soil respiration.

Because budget results were available for comparatively few stands of the same age and standing crop as the *Liriodendron* forest, methods were explored

(Sollins *et al.*, 1973) for comparing standing crop and metabolic parameters across stands of widely differing age, growth rate, and successional stage. Many authors (e.g., Whittaker, 1966; Satoo, 1970; Jordan, 1971) have compared stemwood production (NAI = net annual increment of aboveground components) with either leaf biomass or litterfall. The comparison can be viewed as comparing organic matter immobilized in woody tissues to that made available for decomposition. Ideally wood production both aboveground and belowground should be compared with net primary production less wood production, thus including in the figure for available organic matter such items as branch and bole mortality, insect consumption, root death, and fruit and flower production. Unfortunately these data are lacking in all but a few cases. For comparative purposes, stemwood production (aboveground) was plotted against leaf biomass for eleven temperate deciduous forest stands (Fig. 2). There seemed to be little correlation ($r^2 = 0.43$) between these two variables, indicating that other factors such as water stress or nutrient limitations and not photosynthate production usually determine wood production.

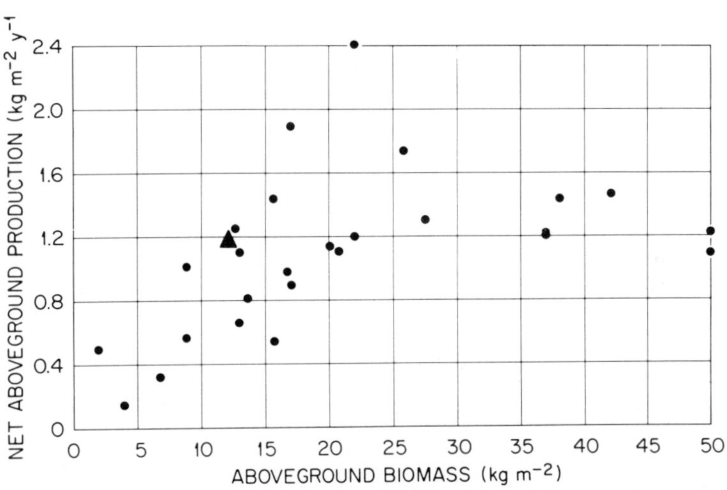

FIG. 2. Relation between stemwood production and leaf biomass in deciduous forests of the world. Data compiled by Sollins *et al.* (1973). Plot described in this study indicated by ▲.

Plotting NPP against aboveground standing crop shows a pattern (Fig. 3) similar to that reported by Kira and Shidei (1967). As biomass increases, NPP first levels off and then appears to decrease. Competition for resources (light, water, and nutrients) is presumably the cause in that respiration increases are not offset by any increase in gross production or nutrient uptake.

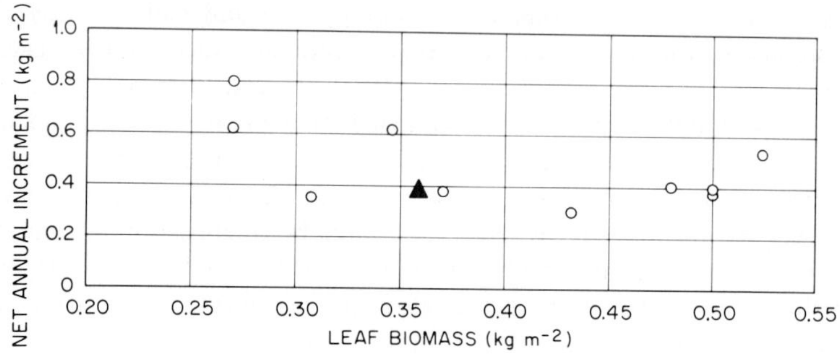

FIG. 3. Relation between aboveground net primary production and aboveground standing crop in deciduous forests of the world. Data compiled by Sollins *et al.* (1973). Plot described in this study indicated by ▲.

B. Forest Floor Organic Matter Budget

Although dynamics of the forest vegetation were studied by first calculating an annual budget of income, loss, and redistribution, a comparable budget could not be calculated for the forest floor. Tremendous spatial variation has prevented us from measuring the slight annual increases in organic matter in the litter horizons. Although litterfall could be measured accurately, organic matter contributed by root death could not. Loss of organic matter through respiration could be measured only for small plots (several dm^2) for short intervals (days). Spatial variation, as well as pronounced effects of temperature and moisture on

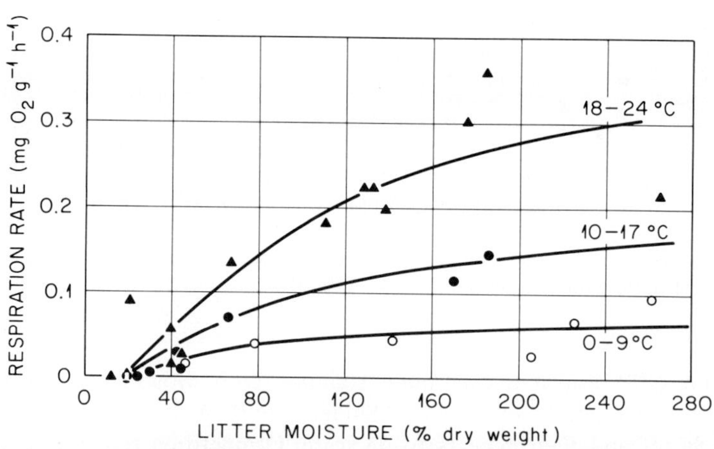

FIG. 4. Comparison of oxygen uptake by O_1 litter as a function of moisture content over three temperature ranges. Curves fitted by eye. ○, 0–9° C; ●, 10–17° C; ▲, 18–24° C.

activity of decomposer organisms, has hindered extrapolation to larger areas or time intervals. Our only method for studying transfers from one litter horizon to another has been to confine a volume of the forest floor either in situ (litterbag studies), or in the laboratory (microcosm experiments, Patten and Witkamp, 1967). In either case, confinement introduces errors into the experiment which are poorly understood (see Witkamp and Olson, 1963).

To study forest floor dynamics, a model was constructed based on measurements of decomposer activity in relation to temperature and moisture (Fig. 4; see also Edwards and Sollins, 1973) and redistribution rates based on litterbag studies (see compilation by Sollins *et al.*, 1973), and a measured rate of increase of soil organic matter (Sollins *et al.*, 1973). The model was used to calculate an annual budget for the system as well as for simulation experiments.

IV. Model Development and Analysis

A. Introduction

We are not aware of a previous attempt to model both forest primary production and decomposition processes simultaneously. Previous efforts to model forest primary productivity have involved two approaches. In one, a single function is postulated to describe stand growth as a function of stand size and site factors. For example, Moser and Hall (1969) described stand growth with a function of the form

$$dB/dt = nB^m - kB, \tag{1}$$

where B is stand basal area and n, m, and k are constants. Leary (1970) and Goldstein and Harris (1973) divided stands into size classes and utilized various growth functions, including Eq. (1), in modeling the system.

In another approach (e.g., Stage, 1973), individual trees in the stand are treated statistically and stochastic functions developed to describe the probability that a given tree or seedling of known size and species will grow at a certain rate or die. Botkin *et al.* (1970) have constructed a model in which these stochastic functions take into account both meteorological and site variables, relying primarily on degree-day accumulation and soil moisture data.

The modeling approach used in this study was unique in that it treated the entire forest ecosystem and developed along physiological lines from the budget analysis of the forest ecosystem. Mathematical functions were formulated to describe transfers of organic matter among compartments in response to key environmental variables such as temperature, moisture, and solar radiation. This physiological conceptualization attempted to provide a mechanism through which descriptions of biological processes could be incorporated into realistic ecosystem models.

As in the budget analysis, overstory vegetation was divided into two components (subsystems A and B) consisting of shade-intolerant and shade-tolerant overstory species, respectively. Subsystems A, B, and C (Fig. 1) were in turn divided into five storage compartments reflecting the physiology of the trees (Fig. 5). These compartments were leaves, buds, fine roots, other metabolically active tissues (including phloem, meristematic tissues, and dissolved carbohydrates), and inactive woody tissues.

The forest floor (subsystem D) was partitioned (Fig. 6) into two litter layers (O_1 and O_2) and two mineral soil layers (0–10 cm and 10–60 cm depth). The O_1 layer was defined as organic matter of identifiable origin (leaves, cones, twigs, etc.), with leaves identifiable to species. O_2 was defined as organic matter sufficiently decomposed such that its origin could not be determined but which had not yet been incorporated into the mineral soil (Soil Survey Staff, 1962,

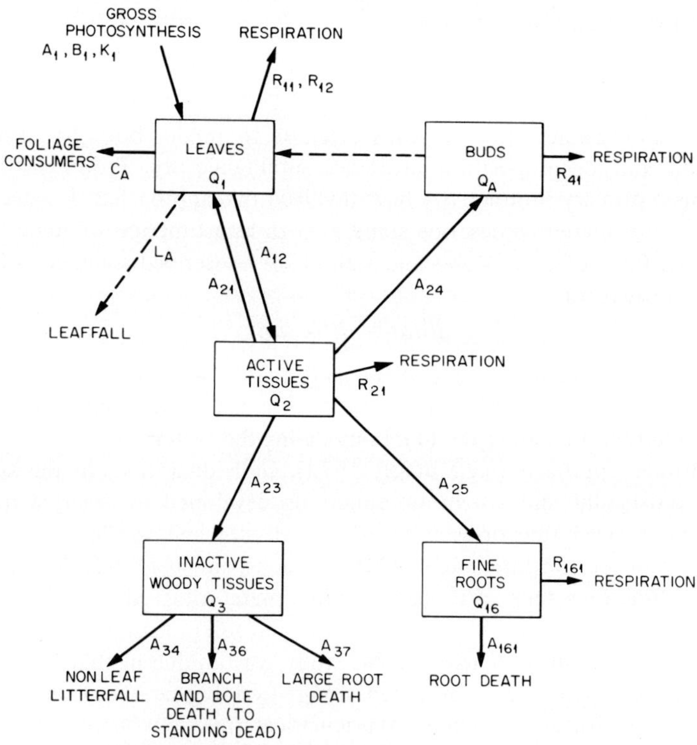

FIG. 5. Arrangement of compartments and transfers in each subsystem of the primary production part of the model. Dotted lines indicate transfers that are discontinuous over time. Fine roots are combined into a single compartment for all three tree subsystems and the herbaceous flora. Compartment and flow indices corresponding to subsystem A have been used for illustration.

pp. 176–177). The O_1 layer (corresponding approximately to the *L* and *F* layers) was divided further into two compartments consisting of leaf and reproductive material in one and fallen branch and bole material in the other. The partitioning of mineral soil was based on the distribution of total organic matter and roots with depth. At 10 cm an abrupt change in soil characteristics and accumulation rates of soil organic matter was apparent (Sollins *et al.*, 1973, based on unpublished data of C. W. Francis).

In this model the decomposers were included with their substrate in defining the five compartments. Decomposer organisms constitute compartments of small storage and rapid turnover whose influences on the system were accounted for in this model through inclusion of temperature and moisture functions in expressions for respiration and transfer between layers. A separate model was developed in which earthworms constituted a separate compartment; this was used to evaluate the contribution of this group of organisms to litter decomposition (Reichle *et al.*, 1972).

Differential equations were formed describing the rate of change of organic matter in each compartment by summing the functions for incoming and

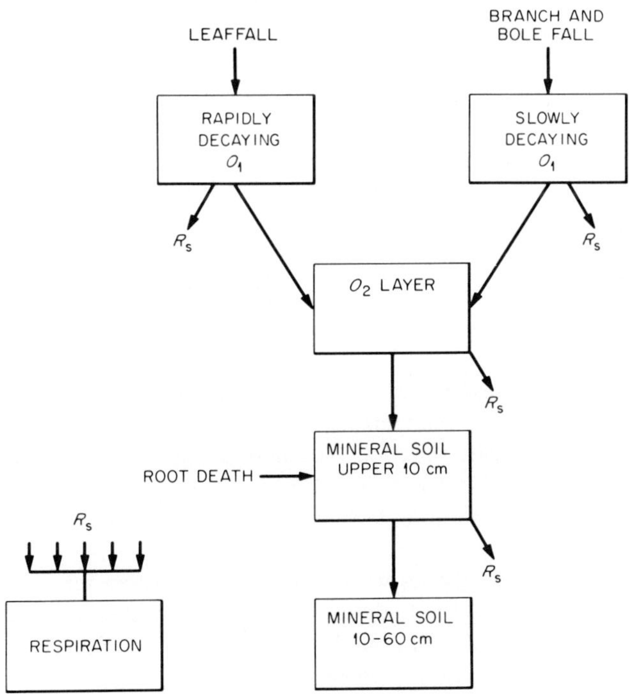

FIG. 6. Arrangement of compartments and transfers in the litter and soil part of the model. R_s refers to soil respiration.

outgoing transfers. The model included seasonal phenomena through temperature and soil moisture and phenological functions; however, daily cycles were not considered. Moisture (percent wet weight) and air temperature (°C) used were based on monthly means, with linear interpolation to provide values at each calculation point. Light intensity was assumed to be a constant 0.7 ly min^{-1} during the growing season. Rate constants were expressed in units of $year^{-1}$, and storages in kilograms dry weight (DW) per square meter. For example, rates of photosynthesis and incident solar radiation were divided by 2 and assumed to hold over each 24-h period rather than only the daylight hours.

Equations were solved using fixed step Runge-Kutta integration. Initial model development depended heavily on use of an interactive time-shared computing system which permitted frequent changes in parameter values and functional forms. Although the original intention was to use a calculation interval of 0.01 y (~4 days), rapid foliage growth early in the growing season required a decrease in step size to 0.002 y.

B. Equations and Inital Conditions for Primary Producer and Consumer Compartments

1. *Phenological Events*

Autotrophic flow expressions (Table V) were all continuous with two exceptions. Transformation from buds to leaves was treated as an instantaneous event occurring at the beginning of the growing season (April 1). Leaffall was assumed proportional to leaf biomass; however, this transfer was set equal to 0.0 during the growing season since less than 7% of leaffall occurs during summer. Any leaves remaining on trees at the start of the growing season were transferred instantaneously to the fast-decaying O_1 compartment.

2. *Respiration*

Except for foliage respiration, respiration functions consisted of a constant multiplied by the source compartment and a temperature function F_T (see Table V). The temperature function used here, in the photosynthesis functions, and elsewhere in the model is an extension of the usual Q_{10} relationship. The function increases to an optimum (32° C) and then decreases abruptly to zero at 44° C.

Respiration of woody tissues was accounted for by transfer out of the active tissue compartment (Fig. 5). Cellulose conversion to starch or sugars, which would have been included in the model by a transfer from inactive back to the active compartment, was assumed to be zero.

Leaf respiration was assumed to have two components (Olson, 1964), a temperature-dependent maintenance respiration proportional to leaf biomass, and a respiration proportional to rate of gross photosynthesis. Leaf respiration was apportioned about equally between the two component processes but more accurate resolution of this may be needed.

3. *Photosynthesis*

Gross photosynthesis G was considered to be the process whereby CO_2 passes from the atmosphere into leaf tissues. In developing a photosynthesis expression for each of the four vegetation groups, our major interests lay in relating G to leaf biomass of that group and to leaf biomass of the entire stand. To simplify the presentation, an expression for G is derived first for a homogeneous stand based on a procedure described by Monsi and Saeki (1953) and in English by Horn (1971).

Consider a thin layer of foliage (thickness dx) at a distance x from the top of the canopy (where $x = 0$). Assume that light absorbed by that layer ($-dI$) is described by the equation

$$-dI = KW(x)I\,dx, \tag{2}$$

where $W(x)$ is leaf biomass per unit area in the layer dx, and K is a light extinction constant. $W(x)$, the leaf density distribution function, is best thought of as the derivative of the curve of cumulative leaf biomass with canopy depth. Equation (2) assumes that light absorption is proportional to this leaf density function (Fig. 7) (Kira *et al.*, 1969; Tadaki, 1966; Satoo, 1970). By rearranging

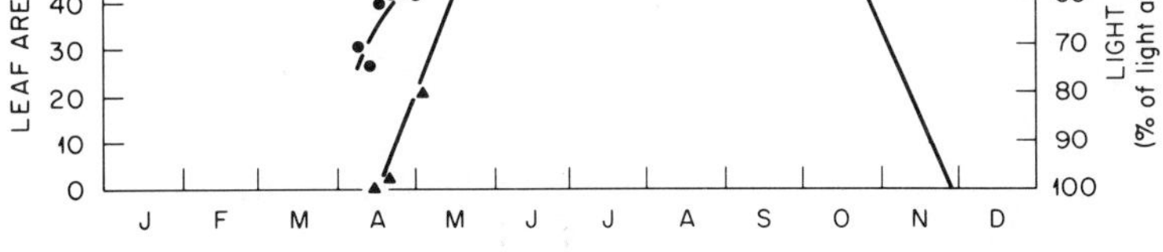

FIG. 7. Pattern of yellow-poplar leaf area during the growing season and relationship to light penetration. Data points were determined from 50 samples collected weekly from the canopy during 1971; each sample consisted of a single stem having from four to 10 leaves each (Reichle *et al.*, 1973c). The straight line representing leaf area decline was derived from 1970 data on litter accumulation for this stand (Sollins *et al.*, 1973). Also shown are percent light penetration to the forest floor and to a height of 12.2 m above the forest floor (Taylor, 1974). Maximum leaf area index was 5.9; average light intensity above canopy was 13.9 kcal cm^{-2} $month^{-1}$. ▲, yellow-poplar leaf area; ●, light penetration to height of 12.2 m; ○, light penetration to forest floor.

TABLE V

Initial Conditions and Equations of Primary Producer Compartments[a]

Compartment (Initial condition, kg m^{-2})	Differential equation
Yellow-poplar leaves (0.0)	$\dot{Q}_1 = dQ_1/dt = (1 - R_{12})G_A + \phi_{21} - (R_{11}F_T + A_{12} + C_A)Q_1 - L_A$ where $G_A = \left[\frac{Q_A{}^* + A_{13}}{A_{13} + 0.005}\right]\left[\frac{-3.914B_1}{A_1K_1}\right] \times \left[\frac{Q_1}{Q_1 + Q_5}\right] \ln \frac{1 + A_1E_1}{1 + A_1I(0)}$ $E_1 = I(0)\exp(-1000K_1(Q_1 + Q_5))$ $\phi_{21} = Q_A{}^*A_{21}Q_2^{0.1}G_A^{-1}$ when $G_A > 0$ $= 0$ when $G_A = 0$ $L_A = A_{14}Q_1$ during dormant season $= 0$ during growing season $F_T = 0.35(40 - T)\exp[-(40 - T)/8]$ where T is temperature (°C) $Q_A{}^*$ = value of Q_A at end of dormant season
Yellow-poplar active tissues (2.67[b])	$\dot{Q}_2 = A_{12}Q_1 - \phi_{21} - (R_{21} + A_{23} + A_{24})F_T - A_{216}Q_2$
Yellow-poplar woody tissues (9.45)	$\dot{Q}_3 = A_{23}F_TQ_2 - (A_{34} + A_{36} + A_{37})Q_3$
Yellow-poplar buds (0.005)	$\dot{Q}_A = A_{24}F_TQ_2 - R_{41}F_TQ_A$
Other overstory leaves (0.0)	$\dot{Q}_5 = (1 - R_{52})G_B + \phi_{65} - (R_{51}F_T + A_{56} + C_B)Q_5 - L_B$ where $G_B = \left[\frac{Q_B^* + A_{53}}{A_{53} + 0.0012}\right]\left[\frac{-3.914B_5}{A_5K_1}\right] \times \left[\frac{Q_5}{Q_1 + Q_5}\right] \ln \frac{1 + A_5E_1}{1 + A_5I(0)}$ $\phi_{65} = Q_B^*A_{65}Q_6^{0.1}G_B^{-1}$ when $G_B > 0$ $L_B = A_{54}Q_5$ during dormant season $Q_B^* = Q_B$ at end of dormant season

[a] Parameter definitions shown in Fig. 5. Parameter values and units given in Table VII.

TABLE V–*continued*

Other overstory active tissues (0.496[b])	$\dot{Q}_6 = A_{56}Q_5 - \phi_{65} - A_{616}Q_6$ $- (R_{61} + A_{67} + A_{68})F_T Q_6$
Other overstory woody tissues (2.816)	$\dot{Q}_7 = A_{67}F_T Q_6 - (A_{74} + A_{76} + A_{77})Q_7$
Other overstory buds (0.0012)	$\dot{Q}_B = A_{68}F_T Q_6 - R_{81}F_T Q_B$
Understory leaves (0.0)	$\dot{Q}_9 = (1 - R_{92})G_C + \phi_{109}$ $- (R_{91}F_T + A_{910} + C_C)Q_9 - L_C$ where $G_C = \left[\frac{Q_C^* + A_{93}}{A_{93} + 0.0012}\right]\left[\frac{-3.914B_9}{A_9 K_9}\right]\ln\frac{1 + A_1 E_1 E_2}{1 + A_9 E_1}$ $E_2 = \exp(-1000K_9 Q_9)$ $\phi_{109} = Q_C^* A_{109} Q_{10}^{0.1} G_C^{-1}$ when $G_C > 0$ $L_C = A_{94}Q_9$ during dormant season
Understory woody tissues (0.918)	$\dot{Q}_{11} = A_{1011}F_T Q_{10}$ $- (A_{114} + A_{116} + A_{117})Q_{11}$
Understory buds (0.0007)	$\dot{Q}_C = A_{1012}F_T Q_{10} - R_{121}F_T Q_C$
Groundcover (0.050)	$\dot{Q}_{13} = (1 - R_{132})G_H - R_{131}F_T Q_{13}$ $- (A_{131} + A_{132})Q_{13}$ where $G_H = \left[\frac{-3.914}{A_{13}K_{13}}\right]\ln\frac{1 + A_{13}E_1 E_2 \exp(-1000 K_{13}Q_{13})}{1 + A_{13}E_1 E_2}$
Standing dead (0.83)	$\dot{Q}_{14} = A_{36}Q_3 + A_{76}Q_7 + A_{116}Q_{11} - (R_{141}F_T + A_{141})Q_{14}$
Canopy consumers (0.002)	$\dot{Q}_{15} = A_{152}C - (M_{15} + R_{15}F_T)Q_{15}$ where $C = C_A Q_1 + C_B Q_5 + C_C Q_9$
Fine roots (0.44)	$\dot{Q}_{16} = A_{216}Q_2 + A_{616}Q_6 + A_{1016}Q_{10} + A_{131}Q_{13} - R_{161}F_T Q_{16}$ $- A_{161}Q_{16}(1 - F_T M)$ where M is soil moisture (% wet wt)

[b] Set at 15% of biomass (see Kramer and Kozlowski, 1960, Chap. 4).

and integrating Eq. (2), we obtain an expression relating light intensity to depth in the canopy:

$$I(x) = I(0) \exp\left[-K \int_0^x W(h)\, dh\right], \tag{3}$$

where h is a dummy integration variable. Now, at any canopy depth assume that gross photosynthetic rate is related to light intensity by the well-known relation (e.g., Monsi, 1968; Hozumi *et al.*, 1972)

$$G(x) = bI(x)\, W(x)/[1 + aI(x)], \tag{4}$$

where b is a constant indicating the maximum possible rate of photosynthesis, and a is a constant related to slope of the curve at low light levels.

Substituting Eq. (3) into Eq. (4) and integrating across canopy depth yields

$$G_T = bI(0) \int_0^D \frac{W(x) \exp\left[-K \int_0^x W(h)\, dh\right] dx}{1 + zI(0) \exp\left[-K \int_0^x W(h)\, dh\right]}, \tag{5}$$

where G_T is gross photosynthesis across the entire canopy depth D in units of biomass/time.

To integrate this expression over depth x we make the following substitution:

$$U = 1 + aI(0) \exp\left[-K \int_0^x W(h)\, dh\right].$$

Then

$$dU = -KaI(0)\, W(x) \exp\left[-K \int_0^x W(h)\, dh\right] dx.$$

Substituting back into Eq. (5):

$$G_T = \frac{b}{-aK} \int_0^D \frac{dU}{U} = \frac{b}{-aK} \ln\left[\frac{1 + aI(0) \exp\ -K \int_0^D W(h)\, dh}{1 + aI(0) \exp\ -K \int_0^0 W(h)\, dh}\right]. \tag{6}$$

But,

$$L = \int_0^D W(h)\, dh, \quad \text{since} \int_0^0 W(h)\, dh = 0,$$

where L is total leaf biomass to a depth D. Thus, the overall expression simplifies to

$$G_T = \frac{-b}{aK} \ln \left[\frac{1 + aI(0) \exp(-KL)}{1 + aI(0)} \right]. \tag{7}$$

It is interesting that we need make no assumption about vertical distribution of the foliage, $W(x)$, as long as Eqs. (3) and (4) are used for light attenuation and the relationship between photosynthesis and light intensity. Changing these assumptions would most likely result in a function that could not be integrated analytically over the canopy depth. Numerical integration would be necessary resulting in a model considerably more complicated and expensive to run.

The gross photosynthesis equations used for subsystems A, B, and C included two additional functions (for temperature and bud effect), and a constant (3.914), which converts from ly min^{-1} and $mgCO_2$ dm^{-2} h^{-1} to kilograms dry weight per square meter per year. The temperature function was the same as that used in the respiration equations (see above). The purpose of the bud function was to set an upper limit on leaf production during a growing season based on the number of buds set in the previous growing season which survived dormancy. The bud function was of the form (using subsystem A as an example)

$$(Q_A{}^* + A_{13})/(A_{13} + 0.005),$$

where $Q_A{}^*$ is the value of (in this case) the yellow-poplar bud compartment immediately prior to bud break, typically about 0.005 kg m^{-2}. The constant A_{13} adjusts the magnitude of the effect. For example, with $A_{13} = 1.0$ the value of $Q_A{}^*$ would be irrelevant; with $A_{13} = 0.0$ the function would increase linearly with $Q_A{}^*$.

4. *Other Transfer Expressions*

A nonlinear function was also used for each of the three tree subsystems to describe organic matter transfer from active tissues to leaves. The transfer was assumed to be inversely proportional to gross photosynthesis rate but to vary directly with the value of the active tissue compartment. A strict proportionality to the active compartment did not provide a realistic simulation. Either transfer to leaves continued at an unreasonably high rate well into the growing season, resulting in final leaf biomass greater than observed, or the initial transfer was not great enough and the initial rate of leaf development was less than actually observed. Raising the active tissue compartment level to a small power (0.1) made the transfer independent of the compartment except at rather low compartment levels and seemed to provide an acceptable *ad hoc* solution. Although a more mechanistic function should be derived, this particular model did not seem to be very sensitive to this transfer coefficient.

Fine root death was made inversely proportional to the temperature and moisture functions so that the majority of root death occurred during the dormant season. The respiratory temperature function F_T was used because temperature effect on root death was unknown.

All of the remaining flow expressions were taken to be linear functions of the source compartment, although in many the rate coefficient was multiplied by the temperature function F_T (see Table V).

5. *Coupling Equations for the Four Vegetation Components*

The three tree subsystems and the groundcover compartment were coupled through their respective equations for photosynthesis, specifically the light intensity terms (Table V), Total overstory foliage biomass ($Q_1 + Q_5$) was used in computing the leaf density variable W_L in the gross photosynthesis expression for the two overstory subsystems. For understory trees understory foliage biomass Q_9 was used, but incident light intensity seen by the understory was taken to be the light penetrating the overstory. Derivation of the gross photosynthesis expression for the groundcover compartment Q_{15} was analogous to that for the understory trees. The forms of these equations (see Table V) are slightly different from Eq. (7), but may be derived using the assumptions and procedure described above.

C. Equations and Initial Conditions for Litter and Soil Compartments

The litter–soil part of the forest organic matter model consisted of five linear differential equations (Table VI). Except for inputs from primary producer compartments (Table V), all transfer expressions consisted of a rate constant, the value of the source compartment, and a temperature–moisture factor F_{TM} all multiplied together. Analysis of the effect of temperature and moisture on respiration of litter and soil (Sollins *et al.*, 1973) showed that over the range actually encountered the product of temperature and moisture (percent wet weight) was an adequate predictor and that more complicated functions did not give significant improvement.

D. Parameter Estimation

Parameters of the vegetation and consumer equations (Tables VII and VIII) were adjusted by trial and error until annual transfers and compartment changes matched budget data (Fig. 1) within about 1%.

For litter compartments, the rate of organic matter input to each compartment from litterfall, the overlying litter layer or soil layer, was known as well as the respiration rate for each compartment. For soil compartments the annual

TABLE VI

Initial Conditions and Equations of the Soil and Litter Compartments

Compartment[a] (Initial condition, kg m^{-2})	Differential equation[b]
Quickly decomposing O_1 layer (0.2)	$\dot{Q}_{17} = L_A + L_B + L_C + A_{132}Q_{13} + (1 - A_{152})C$
	$-(R_{17} + A_{1719})F_{TM}Q_{17} + 0.38D$
	where $D = A_{34}Q_3 + A_{74}Q_7 + A_{114}Q_{11}$
	$C = C_A Q_1 + C_B Q_5 + C_C Q_9$
	$F_{TM} = 0.2TM$
Slowly decomposing O_1 layer (0.12)	$\dot{Q}_{18} = A_{141}Q_{14} + 0.62D$
	$-(R_{18} + A_{1819})F_{TM}Q_{18}$
O_2 layer (0.26)	$\dot{Q}_{19} = A_{1719}Q_{17} + A_{1819}Q_{18}$
	$-(R_{19} + A_{1920}Q_{19})F_{TM}$
Soil organic matter (0–10 cm depth) (4.0)	$\dot{Q}_{20} = A_{1920}F_{TM}Q_{19} + A_{161}Q_{16}(1 - F_{TM})$
	$+ A_{37}Q_3 + A_{77}Q_7 + A_{117}Q_{11}$
	$-(R_{20} + A_{2021})Q_{20}$
Soil organic matter (10–60 cm depth) (9.7)	$\dot{Q}_{21} = A_{21}Q_{21}$

[a] Arrangement of compartments shown in Fig. 6.
[b] Symbols defined in Tables V, VII, and VIII, and Fig. 5.

increase in compartment level was also known and loss rate to the underlying layer was calculated by the difference. Loss rates from litter compartments were calculated by assuming a 1% annual increase in compartment level. The parameters were then adjusted until total soil–litter–decomposer respiration approached values reported by Edwards and Sollins (1973).

Occasionally the model and budget were not comparable. For instance, importance of the bud compartment became apparent only after several unsuccessful modeling attempts. Bud set and survival through the dormant season provided a mechanism whereby net primary production during a given growing season could influence productivity during the next growing season. By making photosynthesis proportional to biomass of buds surviving the dormant season, it was possible to control the growth rate of the various vegetation groups. However, the amounts of organic matter involved in the bud

TABLE VII

Parameter Values for the Primary Production Part of the Forest Organic Matter Model[a]

Input or source compartment	Subsystem parameter values (units are y^{-1} unless otherwise indicated)		
	A	B	C
Photosynthesis			
Light extinction	K_1 = 0.0039[b]		K_9 = 0.02913[b]
Light saturation	A_1 = 21.0[c]	A_5 = 21.0	A_9 = 21.0
Maximum rate	B_1 = 100.9[d]	B_5 = 75.5	B_9 = 133.0
Leaves			
Maintenance respiration	R_{11} = 1.0	R_{51} = 1.35	R_{91} = 1.85
Photosynthesis-dependent respiration	R_{12} = 0.133	R_{52} = 0.15	R_{92} = 0.10
Consumption	C_A = 0.059	C_B = 0.057	C_C = 0.085
Leaffall	L_A = 18.0	L_B = 15.0	L_C = 18.0
Translocation to active tissues	A_{12} = 19.9	A_{56} = 17.9	A_{910} = 20.3
Bud effect	A_{13} = 0.005[e]	A_{53} = 0.005	A_{93} = 0.005
Active			
Translocation to leaves	A_{21} = 890	A_{65} = 600	A_{109} = 500
Wood deposition	A_{23} = 0.563	A_{67} = 0.248	A_{1011} = 1.03
Bud formation	A_{24} = 0.0116	A_{68} = 0.0106	A_{1012} = 0.0170
Translocation to roots	A_{216} = 0.227	A_{616} = 0.276	A_{1016} = 0.55
Respiration	R_{21} = 0.59	R_{61} = 0.57	R_{101} = 1.2
Buds			
Respiration	R_{41} = 2.0	R_{81} = 3.0	R_{121} = 3.0
Inactive			
Litterfall	A_{34} = 0.00415	A_{74} = 0.0046	A_{114} = 0.0076
Branch and bole fall	A_{36} = 0.0	A_{76} = 0.0185	A_{116} = 0.0567
Large root death	A_{37} = 0.0	A_{77} = 0.0025	A_{117} = 0.0066
Light intensity	$I(0)$ = 0.7 ly min^{-1}		
Groundcover			
(Parameter definitions analogous to tree subsystems)	A_{13} = 10.0 R_{132} = 0.42 R_{131} = 0.4	B_{13} = 48.2 A_{131} = 0.73	K_{13} = 0.0334[b] A_{132} = 0.62
Standing deadwood	R_{141} = 0.0168	A_{141} = 0.11	
Canopy consumers	A_{152} = 0.6[f]	M_{15} = 0.5	R_{15} = 120
Fine roots	A_{161} = 0.387	R_{161} = 2.27	

[a] See Fig. 5 for further explanation of symbols.
[b] Units are $m^2\ g^{-1}$.
[c] Units of A_1, A_5, and A_9 are min ly^{-1}.
[d] Units of B_1, B_5, and B_9 are min $ly^{-1}\ y^{-1}$.
[e] Units of A_{13}, A_{53}, and A_{93} are kg m^{-2}.
[f] Dimensionless constant.

compartments were small and no attempt was made to go back and include the compartments in a new round of budget calculations. Such discrepancies will occur in any ecosystem analysis and are evidence that the modeling is producing insights into the structure and functioning of ecosystems which were not evident from the construction of annual redistribution budgets.

For decomposition compartments only total respiration and rate of increase of soil organic matter were known. Rate of increase of O_1 and O_2 compartments was assumed to be 1% y^{-1}. In general, parameter values obtained from local and published sources were used only as guidelines during parameter fitting. Ultimately not a single model parameter corresponded exactly in value to a locally measured rate, although for several (e.g., root respiration, yellow-poplar photosynthesis parameters) only small adjustments were necessary to obtain desired model behavior. Parameter fitting, although unlikely to result in a model that is a unique best fit to the data, is the only recourse until far more rates have been measured locally and their relation to driving variables such as temperature and moisture established.

E. Model Behavior

After a set of parameter values had been chosen, the time behavior of various compartments was examined in detail (Fig. 8). Leaf development matched measured curves (Fig. 7) fairly well. Active tissue underwent a cyclical seasonal variation reflecting utilization of carbohydrate reserves during the dormant season and the early growing season and surplus production later in the growing season. The groundcover compartment experienced the majority of its growth during the early spring before overstory bud break and in midautumn after leaffall.

Time behavior of the decomposition compartments (Fig. 9) showed distinct shoulders in all litter compartment levels reflecting periods of decomposer inactivity during winter and summer. Differences between model results (Fig. 9) and observed curves of weight loss from litterbags (e.g., Olson and Crossley, 1963) are probably due to gradual onset of leaffall in autumn in contrast with the essentially instantaneous placement of the litterbags. Observed patterns of standing amount of litter (Fig. 10) were quite similar to model results. Soil organic matter levels showed little seasonal variation, indicating the ability of overlying litter to insulate the soil from temperature changes and to "even out" seasonal variations in moisture and organic matter input to the forest floor.

F. Forest Floor Organic Matter Budget

Organic matter lost annually through respiration of decomposer organisms was estimated from the model to be 640 g m^{-2}. By adding the value for root respiration from the primary production budget (600 g m^{-2} y^{-1}), the estimate

TABLE VIII

Parameter Values for Decomposition Part of Forest Organic Matter Model

Source compartment	Parameters[a] (y^{-1})	
Rapidly decaying O_1 layer	$A_{1719} = 1.22$	$R_{17} = 1.6$
Slowly decaying O_1 layer	$A_{1819} = 0.3$	$R_{18} = 0.72$
O_2 layer	$A_{1920} = 0.05$	$R_{19} = 0.8$
Soil organic matter (0–10 cm)	$A_{2021} = 0.001$	$R_{20} = 0.027$

[a] Term A refers to transfers to an underlying layer, R to respiratory losses.

of organic matter respired by the forest floor is increased to 1.24 kg m^{-2} y^{-1}. After converting organic matter to CO_2 evolution by assuming that organic matter consists of $C_6H_{10}O_5$, CO_2 evolution was estimated to be 2.0 kg m^{-2} y^{-1}. This value could not be increased to the value calculated from biweekly CO_2 evolution data (3.8 kg m^{-2} y^{-1}) while maintaining the desired rate of soil and litter organic matter increase. Presumably, the role of root death in providing substrate for decomposition was grossly underestimated. More recent results (Harris *et al.*, in press) indicate that root respiration accounts for about 1.4 kg CO_2 m^{-2} y^{-1} and root death for about 1.3 kg CO_2 m^{-2} y^{-1}.

Of the total CO_2 evolution, 19% was attributed in this model to biota of the rapidly decomposing O_1 compartment, 7% to slowly decomposing O_1, 16% to the O_2 layer, 9% to mineral soil, and 48% to various components of the root systems. Of the annual organic matter input of 0.76 kg m^{-2} to the forest floor, 0.215 kg m^{-2} or 28% was calculated to be due to root death. If input due to root death is converted to equivalent CO_2 evolution and added to the value for root respiration, it can be seen that roots provide the substrate directly or indirectly for about 66% of "soil respiration." This large value supports the contention of Wiant (1967) that "the contribution of litter decay to forest 'soil respiration' [has] been overestimated," and illustrates the need for extensive studies of root processes during any analysis of the forest floor subsystem.

G. Sensitivity Analysis

Because of the nonlinear, time-dependent nature of the model, analytic treatment of model sensitivity to changes in parameter values was not attempted. Instead, selected parameters of each subsystem were varied by ±5% and effects on wood production, net production, and soil respiration noted (Table IX). The net production value used consisted of gross photosynthesis less

TABLE IX

Sensitivity of Forest Organic Matter Model to Changes in Primary Production Parameter Values[a]

Paramater[b]	Net production +5%	Net production −5%	Wood production +5%	Wood production −5%	Soil respiration +5%	Soil respiration −5%
	Criterion (% change in parameter value)					
Maximum photosynthesis rate						
B_1	+10.3	−9.3	+ 2.8	− 2.6	+0.42	−0.35
B_5	+ 7.6	−7.2	+ 4.8	− 4.3	+0.057	−0.057
B_9	+ 7.7	−7.3	+33.0	−32.0	+0.085	−0.076
Light saturation						
A_1	− 8.0	+9.6	− 2.2	+ 2.6	−0.30	+0.39
A_5	− 6.1	+7.0	− 1.9	+ 2.4	−0.11	+0.13
A_9	− 4.94	+5.54	−21.8	+24.1	−0.057	+0.057
Light extinction						
K_1[c]	− 0.68	+0.66	− 0.21	+ 0.23	−0.076	+0.085
K_9	− 1.7	+1.7	− 6.9	+ 6.9	−0.038	+0.047
Light intensity						
$I(0)$[d]	+ 1.1	−1.2	+ 0.51	− 0.57	+0.10	−0.095
Leaf maintenance respiration						
R_{11}	− 0.6	+0.6	− 0.14	+ 0.17	−0.028	+0.019
R_{51}	− 0.9	+0.9	− 0.48	+ 0.48	−0.01	+0.01
R_{71}	− 0.8	+0.8	− 3.4	+ 3.4	−0	+0
Leaf photosynthesis dependent respiration						
R_{12}	− 2.0	+2.0	− 0.52	+ 0.55	−0.085	+0.085
R_{52}	− 1.9	+1.9	− 0.97	+ 1.4	−0.02	+0.02
R_{92}	− 1.1	+1.1	− 4.6	+ 4.6	−0	+0.019
Translocation to active pool						
A_{12}	− 7.5	+8.8	− 1.5	+ 1.7	−0.53	+0.65
A_{56}	− 5.7	+6.5	− 1.9	+ 2.4	−0.11	+0.13
A_{910}	− 4.5	+4.9	−12.6	+12.6	−0.076	+0.085
Leaf consumption						
C_A	− 0.02	+0.02	0	+ 0.03	0	0
C_B	− 0.04	0	0	+ 0.48	0	0
C_C	0	0	0	0	0	0
Translocation to leaves						
A_{21}	+ 3.0	−3.0	+ 0.7	− 0.7	+0.18	−0.19
A_{65}	+ 3.3	−3.6	+ 1.9	− 1.4	+0.057	−0.066
A_{109}	+ 2.2	−2.3	+ 6.9	− 8.0	+0.038	−0.028
Wood deposition						
A_{23}	+ 0.2	−0.2	+ 5.0	− 5.0	−0.03	+0.03
A_{67}	+ 0.1	−0.1	+12.0	−11.6	−0.01	0
A_{1011}	+ 0.2	−0.2	+34.5	−35.6	−0.01	+0.02

[a] Values shown are percent change in criterion value corresponding to ±5% change in value of the indicated parameter.

[b] Parameters defined in Tables V and VII and Fig. 5.

[c] Production changes based on sum of subsystems A and B.

[d] Production changes summed across all three subsystems.

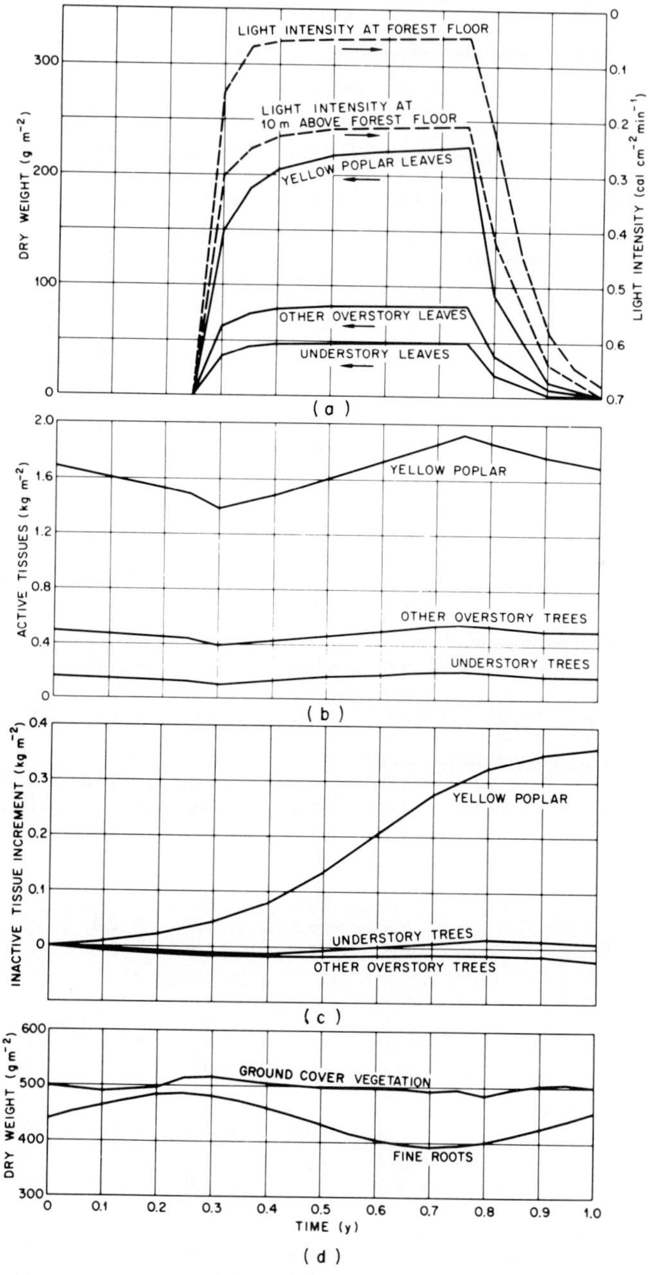

FIG. 8. Simulated seasonal patterns of organic matter in various primary producer compartments. Also shown is simulated light intensity within the canopy and at the forest floor. (a) Light intensity and leaf biomass; (b) active tissues; (c) woody tissues; (d) groundcover and fine root biomass.

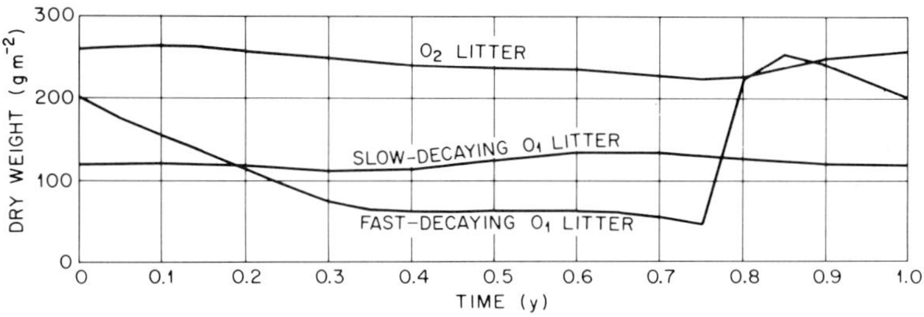

FIG. 9. Simulated seasonal patterns of various litter compartments.

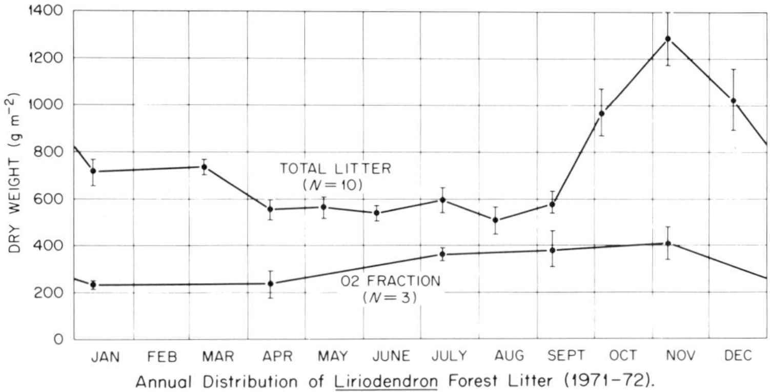

FIG. 10. Annual measured pattern of litter dry weight on a *Liriodendron* forest floor. Data are for 1971–1972 (McBrayer, unpubl.).

leaf, bud, and active tissue respiration. Fine root respiration was not subtracted because of difficulty in partitioning it among vegetation components. "Soil respiration" included respiratory losses from litter, soil, and fine roots.

Photosynthetic parameters had the most effect on primary production and soil respiration, followed by the coefficients A_{12}, A_{56}, and A_{910} for transfer from leaves to active tissues. Parameters with surprisingly little effect on model behavior included all respiration parameters, the light extinction parameters (K_1 and K_9), and the insect consumption rate constants.

Understory wood production was much more sensitive to parameter changes than were overstory wood production or understory net production. However, wood production was, in the unperturbed model, almost balanced by branch and bole fall. The percentage change in this criterion is thus misleading because the denominator, wood production under normal conditions, was nearly zero.

Model response was in general proportional to percent change in parameter value over a wide range although in most cases a 5% increase in parameter value had an effect different from that of a 5% decrease. Only one "interesting" nonlinear response was found. Net production showed a hyperbolic response to changing light intensity (Fig. 11) similar to that observed for single leaves. However, the asymptote was approached by leaves of each subsystem at greater light intensities than for individual leaves, indicating the light attenuating effect of the canopy strata.

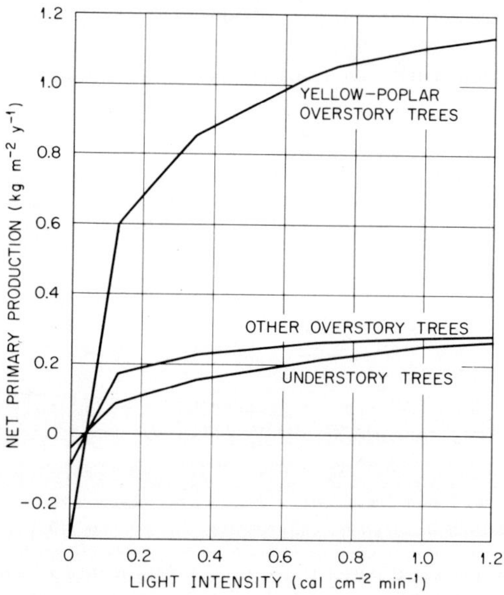

FIG. 11. Simulated effect of light intensity on net production (as defined in text) of overstory and understory components of the vegetation.

V. Simulation Studies

The organic matter model of the *Liriodendron* ecosystem, although crude, permitted an initial assessment of the operation of an ecosystem, both in its natural state of slow change as well as under various perturbation conditions. To demonstrate the usefulness of such a model in studying undisturbed systems, the seasonal course of forest CO_2 exchange was simulated. Three perturbation conditions also were simulated, all involving removal or death of some vegetation component: (i) increased bud kill, a common effect of insect outbreaks and late frost as well as a primary mechanism whereby irradiation affects forest growth (Woodwell and Rebuck, 1967); (ii) stand thinning (selective logging), a common forestry practice although its effects on the entire system, particularly decomposition processes, are poorly understood; (iii) defoliation, an important effect of insect outbreaks as well as the intended result of herbicide applications.

A. Seasonal Course of Ecosystem CO_2 Exchange

A forest is, in many respects, an autotrophic "organism" capable of photosynthesis and respiration. At least two studies (Woodwell and Dykeman, 1966; Odum, 1970) have attempted, at considerable expense, to monitor these metabolic processes for an entire forest over short periods of time. Simulation provides an alternate method of studying ecosystem metabolic processes if the nature of component processes is known. The seasonal course of net ecosystem CO_2 exchange was simulated by combining functions describing its component parts (gross photosynthesis, aboveground and belowground respirations) (see also Bate and Canvin, 1972). Simulation results (Fig. 12) suggest that a deciduous forest "inhales" CO_2 during the growing season and "exhales" it mostly during the spring and fall when conditions for decomposition are more nearly optimum. Photosynthesis occurs during winter, but is very small

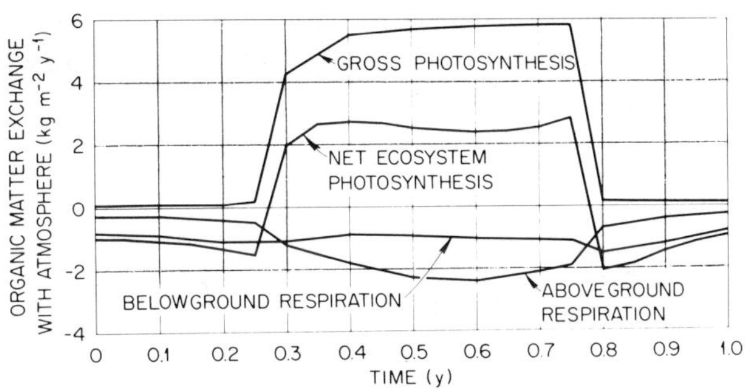

FIG. 12. Simulated seasonal course of CO_2 exchange by a mesic deciduous forest ecosystem.

compared to respiration. During summer, photosynthesis maintains a nearly constant level while respiration increases and then decreases in response to temperature pattern. Net ecosystem photosynthesis (the difference) shows a midsummer depression due to increased respiration. These patterns closely resemble *daily* gas exchange patterns observed for aspen stands (Bate and Canvin, 1972) and many agricultural and aquatic systems. Cycles of carbon dioxide exchange in a deciduous forest appear to operate on two time scales, daily and annual, with a very close analogy between the two.

It will be interesting to compare CO_2 exchange curves for deciduous forests with curves for other forest types. Quite possibly this "inhale–exhale" pattern is unique to the temperate deciduous forest. The mature tropical forest would be expected to show a small but nearly uniform rate of net CO_2 intake throughout the year, while taiga presumably would show almost no activity during the dormant season.

B. Effect of Bud Kill on Primary Production

The effect of various levels of bud kill (0–100%) was simulated by decreasing the bud compartment size on about March 1 of the first year. Bud kill resulting from irradiation was the focus of this simulation because experimental results were available for comparison. Because yellow-poplar is one of the more radiosensitive of the dominant species in our forest (Witherspoon, 1969), only the yellow-poplar bud compartment was decreased. The same simulation procedure could be used to study effects of an organism (e.g., *Odontopus*, the yellow-poplar weevil) that selectively kills buds of *Liriodendron.*

As expected, all levels of bud kill caused decreased production by yellow-poplar, which in turn resulted in increased production by the other vegetation components (Fig. 13). Decreases in net production, leaf production, and wood production followed similar patterns.

Results of the irradiation of the pine-oak forest at Brookhaven National Laboratory allowed some comparison of simulation results with experimental

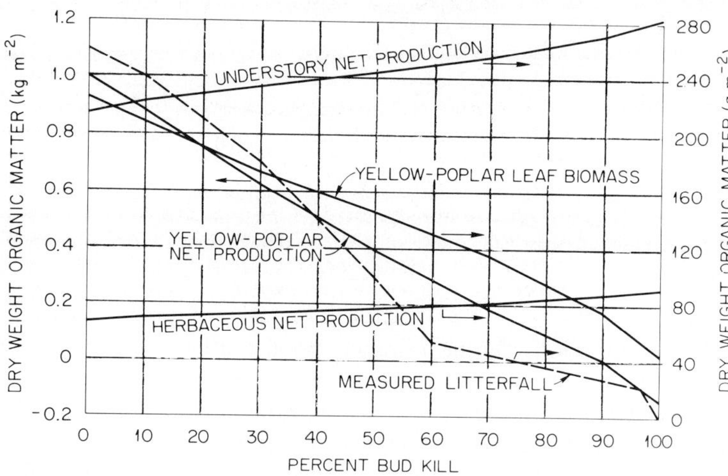

FIG. 13. Simulated and measured effects of bud kill in a mixed deciduous forest. Simulated effects on net production and leaf development are shown by solid lines. Measured effect of bud kill on leaf litter production in an oak-pine forest is shown by the dashed line [based on data of Woodwell and Rebuck (1967), and Woodwell and Marples (1968)]. Bud kill resulted from a range of irradiation levels at varying distances from a gamma radiation source. All data are for the first year following bud kill.

observations. Woodwell and Rebuck (1967) showed that percent bud kill for *Quercus coccinea* increased from 0 to 100% as radiation levels increased from 4 to 150 R day^{-1}. Woodwell and Marples (1968) found that leaffall in the irradiated forest decreased from 250 to 0 g m^{-2} over the same range of increase in radiation level. By combining these results, a curve was constructed of total leaf production in the Brookhaven forest at various levels of bud kill (*Q. coccinea* only) which showed a decrease in leaf production with increased bud kill of the same general form as the present model results (see Fig. 13). Differences may be due to the fact that radiosensitivity of *Q. coccinea* is not typical of the entire forest. *Q. alba* and *Pinus rigida* (Woodwell and Rebuck, 1967), two species which together comprise much of the Brookhaven pine-oak forest, are considerably more radiosensitive. Unfortunately, data on bud kill as a function of irradiation could not be located for either of these species.

Attempts to compare model results with observed results of bud damage due to insect outbreaks were unsuccessful. Rafes (1970) reviewed studies of effects of spring foliage consumption on forest net annual increment. However, none of the studies included measurement of the amount of buds consumed by insects.

Long-term simulation suggested that with 50% yellow-poplar bud kill the forest returns to its previous stage of development after about 5 y; with 90% bud kill some recovery occurs during the second growing season but does not continue thereafter. As carbohydrate reserves continue to be depleted, bud and leaf production decrease again during the third growing season and the yellow-poplar overstory eventually dies.

C. Effect of Selective Thinning on Forest Growth

A 15% thinning of the overstory was simulated by decreasing each of the overstory compartments (see Fig. 1) by 15% on March 1 of the first year. About 2.3 kg m^{-2} was removed from living compartments. As an approximation, 35% of active tissues (0.16 kg m^{-2}) and 25% of inactive tissues (0.46 kg m^{-2}) were introduced into the upper soil layer to simulate root death following cutting. Ten percent of active (0.046 kg m^{-2}) and 20% of inactive tissues (0.36 kg m^{-2}) were added to the standing dead compartment to simulate slash remaining after cutting. The rather large fraction of active tissues added to the mineral soil assumes that during the dormant season most labile carbohydrate is to be found in the roots.

The remainder of material removed from living vegetation compartments (56% or 1.29 kg m^{-2}) was presumed to be removed from the forest as harvest.

The relatively mild thinning regime (15%) was chosen because more drastic manipulations characteristic of silvicultural operations might require changes in model coefficients, functional forms, or even rearrangement of compartments and transfers for accurate simulation.

Results of the simulated thinning (Fig. 14) showed that the yellow-poplar overstory responded more than other overstory species or the understory. Yellow-poplar leaf biomass at the end of the growing season following thinning was 13% less than the value attained without thinning, while the shade-tolerant overstory species decreased only 6%. Herbaceous flora experienced only a temporary increase in biomass following thinning. However, the model did not take into account possible invasion by less shade-tolerant species better able to

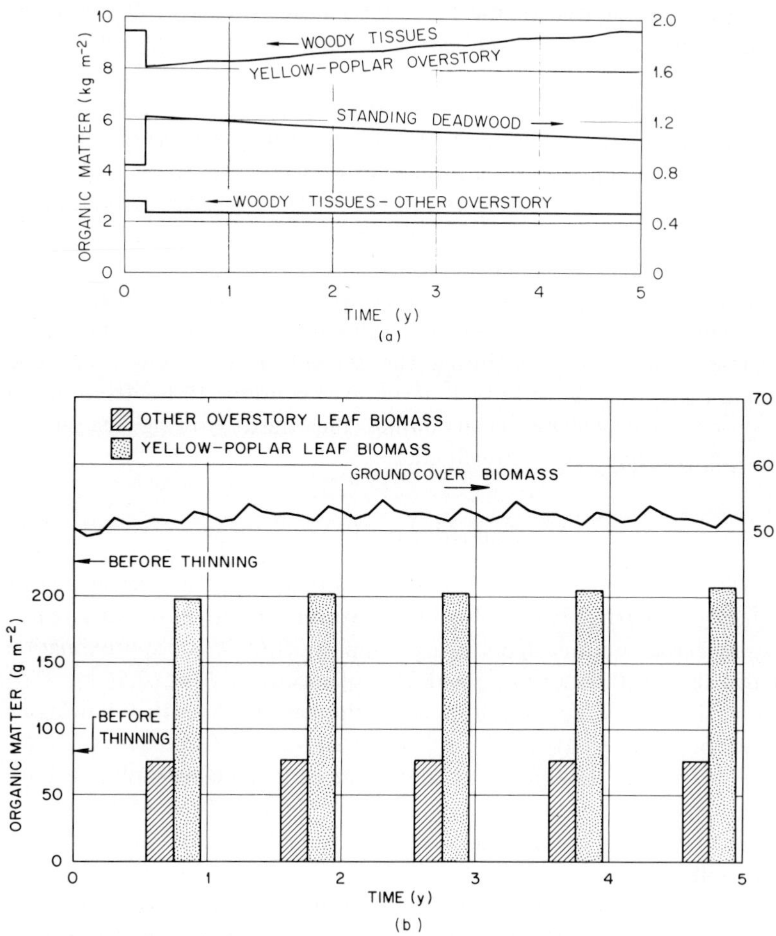

FIG. 14. Simulated response of a mesic deciduous forest ecosystem to 15% thinning of the overstory. Slash was added to the standing deadwood compartment. (a) Behavior of the yellow-poplar and other overstory woody tissue compartments and the standing deadwood compartment. (b) Behavior of the yellow-poplar and other overstory leaf compartments and the groundcover compartment.

take advantage of temporary increase in light levels. Nor did the model allow for stump sprouting which could cause a significant underestimate of the groundcover layer response.

The forest floor showed large increases in both respiration and dry matter accumulation as a result of input of slash and dead roots. If, in the model, litter and soil temperatures were predicted from solar radiation penetrating the canopy, these effects would probably be even more pronounced. Increased light penetration presumably would increase litter temperatures (although probably decreasing litter moisture) and could result in decomposition rates greater than predicted.

D. Effect of Defoliation on Primary Production

In contrast to the previous simulation experiments, where the amount of material in compartments was changed instantaneously at some point in time, defoliation effects were studied by changing a rate constant, the rate of ingestion by canopy consumers. Various rate constants were tried ranging from about 0.06 to 20.0 y^{-1}, resulting in total ingestion of 10–750 g m^{-2} y^{-1}. Consumption rates of all three arboreal vegetation components were varied simultaneously by the same percent change.

During the first year of simulation, active tissue biomass decreased in response to increased insect consumption (Fig. 15). However, woody tissue increment, net primary production (NPP), and leaf production decreased only slightly, even at very high consumption rates (600 g m^{-2} y^{-1} of leaves). Woody tissues were unaffected presumably because of the lag introduced by storage in the active tissue compartment. NPP was unaffected partially because conventionally NPP is defined to include consumption by insects. As long as leaves are regenerated, increased insect feeding would actually increase NPP.

The model predicted nearly complete regeneration of leaves even during severe insect stress (Fig. 15). Rafes (1970) and Franklin (1970) reviewed studies of the effects of insect activity on deciduous forest production and biomass values. Unfortunately, almost all of the studies describe results qualitatively. No studies were located in which amounts of material ingested, biomass of consumers and foliage, and growth rates of vegetation were measured or even guessed simultaneously. To construct realistic models and to test existing ones, sets of such data are needed for various levels of foliage consumption under otherwise similar conditions. Studies must also have been continued for several years. The lack of first-year response to increased ingestion rates predicted by the model may well be accurate. Longer-term simulation suggested that during the second year, although foliage production, woody increment, and NPP continued at the same levels as the first year, active tissues continued to decline.

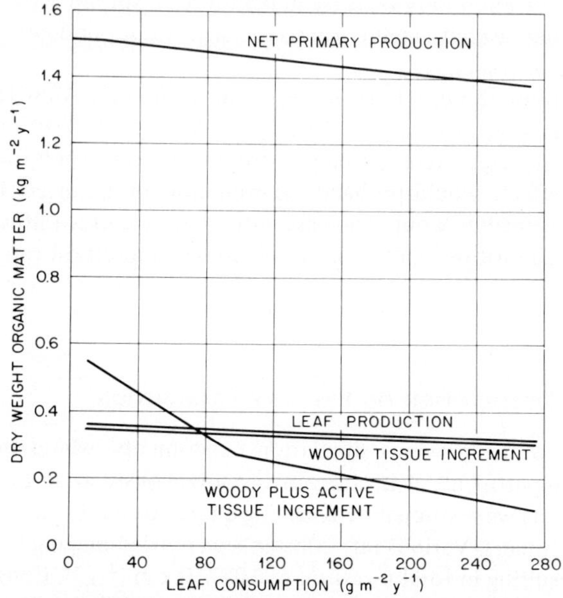

FIG. 15. Simulated effects of increased ingestion levels by canopy consumers on forest net primary production during a 1-y period.

Had longer-term simulations been practical, effects presumably would eventually have appeared.

Although data could not be located for comparison, the prediction that leaf biomass remains unchanged while consumption increases from 10 to 300 g m^{-2} y^{-1} seems unreasonable. A partial explanation of this insensitivity is that the function describing insect consumption is of a form identical to that describing maintenance respiration and translocation to active tissues. The rate constant for consumption is about 0.06 while the constants for translocation to active tissues and respiration sum to about 23.0. Thus, very large changes in the consumption rate constant would be required before any leaf biomass changes would be expected.

Leaf biomass may also be viewed as having a stable steady-state value for a stand of given biomass and canopy structure. The equation for rate of leaf biomass change (see Table V) contains many negative feedback terms which serve to maintain leaf biomass at the stable level at which input balances output. All flows out of the leaf compartments are proportional to leaf biomass; decreases in leaf biomass result in decreased flows out of the leaf compartments. Of flows into leaves, gross photosynthesis decreases nonlinearly with decreasing leaf biomass, but because it decreases more slowly than flows out, leaf biomass tends to return to the stable steady-state value after disturbance. Translocation of carbohydrate to leaves from active tissues was assumed to be inversely related to gross photosynthesis. As leaf biomass decreases, gross photosynthesis decreases and carbohydrate reserves tend to be translocated into the leaves, also tending to replace destroyed leaf material. Perhaps the foliage biomass actually present in the *Liriodendron* forest is a stable value—the forest responds to perturbation by regrowing leaves until the value is once again attained.

Leaf biomass also may be insensitive to insect grazing because the model assumes that leaves destroyed by insects can be replaced as long as light is not limiting. However, many tree species cannot set new buds and develop leaves within the growing season and thus would be unable to replace leaves destroyed by insects. The model also neglects the possibility of nutrient limitations on foliage regeneration. Trees may well be able to replace photosynthate consumed by grazers even at very high levels of consumption. However, protein and pigments may be much more difficult to replace even for species capable of setting new buds and developing new leaves within a single growing season. Assuming that leaves are homogeneous compartments consisting only of carbohydrates and not subject to nutrient limitation may well be an unacceptable oversimplification.

VI. Discussion

We have at this point completed one pass through the entire process of ecosystem analysis—data compilation and analysis, model development, and model application. Rather than continue to refine existing data and the model, we have begun a second iteration based on lessons learned during the first pass. This second pass is producing an improved data base, a more realistic model, independent data sets against which to test the models, and a set of real-life problems to be solved. The days of inventing problems to demonstrate the marvelousness of the model should be numbered.

Many gaps in data reflect processes which cannot be studied with existing techniques. Preliminary simulations suggest that these gaps must be filled before accurate predictions are possible. More measurements of tree diameters and refinements of gas exchange techniques to permit third-digit accuracy will not help as long as processes of equal importance, such as root production and turnover, remain unstudied.

A. Evaluation of the Data Set

After the first year of the Eastern Deciduous Forest Biome Project, root processes were the most critical void in our quantitative understanding of the yellow-poplar forest. Neither their biomass nor turnover rates had been measured satisfactorily, and rates of root respiration were suspect, being based on an inadequate range of root size classes under only a limited range of temperature and moisture. Preliminary experimental results initiated as a result of this study suggest that root turnover could account for as much as 750 g m^{-2} y^{-1} instead of 200 g m^{-2} y^{-1} as estimated previously. No data on temperature and moisture effects on root death or growth could be located.

Associated with lack of data on root processes was the more general lack of information concerning mechanisms of translocation and incorporation of organic matter throughout trees of various species. In fact, existing information on the tree physiology seemed particularly unsuited for model development in an ecosystem context. Many detailed phenomena appeared well understood, such as stomatal control mechanisms, but overall measurements on amounts of carbohydrate produced and transported along various pathways of Fig. 6 appeared to be lacking. Mechanisms regulating the onset of various seasonal phenomena such as leaf emergence, abscission, and root dieback were also not understood.

Even gas exchange processes had not been studied systematically. Detailed information on photosynthesis and respiration was available for few species of deciduous trees (see review by Sollins *et al.,* 1973); almost all work had concentrated on either seedlings or crop plants. Only four studies of woody tissue respiration were located (Møller *et al.,* 1954; Kira, 1968; Woodwell and Botkin, 1970; and Tranquilini and Schütz, 1970), of which only the last two presented respiration data in relation to temperature.

B. Evaluation of the Model

In general, the model performed well. Its deficiencies reflected deficiencies in understanding of mechanisms governing rates of forest processes. These are the same kinds of problems discussed in the preceding section. However, several improvements in the model seem advisable at present, and it is interesting that even a rather uneven model, one in which certain aspects are much better conceptualized than others, permits this sort of evaluation.

The photosynthesis and leaf respiration functions could be simplified tremendously by considering net daytime photosynthesis instead of gross photosynthesis and daytime respiration. Tabulated values of daylength and measured values of solar radiation could then be included easily in the calculations. Also, replacing Eq. (4) with

$$G(x) = bI(x)\, W(x)/[a + I(x)]$$

would simplify the derivation and eliminate the need to modify b when changing light intensity units.

The tendency of the model to postpone wood deposition until fall (Fig. 7) as well as results of the defoliation experiment suggest that the photosynthesis input (less the photosynthesis-dependent respiration) should be added directly to the active tissue compartment (carbohydrate pool) instead of to leaves. This would decrease the lag between onset of net photosynthesis and accumulation of woody material, and also permit distinction between leaf structural material and photosynthate. Photosynthesis could be made a function of leaf structural biomass, and a distinction could be made between insects that feed on photosynthate only (e.g., aphids) and those ingesting structural material as well (e.g., weevils or geometrid moths). The tendency of the model to regenerate leaves during insect outbreaks might be decreased somewhat by interposing the active compartment between the photosynthesis input and the leaf compartments.

A more realistic method of limiting "within season" leaf regeneration after defoliation would be to decrease the maximum bud values (Q_A^*, Q_B^*, and Q_C^*, Table VI) already used in the leaf equations to limit photosynthesis. Species able to regenerate leaves throughout the growing season could be allowed to do so while those unable to set new buds could be constrained from regenerating leaves destroyed by insects or other defoliants.

Preliminary data (Harris *et al.*, in press) gathered after the development of this model suggest that peak root biomass occurs at the beginning and end of the growing season, not just at the beginning as predicted by the model (Fig. 7). The partitioning of root biomass between the O_2 and mineral soil horizons also needs study. Root death was assumed in this model to be an input to the soil organic matter pool only whereas some portion certainly provides input to the O_2. The simulated thinning experiment suggested need for a dead root compartment to introduce lag between tree death and incorporation of roots into O_2 or soil organic pools. Experiments in progress, considering decay rates of roots placed in nylon mesh bags beneath the O_2 layer, should provide data on dynamics of the dead root compartment.

Many other improvements are possible, but most involve coupling of the organic matter model to parallel models of forest water, energy, and nutrient cycling—models not yet in existence. For example, photosynthesis should be made a function of leaf resistance, which could in turn be calculated from soil moisture content. Wood deposition and growth of structural parts of leaves might be made nutrient-dependent and decomposition rates definitely should be. A better function relating decomposition to temperature and moisture could be developed from existing data, but would not be useful without improved data on litter and soil temperature and moisture. However, soil and even litter temperature and moisture could be calculated from precipitation and incident solar radiation data and the light extinction relations used in this study.

In retrospect, the primary production components of the system may have been overemphasized unintentionally both in detail and in realism. Primary producers were divided into 14 compartments, the decomposition part of the system into only six, although total organic matter in the primary producer compartments totalled 17.4 kg m^{-2}, compared to 15.1 for the decomposition compartments. For a better comparison fine roots might be included with decomposers in dividing the ecosystem into functional components. In this scheme the ecosystem would be partitioned into four layers: canopy, fixing carbon and energy; forest floor, taking up water and nutrients; a massive but functionally unimportant layer of structural and conductive tissue interposed between; and a layer of dead, relatively unavailable organic matter in the soil below the zone of root uptake. Fine roots, mycorrhizal flora, and free-living organisms function as a coherent unit converting organic matter (and a small amount of inorganic matter added from the atmosphere and parent bedrock) into a form suitable for rapid utilization by the entire system at the start of the growing season. Alternative models should be constructed which better reflect this inherent structure, and their behavior studied under a variety of conditions.

C. Evaluation of the Approach

The decision to study physiology of the forest with a material balance model rather than to simulate tree growth empirically appears reasonable. Box and arrow diagrams provided a logical and systematic framework for displaying available data, for identifying voids in the data base, for approximating missing values or calculating them by assuming conservation of mass, and for insuring that important processes were not neglected. The box and arrow approach proved less prone to error and confusion than more artistic representations (e.g., Duvigneaud and Denaeyer-DeSmet, 1970), and more amenable to biological interpretation and model conceptualization than a matrix representation (e.g., Olson, 1965), although the matrix approach may in some cases permit greater insight into the mathematical nature of the system.

Although other approaches, for example tree-by-tree simulation (e.g., Botkin *et al.*, 1970; Pienaar and Turnbull, 1973; Stage, 1973), may be advantageous when the goal is to study spatial variation within the forest, a physiological approach is still needed if causal relations are to be studied. Effects of disturbances such as defoliation, bud kill, and stand thinning are propagated through the entire ecosystem through causal chains which are understood poorly. Particularly when mechanisms are not well understood, it is essential that we incorporate a minimum of assumptions into the analysis. This means, for example, that no system component should be treated in more detail or more mechanistically than any other else one make a priori assumptions about the contribution of that component to the ecosystem response.

As an example, tree-by-tree simulation begins by assuming that differences between individual trees are important; this immediately introduces considerable, and perhaps unnecessary, complexity into the simulation. We prefer to portray the ecosystem as a functional unit, modeling primary and secondary production and decomposition processes before concentrating on differences between individual primary producers.

Using sensitivity analysis we can then identify factors critical in predicting the effect of some disturbance and improve our data base and equations with respect to them while hopefully simplifying the model with respect to less critical processes. If, for example, initial simulation studies of a particular problem in primary production indicate lack of any effect of decomposition processes on primary production, the decomposition compartments can be ignored and incorporation of differences between individual primary producers might be considered. But, such a step should be considered a posteriori after experimentation with simple ecosystem models, and not decided a priori after only initial examination of objectives, premises, and available data.

The physiological approach also makes possible coupling of an organic matter model such as this with parallel models of water and nutrient flux through the system. For example, in addition to making photosynthesis and leaf respiration functions of leaf resistance and calculating this from soil moisture content, transpiration could be made a function of leaf resistance, leaf biomass, and perhaps root biomass. Interception could also be related functionally to leaf biomass. As part of a nutrient cycle model, leaf nutrient concentrations could be calculated and used to predict leaf gas exchange efficiency and perhaps stomatal control. Uptake by the fine root–mycorrhizal compartment could be made a function of its biomass, nutrient concentration in the soil solution and perhaps transpiration rate. Alternatively, uptake could be considered independent of transpiration and, instead, a plant nutrient pool could be imagined with redistribution from fine roots to this pool dependent on transpiration. The possibilities are both exciting and staggering.

Now that we have completed an initial ecosystem model of organic matter, the highest priority should be given to development of initial nutrient (particularly nitrogen), water, and energy models rather than to continued refinement of the organic matter model. The coupling of these models through mutually dependent rate expressions has not been attempted often in the past and may present mathematical and programming problems of considerable challenge. These problems should be considered before large amounts of time and money are committed in the development of detailed single substance models.

This study, like its predecessors, of necessity gave insufficient attention to the testing and application of ecosystem models. Models are not an end in themselves. They can be a valuable tool for presenting and testing hypotheses. Once validated, their accuracy assessed, they can play a vital role in rational management of limited natural resources. Unfortunately, it is much easier to postulate models than to evaluate their parameters or collect data to test them. Much tedious data collection is still needed and new techniques must also be invented. Even after practical and accurate models are available, the difficult tasks will remain of devising alternate resource management strategies to test with the models, and of translating recommendations into actions.

VII. Summary

A conceptual model was first formulated to aid in data collection and analysis. Data were compiled on standing crops of the various compartments, their annual increments and patterns of change through the years, and various transfers. Local data were used whenever available, but much data had to be obtained from literature and unpublished sources. Average values for the driving variables, air temperature and solar radiation were compiled in tabular form.

A budget was calculated describing the production, allocation, and decomposition of organic matter throughout the ecosystem (Fig. 1). This facilitated model construction and permitted instructive comparisons of total annual transfers. Comparisons between estimates of annual gross production based primarily on harvest data (4.46 kg CO_2 $m^{-2} y^{-1}$) and primarily on gas exchange data (3.33 kg CO_2 m^{-2} y^{-1}) eventually may provide a measure of the role of root uptake in organic matter accumulation. The annual input of organic matter to the forest floor, primarily via litterfall and root death (0.76 kg DW m^{-2}), was lower than estimates of forest floor respiration plus soil organic matter accumulation (1.34 kg m^{-2}). Presumably, the contribution of root death and excretion of organic material was again underestimated (see more recent data by Reichle *et al.*, 1972).

Metabolic and structural parameters for the *Liriodendron* forest were compared to values for 28 deciduous forest stands throughout the world. The local stand showed slightly higher net ecosystem productivity (NEP = 0.50 kg DW m^{-2} y^{-1}) for its age (50 y) and standing crop (16.1 kg DW m^{-2}), although differences may be due to use of sampling techniques more accurate and detailed than those used previously.

A model of the *Liriodendron*-dominated ecosystem was constructed by considering the physiological mechanisms responsible for production, distribution, and decomposition of organic matter. In some cases, a plausible mathematical derivation of a function describing the process was possible (e.g., gross production as a function of incident solar radiation and leaf biomass). In many cases (e.g., translocation or root death), lack of data permitted only *ad hoc* (though perhaps provocative) postulates of functional forms. Parameter values not based on local measurements were adjusted by trial and error until annual transfers and compartment increments matched budget calculations.

Although data were lacking for model validation, model behavior was studied under a variety of quiescent and disturbed conditions and was compared to published descriptions of system behavior. Some model predictions, such as the simulated pattern of ecosystem CO_2 exchange and effects of various levels of bud kill, seemed very realistic when compared with patterns reported in the literature. Others, such as the simulated defoliation experiment, produced unexpected and perhaps incorrect results. In the defoliation example, although effects appeared to be simulated correctly, defoliation itself required very high levels of consumer activity. The foliage biomass actually present in the forest seemed, from the simulation, to represent a stable value. The model responded to attempts at decreasing foliage by replacing it from current photosynthate or from carbohydrate reserves.

Even in cases of unreasonable model behavior, simple changes in the arrangement of compartments and flows, or nature of the functions describing the flows, seemed to offer solutions. The basic assumption that the forest could be modeled as a homogeneous system on a unit area basis appeared justified.

Not only were the simulation experiments instructive and reasonably successful, but alternatives to the unit area approach appeared to offer disadvantages when responses to be expected were only poorly understood.

Although a proper ecosystem analysis and simulation should have considered nutrient and water cycles as well as the flow of organic matter, time and manpower limitations precluded this. Work in progress at Oak Ridge IBP site (see Henderson *et al.*, 1972) is directed toward the development, validation, and application of coupled models of organic matter, water, nutrients, and energy in a deciduous forest ecosystem. When complete, these should permit simulation of effects of fertilization on forest growth and nutrient yield in runoff, problems which clearly cannot be studied with an organic matter model alone.

Acknowledgments

A synthesis study such as this draws upon ideas and data, much of it unpublished, of many people. We wish particularly to thank D. E. Reichle, site coordinator during this study, B. E. Dinger, B. A. Hutchison, C. E. Francis, J. F. McBrayer, and B. S. Ausmus (all of ORNL) for sharing with us their ideas and data. J. S. Olson (ORNL) provided unlimited access to data gathered during the ^{137}Cs experiment conducted earlier on the same study plot. The IBP Information Center at ORNL helped with much of the data processing and in particular were responsible for preparation of Table II.

This research is based on a dissertation submitted by the senior author to the Department of Botany, University of Tennessee, in partial fulfillment of the requirements for the degree of Doctor of Philosophy.

Research supported in part by the U.S. Energy Research and Development Agency under contract with the Union Carbide Corporation, and in part by the Deciduous Forest Biome Project, International Biological Program, funded by the National Science Foundation under Interagency Agreement AG-199, 40-193-69 with the Energy Research and Development Agency, Oak Ridge National Laboratory. This is Contribution No. 94 from the Eastern Deciduous Forest Biome U.S. IBP.

REFERENCES

Andersson, F. (1970). *Bot. Notiser* **123,** 8.

Art, H. W., and Marks, P. L. (1971). *In* "Forest Biomass Studies" (H. E. Young, ed.), pp. 2–32. Univ. Maine Orono, Maine.

Auerbach, S. I., Olson J. S., and Waller, H. D. (1964). *Nature (London)* **201,** 761.

Bate, G. C., and Canvin, D. T. (1972). *Can. J. Bot.* **50,** 205.

Botkin, D. B., Janak, J. F., and Wallis, J. R. (1970). The Rationale, Limitations, and Assumptions of a Northeast Forest Growth Simulator. IBM Res. Rep. RC 3188 (#14604), IBM Thomas J. Watson Res. Center, Yorktown Heights, New York.

Botkin, D. B., Woodwell, G. M., and Tempel, N. (1970). *Ecology* **51,** 1057.

Dinger, B. E. (coord.) (1972). Gaseous Exchange, Forest Canopy, Meteorology. Ecol. Sci. Div. Ann. Progr. Rep. (1971), USAEC Rep. No. ORNL-4759, pp. 49–56. Oak Ridge Nat. Lab., Oak Ridge, Tennessee.

Duvigneaud, P., and Denaeyer-DeSmet, S. (1970). *In* "Analysis of Temperate Forest Ecosystems" (D. E. Reichle, ed.), pp. 199–225. Springer-Verlag, Berlin and New York.

Edwards, N. T., and Sollins, P. (1973). *Ecology* **54,** 406.

Franklin, R. T. (1970). *In* "Analysis of Temperate Forest Ecosystems" (D. E. Reichle, ed.), pp. 86–89. Springer-Verlag, Berlin and New York.

Goldstein, R. A., and Harris, W. F. (1973). SERENDIPITY–A Watershed Level Simulation Model of Tree Biomass Dynamics, *Proc. 1973 Summer Simulat. Conf., July 17–19, Montreal, Canada.*

Harris, W. F. (coord.) (1972). Terrestrial Primary Production. Ecol. Sci. Div. Ann. Progr. Rep. (1971), USAEC Rep. No. ORNL-4759, pp. 56–68. Oak Ridge Nat. Lab., Oak Ridge, Tennessee.

Harris, W. F., Sollins, P., Edwards, N. T., Dinger, B. E., and Shugart, H. H. (1975). Analysis of Carbon Flow and Productivity in a Temperate Deciduous Forest, *Symp. Productivity World Ecosyst., Proc. IBP V General Assembly,* pp. 116–122. National Academy of Sciences, Washington, D.C.

Harris, W. F., Kinerson, R. S., and Edwards, N. T. (in press). Comparison of Belowground Biomass of Natural Deciduous Forests and Loblolly Pine Plantations. *In* The Belowground Ecosystem: A Synthesis of Plant Association Processes, September 5–7, 1973, Ft. Collins, Colorado.

Henderson, G. S. (coord.) (1972). Walker Branch Watershed: A Study of Terrestrial and Aquatic System Interaction, Ecol. Sci. Div. Ann. Progr. Rep. (1971), USAEC Rep. No. ORNL-4759, pp. 30–48. Oak Ridge Nat. Lab., Oak Ridge, Tennessee.

Horn, H. S. (1971). "The Adaptive Geometry of Trees." Princeton Univ. Press, Princeton, New Jersey.

Hozumi, K., Kirita, H., and Nishioka, M. (1972). *Photosynthetica* **6,** 158.

Jordan, C. F. (1971). *Amer. Sci.* **59,** 425.

Kramer, P. J., and Kozlowski, T. T. (1960). "Physiology of Trees." McGraw-Hill, New York.

Kira, T. (1969). *In* "Functioning of Terrestrial Ecosystems at the Primary Production Level" (F. E. Eckardt, ed.), pp. 399–407. UNESCO, Paris.

Kira, T., and Shidei, T. (1967). *Jap. J. Ecol.* **17**, 70.

Kira, T., Shinozaki, K., and Hozumi, K. (1969). *Plant Cell Physiol.* **10**, 129.

Leary, R. A. (1970). System Identification Principles in Studies of Forest Dynamics. U.S. Forest Serv. Res. Paper NC-45, North Central Forest Exp. Station, St. Paul, Minnesota.

Møller, C. M., Muller, D., and Nielsen, J. (1954). *Det Forstlige Forsøgsvaesen i Danmark* **21**, 273.

Monsi, M. (1968). *In* "Functioning of Terrestrial Ecosystems at the Primary Production Level" (F. E. Eckardt, ed.), pp. 131–148. UNESCO, Paris.

Monsi, M., and Saeki, T. (1953). *Jap. J. Bot.* **14**, 22.

Moser, J. W., and Hall, O. F. (1969). *Forest Sci.* **15**, 183.

Odum, H. T., and Pidgeon, R. F. (1970). A Tropical Rain Forest: A Study of Irradiation and Ecology at El Verde Puerto Rico. USAEC Rep. No. TID-24270, Nat. Tech. Inform. Serv. Springfield, Virginia.

Olson, J. S. (1964). *J. Ecol.* **62**, 99.

Olson, J. S. (1965). *Health Phys.* **11**, 1385.

Olson, J. S., and Crossley, D. A. (1963). *In* "Radioecology" (*Proc. Nat. Symp. Radioecol., 1st*) (V. Schultz and A. W. Klement, eds.), pp. 411–416. Van Nostrand-Reinhold, Princeton, New Jersey, and AIBS, Washington, D.C.

Patten, B. C., and Witkamp, M. (1967). *Ecology* **48**, 813.

Peters, L. N., Olson, J. S., and Anderson, R. M. (1969). Eight-Year Summary of Plant–Soil Data on Cesium-137 in a Tagged Appalachian Forest Dominated by *Liriodendron tulipifera*. USAEC Rep. No. ORNL-TM-2737, Oak Ridge Nat. Lab., Oak Ridge, Tennessee.

Pienaar, L. V., and Turnbull, K. J. (1973). *Forest Sci.* **19**, 2.

Rafes, P. M. (1970). *In* "Analysis of Temperate Forest Ecosystems," pp. 100–106. Springer-Verlag, Berlin and New York.

Reichle, D. E., Crossley, D. A., Edwards, C. A., McBrayer, J. F. and Sollins, P. (1972). *In* "Radionuclides in Ecosystems" (*Proc. Nat. Symp. Radioecol., 3rd*) (D. J. Nelson, ed.). USAEC Conf. 710501, Nat. Tech. Inf. Serv., Springfield, Virginia.

Reichle, D. E., Dinger, B. E., Edwards, N. T., Harris, W. F., and Sollins, P. (1973a). *In* "Carbon in the Biosphere" (G. M. Woodwell, ed.), pp. 345–365. USAEC Conf. 720510. Nat. Tech. Inf. Serv., Springfield, Virginia.

Reichle, D. E., O'Neill, R. V., Kaye, S. V., Sollins, P., and Booth, R. S. (1973b). *Oikos* **24**, 337.

Reichle, D. E., Goldstein, R. A., Van Hook, R. I., and Dodson, G. J. (1973c). *Ecology* **54**, 1076.

Satoo, T. (1970). *In* "Analysis of Temperate Forest Ecosystems" (D. E. Reichle, ed.), pp. 55–71. Springer-Verlag, Berlin and New York.

Soil Survey Staff (1962). "Soil Survey Manual," U.S. Dept. Agric. Handbook No. 18. U.S. Govt. Printing Office, Washington D.C.

Sollins, P., and Anderson, R. M. (eds.) (1971). Dry-Weight and Other Data for Trees and Woody Shrubs of the Southwestern United States. USAEC Rep. No. ORNL-IBP-71-6, Oak Ridge Nat. Lab., Oak Ridge, Tennessee.

Sollins, P., Reichle, D. E., and Olson, J. S. (1973). Organic Matter Budget and Model for a Southern Appalachian *Liriodendron* Forest. USAEC Rep. No. EDFB-IBP-73-2, Oak Ridge Nat. Lab., Oak Ridge, Tennessee.

Stage, A. R. (1973). Prognosis Model for Stand Development. U.S. Forest Serv. Res. Paper INT-137. Intermountain Forest and Range Exp. Sta., Ogden, Utah.

Tadaki, Y. (1966). *Bull. Forest Exp. Station Tokyo* **184**, 135.

Taylor, F. A. (1974). In "Phenology and Seasonality Modeling" (H. Leith and F. Stearns, eds.) pp. 237–254. Springer-Verlag, Berlin and New York.

Thornthwaite, C. W. (1948). *Geogr. Rev.* **38**, 55.
Tranquilini, Von W., and Schütz, W. (1970). *Cbl. Ges. Forstwesen* **87**, 42.
Whittaker, R. H. (1966). *Ecology* **47**, 103.
Wiant, H. V. (1967). *J. Forest.* **65**, 408.
Witherspoon, J. P. (1969). *In* "Symposium on Radioecology" (*Proc. Nat. Symp. Radioecol. 2nd*) (D. J. Nelson and F. C. Evans, eds.), pp. 120–126. USAEC Conf 670503, Nat Tech. Inform. Serv., Springfield, Virginia.
Witkamp, M., and Olson, J. S. (1963). *Oikos* **14**, 138.
Woodwell, G. M., and Botkin, D. B. (1970). *In* "Analysis of Temperate Forest Ecosystems" (D. E. Reichle, ed.), pp. 73–85. Springer-Verlag, Berlin and New York.
Woodwell, G. M., and Dykeman, W. R. (1966). *Science* **154**, 1031.
Woodwell, G. M., and Marples, T. G. (1968). *Ecology* **49**, 456.
Woodwell, G. M., and Rebuck, A. L. (1967). *Ecol. Monogr.* **37**, 53.

PART III

MODELS OF HUMAN ECOSYSTEMS

The three chapters of this section treat several different aspects of and approaches to modeling ecosystems with man at the center of focus.

Chapter 7 by Dr. Koenig describes a "sociocybernetic" approach to human ecosystem design and management. The author asks what options exist for dealing with ecological and related sociological problems, and what is the scientist's role? Computer simulation models on regional scales have served to organize multidisciplinary research activities as a needed first step, but to make them more useful for management and decision making, a consistent theoretical and conceptual framework is needed. This chapter offers such a framework in a model of how industrialized man relates to the natural environment. Problems of technological planning and regional economic development are presented in a cybernetic perspective to identify states to be avoided and means of avoiding them. The model relating man to physical resources and natural environment consists of a life-support system with energy generating and material processing activities, and a cybernetic control system of institutions for social regulation. The former is a "template" to which the latter adaptively adjusts. Two modes of life-support behavior are short-term mass-energy dynamics, and long-term successional changes in technological and physical structure. The planning and development problem is to control both. The author's approach is one of design: to match social mechanisms to physical/ecological constraints. Dynamic characteristics of the controlled object dictate those of decision processes, including allowable time delays. Virtually all social control instruments were formed without an ecological perspective at a time when resources were not limiting and regional environments not overtaxed. Modifications now must be made and adjustments are in progress. Questions of who bears the costs of such adjustments and at what rates lie outside the domain of science. But science can provide models to be employed according to principles of on-line control to aid the political decision processes aimed at bringing man's life-support activities into harmony with his natural environment as a closed system.

Chapter 8 by the editor emeritus of *Simulation* continues a long-time advocacy of simulation as a principal means of understanding and predicting dynamics of complex systems. Long concerned with physical and engineering systems, John McLeod in recent years has focussed on social, economic, and ecological systems as problems with these materialized. In this chapter he considers simulating the world. In 1969, while he was forming the World Simulation Organization to do this, J. W. Forrester was beginning the Club of Rome projects which led to *World Dynamics* and *The Limits to Growth*. The latter work became famous as computerized verification of the consequences of human population trends, and infamous for its simplicity, untenable assumptions, and inadequate data base. Criticisms of the model are enumerated and an assessment of its ultimate impact offered. The author holds that it drew needed attention to the problem and this is sufficient to justify the model even if it were not valid. World simulation models may not be capable of predicting what will happen, but they can project what may happen given the present course of events. "It is characteristic of simulation prophecies," says McLeod, "that they are self-defeating. Just as surely as it is predicted that something will happen, it is almost certain that someone will take steps to see that it does not."

Chapter 9 by Dr. Odum prescribes small-scale models of large-scale systems of man and nature. Complexity is in the way of understanding, and must be reduced if man is to see the essence of his interactions with natural ecosystems. "Macroscopic minimodels" are one way to reduce complexity; models are the means for the scientist to form impressionistic views of reality. A number of analog computer-simulated models are described to illustrate the author's approach to ecological impressionism. His method involves energy circuit diagrams, now well familiar in ecological literature. This pictorial language forces change from reductionist to synthetic thinking. In analog computer simulation, a number of shortcuts have been devised to save time, decrease errors, and increase facility of interaction with minimodels. Among these, methods for quick scaling, and debugging and explanation are described. Nine minimodels of human ecology are discussed by way of illustration. The chapter concludes with a comment on the need for symbolic languages in ecology to depict man's complex relationships to nature, abstractly but with scientific accuracy.

7

Human Ecosystem Design and Management: A Sociocybernetic Approach

HERMAN E. KOENIG

DEPARTMENT OF ELECTRICAL ENGINEERING AND SYSTEMS SCIENCE
MICHIGAN STATE UNIVERSITY, EAST LANSING, MICHIGAN

I. Introduction

The ecological and related sociological problems (including the energy crisis and the "urban crisis") facing developed nations, and indeed, the whole world, need little elaboration. These are the remaining questions: what are the options for dealing with their many dimensions and how are these options to be effected, and what are the potential contributions of the scientific community and how are these contributions to be given their full weight without infringing on those dimensions that are intrinsically extrascientific?

Many hopes have been invested in large-scale, empirically based computer simulation models of the dynamic ecological, sociological, and economic factors of major regions, both as a mechanism for organizing multidisciplinary research activities and as a critical first step in identifying alternative solutions. It is increasingly evident, however, that such large-scale simulations are in themselves of limited value in dealing with many of the most important aspects of our ecological and sociological problems. In themselves they provide little in the way of an organizational framework that is scientifically useful.

Reassessment of the problem reveals that, where simulation models have been useful as tools in the design and management (control) of physical processes, they depend critically upon a logically consistent theoretical and conceptual framework to give them form and substance and to provide a basis for decomposing the overall system into manageable subproblems.

So it would also appear that in ecological and sociological problems a framework of theoretical principles and constructs is a prerequisite both to identifying meaningful subquestions and to giving substance to the conceptual, mathematical, and computer simulation models we construct to deal with these subquestions (at appropriate levels of functional and spatial organization). This chapter presents the basic conceptual features of a sociocybernetic model or paradigm which characterizes the essential processes by which industrialized man relates to his physical resources and natural environment. The problems of technological planning and regional economic development are presented as fundamental problems in the control of the short-term mass–energy and long-term structural dynamics of a physical–biological system (the human life-support system) by collective man (i.e., by society). From this cybernetic perspective the crucial questions are these: what family of states are to be avoided and, given this information, what kind of social instruments are required to carry out the essential functions of control and how shall they be deployed? The answers to these questions involve a mixture of purely technical matters of design and management and matters of collective social judgment which are intrinsically a part of the sociocybernetic control process. The spectrum of systems–theoretic problems required to deal effectively with the technical matters are discussed along with some of the emerging principles of structural design and sociocybernetic regulation.

These concepts and principles have proven useful to the author and his associates in identifying, coordinating, and managing multidisciplinary scientific efforts toward the solution of important subproblems involved in bringing the life-support activities of industrialized nations into harmony with the natural environment.

II. A Sociocybernetic Perspective

The structural features of a model characterizing processes by which industrialized man relates to his physical resources and natural environment are illustrated in Fig. 1. This model consists of two major subsystems: (i) a life-support system with its physical energy generation and material processing activities, and the associated material and energy exchange rates with the

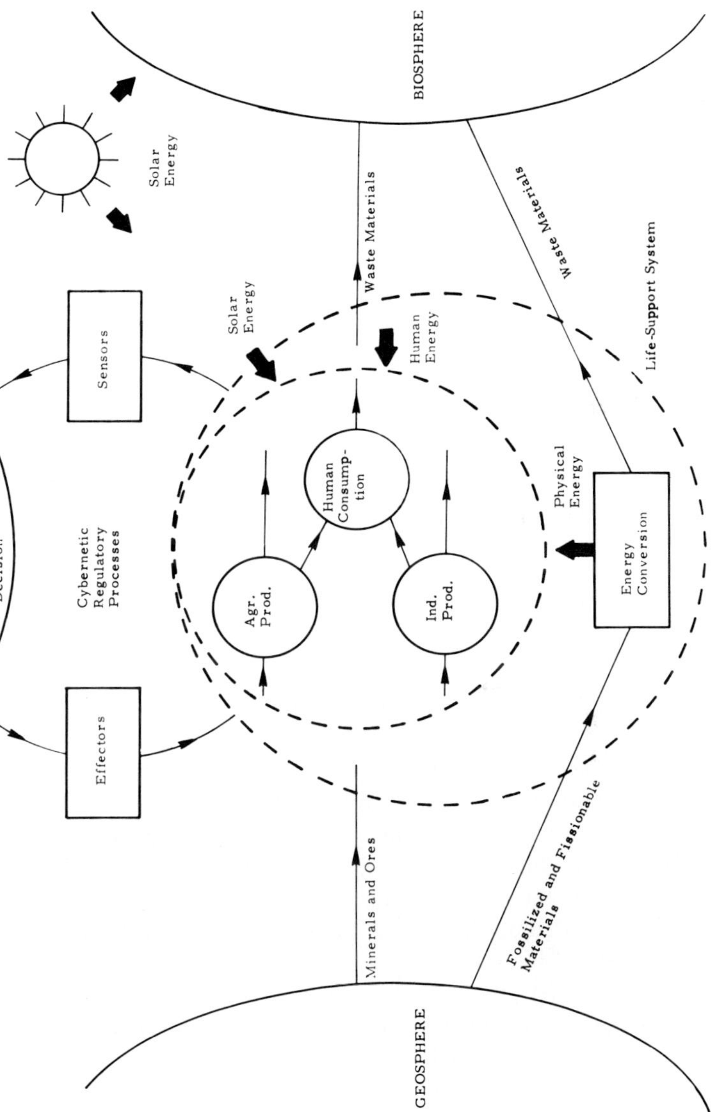

FIG. 1. Cybernetic control of human life-support system by decision-making institutions.

geosphere and biosphere, and (ii) a cybernetic regulatory and control system with its economic, political, legislative, religious, educational, and other institutions of social regulation. The basic structural and functional features of each of these subsystems and their interaction can be summarized as follows (Koenig *et al.*, 1972; Koenig and Tummala, 1972):

The essence of the life-support system is a set of interconnected transformations on the structural state of materials, their spatial transportation, and their physical and biological storage. Physical, solar, and human forms of energy are required to carry out each of the processes in industry and agriculture according to known physical and biological laws that, in principle, are known as a function of the technology used. The laws of material and energy balance govern interaction of the components as a closed ecological process.

Each component of the natural environment (each lake, stream, airshed, coastal region, and so forth) has a limited capacity to accept identifiable classes of materials and energy discharged as wastes from the life-support processes without altering its physical and biological state beyond preassigned limits. These processing capacities are *regionally specific* and they depend upon the "quality" of the environmental component to be maintained, quality being a subjective judgment based on the physical and biological state of the environmental component.

The "effectiveness" of the life-support system in providing the physical needs and wants of man is determined ultimately by the *mass–energy* characteristics of the system, the availability of the requisite forms of energy to drive it, and the temporal stability and reliability of the functional system as a whole. Clearly man lives by materials and energy, not dollars.

The physical and technological structure* of the life-support system can be viewed as a "template" to which social, cultural, and service activities adjust through homeostatic or self-adaptive processes. To this extent, the potential for individual human and cultural fulfillment is deeply rooted in the physical and technological structure of the system. Indeed, the physical structure of the life-support system is the "house" in which man lives, works, and recreates on the surface of the earth. In this respect many of its physical features can and must be evaluated on the basis of subjective judgments or social preferences.

Two general modes of dynamic behavior of the life-support system must be distinguished: (i) the short-term *mass–energy dynamics,* and (ii) long-term *successional changes in its technological and physical structure.* The first mode of behavior has to do with temporal and spatial changes in the flow rates of material and energy for an essentially fixed physical structure. The second mode

* Physical structure includes all aspects of technological form and spatial distribution and location of the production processes, transportation, and human habitat. The degree of spatial centralization and specialization of processes is an aspect of physical structure of particular concern.

deals with temporal changes in the technological form and physical structure, including the development of capital production facilities. The two modes can be distinguished, in part, by time constants that differ by orders of magnitude. Both models are regulated by a variety of man-made social instruments of control, including in particular the economic system.

In principle, all social decision-making processes should act on the information provided by the "sensors", information from which the dynamic *states* of the life-support system can be computed or inferred. Effective and flexible sociopolitical mechanisms must contain a variety of separate but coordinated decision-making mechanisms with a corresponding spectrum of time constants and effectors for transmitting instructions of the decision-making mechanism back to the life-support system. The specific mechanisms selected depend on the choice of behavioral variables in the life-support system to which the effectors are to be coupled, and through which control is to be exerted. The choice of sociocybernetic control loops and associated couplings to the life-support system is, in part, a technical matter governed by objective dynamical properties of the controlled system (the life-support system), and in part a matter for social, economic, and political judgments. Certain technical performance features, such as time delays, reliability, and compatibility with the controlled system, must be built into the design of the control loops, but the decision mechanisms need not, and indeed *cannot,* be characterized in advance.

From this sociocybernetic perspective, the problem of technological planning and regional economic development is essentially that of controlling the short-term mass–energy *and* long-term structural dynamics of a *physical* system (the life-support system) by collective man (i.e., by society). These are the crucial questions:

1. What family of states of the controlled system are most desirable? Alternatively, what family of states are to be avoided?
2. Given this information, what kinds of social instruments are required to carry out the essential functions of control and how shall they be implemented?

In both of these questions we are as much, or more, interested in "what might be" than simply "what is," and they involve a mixture of strictly technical matters, which may be decided purely on the grounds of science and technology, and matters that are intrinsically extrascientific. In this sense they are questions of *design* rather than questions in the simulation of an existing system or projecting present trends.

One of the central objectives in any modeling effort based on this perspective is to separate carefully "mechanistic" aspects of the system from all decision-making aspects, so that purely technical, scientific analyses may be given their full weight without encroaching on those aspects that fall outside their domain. Stated in another way, the objectives are to provide (by redesign, if necessary) effective social mechanisms and institutions for (i) collective social control or management of regional development, and (ii) sound characterizations of the full range of alternatives or options in the physical structure of the life-support system that exist within the physical and ecological laws that govern existence.

In general, each of the social decision units relates to a particular spatial and/or functional subunit of the total life-support system as an object of control. The set of such sociocybernetic control loops must then be coordinated between levels of organization and across spatial and functional subsystems in a way that is not as yet well-understood. One dimension of the interaction (and hopefully coordination) clearly takes place through interactions between the objects of control as components of the life-support system. Indeed, control actions taken in any given control loop, say loop A, are determined by the states of object A. But the states of object A are also conditioned by interactions between object A and other components of the life-support system, objects B, C, D, and so forth. These interactions are in turn conditioned by responses of objects B, C, D, and so forth, to their respective controllers. One of the most interesting systems theoretic questions is, how shall such a system be organized so as to be consistent and stable as a "goal-seeking" system?

Models that effectively deal with such questions must be structured so as to separate the dynamics of the controlled object from the decision-making functions. Indeed, dynamic characteristics of the controlled object dictate virtually all the technical characteristics of the sensors and effectors and the allowable time delays that can be tolerated in the decision-making processes. As an expedient one may wish to simulate the time delay, reliability, and other performance characteristics of existing or proposed control instruments under a variety of decision alternatives. But the particular decision that will be made under a given set of contingencies need not and *cannot* be characterized in advance. A more explicit conceptualization of the structure of major cybernetic control loops in relation to the controlled object as they currently exist in the present system would in itself be a major contribution.

Since there are many controllers, we need not one model of the controlled object but many models, each of which serves as a "feed forward" mechanism in the context of decisions to be made in any given control loop. Further, as discussed in a later section, the range of available alternatives in the physical structure of the life-support system alone extends beyond what conceivably can be included in any one computer program or input–output format. Systems as simple (relatively speaking) as the electrical power generation and distribution system, the telephone communication system, or indeed the computer itself, have never been simulated in total. Yet these systems have been designed, built, and managed to operate effectively and reliably as a system of interrelated cybernetic control loops. So it would also seem that our goal in regional planning is to design a functional sociocybernetic system for management of the short-term mass–energy and long-term structural dynamics of the regional life-support system. Like the computer, the electrical power system, and the communication system, it defies total simulation; it is, in effect, its own computer.

We must look at the overall life-support system from a variety of perspectives and levels of organization, each with its own degree of resolution, time horizons, and specific questions. At one end of this spectrum we have theoretical studies of the general mass–energy and economic characteristics of the life-support system as an abstract material processing "machine" constrained in its design and operation by physical and ecological dynamics of the environment. At the other end we have more specific in-depth problems dealing with quantitative analysis and design of the detailed physical and technological structures of an agricultural or industrial production process, an urban community, or a lake or stream as real-world components of the total life-support system. Examples of such in-depth analysis and design efforts are cited in later sections. It is largely through such efforts that the "cutting edge" of action is generated. But developments in sound theoretical foundations and basic principles of overall design and management of the total life-support system as a closed ecosystem* are the essential elements of the evolving science which must direct, integrate, and coordinate these more detailed cutting-edge activities into something more than a random set of disjoint activities. It is the key to divided but coordinated individual efforts that ultimately leads to a stable, functioning sociocybernetic system. Some basic principles have already begun to emerge. They are presented in the sections following, along with a discussion of the perspective and direction they provide to the more detailed lower-level studies of important subquestions.

III. Ecological Constraints

As a prerequisite to sound ecological planning we must know explicitly the time–space capacities of the various components of environment (lakes, streams, coastal regions, airsheds, etc.) to process specific classes of materials (heavy metals, oxides of sulfur, organic compounds, etc.) as a function of the "quality" (ecological state) to be maintained in the environmental component. What, for example, are acceptable input rates of organic and inorganic nitrogen and phosphorus to particular lakes and streams, or the acceptable rate of application of urban wastes (spray irrigation) to terrestrial areas as a function of soil type, plant communities, climate, and desired ecological state?

* The word ecosystem is used here in a somewhat more inclusive sense than usual, to refer to man *with* his natural environment.

These mass–energy processing capacities stand as unremovable physical constraints on regional developments, population densities and distributions, and technological activity. The family of structural states that are not ecologically feasible in this sense must be avoided. Models developed to answer such questions must be able to predict future states of systems that are subjected to input levels not yet experienced by the system. Experiences in current efforts by the author and his colleagues to answer questions of this general class for fresh-water lakes and streams and terrestrial regions indicate that models based on statistical data from natural unstressed systems are of limited value, unless they are in fact based on behavioral laws of the basic structural components of the system and their associated laws of interaction. Unfortunately, it is not yet clear that a complete scientific base from which such models can be developed is available. It is clear, however, that research directed at answering such questions should lead ultimately to the identification of material and energy rate capacities of each major aquatic and terrestrial component of the region as a function of the ecological state that the planner might wish to implement in response to social preferences. In principle, these rate capacities must be regarded by the planner as limited resources of the region in much the same way that mineral deposits and forests are regarded as resources.

IV. Physical and Technological Structure of the Life-Support System

The problem of identifying states of the life-support system that are to be avoided for reasons of long-term incompatibility with the environment is fundamentally a problem in the analysis of mass–energy characteristics of alternative technological and physical structures. Such analysis must identify classes of structures for which the *steady-state* material and energy exchange rates with the natural environment are inconsistent with constraints referred to in the previous section. In contrast to natural ecosystems, the laws which govern production, transport, and storage processes in agriculture and the physical industries as components of the life-support system are in principle known since these activities were designed and developed by man. Likewise, the laws of material and energy balance which constrain the interactions of these components to form a closed system with the environment are well understood in the physical and economic sciences.

Within the framework of these general principles, well-informed economic pricing policies and other potential regulation functions require considerably more detailed and complete knowledge of what the potential tradeoffs really are, for example, between technological recycling and environmental recycling of the externalities of production. Such tradeoffs, in general, depend upon the product type, the geographic region, and the alternate technological and spatial organization of the industry producing it. For example, beef and other meat products can be produced through a system of producers in which various phases of production are coordinated and diversified under the management of each production unit. Alternatively, they can be produced by a highly centralized, specialized system in which the individual producer allocates a major land region to one particular phase of production (such as corn production or cattle feeding, for example) for an indefnite period of time. The ecological and sociological tradeoffs implicit in these alternate physical structures are of major concern to regional planning, as are the implied differences in the physical distribution of human population over the landscape.

A similar set of ecological and sociological tradeoffs pertain to the counterpart structural features of physical production and the marketing and distribution activities in urban regions. For example, the generating capacity and location of power plants can be specifically designed to match the capacity of natural bodies of water and airsheds to process the thermal and material externalities of generation. Alternatively, artificial means (spray canals, cooling towers, SO_2 scrubbers, and so forth) may be used to extend the generating capacity of the plant beyond the limits imposed by local environmental components. What are the tradeoffs between the physical, ecological, sociological, and economic factors associated with these alternative physical structures? In reference to urban regions it is important to know to what extent the socioeconomic conditions and regenerative characteristics of the community are stabilized by a family-owned retail system in contrast to absentee ownership of a large-scale centralized marketing system. What are the tradeoffs in energy requirements, transport requirements, and waste concentrations?

It is suggested in a related paper (Koenig and Tummala, 1972) that many of the physical, ecological, and economic tradeoffs important to such questions can be evaluated from relatively simple *static* or steady-state models characterizing in a coordinated way the land, labor, physical energy, and monetary costs required to produce a given set of products with alternative system structures. These models represent a modified form of the classical economic input–output models appearing extensively in the economic literature. Interestingly enough, they can also be viewed as a generalization of the mass–energy models developed in the physical sciences. Such analyses have proven to be relatively simple to apply to problems in agricultural production and power generation. They provide at least one important perspective on the classes of physical and technological structures that are ecologically feasible as "targets" for regional development.

Social and cultural considerations in general impose additional constraints on the class of allowable structures which may be more restrictive than ecological considerations. Some of these considerations are discussed by Koenig *et al.*

(1972), but the scientific literature on the subject is virtually nonexistent. Investigation of these questions is not yet an established area of research in the social sciences, and it is not clear to what extent such constraints can be quantified or the major factors even identified. Most social and cultural considerations must, therefore, be regarded as extrascientific, at least for the present. In principle they must be dealt with on a subjective basis as part of the decision process in the sociocybernetic control loops.

V. Regulatory Processes

Given the present physical and technological state of the life-support system in relation to the class of states that are ecologically and physically feasible in the sense referred to above, the critical questions are (i) how are the social options implicit in the choice of target states to be expressed, and (ii) what adjustments or new developments in social instruments of control are required to bring about necessary changes in structure? Given the tradeoffs in ecological, sociological, and economic factors associated with various scales of mechanization and levels of spatial diversification in agriculture, for example, how are choices between alternatives to be made and how are they to be deployed? The decision-making unit is by no means a single entity and, because of spatial variations in environmental capacities, the options vary from region to region. The choice among alternatives hopefully will represent a collective social decision and reflect regional variations.

The instruments of control range from the life style and behavioral patterns of individual participants in society to the economic system itself. These decision units, of course, must be coordinated at all levels of aggregation of political and ecological boundaries. It appears essentially self-evident that one of the greatest contributions that can be made to regional planning is to identify a structure through which these regulatory functions can be carried out. Indeed, present political boundaries, and essentially all existing regulatory institutions were formed at a time when resources were not a limiting factor, that is, at a time when we did not have the industrial capability to seriously overtax the limited reserves of the geosphere or to overdrive the limited capacities of regional environments. Consequently, virtually all instruments of regulation and all social attitudes were formed without the benefits of an ecological perspective.

Increased understanding of the limitations imposed on a highly advanced technological society by ecological constraints is clearly an important and critical element of change that must take place in virtually all social instruments of control. But it is increasingly evident that interregional decision-making units are essential to negotiate socioeconomic tradeoffs and implement policy. The economic system itself is potentially a very effective instrument for regulating the structural state of the life-support system. Some insight as to how nonuniform regional taxing policy, for example, might be used in regulation can be gained from a careful study of the mass–energy and economic characteristics of a model of the life-support system (Koenig and Tummala, 1972).

Through the economic system we associate a monetary value with each material flow in the life-support system, including material exchange rates with the environment. These monetary values form the basis for at least one major component of a countless number of priority decisions by individuals and groups of individuals at all levels of social organization. Insofar as the technological and physical structure of the life-support system is concerned, the economy can therefore be regarded as a regulatory mechanism. The monetary values assigned to material exchange rates with the environment have a direct influence on the monetary values of structured materials in the life-support system. In this sense they serve as weighting "signals" used by the engineer, the developer, the businessman, the farmer, and all other social decision-making units in assessing "value functions," whether these value functions are definable mathematically or not.

One of the fundamental difficulties, if indeed not *the* fundamental inconsistency, of present pricing mechanisms in our economy is that they are based on the precept that regional limitations of the environment for processing wastes are unbounded. The monetary costs, for example, now assigned to materials discharged in the natural environment are all taken as zero, and costs assigned to materials extracted from the geosphere are based on short-term opportunities rather than long-term ecological considerations. To assign a zero value to material exchange rates with the natural environment is to assume that the environment has an unbounded capacity to process externalities of life-support activities, or that we do not care about it. Neither assumption, of course, is valid. The economy as a cybernetic control process is manipulating successional changes in structure that may not be ecologically sound because it receives the wrong weighting signals. In less abstract terms, the day-to-day economic feasibility analyses and cost-effectiveness studies made by engineers, regional developers, planners, and other in the context of their individual problems simply do not include a sufficient set of the important socioecological factors, and the rules of economic survival were not designed for a game of life in a finite world with a highly sophisticated technology.

Modern technology has greatly expanded the scale economies to human forms of energy (labor) in both physical and agricultural production through automation and large-scale mechanization. In general, modern computers extend these scale economies further through mechanization of the management function of the firm. It can be anticipated that new technological innovations will extend these scale economies beyond the present range. From the structure of the material and energy balance model of the life-support system it follows that structural changes toward centralization and specialization will persist as long as competition and decision making at the firm level is dominated by monotonically increasing scale economies (Koenig and Tummala, 1972). Empirical evidence seems to indicate that, after the firm reaches a critical scale of operation, economic and political advantages add to the competitive advantages. These advantages frequently lead to reduced economic efficiencies in the long run. If these competitive advantages are interpreted as being equivalent to scale economies that are strictly increasing, the system will not have a favorable climax state.* The scale of mechanization in agriculture has already progressed to the point where only 4.8% of the nation's working population is involved in agricultural production. Under present trends it is projected that this number will drop considerably. In relation to industrial production, it can be noted that twenty years ago half of the productive capability was in the control of 100 firms. Under present trends the number of people involved in physical production is expected to drop to about 13% of the working population.

The central impact of technology on society appears to be this: The economic forces of competition generated by a highly advanced technology in the context of present economic price mechanisms is creating structural changes toward what appears to be essentially "unbounded" concentration, specialization, and centralization of production processes. Since the physical structure of the life-support system serves as a "template" for many of the service, social, and cultural activities of man, these successional changes are generating unbalanced spatial distribution of population, spatial concentrations of waste, extreme job specialization and work activity, and centralized control of production facilities. Large-scale mechanization and specialization in agriculture has a particularly profound impact on the spatial distribution of people and the distribution of organic wastes. In a sociological context one must ask what percentage of the population would prefer to be engaged in agriculture production? Is it 2% or 20%? Can a society with only 13% of its working population involved in physical production provide meaningful work activity for all of its people? At what point does leisure time become socially degenerating idle time?

Finally, some of the most important questions of sociocybernetic control center around the relationship between degree of centralization in the control processes and level of specialization and centralization of the controlled object (the life-support system). The problems of realizing effective, responsive, coordinated, and reliable social instruments of management in themselves apparently place an upper bound on the degree of centralization and specialization that can be tolerated in the physical structure of the life-support system. These bounds may be more restrictive on the physical structure than the

* In reality social forces other than economics will eventually bring about a climax state that may or may not be orderly.

constraints imposed by ecological and sociological considerations. The dominant limitations apparently are to be found in problems of decision reliability and interinstitutional compatibility, the information required to support them, and the time delay involved in their implementation. Wrong decisions or correct decisions with excessive temporal delays can and do lead to regenerative feedback and unstable loops within the system. Two contemporary examples illustrate these points:

1. The "time-honored" strategy of the central government of the United States to deal with unemployment problems has been to stimulate economic expansion through construction of new production facilities and promotion of new products. The business community augments this strategy by promoting obsolescence and consumerism. The combined result in general speeds up the production machinery. Increased flow rates in the system not only increase the waste-processing requirements of the environment and the physical energy requirements but they make increased specialization, centralization, and automation economically feasible with a subsequent cycle of unemployment.
2. The population of rural America displaced by large-scale mechanization in agriculture moves to large urban areas (approximately 20 million in 20 years) with employment expectations that appear to be essentially proportional to the size of the physical production complex of the urban community. The urban planner, the Chamber of Commerce, and other well-meaning organizations, on the other hand, attempt to provide job opportunities for the unemployed by promoting further industrial expansion in the urban community. The industrial expansion, in turn, raises the employment expectations and with it increased migration to the large urban community. The whole process is a regenerative spiral with no apparent equilibrium point.

If the concept of regional planning embraces such ecological and sociological dimensions as those discussed above, then it would appear that one of the most critical areas of research and development centers about the problem of identifying effective instruments of sociocybernetic control. In this context the use of regionally specific tax policies and other pricing mechanisms stands as one option open to the planner for identifying social preferences and for effecting structural regulation. It appears, for example, that changes in technological and physical structure of the human life-support system can be constrained within an ecologically feasible class of structures through monetary costs assigned to material exchange rates with the natural environment (Koenig and Tummala, 1972). But as a general principle, the monetary costs assigned to these exchange rates must increase exponentially as the steady-state material processing capacity of the *recipient* environmental component is approached. Such costs in effect reflect environmental diseconomies as the assigned threshold of capacity is approached. Under such pricing mechanisms it can be shown that successional changes in the structure of the life-support system will theoretically approach an ecologically feasible climax state.

There are clearly many practical difficulties in implementing such pricing mechanisms. Some of the practical difficulties inherent in the direct application of graduated taxes or other pricing policies to environmental exchange rates can perhaps be overcome by indirect application. For example, for any given product type and production technology the waste discharge rates into the environment are essentially proportional to production rates (outputs). Consequently, the same objective can be realized (perhaps with relatively minor loss of precision) by nonuniform taxes based on the potential productive capacity (output) of the processing unit. Such taxes must vary with product type, production technology used to produce the product, productive capacity of the unit, and geographic region to account for regional variations in environmental capacities. In agriculture such pricing policies reduce essentially to the simple concept of a tax on agricultural production land that is specific to the type of production technology in use, but *graduated* with respect to the potential productive *capacity* (size) of the unit. The rate of graduation should be based on ecological and sociological considerations rather than on "pollution damage" as such.

It can also be shown that, as a general principle, uniform environmental standards (physical regulations) on waste discharge rates imposed on the polluter *cannot* regulate the successional changes in physical structure referred to above. In fact, such policies will force centralization in the context of present pricing mechanisms. It is easy to show that uniform standards on SO_2, for example, are counterproductive in the successional dynamics they produce–technological removal is economically feasible under present pricing structures only for large-scale centralized operations. As a basic principle, it is critical to recognize that economic and physical regulations that are uniform with respect to material rates and spatial regions *cannot* regulate successional changes in the spatial distribution of production processes or their scale of operation (centralization and specialization).

Deployment of temporal changes in economic and/or physical regulations designed to alter the structural state of the life-support system will also, in general, impose changes in the socioeconomic state of the region through such factors as relative income and wealth distribution and distribution in the control of production facilities. Appropriately structured models (conceptual, analytical, or computer simulations) characterizing these dynamic changes are helpful in evaluating the impact of proposed changes in regulatory policy. But the questions of who bears the cost of adjustment and at what rates are political and sociological questions that are clearly outside the domain of science. In principle they are settled by political decision processes of the sociocybernetic control loops. Just how and at what levels of functional and spatial organization models of dynamic changes in state are to be structured and deployed as aids to these political decision processes is not clear. However, the principles of on-line control do provide important guidelines to the general nature of the questions to be answered and the relative accuracy required of such models. Indeed, in view of the cyclic nature of the sensing–decision–implementation process and the relative "conservative" nature of the decision processes in making policy changes, projections over relatively short time periods under relatively small perturbations may be sufficient. If the frequency of the sensing and decision cycle is properly adjusted to the time constants of the controller, the "feed forward" models of decision making need do little more than indicate the direction of change implied by proposed policy changes.

The fundamental nature of these considerations should not be overlooked. Is it not, in fact, just this consideration which makes it possible for the central government to maintain relatively effective control over the short-term economic dynamics of the nation through a board of economic advisers meeting once a month? This is not to say that such control cannot be improved with improved "feed forward" models. It only recognizes that improvement in this aspect of the control loop alone is not sufficient. The perceived *structural* state as a target for regional planning, the general class of feasible regulation policies, and the organization of social instruments for "on-line" deployment of these regulatory policies are certainly of no less fundamental concern.

VI. Conclusions

The sociocybernetic perspective discussed above points to many deep and important problems of major significance in planning and regional development which are directly or indirectly related to the physical and technological structure of our life-support system and the economic and other social institutions that direct its trajectory of structural change. The principles of structural design and sociocybernetic regulation suggest that the basic issues center about the tradeoffs among ecological, sociological, and economic factors of a highly centralized and specialized life-support system in relationship to a more decentralized, diversified system. The scale of mechanization (size of operating unit) and the degree of product specialization in agriculture apparently have a particularly profound effect on both ecological and sociological factors. At what point on the trajectory of current development should centralization and specialization be curtailed for ecological and/or sociological reasons, or has the successional development already progressed irreversibly in this direction? What social institutions are required to bring about the necessary adjustments and how are they to be deployed?

To be sure there is no simple, universal, time-independent answer to these questions across all products, geographic regions, or production technologies, and they include many elements that are extrascientific in nature. It is not likely that large-scale simulations of the ecological, sociological, and economic factors associated with any region in existence will be of any significant value in answering these important questions. Indeed, simulation models based on empirical data from existing systems can only reflect social attitudes, economic values, and physical structures that are known to be ecologically unsound; that is, they are by their very nature based on spurious information.

Responsible scientific inputs to these questions raise conceptual and methodological questions of the greatest possible importance. They require a framework of theoretical principles and constructs to provide a basis for decomposing the overall system into manageable subproblems and for organizing divided but coordinated research efforts to deal effectively with their solutions. The sociocybernetic perspective presented in this paper has proven effective to the author and his colleagues as a framework around which to coordinate and manage multidisciplinary scientific efforts toward solution of important subproblems. The specific subproblems included in these efforts are listed below as an illustration of these organizational questions:

1. What are the rate capacities of fresh-water lakes and streams to process organic and inorganic forms of nitrogen and phosphorus as a function of physical and ecological characteristics of environmental components, and the quality to be preserved?

2. What are the rate capacities of terrestrial regions to accept urban, agricultural, and industrial wastes (spray irrigation) as a function of the physical and ecological structure of the region?

3. What are the tradeoffs in the mass–energy and economic factors of alternative technological organizations and scales of operation at the firm level in the production of feed beef as a prototype of levels of diversification and scale of mechanization in agricultural production? What are the regulatory mechanism and policy options for promoting changes toward these alternatives, and what are the transient socioeconomic adjustments implicit in their temporal deployment?

4. What are the tradeoffs in mass–energy couplings of power generating units with local natural environment as a function of technological form and scale of the operating unit as a prototype of industrial production processes? What are the relative socioeconomic merits of alternative regulatory policies in promoting alternative structural forms?

5. What specifically are the critical technical, economic, and sociological factors involved in implementation of on-line integrated pest management systems in agricultural production as a prototype of sociocybernetic control processes? What are the options in sociotechnical instruments for periodically sensing the critical biotic, abiotic, and economic states of the controlled region

and for effecting alternative control strategies? Can these regulatory processes be carried out, for example, by readjusting the structure of agricultural extension services of land-grant universities, or is this sociocybernetic regulatory function best carried out within the framework of rewards and motivations of a pest management industry specifically developed for this purpose?

It is to be noted that each of these subproblems is a prototype of a basic class of subproblems central to the overall problem of bringing man's life-support activities into harmony with his natural environment as a closed system. Although there are many concepts, techniques, and computational procedures common to these subproblems, the dominant coordination takes place at a level of conceptual organization *above* the subproblems. Each of the subproblems, in fact, must operate as a functional component of a higher-level system over a range of contingencies imposed by the higher level of organization. Is this not one of the central features of the "systems approach" that makes it possible for collective man to construct and manage complex functional systems through divided efforts, systems that in fact are too complex to be simulated by another machine?

Acknowledgments

The research leading to this paper was supported by the National Science Foundation under Grant GI-20. The author is deeply indebted to his colleagues, especially Drs. W. E. Cooper and R. Rosen for their many contributions to this paper.

REFERENCES

Koenig, H. E., and Tummala, R. L. (1972). *IEEE Trans. Syst. Man Cybernet.* **SMC-2,** No. 4, 449.

Koenig, H. E., Cooper, W. E., and Falvey, J. M. (1972). *IEEE Trans. Syst. Man Cybernet.* **SMC-2,** No. 3, 319.

8

Simulating the World Ecosystem

JOHN McLEOD

SOCIETY FOR COMPUTER SIMULATION
(SIMULATION COUNCILS, INC.)
LA JOLLA, CALIFORNIA

I. Introduction

I am no ecologist; indeed I have only a concerned layman's knowledge of the subject. But I do know that ecology is that branch of biology which treats of the relations between organisms and their environment. And because organisms are many and varied, and the environment is the aggregate of all surrounding things, conditions, and influences, it is obvious that the world ecosystem is large, complicated–and extremely important! Add to that the fact that it is less than perfectly understood and we have a system that the unaided human intellect cannot fully understand. Yet, if man is to exert any control over his environment and thus the quality of his life, he must have a better understanding of the world ecosystem. Computer simulation is one of the most powerful tools modern technology has placed at our disposal to study and gain insight into the dynamic interactions of large, complicated, ill-defined systems. We might be well-advised, therefore, to consider the experience of others who have used simulation to study such systems.

My specialty on such occasions as this is to give a broad-brush tutorial on the power and limitations of computer modeling and simulation. However, I note that 70% of the chapter titles in this volume include the words "model" or "simulation." I take it, therefore, that this is no place to indulge in generalities. So I will restrict myself to two subjects that are of particular interest to me: the widespread criticism of the Forrester–Meadows world models, and the capability–demand relationships of a wide spectrum of simulations.

II. Toward a Meaningful Discussion

The words that I write and the words that you read may be the same, but if these words mean different things to you and to me, confusion will be compounded and no useful communication can ensue. So with slight apology to Lewis Carroll I will, as Humpty Dumpty said, make the words mean what I choose them to mean. For the purpose of this chapter, therefore, I will assign meanings to key words as follows:

Computer modeling is the programming of a computer to exhibit selected characteristics similar to those of an existing or hypothesized system or situation.

Simulation is the use of a model (not necessarily a computer model) to perform experiments designed to reveal certain dynamic characteristics of the model and, by implication, of the system, idea, or situation modeled.

Simuland is a generic term of convenience used to refer to that which is modeled, whether it be real or hypothesized, a system or a situation.

Before continuing I would like to call attention to two important phrases in the foregoing definitions. Note "*selected* characteristics" in the definition of modeling. In most cases it is necessary to leave out of the model all those characteristics of the simuland that are not important with respect to the study at hand. Otherwise the model would be as complex and as difficult to understand as the simuland. Of course, it is the responsibility of the investigator to determine what is and what is not important.

The other phrase is "experiments *designed*" in the definition of simulation. In simulation, as in any other scientific investigation, the experiment must be designed specifically to yield the kind, quantity, and quality of data necessary to draw valid conclusions relative to the purpose of the investigation. Again, it is the responsibility of the investigator to see that the experiment is properly designed.

I call attention to the foregoing points for the benefit of those who might be inclined to put too much credence in simulation. Simulation is only a tool; as such it can multiply manyfold the ability of a competent investigator. But simulation can also propagate personal prejudices and/or mask sheer stupidity. Credence should be placed in the investigator rather than in simulation results per se!.

III. "The Limits to Growth"–A Case in Point

About the time (mid-1969) that I published a statement (McLeod, 1969) that I intended to work toward development of a model of the world, and began forming the World Simulation Organization (WSO) to do just that, Professor Jay Forrester and colleagues at MIT were getting underway with world models of their own. The WSO was unable to obtain funding in this country for such a "far-out" project, but the Club of Rome, which was sponsoring the MIT work, got a quarter-million-dollar grant from the Volkswagen Foundation in Germany. Thus, they were able to produce the models World II, on which "World Dynamics" (Forrester, 1971) was based, and its successor, World III, upon which was based "The Limits to Growth" (Meadows *et al.,* 1972) (hereafter referred to as "LTG").

The Club of Rome has printed fifty thousand copies of "LTG" and has attempted to place copies in the hands of decision makers, or the influencers of decision makers, throughout the world. Most readers probably are familiar with that controversial work, but if not, a good idea of its contents will emerge from the selected remarks of critics that I will discuss.

A. The Fury and the Hope

I have collected and studied more than 30 reviews and criticisms of the Forrester–Meadows works. Some may say–considering the great number that have been written–that 30 is not a statistically significant sample. However, I believe the opinions I have read are probably representative. They vary from furious attacks verging on the personal to apparent willingness to accept as gospel anything that comes out of a computer. In my opinion, both extremes are unjustified–and wrong. But between the extremes there is hope, a conviction on the part of some that with all its admitted shortcomings these first attempts at modeling the world have pointed the way to a better understanding. And, as noted, understanding is a prerequisite to corrective action.

There are several reasons for the adverse reaction to the Forrester–Meadows work:

1. Many established experts in the social sciences resent the intrusion of Forrester and Meadows, both with backgrounds in the "hard" sciences, into their field–especially when the findings do not agree with their own.

2. The publications present predictions for our future that people simply do not want to believe. Critics somehow miss the point that the predictions concern what we can expect *if* no action is taken to change the present course of events. But some action will undoubtedly be taken and, in part, as a result of the warning the Forrester–Meadows work has given us.

3. The uninitiated do not understand, and therefore do not trust, simulation. Nor do many who actually use simulation for certain purposes realize that there are other kinds of simulation that are useful for other purposes.

There is, in fact, a continuum of simulations from the fuzziest to the most rigorous. And it is the place in that spectrum where a particular simulation fits that determines what one must put into it and what one can expect to get out. At one extreme there is a tendency to look upon computers as oracles, which stems from a lack of understanding of the computer–investigator relationship mentioned earlier. In one respect, however, this view is justified: The pronouncements of the Oracle at Delphi required interpretation by the priests; simulation results also require interpretation by the experts!

I believe we should ignore the extremists and focus attention on the points most often raised by more rational critics. In general these are of three kinds, those concerned with (i) validity of the model, (ii) interpretation of the results, and (iii) acceptability of the recommendations.

B. The Model

The critics complain that the model World III, upon which "LTG" is based, is too highly aggregated, that some of the assumptions are untenable, and that the input data are unreliable. Furthermore, the model is strictly materialistic, having no provision for taking human values into account. Furthermore there are no provisions for including the impact of technological advances. At least one critic points out that the model is deterministic whereas the phenomena with which it deals are stochastic in nature.

These are important points, and deserve serious consideration. I will comment on them in the order mentioned.

1. *Aggregation*

There are at least three ways to undertake development of a model of a large complicated simuland. One can try to include at the outset all details which might be pertinent to the proposed study. As there is usually no reliable way of determining what degree of detail should be included, this approach will most likely lead to an overcomplicated model, confusion, and wasted effort.

A better way is to start by modeling subsystems and, after they have been simplified and validated to the greatest extent possible, to interconnect them to complete the overall model. Even then it may be found that some of the subsystems have negligible influence on operation of the overall system, so again effort may have been misdirected. Worse, interconnecting optimized subsystems does not necessarily lead to an optimum overall system—in fact, the resulting model might even prove to be unstable!

A third way is to begin with a "skeletal" model which can be fleshed out with details as the situation warrants. Such a model can give insight into the operation of the overall system and some indication where further work is most needed. But it is not, of course, adequate to predict the future behavior of the simuland to any degree of detail greater than that included in the model. The authors of "LTG" are well aware of this, and Dennis Meadows—now at Dartmouth—is busily at work on disaggregation by way of developing more detailed submodels of population and resources.

2. *Assumptions*

Some critics have presented telling arguments to the effect that many explicit relationships in the "LTG" model are not correct, at least quantitatively. As assumptions concerning these relationships make the model, the model is valid only to the extent that the assumed relationships hold. This can be tested by modifying the individual assumptions to the greatest reasonable extent. If the modification causes profound changes in model behavior, the assumptions must be restudied and their values refined (of course the experimenter must determine what is "reasonable" and what is "profound"!). But even an invalid model is not necessarily useless—a point to which I will return later.

3. *Data*

Critics object to the authors' apparently "pulling numbers out of the blue" when "hard" data are not available. Forrester's answer is that the models are not very data-sensitive. Any reasonable (that word again!) numbers will give similar projections; only the timing and magnitude of responses to alternative courses of action will change somewhat.

So, considering all of the foregoing, it would seem that the models are *not* valid for accurate predictions. But, as one critic puts it, the results of the experiments described are not predictions but projections. So be it. But the problem is that too many people will take them for predictions. That, I believe, is the most dangerous of all the charges that have been or may be leveled against "LTG." The book is replete with caveats, yet the general tone is such that somehow the importance of the disclaimers is overlooked by most readers. It is as though the authors had felt obliged to write: "Believing this might be dangerous to your perspective—but isn't it exciting?"

4. *Materialism*

It is true that the model is materialistic; there *are* no provisions for taking human values into account. But the authors do not claim otherwise. They realize, as do all thinking people, that things such as food, an uncrowded and relatively clean environment, and material possessions are not everything. But these factors are easier to deal with in the development of a model than the possibly more important, intangible ingredients such as religion, recreation, and the arts, which must ultimately be considered in determining the "quality of life." And is not such simplification justified in any early attempt to simulate a complex system like the world?

5. *Technology*

Certainly technological advances can be expected, and they will have their impact—probably a tremendous impact. But it is not true that the "LTG" model lacks provision for taking such changes into account. It does not do so automatically, but if any investigator, decision maker, or student of the future wishes to study the probable impact of a technological change taking place at any given time, the expected change in the value of any of the model's parameters can be introduced at the selected time, and the resulting impact on all variables measured.

6. *Determinism*

When is a deterministic model of a stochastic simuland justified? That is a matter of judgment and again it is the responsibility of the investigator to decide. But it would seem to this writer that the introduction of noise to make an otherwise deterministic system behave stochastically is justified only when the uncertainty introduced by the noise is greater than that introduced by questionable assumptions and uncertain data. So, insofar as the subject model is concerned, "who needs it?"

If I seem to damn with faint praise the model upon which "LTG" is based, that is not my intention. I am only developing ammunition for my contention that a model may have merit in some respects, yet be completely invalid and even dangerous if the results are not interpreted in light of the model's limitations.

C. The Results

I contend that the interpretation of results of any experiment should be the responsibility of the experimenter. "Let me not judge a man until I have walked a mile in his moccasins"—or struggled with the complexities of his simulation. As

I am not intimately familiar with their struggles, I am of necessity willing to go along with the meaning that the authors of "LTG" ascribe to the various curves their computer has produced, even though some of their critics are not.

I believe this points up another real danger of simulation: People who do not understand sufficiently the model or the experiment—sometimes influential people—often *do* give *their own* interpretation to the results. For that reason decision makers, or people who influence them, should be intimately involved in the simulation process, preferably from the concept, certainly in problem elucidation, hopefully in the development of the model and planning of the experiments, and necessarily in the interpretation of results.

I do not believe that those who criticize the interpretation of results in "LTG" qualify in the above respect.

D. The Impact

There can be no doubt about the impact of "LTG"—it has been sensational. But how much more than sensational? Will it really change things? I believe that it will!

You may remember the story of the mule trainer who started his course of instruction by breaking a two-by-four across the mule's head. When asked why he had to resort to such drastic measures he replied: "First you have to get the critter's attention!"

Well, "LTG" has now got the attention of a goodly number of us critters. Even if the models are faulty, the results misinterpreted, and the recommendations unacceptable, the publication of "LTG" has caused many people to begin thinking in ways they had not thought before. I believe that is salutary; man must *think* his way out of the mess that thoughtlessness, carelessness, and lack of understanding have gotten him into. Simulation won't do it, but it certainly can help.

A number of critics argue convincingly that the recommendations of "LTG" will not be acceptable to those in a position to act on them. Unfortunately, with few exceptions, they are all too right!

Can you imagine a politician supporting measures that a computer simulation indicates will be beneficial to his constituency 20 years from now, when his problem is to get elected next year? No, we have a long uphill fight to sell the usefulness of simulation as an aid to understanding. But "LTG" has attracted the necessary attention. Now it is up to those of us who are concerned to take advantage of the positive impact and at the same time counter the undeniable negative impact by presenting the capabilities and limitations of computer modeling and simulation in proper perspective. We can do this in part by employing the same scientific approach and professional objectivity to our simulation work and its reporting that we use in connection with our other work. But we must go further. We must overcome any remaining aura of mysticism by getting the "customer" involved. If he can be made to feel that simulation is a tool that *he* can use to gain a better understanding of *his* problems he will use it. But so long as we take the attitude—or even appear to take the attitude—of "bring us your problems and we'll give you solutions," the potentially favorable impact that simulation could have on our way of life will not be realized.

IV. Models: From Fuzz to Fact

Dennis Meadows has said that the question should not be "Is the model valid?" but "Is the model *useful*?" I will go further and say that I believe that just the attempt to develop a model is useful. But usefulness is a relative term. I like to think of it as a ratio of the capability of a simulation to the demands placed upon it; that is, what it can do with respect to what it is asked to do.

Ideally the capability/demand ratio should be unity—if it is greater the simulation is not being used efficiently, and if it is less it should not be used. But it is difficult to place numerical values on either capability or demand. Capability is a function of the weighted confidence factors that can be assigned to the validity of assumptions, the accuracy of data, and the competence of the investigator—among other things. Quantifying the demand is just as difficult. It is some function of the purpose of the simulation project, but more than that I cannot say. However, I find it helpful to think of a simulation in these terms, even if intuitive values must be assigned.

Now if we apply this line of thinking to "LTG," it seems possible that the authors were working with a capability/demand ratio of less than one. It seems they might have read more into the results, that is, demanded more information than the model was capable of substantiating. But of course I cannot be sure because I am not privy to all the facts. But I *am* sure that many of the critics read more into the published results than the model was capable of giving.

To the best of my knowledge World III and its immediate predecessors represent the first serious attempt to model the world. Simulations using these models are not capable of predicting what *will* happen to the world in the future. They only indicate what will probably happen sooner or later *if* nothing changes the present course of events. However because of, to some extent, the warning that "LTG" has given us, things *will* change.

It is a characteristic of simulation prophecies that they are self-defeating. Just as surely as it is predicted that something will happen, it is almost certain that someone will take steps to see that it does not. Let us hope that will be true of the "LTG" projections.

V. Conclusion

Although I have discussed critics' reactions to "LTG" and injected many of my own thoughts, I believe that the considerations outlined are pertinent to modeling and simulation of all large, complicated, ill-defined systems, especially the world ecosystem.

The primary points that I hope I have made are as follows:

1. A simulation does not have to be "valid" to be useful.
2. One should feel confident that the capability–demand ratio is one or greater before drawing conclusions or recommending action based upon any simulation results.

REFERENCES

Forrester, J. W. (1971). "World Dynamics." Wright-Allen Press, Cambridge, Massachusetts.

McLeod, J. (1969). *Simulation* **13**, 7.

Meadows, D. H., Meadows, D. L., Randers, J., Behrens, W. W. (1972). "The Limits to Growth." Universe Books, New York.

9

Macroscopic Minimodels of Man and Nature

HOWARD T. ODUM

DEPARTMENT OF ENVIRONMENTAL ENGINEERING SCIENCES
UNIVERSITY OF FLORIDA, GAINESVILLE, FLORIDA

I. Introduction

Man is embedded in a complex world of confusing cues threatening to overwhelm him psychologically as much as physically. His finest role in the mechanism of this planet may be to become its steersman—*if* he can learn to cut through the plethora of detail his knowledge has brought and "see" the essence of man–nature interactions. Science is unlikely to help if it continues to focus on atoms and the short term. But science is beginning to show some adaptation in the recent shift to systems thinking and methods.

The thesis here is that complexity must be reduced to essentials if complexity is to be overcome as an impediment to understanding and correct action, and this means modeling. The specific tool envisioned is overview models that are macroscopic in viewpoint but minidimensional in complexity—"macroscopic minimodels."

There is a type of orthogenesis within our scientific system that sometimes says a proposed new outlook, theory, or process is more advanced, more sophisticated, and thus better if it is more complex. This is false whenever the new thing is not as energy conserving as a simpler one. It is also false if the complexity causes a communication failure with the human being with which the new science must interface. A science of detail must in the long run be self defeating if, from time to time, useful means of synthesis and summary are not sought. Just as an artist seeks to capture an impression of what he views, scientists also must find ways to suppress detail and formulate the subjective, qualitative essence of facts and figures.

Macroscopic minimodeling is seen as one way to do this, a way that can be freely operationalized on modern, high-speed computers. Such abstract condensations of the real world, it is held, will be useful if they include most of the energy forcings and dominant pathways pertinent to particular considerations. A simulation minimodel of a large system may be the only way to assess consequences of human acts when the system is too complex and ill-defined for direct, controlled experimentation. This chapter presents an approach to such models.

Examples will be given for some of the main issues of uncertainty about systems of man and nature. Generalization to other systems where energy flow and structure are involved will be self-evident. The approach is modular: the main configurations of macro-minimodels have characteristic performances which are implied by systems diagrams in energy circuit language. The diagrams become, in effect, a new kind of mathematics, or at least a substitute for existing mathematics.

The chapter presents performance characteristics of certain minimodels as determined by analog computer simulations. The approach is abstract and general, and related to applications important to current survival issues of man and nature. The models discussed include: money and energy; order and disorder; producer–consumer symbiosis and competition; recycling and mining; succession with declining energy reserves; power titration with war, disease, and diversity; regional development optima; auxiliary power and diversity; and systems with mixtures of constant force and constant flow energy sources.

Included is a section describing basic module performance as an aid to interpreting the energy circuit diagrams.

II. Methods

A. Formulation of Basic Minimodels

There are two approaches toward realizing basic macro-minimodels. In the first, large complex diagrams are drawn representing in detail all relevant components of the system and all outside forcing functions (energy sources). In our experience, this diagram was often done by a group of people with varied backgrounds, a class, or a panel of decision makers. Then, the large model of perhaps 50 units and many pathways is redrawn over and over again, condensing and compartmentalizing, but retaining the key classes of mechanistic configurations, the main energy sources, the state variables of concern, and all constraints of energy, matter, and information.

In the second procedure, a priori interesting configurations for generalizing about systems theory are considered, scanning the varied world of man and nature for examples of their application. Such simple models constructed in advance aid in the recognition of central mechanisms and driving forces that tend to be hidden from the human eye in all the confusions of too much data, too much macroscopic noise, and too close proximity of view from within.

Once a model is defined in simple form, it can be quantitatively parameterized from whatever data are available, translated from energy circuit language into other (mathematical and/or programming) languages, and simulated. Simulations, providing contact with computer software and hardware, facilitate communication between individuals who do not think naturally in terms of energy flow language. However, the prime thrust of the modeling effort is to establish macroscopic minimodels and their diagrammatic language. Energy circuit diagrams, once understood, can be the focal point of communication between scientists and laymen, and in many respects they are more precise than some other systems languages that may be used to represent the same model.

B. Energy Circuit Language

The diagrammatic energy circuit language evolved for macro-minimodeling serves a variety of purposes. It provides a symbolic general systems lexicon with built-in attention to energy constraints and to dual force–flux laws of single force and population drives. It allows combining energetics and kinetics, formalizing maximum power selection, and generally extending features recognized in ecosystems to other systems, and vice versa. As such, it makes possible unification of principles known in many fields under different names and expressed in different formalisms, it reduces semantic arguments, and facilitates the general process of discourse. Above all, it forces a change in reductionist thinking to synthetic thinking.

Published discussions of different aspects of energy circuit language include the following:

1. description of basic modules and models (Odum, 1967, 1968, 1972a);
2. a general book surveying applications (Odum, 1971);
3. nonmathematical introductions (Wetterquist *et al.*, 1972; Leaird, 1972; Jannson *et al.*, 1971);
4. geochemical and marine models (Odum, 1971c, 1972d, 1975);
5. rainforest minimodels (Odum and Pigeon, 1971);
6. climax man (Odum and Peterson, 1972);
7. economic model (Odum and Bayley, 1975);
8. environmental impact statements (Odum, 1972b; Odum *et al.*, 1972; Littlejohn, 1973; Gilliland, 1975; Zucchetto, 1975).
9. a book for junior college that uses the energy language to explain energy, economics, and environment (H. T. Odum and E. C. Odum, 1976).

An excellent example of application to ecosystem analysis appears elsewhere (Lugo *et al.*, this volume).

Since one of the main technical discussions of basic energy circuit modules is in a previous volume of this series, the reader is referred there (Odum, 1972a). Simple basic configurations not included there are presented here to help in understanding the minimodels that will be described. Figures 1–4, pp. 255–257, summarize the modules and their translations into equation form. It is noted that the equations do not necessarily carry all the information in the diagrams. For example, there may be more than one module possible for a given equation and, in general, the equations do not possess the energy constraints, or exact details of force–flux relationships, or immediate perspective of function beyond the simplest (often linear) cases. A set of differential equations is another way of writing an energy circuit diagram, but the translation is often qualitatively incomplete. In this chapter, mathematical representations are provided on the energy circuit diagrams for clarity, but this is redundant since the language already specifies these quantities implicitly. Each module and pathway has a function transfer coefficient, derivable from data on storage and flow.

C. Simulation

At the time *Environment, Power, and Society* (Odum, 1971) was written in 1968, most of the models described therein had not been simulated. Discussions were based on perspective thinking facilitated by the energy circuit language.

Now, however, most of those models have been simulated, and in addition approximately 200 others contrived by various scientists, engineers, and students using the approach. At the University of Florida, we use an Electronic Associates, Inc. (EAI) Miniac analog computer for small models, and for larger ones a variety of machines including EAI 580 and 680, and Applied Dynamics AD-30 and AD-80. Some of our group use digital simulation, implemented by languages such as Fortran, Dynamo, and CSMP, but for macroscopic minimodels most prefer the advantages of hands-on contact and instant turnaround provided by analog computers.

In our use of analog simulation we have developed a number of shortcuts that save time, reduce errors, and increase interaction with the minimodel itself, as opposed to its machine intermediary. Two of these techniques are described below.

1. *Quick Scaling*

Voltage and time scaling are the most difficult parts of analog simulation. They impose a mechanical rigidity on the process that the minimodeler does not want. One approach is to bypass scaled equations (and associated errors) entirely and scale directly on the analog computer diagram. The diagram is arranged so that all potentiometers are arrayed in a column along the left margin of the paper. Opposite each of these a quotient is constructed, with two sets of parentheses in the numerator and two sets in the denominator. In the numerator (left) is written the coefficient value as determined from data. In the numerator (right) the scale factor of the amplifier upstream to the potentiometer is written. In the denominator (left) is put the scale factor of the amplifier (integrator) into which the potentiometer is directed. The remaining set of parentheses is left blank at first (value = 1).

The quotient value is the potentiometer setting. If all the potentiometers fall within a settable range (0.001–1), then scaling is complete and no time scale factor is necessary. A unit of time in the data becomes one second on analog output. If, however, one or more potentiometers cannot be set, all are divided by an appropriate factor to make this possible. This factor, entered in the denominator (right), becomes the time-scaling constant. That many units of computer seconds corresponds to one time unit in the data.

2. *Use of Energy Circuit Diagrams for Debugging and Explanation*

After patching is done energy circuit diagrams can be used directly in place of standard analog computer programs. This increases contact with the primary model and speeds debugging enormously. Amplifier and potentiometer addresses are written directly on the energy diagram and the analog program put aside.

Simulation manipulations proceed directly with reference to the energy circuit model. The latter, furthermore, is simpler since half the pathways are eliminated; there are two analog wires for most single pathways on the energy circuit diagram.

Analog simulation has another benefit specific to the minimodeling approach. Even with large analog or analog–hybrid installations, models ought to be kept to a small size (relative to the complexities of macrosystems). This ensures that real tests of essential system properties can be made in a small amount of time and at minor computer cost. Even a moderate-sized analog computer model, with 20 storages, 15 multipliers, and perhaps 100 coefficients, would have of the order of 10^4 combinations of meaningful manipulations. This is not so easily explored digitally. Often, complex digital computer models are only explored in a few combinations. The exercise becomes a computational rather than a theoretical and predictive endeavor.

All models and modules in this chapter have been simulated and explored to determine coefficient sensitivities in varying combinations, tendencies to display oscillatory or unstable behavior, and the relative importance of different pathways. Many of the configurations appear frequently in different applications so that generalizations are beginning to emerge. Most of the time behavior shown was traced from oscilloscope displays.

D. Basic Modules and Configurations

A sampling of some (not all) basic modules and configurations of the energy circuit language, and associated behavior, is shown in Figs. 1–6. To the extent that basic performance curves become mentally associated with network configurations do larger diagrams of systems begin to take on a dynamic perspective.

Figure 1 has force and flow sources, heat sink, and basic pathway equations. Figure 2 defines the storage module and the storage capacity factor. Also given there is a derivation of the interesting truth that the two-thirds law of metabolism and size can be derived from the normal force–flux of first-order systems. The laws of metabolism turn out to be based on the same basic relationships as Ohm's law. Figure 3 has many of the intersection functions commonly used. Figure 4 shows some of the simpler autocatalytic modules (with work looping back from storage to pump in additional energy) and their time dynamics. Depending on the nature of the source (input), some of these will level while others will grow exponentially.

Pairs of self-maintaining units (autocatalytic pairs) are shown in several designs in Fig. 5. Some of these are competitive and some cooperative. Well-established in microbiology and population ecology is the competitive exclusion principle by which exponential growth at high energy levels leads to

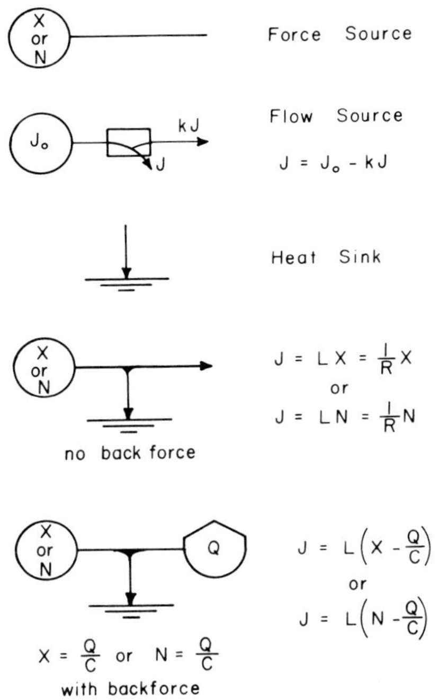

FIG. 1. Basic energy source modules and pathways.

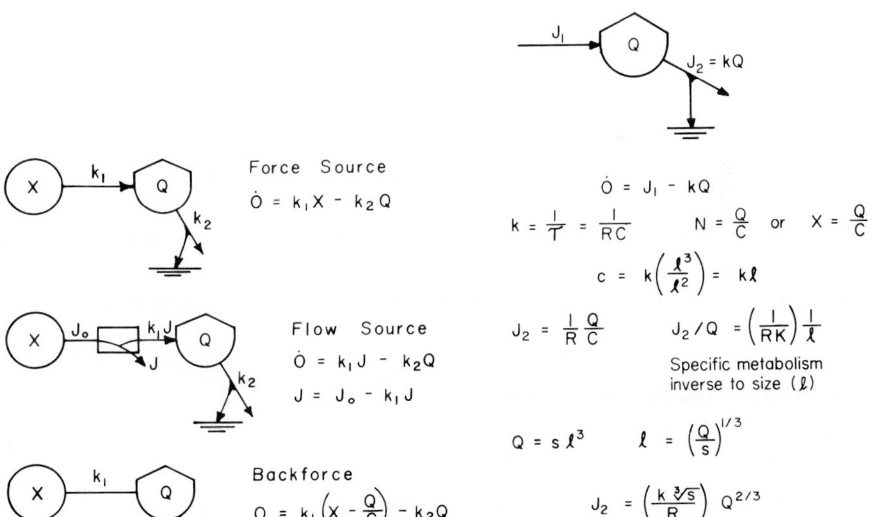

FIG. 2. Basic storage module and its relationship of (a) force and flux and (b) size and metabolism.

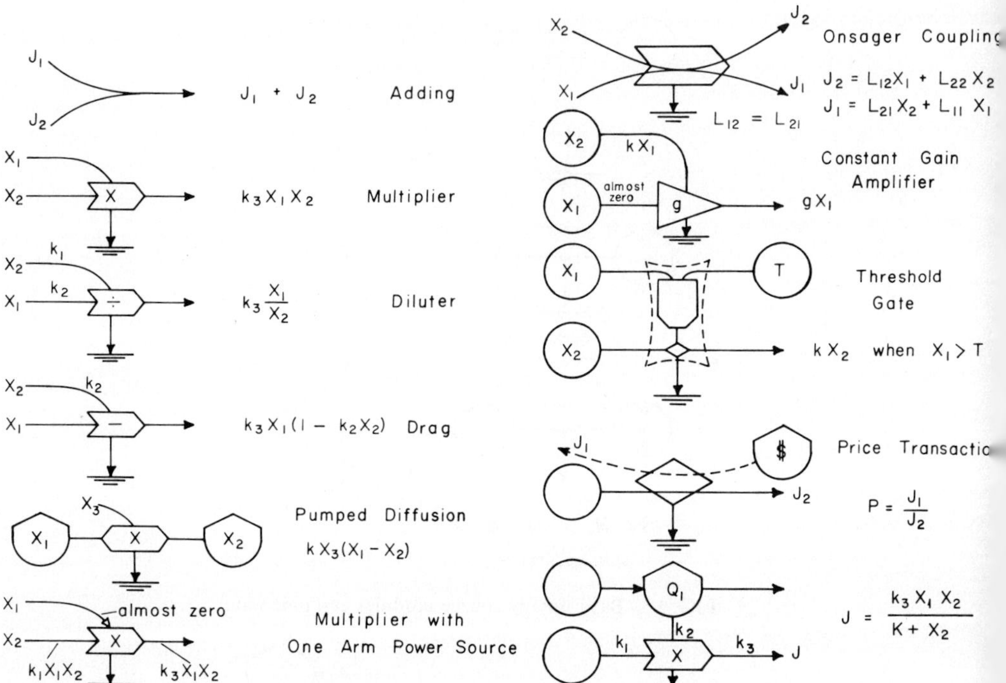

FIG. 3. Basic intersection modules.

only one competitor surviving and absorbing resources. Two designs for this are given in Fig. 5. A pattern that allows continuance of both interactive units is also shown. Most of the more complex designs of minimodels avoid competitive exclusion (i.e., are stable). Finally, in Fig. 6 are several configurations whose behavior is derived from the manner of coupling flow and force-type sources.

III. Macroscopic Minimodel Examples

A. Symbiosis of Order and Disorder

If order is defined as a higher state of stored potential energy and lower entropy (less randomness) than average in a system, and disorder as the opposite, then a general symbiosis can be said to exist between ordered and disordered storages. Figure 7 illustrates the essence of this property, showing disordered parts as reactants to incoming potential energy that interact to generate order. The pattern of order is selected that maximizes inflow by feeding back

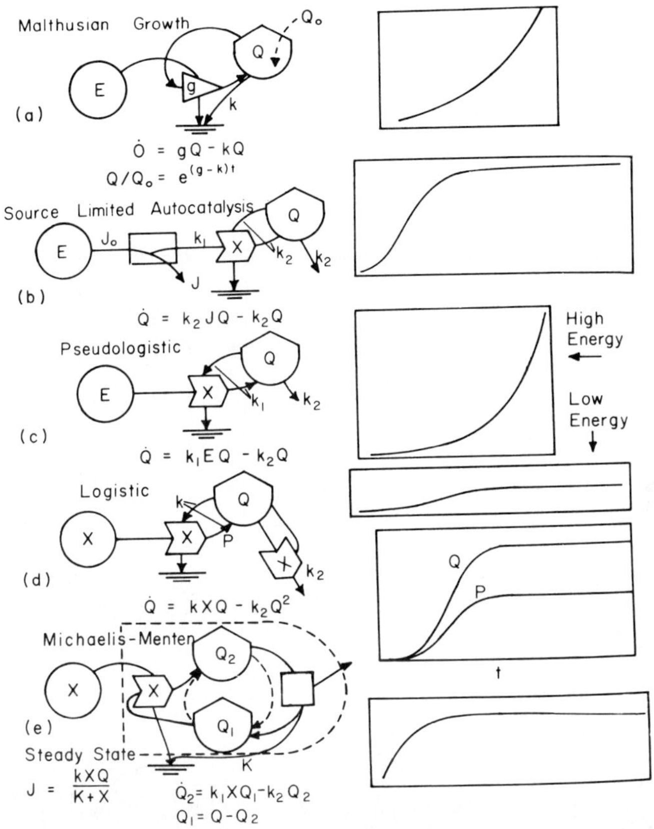

FIG. 4. Basic loop modules.

high-grade, organized potential energies; the power of the system is maximized together with its competitive advantage.

However, as storage energies rise and power flow increases, a shortage of disordered parts to interact with the high power develops and becomes limiting. Increasing disorder is generated as power flow increases, due to entropy-generating tendencies inherent in any high energy state or flow. Thus, disorder stimulates reordering, and reordered energies stimulate disorder. Shown in Fig. 7 (bottom) is a representative response of a system diverted from its steady state. It promptly returns. In some ways the model simulates both Schrödinger's "ordering by disordering" and Lotka's maximum power principle of selection, choosing order from disorder. The model is really a generalization of the P and R ecosystem model in which organic structure is the organized state and regenerated nutrients are among the disordered states stimulating reordering.

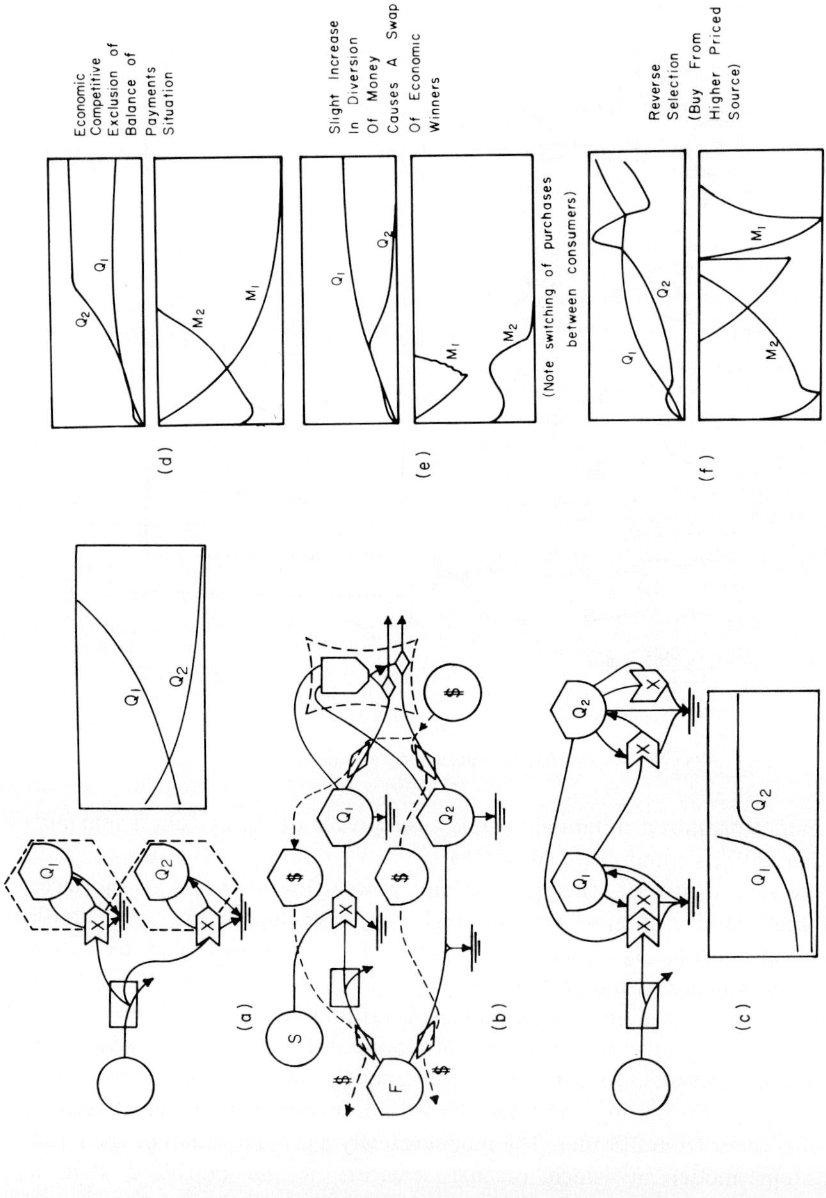

FIG. 5. Configurations of paired subsystem modules.

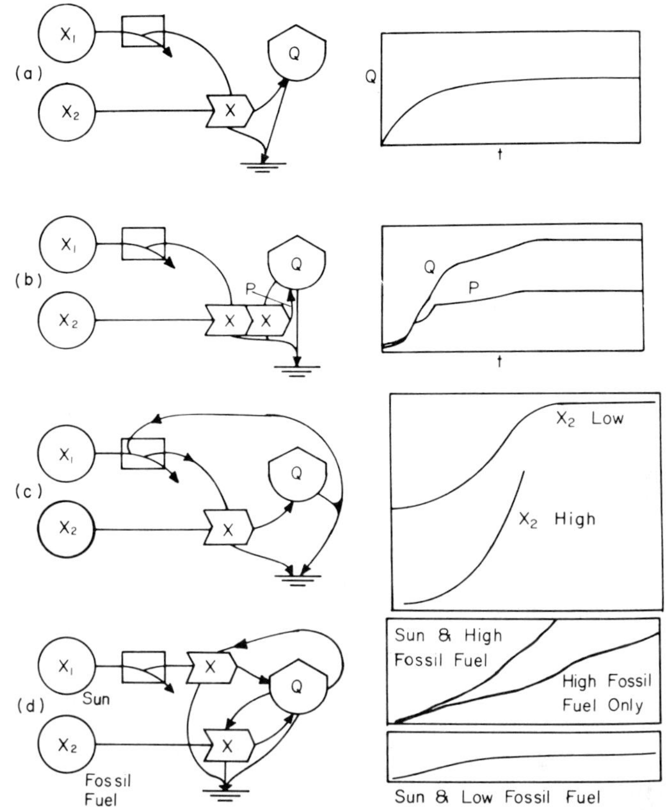

FIG. 6. Interplay of dual energy supports.

Another example of great recent interest is the war states in Vietnam in which the organizing energies of sun, wind, waters, and tides develop vegetation, mangroves, marine communities, and, with fossil fuel inputs, towns and agriculture. With this there also develops a normal disordering tendency that keeps lands, peoples, and materials in slow recycle. Superimposing on this scenario the organized disorder of war, the steady state shifts to a higher disorder–order ratio. It becomes progressively more difficult to disorder further as the ordering potential is increasingly stimulated by more fallow lands, loose building materials, photosynthetic sunlight, and people without occupations. If the disordering energies are turned down, reconstruction is very rapid. We saw this in Germany, Japan, and Italy after World War II, and it will happen in Vietnam also if disordering energies do not continue to be applied.

Another example is the effect of noise, fire, explosions, and so forth, on

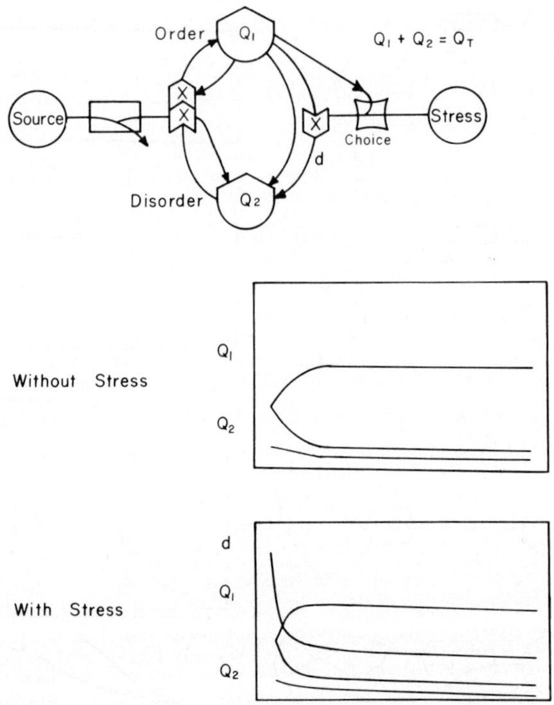

FIG. 7. A model of the symbiosis of order and disorder.

ecosystems. Richey (1970) showed experimentally that relatively small energies are required to divert order to disorder, but the disorder stimulates very prompt recovery of the more ordered condition. Photorespiration is a further example, disordering in this case being aided by the sun (Odum *et al.*, 1971). Kelly (1971) simulated models of high-temperature disordering that stimulated recycling.

The following dual view of order–disorder seems appropriate. An anthropocentric view of the world generally favors paths from disorder to order. To man, entropy-generating tendencies are "bad." This shows up culturally in various forms: in religion disordering activities are actions of the devil, in economics the antithesis of goods is "bads", in ecology perturbing influences are stresses, and so forth. One wonders if there is not a true general symbiosis of order and disorder, with both essential to the other. Is it too human to regard as good only the ordered half? Is the word "order" suspect since disordered states ultimately have as much amplifier value as ordered states in the total process? A principle of order by energy selection may be as fundamental as the second thermodynamic law of energy degradation, as suggested by Lotka (1922).

B. War Titration of Disordering Energy*

Whereas the essence of war's action as a special disordering energy is embodied in Fig. 7, there are additional roles of war such as in power titration for setting of boundaries between two regimes. Given in Figs. 8a and 9a are models that have some features of the Viet Nam situation, such as two large energy sources that tend to match each other in terms of energy effectiveness impinging on an area (Viet Nam). The essence of the order–disorder recycle symbiosis of Fig. 7 is also included. From one side money is added to purchase material and fuel for reconstruction, but added energy may also bleed off into further war. Figure 8b shows adaptation to war and Fig. 9b a simulation in which a 5-year pulse is introduced, representing the surge of energies, manpower, and money of the U.S. war effort; a matching effectiveness is injected for the opposition, accomplished on the analog computer by a comparator function. The simulation shows a surge of disorder and a new order–disorder balance achieved, which subsequently returns to an older state as the energy pulse fades. It is a property of this model that it is quite impossible to drive the order to zero, a characteristic commented upon by many in the adaptation of the countryside to war.

There are many situations and sizes of systems to which this macro-minimodel may apply: for example, fire fighting of brush in pine ecosystems, cops and robbers, and competition between two ecosystems, each of which has soldier species whose preferences in behavioral action are programmed for antistatic actions against foreign members.

C. Leveling Tendencies of Systems Receiving Combinations of Constant Flow and Constant Force Energy Sources

Summarized in Figs. 2 and 4 are basic energy consumption modules and their performance with time when supplied from constant force or constant flow energy sources. First there are the growth patterns of a single tank storage (von Bertalanffy module; first-order system) in response to both kinds of source. Steady states are reached as the linear density-dependent outflow equals inflow. Another variation of this has a backflow pressure included. Figure 6a shows the effect of multiplicative interaction of both kinds of sources, as in photosynthesis on an infinite nutrient pool. It also levels.

More realistic for self-maintaining modules are models that include one or more reward loop feedbacks under conditions of competition and maximum power selection. Stored energies have reward loop actions as required to offset

* This section by C. Swallows and H. T. Odum.

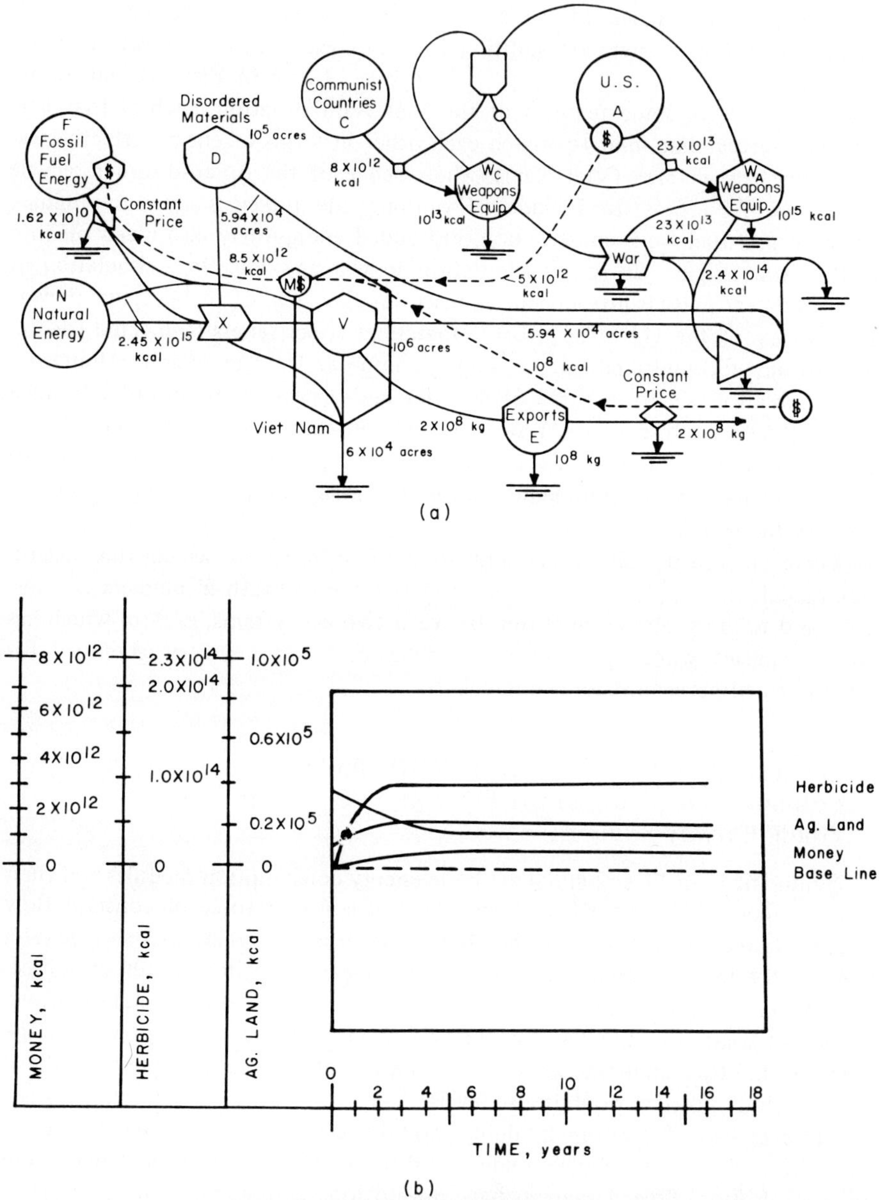

FIG. 8. (a) Minimodel of a steady-state Vietnam-type war. (b) Representative simulation.

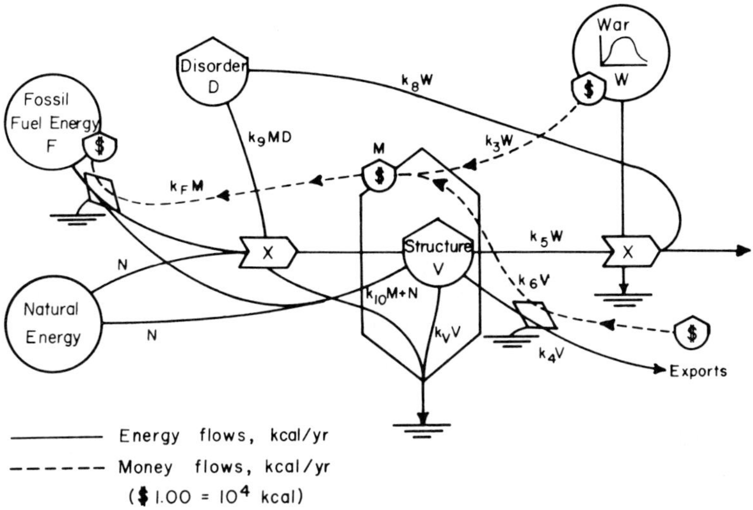

(a)

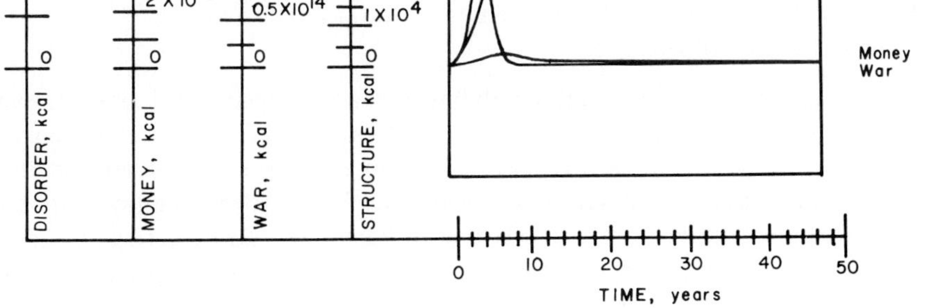

(b)

FIG. 9. Pulse of war: (a) A minimodel of disorder–order symbiosis upon structure as stressed by a Vietnam-type war. (b) Representative simulation.

their own drain. The behavior of these units drawing on constant flow and constant force sources is shown in Fig. 4. The constant flow source automatically levels, but the constant force source drives the basic self-maintaining module into exponential growth once power levels exceed a low threshold value. By analogy, when man shifted from solar energy (constant flow source) to fossil fuel (that behaved at first like a constant force source), his system shifted from a relatively steady-state pattern to exponential growth.

If, however, as shown in the Fig. 6 models, the system requires some of each type of input, then the two energy sources are not additive and alternative, but multiplicative and thus essential. In the latter case the system levels to steady state, with limiting energy from the second (fossil) source. But, if the unlimited source can be fed back to do the work of the necessary constant flow source (Fig. 6c), then the system can maintain exponential growth again. Such a system would actually inhibit the steady flow source by preempting the lands and overgrowing the mechanisms.

In nature, systems compete for service and feedback selection control. The system that gets the energy is the one that uses it most effectively. The most effective use is feeding back to auxiliary energy sources, developing energy sources in addition to the primary source. Thus, the first two systems of Fig. 6 get more energy than the third. The power-maximizing system is one that taps all energy sources, not just one.

Suppose the system is one such as man's civilization in which money flows and energy is purchased with a balance-of-payments mechanism that fine tunes the maximum power principle. In this case (Fig. 5b) a system increases its balance of payments for its work if it best uses auxiliary sources of energy. Such a system has a selective advantage in diversifying the energy sources upon which it is dependent. Making good use of auxiliary energies, such as environmental subsidies, levels growth.

D. Diversity as an Auxiliary Energy Generator and High-Power Regulator

Whereas autocatalytic modules with linear, density-dependent drains can level at steady state with low energy inflows, they switch into exponential mode with high energy inputs (Fig. 4c). However, as energy storages rise, square function outflow terms begin to develop. These include backforce interaction with the feedback loop (Odum, 1971, p. 316), and drains due to interactions of dense populations (including conflict, frictions, wrecks, noise, organized waste removal, and so forth). It may be a general rule that high energy levels generate square terms.

In many energy storage devices such as water tanks, capacitors, and flywheels, storages increase as the square of the quantity carrying the energy storage. High levels of such energies may generate proportional losses. In other words, higher storages have higher maintenance costs, proportionally, expended to prevent losses. In simulating minimodels, autocatalytic subsystems with unlimited energy inputs and no square terms in their drains are soon recognized for their explosive growth characteristics. Such systems are adaptive when rapid growth is adaptive, for example, during successions or epidemics.

When the drain on an autocatalytic module is a square function of the storage quantity Q, the module's growth is that of the standard logistic model (Fig. 4d). To some extent, the wide use of such models indicates some generality of quadratic terms in units leveling on high energy support. However, if such terms merely drain a system, it may be selected against in competition with systems using square terms to gain energy from auxiliary sources or to increase efficiencies.

One kind of square term that improves efficiency and taps additional energies is diversity. The maintenance cost of diversity may be in proportion to interaction between the diverse members, and this is a function of the square of the number of types. As shown in Fig. 10, diversity feeds work back to tap other energies, but with increasing costs in the form of regulatory structures, Parkinsonian losses, and general noise.

In Fig. 10 the diversity unit serves as a quality index, a bellwether of the energy condition of the system, and an indicator of high net energies flowing into informational types of processes. The system that grows until leveled by the square costs of diversity outcompetes one that grows until limited by more destructive square terms characteristic of less organized or younger systems. The climax system thus maximizes its total power while using it all in the process, being ultimately the most efficient input processor and least efficient net storage or output processor. The Fig. 10 simulation model develops mass and diversity alternatively during succession, and levels at a diversity allowed by the balance between energy inputs and losses (which are not a part of diversity).

In human systems, with rapid growth based on a succession of new fossil fuel sources, there has been diversification of occupations, specialization, organization, and a trend toward a diversity structure that captures additional sources of energy. Super strains of crops capture more light. Super technology is developing nuclear power. There is a trend toward better land use, better use of the natural energies of ecosystems that support man. The excess energies of wars may draw more energies into the economic and political spheres of influence. Efforts to couple these auxiliary energies may cost more energy than is gained, but at least frivolous expenditures are minimized. Care must be exercised in identifying which parts of human endeavors induce energy savings and auxiliary energy capture. Too little is known, for example, about how activities such as sports, the arts, hobbies and leisure pleasures couple with the total energy inputs and conservation.

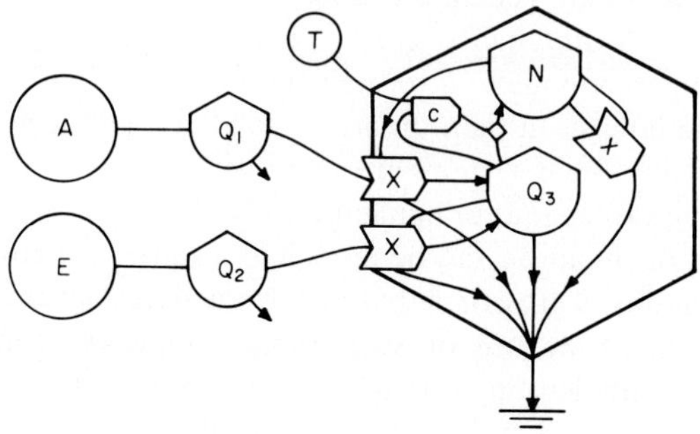

$$\dot{Q}_1 = A - k_2Q_1 - k_8NQ_1$$

$$-\dot{Q}_2 = -E + k_4Q_2 + k_5Q_2Q_3$$

$$-\dot{Q}_3 = -k_{12}Q_2Q_3 + k_{10}Q_3 + \overbrace{k_3Q_3}^{\text{IF } Q_3>T} - k_{11}NQ_1$$

$$\dot{N} = k_{13}Q_3 - k_9Q_1N - k_6N^2$$

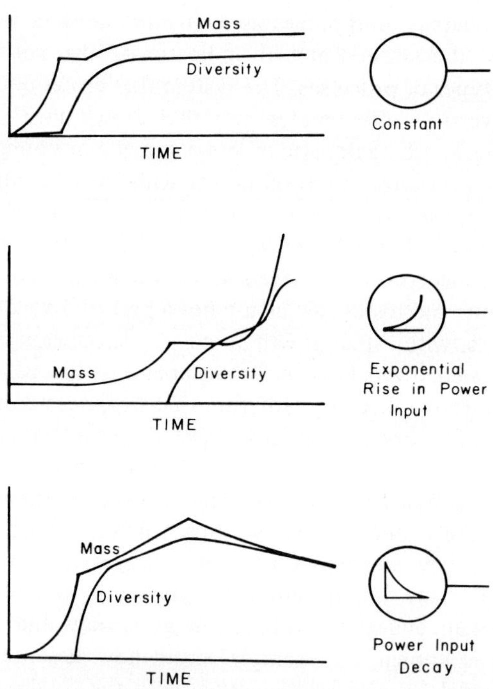

FIG. 10. Diversity and succession (Odum and Peterson, 1972).

E. SUCCESSION AND CHANGE IN DECLINING ENERGY SOURCES, NUCLEAR ENERGY, AND RECYCLING

Figure 11 simulates the main features of succession to steady state on solar energy. The fossil fuel world of man is like the fuel supply of a dead log or hay infusion microecosystem. Available fuel declines from the start, firing the successional development of a consumer (heterotrophic) order that builds structure, develops diversity, and finally recedes as the energy disappears. The system that ultimately remains is one that shifts to other energies. The nature of such patterns is given in the minimodels of Figs. 12–14. Each of these shows patterns of exponential growth, sigmoid decay of reserves, and finally decline of processes leaving only some of the remnant structure of the system.

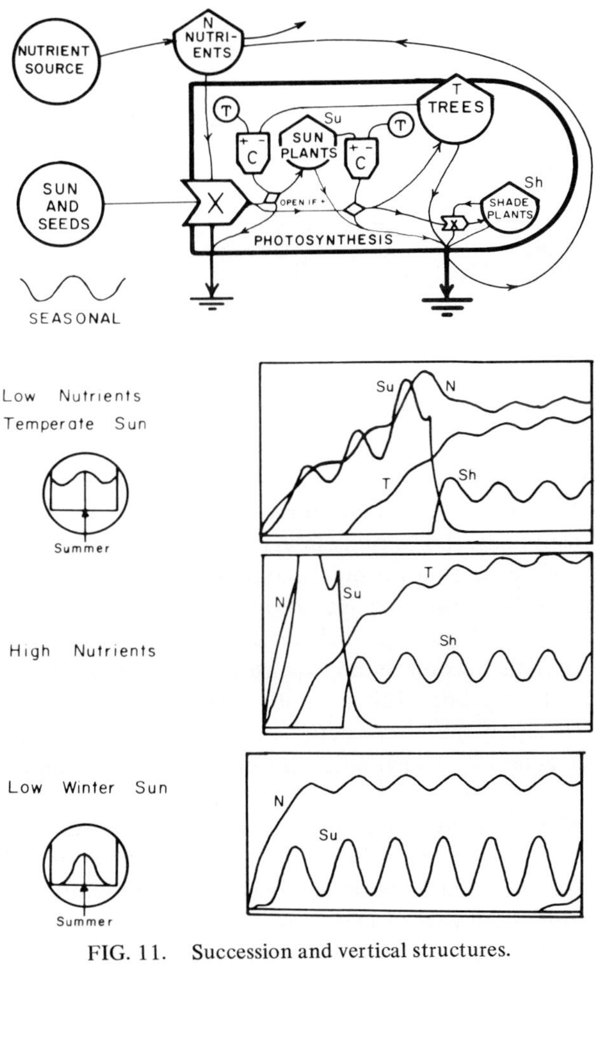

FIG. 11. Succession and vertical structures.

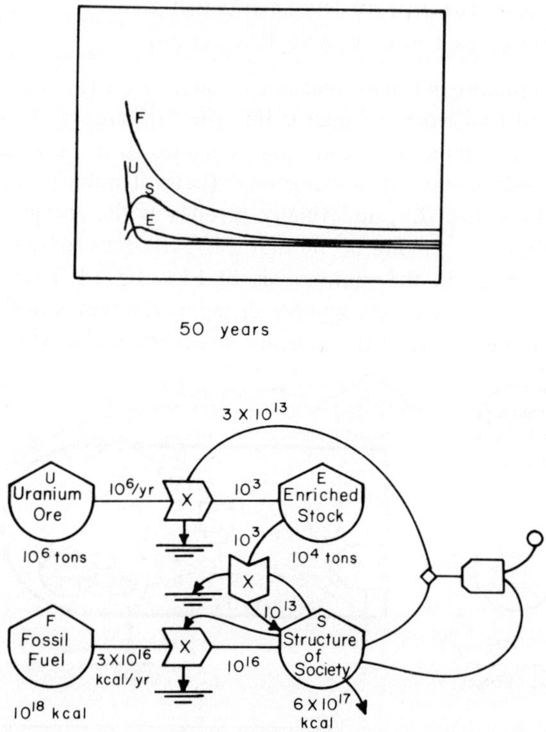

FIG. 12. Nuclear energy and auxiliary energy.

Figure 11 shows the progress of diversity during the rise and fall of a seasonal pulse. Figure 12 shows the pattern for nuclear power as it is now heavily subsidized by fossil fuels. If new processes (such as breeder reactors) with wholly different levels of conversion do not become new yielders, the nuclear power will run out before the fossil fuel, just as diversity disappears before all flow of life in a decaying log. Figure 13 includes material recycling as an alternative when scarce materials become more expensive to obtain from primary sources than the costs of recycling. This is another example where the main energy source is used to supply auxiliary energies, and where scarce materials have the amplifier release action by which necessary materials are recognized as auxiliary energy sources.

A word about net energy. Familiar to ecologists are the contrasting concepts of gross and net production, gross being input from synthesis and net being what remains after some feedback work has been done. The difference is not so familiar to the public, nor to leaders now showing concern for power policy. Usually when large fossil fuel reserves have been estimated gross energy was

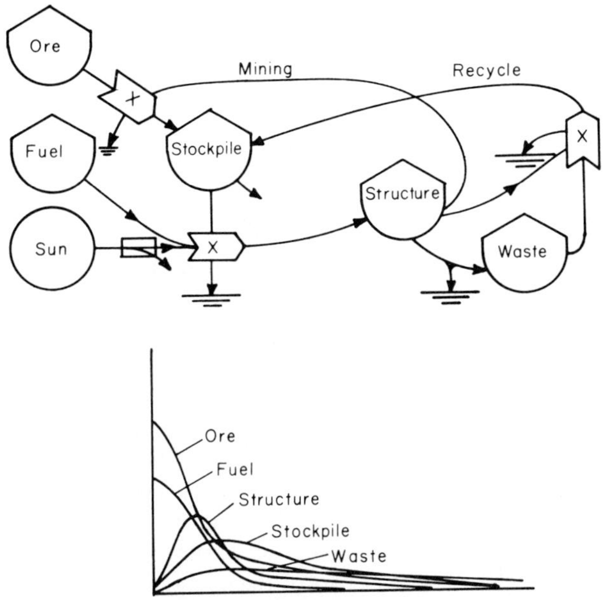

FIG. 13. Recycling and mining.

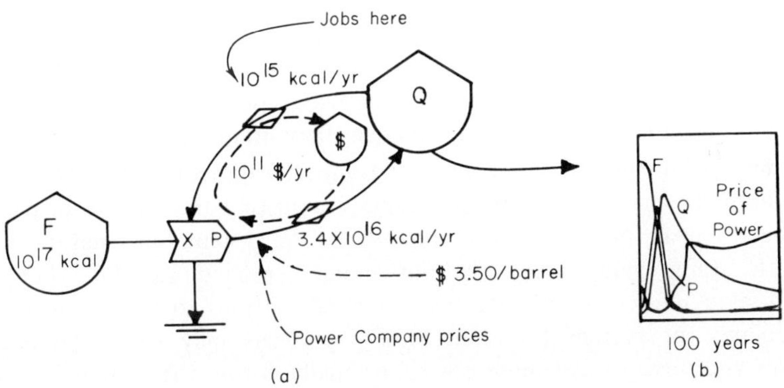

FIG. 14. Basic money–energy relations.

measured; the costs of concentrating and distributing were not subtracted. Much of the presently discussed reserves may actually be negative when the costs of delivering them for use are read in. When there are excess supplies of rich energy it may pay to expend some to develop less rich sources. But as rich energies decline the non-net paying energies must be dropped. Thus, if the new nuclear processes cannot be brought to new levels of magnitude of *net* yield in time, they may have to be dropped entirely.

F. The Reward Loop as a Constant Value Circle, and Countercurrents of Money as Power-Maximizing Mechanisms

The energy theory of value described in *Environment, Power, and Society* (Odum, 1971) apparently caused some confusion because there are two values to be associated with any energy flow, both correct. Figure 14 shows a reward loop feedback system in which entering energies are gradually dispersed into heat sinks according to the second law of thermodynamics. If the feedback flow of energy from Q were diverted and its heat content evaluated, 10^{15} kcal would be found if it were used alone. However, reward loop feedback energies are of very high quality. The energy diverted from Q has the special and unique property of being able to release input flow of raw fuels at P. The flow may be a special nutrient, some information, a technological development, and so forth. If the feedback flow acts as P it becomes part of the energy flow of that multiplicative process and is as much responsible for the 10^{16} kcal output as the fuel flow itself. Thus, the feedback path shares a potential energy value of 10^{16} kcal with the fuel flow as it acts and while it is en route. The energy lost to heat sinks in developing the high quality of energy is needed to maintain the action releasing value constant within the closed loop value circle. Since other, open loop pathways handicap the energy sources they draw upon, natural selection will tend to retain those path loops that provide equal multiplier value to the upstream action in exchange for the energies they drain.

Money in human systems flows in the opposite direction of energy around a value loop. It serves as a lubricant, rapidly adjusting the choices of pathways (feedback loops) for total maximum power of the system. Since natural systems have the same value loops without money, money can be taken as an evolutionary refinement of effective value calculations to harness man's mind and energy for system function. Money patterns may follow from energy patterns and serve to distribute power optimally within the system for survival. Money causes those energy paths to interact that best release more power loop values.

Within the constant value loop there is an overall constant ratio of potential energy value to money. The simplest such loop is shown in Fig. 14, with a countercurrent of money. In general, money flows from a site in proportion to its storage, and is spent soon after it is acquired. Its turnover rate may be two to four times per year, representing the tendency to reinvestment. This is a fast turnover relative to the system in which the flow is a countercurrent, and thus money may show noisy fluctuations in the same way that scarce nutrients do in plankton systems.

1. *Variable Price Models*

Equations for a system of energy and money flow may be forced on a model system using empirical inverse curves for the relationship of price (energy per unit money) and supplies of energy stocks. A simulation with this kind of model was given by Odum and Bayley (1975). However, the model discussed next suggests that these properties emerge from energy flow loops without being forced into the model.

2. *Prices Determined by Energy Flow*

Consider a world system in which the feedback goes to the fuel-processing industry and the storage is the main structure of society. Suppose the amount of money circulating is constant, and the system is putting all its work into developing feedback multipliers to maximize power. Instead of forcing a price relationship into the model, let the price emerge as the ratio of the energy flow and money flow systems. In effect, this assumes expenditures in proportion to concentration. In the Fig. 14 minimodel, price emerges as an inverse function of energies available for purchase and monies available to be spent. With both money and energy flowing according to their stock, their ratio (price) develops in relation to stocks, as in classical (supply–demand) economics. Man, with "tunnel vision," may actually achieve the macroenergetic needs for system survival through greed about money, and culture is apparently selected to do so.

3. *Inflation from Declining Energy*

Now consider the money–energy model of Fig. 14, with a constant supply and turnover rate of money, but with declining energy reserves. When energy is (initially) plentiful, fuel prices are low and money tends to accumulate with the processors of power. Few work services are needed to get the fuel and so the money paid to feedback services is large. People have high salaries and wages, and there is surplus energy to put into higher living standards and seeking auxiliary energies. As fuels decline, feedback work to maintain the system is required. Gross energy consumption may increase but net energies soon begin to drop. Then fuel prices become high and payments for services low. The net energy circulating within the loop is less and the ratio of money to energy is high. This is inflation.

How much of world inflation is due to the declining concentration of fuels? Many economic indicators seem to show economies expanding because gross energy input is increasing. But, as more and more resources go back into securing power inputs, the net effect is less work value for the same energy. Net energy and standard of living depend on energy concentrations.

4. *Economic Competitors in World Fuel Monopoly*

The model shown in Fig. 15 features two economic competitors for fuels, each with a balance of payments with which to buy the fuels. A comparator function selects the pathway with highest fuel prices for the fuel processor to sell to. Another comparator selects lowest prices for purchase of feedback goods and services. The two systems have auxiliary energy sources such as environmental inputs. The one with the greater environmental subsidy is able to sell its services more cheaply, thereby gaining more money for fuel purchase. This element should win the competition. But, later in time, the competitor with higher levels of structure based on the earlier preferred energy position has increased energy losses and less cash. The other system begins to take the

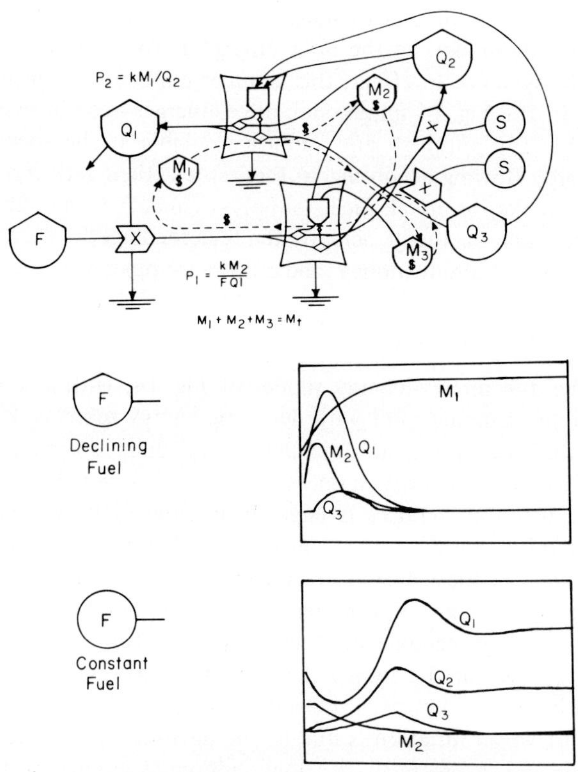

FIG. 15. Competing economic systems.

energies. There is an alternating action of first one buying and then the other (as indicated on the computer by flashing of comparator lights). The net effect is to maximize the coupling of both fuel energy cycles, giving maximum power flow for the whole system.

In this model the monies end up with the monopoly fuel supplier. The model was run with a constant energy source and also one of declining concentration.

G. Disease and Diversity

Diseases often regulate populations. Collectively, diseases generate diversity by eliminating dominance of energy flows. In steady-state situations diseases

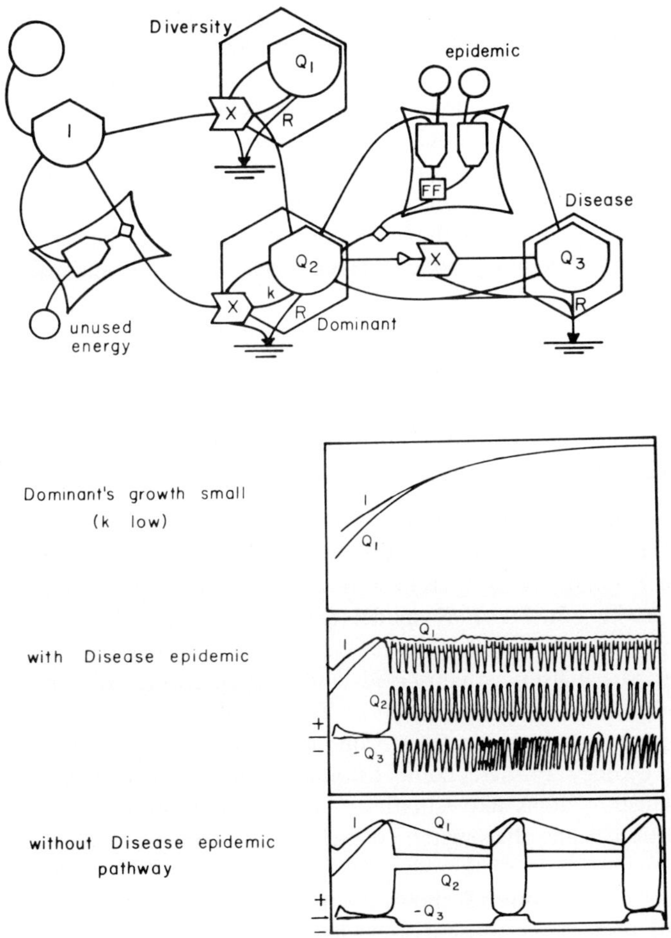

FIG. 16. Disease and diversity.

take relatively small energies, but enough to suppress populations to levels where they are present and available. When host stocks become large, the disease becomes epidemic and repressive square terms begin to operate through such mechanisms as increased probability of infection and increased virulence due to mutation. Diseases, like carnivores, may provide continual energy tests to organisms. A host organism with energy reserves generally dissipates the disease so long as its energy inputs are large. Old age, starvation conditions, or unbalanced density–resource ratios are properties that may set disease into nonequilibrium action.

In Fig. 16 some properties of disease and diversity interactions are incorporated into a minimodel that has provision for epidemic response when stock thresholds of a dominant host are high. There is also a threshold for disease disappearance when host levels become very low. The model can develop dominance first, followed by diversity, with disease acting to ensure that the diverse stocks replace the dominant. By this action the total energy budget of the system is increased through greater efficiencies. The disease is part of the energy-augmenting auxiliary work. The Fig. 16 simulations show oscillations, with period determined by energy flows into the dominant's growth. Disease action produces a faster system with more stability, and the highest energy levels are without the tendency to overdevelopment with the energies going into diversity.

With declining energies available to human systems, disease can be expected to increase as less energy is invested in eliminating the population control role.

H. Designs for Simultaneous Competition and Symbiosis, Environment, and Fossil Fuel

If competition is defined as two systems drawing on the same energy source in parallel, and symbiosis as two systems that are serial parts of a constant value loop, then a number of systems become easily distinguishable, as represented by examples in Fig. 5. The *P–R* model (Fig. 17) is a cooperative, reward loop linkage that has been much used to simulate diurnal and seasonal patterns in ecosystems. If matter is conserved and recycled, the system becomes a Michaelis–Menten module (Fig. 4e). The parallel competing models and predator–prey models in Fig. 5 were among the first models (Lotka–Volterra) in population ecology. In these, competition is distinct from symbiosis.

When larger systems like estuaries or world energy sources are diagrammed, opportunistic designs that switch to different combinations of energy sources begin to emerge. These systems may have cooperative and competitive pathways at the same time. Figure 18 shows a simple case of two parallel (competitive) modules in respect to nutrient use when there are only two energy sources. When there is only one source the units become symbiotic and recycling.

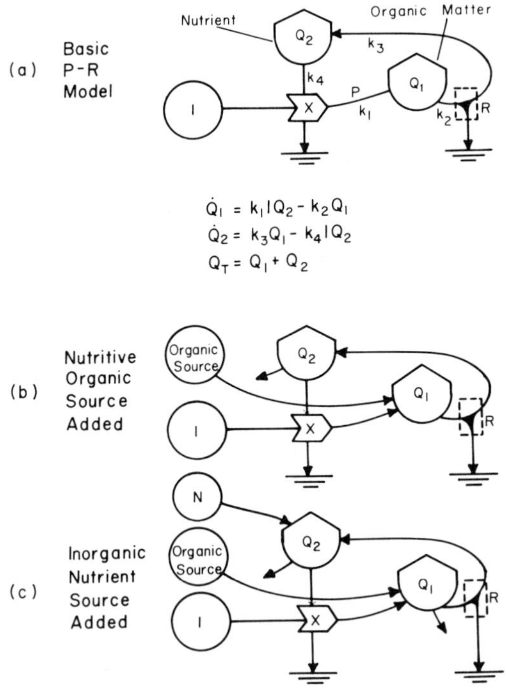

FIG. 17. (a) Basic *P* and *R* recycle model for closed system and its variation with (b) organic and (c) inorganic advections.

Estuaries of the central Texas coast sometimes receive river organic inputs from the land, which leads to bacteria–plant competition for nutrients. When these inflows are zero, the heterotrophs use only plant output and become recycling consumers. With converging food chains higher animals in these situations switch to exploit both chains as the energy source ratios change from plants to microconsumers.

The worldwide man–nature system is similar, with organic inflows from fossil fuels supplementing solar energy, both contributing to man's urbanism as a switch consumer mechanism. Without fossil fuel man is symbiotic with the solar-based environment; with it his relationship takes on competitive aspects. Viewing the system's overall maximum power, however, the design with both competitive and cooperative pathways allows flexibility in drawing from many energy sources, thereby optimizing survival. Switching between energy sources when competing for scarce nutrients in effect allows a continuous selection for maximum power.

The time dynamics curves in Fig. 18 indicate consequences of the system

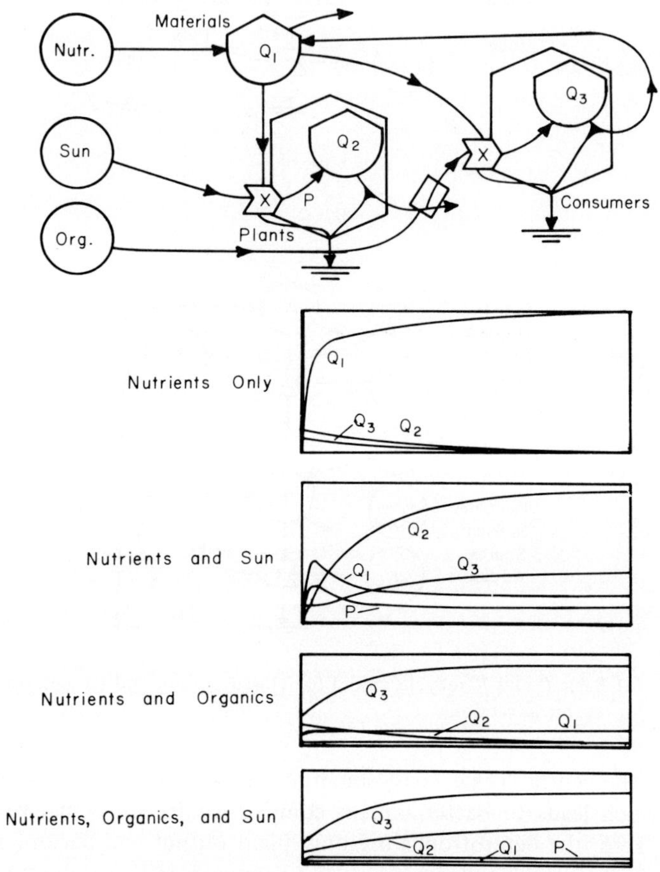

FIG. 18. Mixture of parallel and series.

(same parameter values) operating with energy supplied in four combinations from two sources: on, off, and each alone. Whereas the model is simple and clear enough, attempts to describe it in terms of symbiosis and competition concepts lead quickly to semantic problems. The history of ecology is replete with semantic debates caused by lack of adequate models to expose essential relationships. Language clarification is another benefit of macroscopic mini-models.

The flexibility of serial–parallel designs in switching a system between different inputs in whatever combinations they occur makes for optimality. The system is doing something like automatic linear programming for maximum power.

I. Energy Value Tabulations

Macro-minimodels can also be used to calculate value in large-scale applied problems involving public decisions. We refer to this procedure as an energy cost-benefit method because all flows in the system (matter, information, money, and so forth) are converted to energy equivalents.

Figure 19 shows a simple example of energy evaluation in which parts of the system's transactions are accompanied by money flows and other parts are not.

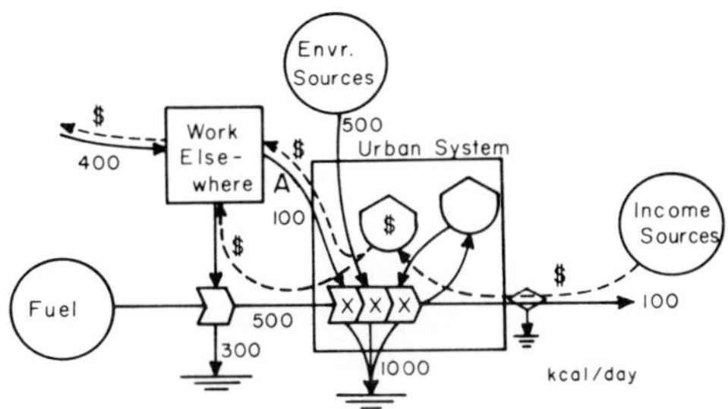

FIG. 19. Diagram for energy analysis.

As in Fig. 19, flows of various kinds of energy are shown in Calories of heat they release when degraded. According to the first energy law, the inflow energies ultimately balance outflow energies. However, the value of these energy flows in causing work differs because the kinds of energy flow are different (sunlight, food, human service, and so forth).

Energy analysis is done by converting the Calories of energy flow of various types and quality into equivalent Calories of the same type such as coal. Money flows as a countercurrent to some of the flows of energy. Money value is believed to equal the energy flow in the opposite direction when that energy has been expressed in units of equal quality. On the average, in 1975 20,000 Calories (Kilocalories) of coal equivalent energy was used per dollar spent. Converting money data to energy data allows one to evaluate high quality goods, services, and labor contributing to the work of a system.

Next, energy flows of the first diagram are multiplied by energy quality factors to put them all on the same basis such as Calories of coal equivalents. For example, sunlight is divided by 2000, because 2000 Calories are required to develop a Calorie of coal. A car built in Detroit with energy flows there brings high energy value with it elsewhere. Its value is measured by the energy required to replace it. Where several energy flows interact energy value is that of their collective action.

After the energy evaluations are tabulated they may be compared with diagrams for alternative plans. Systems with more energy flow can prevail. More energy flow results if more income can be attracted. More income is attracted if there are good, free, environmental energies as matching energy. Therefore, the level of investment that can be economic is determined by the level of environmental sources. Maximum energy flow occurs when economic development is at an optimum intermediate density.

IV. A Note on Symbolic Languages

Most scientists think pictorially, and ecologists are no exception. The diagramming of energy flows as a means to represent ecological systems started early in this century when vague food chain arrow diagrams were first drawn. Gradually, as ecology became more quantitative, these diagrams were made more quantitative. Such evolution is not unique to ecology; dozens of diagramming languages have developed in the highly mathematical engineering sciences (signal flow graphs, block diagrams, electrical networks, etc.).

In 1963, Forrester (1963) introduced a symbolic simulation language in connection with Dynamo, a digital computer language now widely established in programming libraries. The initial applications were industrial and urban problems, but this has now extended to world simulation (see preceding chapter), culminating recently in the best-seller status of "The Limits to Growth." The symbol diagrams of Forrester's language are now extensively used in systems ecology. The basic modules are illustrated in the right-hand side of Fig. 20.

Energy circuit language may be compared to the Forrester scheme. Correspondences between symbols are indicated in Fig. 20. Energy circuit language is less empirical and based more on certain physical and biological principles. Conservation of energy, for example, is built in as heat sinks. Forrester's language visualizes a system as a set of material subsystems linked with cross system coefficients or functions, and influenced from outside through such functions. Energy circuit language shows all paths in a system as energy flows explicitly governed by energy laws. The natural selection principle for maximum power is represented by feedback multipliers on all pathways, making linear models unreal for most purposes (next chapter). Energy force constraints prevent developing a model with more or less action than the energies involved. Switching processes are explicitly represented. In short, energy circuit language is consistent with important principles under which macroscale natural systems

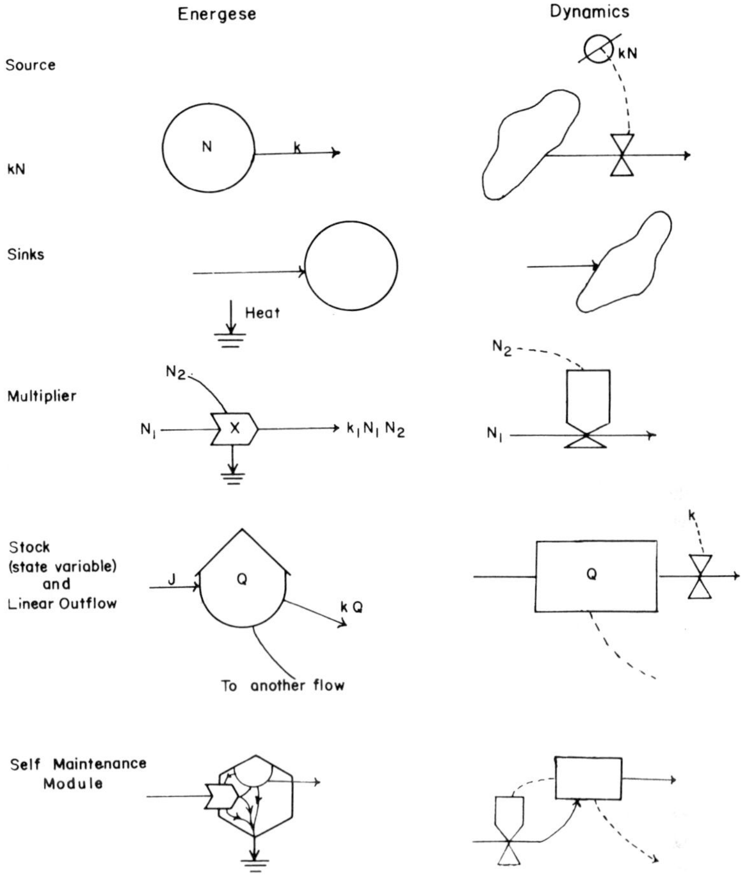

FIG. 20. Forrester translations.

operate. Since these principles are built into the language, the language can be used without conscious attention to them.

As such, it becomes a medium for rapid broad-brushing of the essential aspects of complex systems. It is a tool for suppressing detail and expressing the essence of systems' global operating features. It is a means of gaining quick insights at the whole system level by minimizing the scientist's reductionist tendency. Operationalized on computers, energy circuit diagramming of man–nature systems offers the possibility of piercing complexity, visualizing questions important at the system level, evaluating alternative human actions also at the system level, and in general makes possible a macro approach to man's energetic relation to this planet. Such is the role visualized for macroscopic minimodels.

REFERENCES

Forrester, J. W. (1963). Industrial Dynamics. MIT Press, Cambridge, Massachusetts.

Gilliland, M. W. (1975). *Science* **189**, 1051.

Jansson, B. O. (1971). Ecosystem approach to the Baltic problem. *Bull. Ecol. Res. Comm.* **16**, 1–82. Swedish Natural Research Council.

Kelly, R. A. (1971). The Effects of Fluctuating Temperature on the Metabolism of Laboratory Freshwater Microcosms. Doctoral dissertation, Dept. of Zoology, Univ. of North Carolina, Chapel Hill, North Carolina.

Leaird, J. (1972). What is the dollar value of aesthetics? *Fl. Nat.* **45**, 18.

Littlejohn, C. (1973). An energetics approach to land-use planning in a coastal community of Southwest Florida. Master's thesis, Dept. of Environmental Engineering Sciences, Univ. of Florida, Gainesville, Florida.

Lotka, A. J. (1922). *Proc. Nat. Acad. Sci.* **8**, 147.

Odum, H. T. (1967). *In* "Pollution and Marine Ecology" (T. A. Olson and F. J. Burgess, eds.) pp. 99–157. Wiley (Interscience), New York.

Odum, H. T. (1968). *In* "Primary Productivity and Mineral Cycling in Natural Ecosystems" (H. E. Young, ed.) pp. 81–138. Univ. of Maine Press, Orono, Maine.

Odum, H. T. (1971). Environment, Power and Society. Wiley, New York.

Odum, H. T. (1972a). *In* "Systems Analysis and Simulation in Ecology" (B. C. Patten, ed.). Vol. II, pp. 139–211. Academic Press, New York.

Odum, H. T. (1972b). *In* "Tools for coastal management", pp. 197–213. Marine Technology Society. Washington, D.C.

Odum, H. T. (1972c). *In* "The Changing Chemistry of the Oceans" (D. Dryssen and D. Jagner, eds.) pp. 223–259. Nobel Symposium No. 20. Wiley, New York.

Odum, H. T. (1972d). *In* "Thermal Ecology" (W. Gibbons and B. Sharitz, eds.) p. 628–649. CONF-730505, Tech. Inf. Service, Springfield, Virginia.

Odum, H. T. (1975).*In* "Modelling of Marine Systems" (J. Nihoul, ed.) pp. 127–151. Amer. Elsevier, New York.

Odum, H. T., and Bayley, S. (1975). *In* "Economics for Decision Making for Environmental Quality" (J. R. Connell and E. Loehman, eds.) pp. 284–299. Univ. of Florida Press, Gainesville, Florida.

Odum, H. T. and Odum, E. C. (1976). Energy Basis for Man and Nature. McGraw-Hill, New York.

Odum, H. T., and Peterson, L. (1972). *Archit. design,* **42**, 624.

Odum, H. T., and Pigeon, R. (1971). A tropical rainforest. Division of Technical Information, U.S. Atomic Energy Commisssion, TID 24270. Clearinghouse for Tech. Inf., Springfield, Virginia.

Odum, H. T., Nixon, S., and DiSalvo, L. (1971). *Symp. Amer. Micro. Soc. In* "The Structure and Function of Freshwater Microbial Communities" (J. Cairns, ed.) pp. 1–29. Research Division, Monograph 3, Virginia Polytechnic Inst. Blacksburg, Virginia.

Odum, H. T., Littlejohn, C., and Huber, W. (1972). An environmental evaluation of the Gordon River area of Naples, Florida, and the impact of development plans. Rep. to County Commissioners of Collier Co., Florida.

Richey, J. E. (1970). The role of disordering energy in microcosms. Master's thesis, School of Public Health, Univ. of North Carolina, Chapel Hill, North Carolina.

Zucchetto, J. (1975). Energy basis for Miami, Florida, and other urban systems. Doctoral dissertation, Dept. of Environmental Engineering Sciences, Univ. of Florida, Gainesville, Florida.

PART IV

SPECIAL PROBLEMS IN ECOSYSTEM MODELING

Complexity and large size are attributes of ecosystems that do not confront physical system modelers to the same degree as they do ecological modelers. Many workers, such as H. T. Odum (Chapter 9), advocate minimodels. Others believe that complexity is of the essence and to lump it away may be to destroy what is unique about ecological levels of organization. Some systems ecologists consider that only technical details need be worked out to bring ecosystem modeling to an advanced state. Others are inclined to the view that significant theoretical development will be necessary—theory of modeling, theory of ecosystem as a special class of formal general system, etc. The chapters of this section undertake to examine three basic problems of ecosystem modeling that have become fairly well identified in contemporary discussion. They are linearity versus nonlinearity, aggregation, and validation.

In *Chapter 10*, L. J. Bledsoe discusses the issue of linear versus nonlinear differential or difference equations for representing ecosystem dynamics. The general rationale for modeling real systems and criteria for methodological decisions are first discussed. Man must abstract reality to comprehend it. The abstraction may be symbolized in words, diagrams, programming language, or mathematics. In the latter case, choice of the kind of mathematics used depends on the modeler's view of the system and objectives, and is conditioned by historical factors. Linear models were introduced into ecology to represent dynamics of radionuclide tracers. This choice, justifiable in terms of physical processes of intercompartmental tracer transfers within individual organisms, is less defensible at the ecosystem level. Processes are rarely represented with biological realism in a linear model. Such models provide only a sketch of behavior, with little mechanistic basis. Consequences of linearizing a nonlinear model are illustrated by linearizing the logistic growth model. Linearized models are limited to depiction of equilibria and approaches to equilibria from small deviations. The severe abstraction that results from linearization is illustrated by comparing a linear version to an original representation of primary production in an intraseasonal terrestrial ecosystem model. Finally, five conclusions of linear versus nonlinear modeling are reached: (i) Technical difficulties of model construction are comparable in both cases when the linear model is time-varying; the linear modeler, however, labors under additional constraints of not being able to accept obviously desirable and realistic mathematical analogs of biological mechanisms. (ii) Flexibility to fit available data to model parameters is much greater in the nonlinear case. (iii) Accurate reflection of existing empirical knowledge requires the full range of mathematical functions available only in a nonlinear approach. (iv) Linear models contribute little to understanding of ecosystem processes, although macroproperties may be reasonably well characterized. (v) Model solutions and analyses by numerical methods enable nonlinear ecosystem models to be investigated as fully as linear ones, although analysis procedures specific for linear systems are precluded.

Chapter 11 by Dr. Zeigler considers the aggregation problem from a theory-of-modeling standpoint. Analytic, and to a lesser extent computer simulation, models are said to be limited in descriptive power considering the wealth of known detail about ecosystems. Aggregation of variables, the usual response to this problem, is not very well understood as a formal principle of modeling. The chapter outlines a framework for better understanding. The ecosystem is characterized by a *base model* consisting of input time segments, state space, and a state transition function, and represents a formalization of the modeler's conception of reality. The base model is narrowed to a *lumped model* by a restricted scope of enquiry called the experimental frame. The lumped model, in addition to the three elements of the base model, contains an output space and a mapping (output function) from state space to output space. Thus, the experimental frame supplies observer-determined aspects of the system. The state space of a base model can be reduced to that of the lumped model to achieve a state space *homomorphism*; both transition and output functions are preserved and base and lumped models are behaviorally equivalent. Thus, it is logically possible for a base model to have a reduced lumped model behaviorally equivalent to it. The construction of such base and lumped model pairs by "aggregation of variables" is illustrated. Transition function and output function preservation, and the crucial role of experimental frame in determining existence of valid lumped models given a base model, are demonstrated. Relevance to formulation of ecosystem research, construction of models, model validation and structural inference, correlation and organization of models and model retrieval and application are then discussed. Finally, an example of the concepts developed is given for linear systems in an Appendix.

Chapter 12 by Dr. Caswell analyzes the problem of model validation as it confronts the ecologist constructing a large, complicated ecosystem model. How does a scientist assess how his model conforms to reality? The normal concept is agreement of simulated behavior with a set of time-dependent data. The author examines this concept applied to a model (von Foerster) of human population growth and arrives at a paradox. The model agrees well with existing past and present data and makes reasonable predictions of 2000 A.D. population size, but it projects infite size at 2027 A.D. Investigation of the "validity" of this model leads to a distinction between models for understanding and models for prediction. Validation of a predictive model means only to evaluate its accuracy and the range of conditions over which it is useful. Truth or falsity is not an issue. An explanatory model is validated by a falsification procedure involving a sequence of conjectures and refutations. As long as the model resists falsification it is said to be corroborated. For Caswell, models for understanding are "corroborated" while those for prediction are "validated." The von Foerster population model is valid for a range of time as a predictive model, but uncorroborated as a theoretical model. Ecological models, which frequently combine both predictive and explanatory purposes, must be evaluated by both validation and corroboration criteria and procedures. Validation criteria are straightforward: prediction error, proportion of variance explained, etc. Testing of theoretical models is more difficult, and the procedure of "strong inference" (Platt) involving iterative falsification and refinement of alternative hypotheses is recommended. The chapter concludes with an appeal that ecological models presently need to be more than descriptive. They need to provide steps toward a theory of ecosystem dynamics. As such, they represent theories and should be rigorously tested. Failures have as much to contribute at this point as successes.

10

Linear and Nonlinear Approaches for Ecosystem Dynamic Modeling

LEWIS J. BLEDSOE

CENTER FOR QUANTITATIVE SCIENCE IN FORESTRY, FISHERIES, AND WILDLIFE
UNIVERSITY OF WASHINGTON, SEATTLE, WASHINGTON

I. Introduction

The problem addressed here is the choice of linear versus nonlinear systems of differential equations* for purposes of modeling the dynamics of an ecosystem. First, the general rationale for modeling of real systems and the criteria for general methodological decisions are discussed. Next, the historical development of ecosystem modeling is considered and related to linear and nonlinear

* In this chapter, the term "difference equation" may be substituted for "differential equation" with no change in the arguments presented.

approaches. The choice of a linear or nonlinear approach is then related to the general goals of ecosystem modeling. The mathematics of representing a nonlinear system by a linear approximation are developed and an example for a simple nonlinear differential equation is presented to illustrate the method and its consequences. Finally, five points of modeling consideration are evaluated for the linear and nonlinear approaches in the context of three distinct modeling goals.

II. Modeling Rationale

In attempting to understand the functioning of the ecosystem in which he lives, man is forced to abstract from reality sets of symbolic interactions that he perceives as analogous to some subset of his environment. This must be done because he cannot comprehend directly the interactions of but a few real-life variables at a time. Such a symbolic abstraction of reality is called a model. The symbols used to form a model generally are taken from a system containing general sets of symbols and relations with rules for manipulating the symbols. In a word model, this system is called a metalanguage, such as English; in a diagrammatic model, the symbols are geometric shapes. In a computerized model, a computer programming language is used.

Because of the possibility of making conceptually objective measurements on reality and relating the measurements directly to symbols in the model, and, because of the generality of manipulation methods that are available, mathematics is one of the most useful symbol systems in which to imbed an abstraction of reality. Because of its generality the specification of a mathematical model leaves many decisions to be made about the type of mathematics to be used.

Most mathematical ecosystem models to date have been expressed in the mathematical language of algebra and calculus. A subset of these models includes dynamic models in which ecosystem properties are abstracted as a set of (conceptually) measurable state variables. In this approach, rates of change of the state variables, either over continuous or discrete time intervals, are expressed in terms of basic processes abstracted from the modeler's knowledge of the real system. At this point, however, it is necessary to specify the system of mathematical techniques to be used before further system modeling can occur. Common dichotomies between which choices must be made include discrete or continuous variables (either dependent or independent), deterministic versus stochastic techniques, spatial versus point techniques, and so on. The major considerations in this series of choices are that the decision criterion at each point depend upon the way in which the modeler views the real system, and the purposes to which the model will be put once it is developed (model objectives). These are the primary considerations.

III. Historical Development

Since ecosystem modeling is a fairly new science, the methods of modeling that are open to the systems ecologist are continually increasing and have been increasing for some years. One pattern of development of mathematical modeling in ecology coincides historically with attempts to model movement of radionuclide tracers placed in various parts of ecosystems and allowed to move according to natural processes (Olson, 1964). This type of experiment may have been suggested by similar placement of tracers in animal metabolic systems. The model generally used to describe radionuclide movement in various parts of an animal or plant usually conceived of these parts as being homogeneous compartments in which the nuclide would freely and perfectly mix with other naturally occurring isotopes. The mechanisms by which the nuclides and their naturally occurring isotopes move from one compartment to the other are usually abstracted in terms of diffusion laws in which movement from a compartment surrounded by a semipermeable membrane is governed by concentration differences (actually, entropy differences) between solutions on the two sides of the membrane. In modeling such a system, a mathematical system analogous to this diffusion process was developed. The mathematical system turned out to be a set of linear differential equations. The basic rationale for such a mechanism is that movement out of a compartment is proportional to the concentration of material in that compartment (see, for example, Bergner, 1965). Notice here that the rationale for development was to choose mathematics analogous to physical processes which were occurring.

IV. General Criteria for Modeling Decisions

With this historical and philosophical perspective, let us look at the current state of ecosystem modeling, and particularly the modeling of intraseaonsal dynamics of ecosystems with respect to the choice of a linear versus any other type of differential equation system to define fluctuations of the state variables in time. First, let us define a set of linear, first-order, ordinary differential equations. Equation (1) below defines such a set, where the coefficients a_{ij} are, in general, variables dependent only upon time (i.e., it is not necessary to know the state of the system to determine the value of the variables a_{ij} at time t):

$$dx_i/dt = \dot{x}_i(t) = \sum_{j=1}^{n} a_{ij}x_j + h_i(\mathbf{z}), \qquad i, j = 1, \ldots, n, \tag{1}$$

where $\mathbf{x}(t)$ is the vector of state variables having components $x_i(t)$, and $h_i(\mathbf{z})$ is a function of m system driving variables $\mathbf{z} = [z_k, k = 1, \ldots, m]$. It is important to note that if one element in the set of a_{ij} coefficients is state dependent in any way, even to the extent of a simple step function as a state variable exceeds a threshold, then the set of differential equations can no longer be said to be linear. Methods that mathematicians have devised for analyzing linear systems no longer apply.

As ecologists began to describe the movement of nuclides through systems of organisms and their environment, that is, ecosystems, it seemed a natural extension to apply the same type of mathematics to these systems as had previously been applied in metabolic tracer studies within a single organism.* Notice, however, that in modeling by analogy the ecologist had forsaken the principle of making his modeling decisions based on the realities of the modeled system and the purpose for which the model was intended. This is because it is easier to choose mathematics by analogy rather than exert the mental effort necessary to hypothesize (or determine through literature or data collection) what the actual biological mechanisms might be and translate these into mathematical notation. By borrowing a previously defined mathematical form, the modeler imposes the mathematics upon the biology rather than letting the ecosystem itself dictate the way in which the model is to function.

It might happen, and certainly does in some cases, that the appropriate mechanism based on ecological principles is linear. However, it is certainly not going to be the case that every mechanism will be linear in any ecosystem. Use of the linear approach might appeal to an ecologist because its solution has all the trappings of a good ecological model without requiring a lot of mathematical thought. Involved mathematical techniques are required to solve the equations, the solution has the immediate appearance of giving the dynamic characteristics of a complex system, and yet the model is rather simple to formulate. On the other hand, disadvantages in the use of linear models include the fact that one is not able to relate the predicted phenomena, that is, fluctuations of the state variables, to specific characteristics of the processes that contribute to and make up the structure of the ecosystem. This is because processes are rarely represented with any biological reality in a linear model. Indeed, it would seem that the agreement of the predicted state variable fluctuations with observations in some cases where it has occurred is purely fortuitous and that this agreement can rarely be maintained if the model results are compared with the ecosystem under a widely varying set of driving conditions. Basically, such a linear model then is not more than a pictorial display of ecosystem variability, having no true dynamic characteristics in the sense of being responsive to modifications of driving functions and ecosystem parameter changes. Similar depictions can be built with much simpler mathematics if such is the goal of the model.

Consider the alternative approach to system modeling via sets of ordinary

* This is in accordance with a well-precedented principle in biology: that of borrowing techniques and terminology from other sciences by analogy. For example, since physicists have cyclotrons and bevatrons, biologists must have phytotrons. The curse of biology is physics envy.

differential equations. The general symbolic approach to such a set of equations is not as aesthetically pleasing as Eq. (1) because it does not specify exactly the algebraic form that such equations take. However, whenever a system requires generality, specific directions for accomplishment must be sacrificed. To substitute for this lack of specific direction, the modeler must use his intellect to determine the appropriate form for the functions in each modeling instance. That is, he must determine how each term in each differential equation relates mathematically to the other state and driving variables of the system. At this point, however, little or no mathematics is required.

Ecologists and biologists have for quite some time utilized graphical representations to display relationships between variables, and this approach can again be used to define empirically the functions to be substituted into the above equations. The conversion of these graphs into mathematical notation is strictly a mathematical problem and quite an easy one at that. (The simplest and most straightforward approach, and frequently a satisfactory one, would be to approximate the graphs by a series of connected line segments.) If one does not have enough insight or knowledge about the system to hypothesize graphically, with the aid of a sympathetic mathematician, relationships to go into the terms of the differential equations, then it is questionable whether or not it will be possible at all to enhance one's knowledge of the system by modeling.

V. Effect of Linearizing a Nonlinear System

It is generally conceded by systems ecologists that the true behavior of ecological systems exhibits nonlinear characteristics. Let us examine the mathematical consequences of utilizing a linear model to display the dynamics of a system truly represented by a series of general differential equations, that is, either linear or nonlinear. The usual method of converting an equation to linear form is via a Taylor expansion utilizing the first derivative of the function taken at a point in the state space of the system. Equation (2) represents one differential equation in a set of n in which the m terms g_{ij} are dependent upon time, the state of the system (represented by the vector $\mathbf{x}$), and the driving forces (given by the vector $\mathbf{z}$):

$$dx_i/dt = \dot{x}_i(t) = \sum_{j=1}^{m} g_{ij}(t, \mathbf{x}, \mathbf{z}) + h_i(\mathbf{z}), \qquad i, j = 1, \ldots, n. \tag{2}$$

The functions $h_i(\mathbf{z})$ represent driving or forcing terms.

Suppose we wish to develop a linear expansion about the point in space represented by the constant vector $\mathbf{x}(t_0)$. Equation (3) gives the appropriate linearization of the model:

$$\dot{x}_i(t) = \sum_{j=1}^{m} \Bigg\{ g_{ij}(t, \mathbf{x}(t_0), \mathbf{z})$$

$$+ \sum_{k=1}^{n} \left[\frac{\partial}{\partial x_k} g_{ij}(t, \mathbf{x}(t_0), \mathbf{z}) \right] [x_k - x_k(t_0)] \Bigg\} + h_i(\mathbf{z})$$

$$= \sum_{j=1}^{n} a_{ij}(t, \mathbf{z}) [x_j - x_j(t_0)] + p_i(t, \mathbf{z}); \tag{3}$$

$$\left[\text{notational convention:} \qquad \frac{\partial}{\partial x_k} g_{ij}(t, \mathbf{x}(t_0), \mathbf{z}) = \frac{\partial}{\partial x_k} g_{ij}(t, \mathbf{x}, \mathbf{z}) \Big|_{\mathbf{x}=\mathbf{x}(t_0)} \right].$$

Equation (4) shows the relation between the time-varying coefficients of the linearized model and the original functions, g_{ij} and h_i, of the general nonlinear model of Eq. (2):

$$\begin{aligned} a_{ij}(t, \mathbf{z}) &= \sum_{k=1}^{m} \frac{\partial}{\partial x_j} g_{ik}(t, \mathbf{x}(t_0), \mathbf{z}), \\ p_i(t, \mathbf{z}) &= h_i(\mathbf{z}) + \sum_{k=1}^{m} g_{ik}(t, \mathbf{x}(t_0), \mathbf{z}). \end{aligned} \tag{4}$$

Notice that the coefficients in the linearized version do not depend upon the state variables (except at a constant point in time), and thus the equation system described in Eq. (3) is purely linear.

The procedure outlined above is straightforward and can be applied to any nonlinear model provided only that the functions g_{ij} are differentiable with respect to the state variables in a neighborhood about the point $[\mathbf{x}(t_0)]$ of expansion in the state space of the system. However, it is a consequence of this linearization that the behavior of the linearized system will resemble that of the original system only in a neighborhood about the point in which the linearization is made (and only then with some restrictions on the original nonlinear system). If this is a nonequilibrium* point of the nonlinear model at which positive feedback occurs, then the linearized version will be, in many cases, unstable.

In any event, if the linearized version is stable, there is absolutely no guarantee that the equilibrium point for the linearized equations will bear the slightest resemblance to equilibria in the nonlinear system. The exception to this rule is when the point of expansion is an equilibrium point of the nonlinear model, assuming such points exist in the domain of practical solutions to the system. In this case the linear version might describe the approach to an

* Let an equilibrium point of a system be defined as a point in state space with uniformly zero time derivatives.

equilibrium point of the nonlinear model, but there would be no guarantee that this approach was real for any starting point except one very close to the equilibrium. Such a study could be made with the original nonlinear model, utilizing numerical techniques.

Consider the following example in which the nonlinear differential equation that gives the logistic growth curve, frequently used to describe the growth of one-celled and other organisms, is linearized. The logistic equation is a perfectly legitimate description of one aspect of some ecosystems. However, it is a nonlinear description as can be seen from examination of Eq. (5):

$$\dot{x} = rx(x_{max} - x)/x_{max}, \tag{5}$$

where x is a state variable and r and x_{max} are parameters. Let us use the form of the linearized version as shown in Eq. (6), where the constants a and b are not dependent upon time since the original equation has no time-dependent coefficients or driving functions:

$$\dot{x} = a(x - x_0) + b; \tag{6}$$

$$[\text{notational convention:} \quad x_0 = x(t_0)].$$

Let us examine the resulting equation and the values of a and b when we expand the original nonlinear function given in Eq. (5) about two different points in the state space of the dependent variables. First, expand about the equilibrium point for the nonlinear equation $x_0 = x_{max}$. Obtaining the derivatives of $\dot{x}$ with respect to x we get the results shown in Eq. (7):

$$\begin{aligned} \frac{\partial}{\partial x}[rx(x_{max}-x)/x_{max}]\Big|_{x=x_0} &= r - 2\frac{r}{x_{max}}x_0 = a, \\ rx(x_{max}-x)/x_{max}\Big|_{x=x_0} &= rx_0 - \frac{r}{x_{max}}x_0^2 = b. \end{aligned} \tag{7}$$

Substituting this into Eq. (6) gives the result in Eq. (8):

$$\dot{x} = \left(r - 2\frac{rx_0}{x_{max}}\right)(x - x_0) + rx_0 - \frac{r}{x_{max}}x_0^2. \tag{8}$$

Finally, letting x_0 equal x_{max}, we get the resulting Eq. (9):

$$\dot{x} = -r(x - x_{max}). \tag{9}$$

It can be seen that the constant term b drops out and we have the equation of a simple exponential decay to equilibrium.

Equation (9) shows that the equilibrium state of the linearized model is the same as for the nonlinear system, as expected since the expansion is about an equilibrium point. Since the original system has no driving forces, a more realistic model would have x_{max} as a function of environmental driving variables and other state variables that change the "carrying capacity" of the system. The expansion into a linear model must be made about a fixed point (with respect to the state of the system) and could not reflect the way the system modified its own "carrying capacity" via feedback relations. Thus, the linear version would be valueless for management purposes and is limited to describing an ecosystem at equilibrium, and at approaches to equilibrium from very slight perturbations.

To expand about a different point, let x_0 equal half of x_{max}, the inflection point of the system. Again, substituting this into Eq. (8) we get the result in Eq. (10):

$$\dot{x} = \tfrac{1}{4} r x_{max}. \tag{10}$$

Here, the resulting differential equation no longer depends upon the dependent variable at all, and the resulting solution is a straight line with a positive slope since both r and x_{max} are positive numbers. Biologically, this is an impossible result since it would predict an indefinite* increase in the population at a constant rate.

The purpose of the above exercise is to illustrate that blindly imposing a mathematical procedure upon a model that is supposed to describe, in some sense, a biological reality can give absurd results. The procedure used by the modeler to avoid absurd results when he linearizes a nonlinear system may consist of intuitively juggling the coefficients in the linear model so that the results, that is, equilibrium positions and state variable dynamics, do not counteract biological intuition. However, the consequence of this is that the terms on the linear model bear no resemblance to the processes that actually occur in the ecosystem, and it is impossible to draw from such an approach any inferences about the way the ecosystem functions.

VI. Application to an Intraseasonal Model

As a further exercise into the possibilities of using linear models to depict ecosystem dynamics, let us consider the state variables and equations defining their rates of change for a model of intraseasonal herbage dynamics in a terrestrial ecosystem, such as the producer submodel of Bledsoe *et al.* (1971). The state variables of such a model would include the biomass density of aboveground producer organisms. A simple set of driving functions for such a model would consist of average temperatures and total precipitation amounts over some time span of interest. The equation for the rate of change of the primary production variables should include a dependence upon these two driving forces as well as the amount of biomass present, the harvest rate by

* "Indefinite" within the neighborhood of the inflection point. EDITOR.

consumer organisms, the mineral nutrition factors (assuming that a whole ecosystem approach having a decomposer and/or nutrient cycling section is employed), and possibly upon the phenology of the species to be considered. The other state variables of the system, that is, biomass density of consumers, enter into the consideration in determination of the harvest intermediate variable. The processes contributing to and detracting from the rate of biomass accumulation might be photosynthesis, translocation, death, and harvest.

Now, it is unfortunately a characteristic of linear models that the terms in the differential equation can depend on only one state variable. There is no way to incorporate the dependence of a term, for example, photosynthesis, upon more than one state variable of the model without introducing a nonlinearity. Thus, the modeler must choose among the various state variables, one on which the photosynthesis terms in the primary production equation are to depend. In one approach to linearizing a nonlinear model, Patten (1972) chose to eliminate the dependence of photosynthesis upon the biomass in the primary producer compartment, to convert the soil moisture state variable in the original nonlinear model into a driving function, and to make photosynthesis depend only upon the relative nutrient level given by a state variable in the decomposer section of the model. This approach constitutes a violation of the scientific respectability* of the original model for the purpose of achieving mathematical simplicity. It should be apparent from the foregoing that the outcome is to produce an absurd abstraction of reality, even though the resulting parameters may end in a fortuitous and cursory agreement between the model predictions and the real data of the system. The same would be true for the other terms in the producer equations as well as equations for other sections of the linear model.

VII. Criteria in the Linear–Nonlinear Controversy

In order to make the comparison of linear versus mechanistic approaches to ecosystem modeling more concrete, five specific points, each of which must be considered in the process of ecosystem model building, will be covered below.

A. Technical Difficulties in Model Construction

This subject has to do with problems associated with model building that might be alleviated by the mathematical background of the modeler. One of the overriding attractions of linear models is that a conceptual mathematical

* For purposes of this chapter, "scientific respectability" means the dependence of one scientific work upon previous works. In this case, the previous works are the general scientific knowledge of external environmental factors governing net photosynthesis rate. See Section VII.C.

framework is given a priori and little mathematical ability is required on the part of the modeler to insert coefficient values into the system matrix. On the other hand, frequently it is desirable to have a nonconstant coefficient matrix in a linear model; in fact, this is essential if the resulting linear equations are to display anything but the most trivial types of behavior. Formulation of the time-varying coefficients in terms of driving functions requires some basic mathematical manipulative ability. This same mathematical manipulative ability requirement on the part of the modeler for a linear modeling approach with nonconstant coefficients could be put into the development of more realistic mechanisms for a nonlinear approach with a resulting increase in realism.

In summary, the argument on this point is that the modeler must have a basic mathematical capability and the imposition of the constraint that he must use only a linear model will simply be a hindrance to his modeling activities since it will restrict his freedom to insert the most desirable and realistic mathematical analog of the biological mechanisms into the model.

B. Data Availability

This subject has to do with the relative ease of incorporating data into the model-building process, including model validation as well as development of equations. There are frequently restrictions on the types of data available to the modeler. He must supplement available data using intuition and ideas from the literature about the way the system operates.

In a linear, constant coefficient modeling approach, the interpretation of constants in the coefficient matrix as "rate constants," having units of reciprocal time, enables one to estimate their value quite easily from time series information on state variable dynamics, provided one also has information on the rates at which the basic processes occur (say, units of biomass, energy, or nutrient density per unit time). One simply divides the rate of change of the process by the size of the state variable to come up with a rate constant at any particular time. However, of course, this value will change with time in any real ecosystem; a measure of whether or not the system can truly be described by a linear, constant coefficient model would be the relative variability of the resulting rate constant. For a nonconstant coefficient approach the same procedure could be used if the resulting "rate variable" were considered a function of driving forces only and not of the system state. Otherwise, the linear constraint on the model would be lost.

On the other hand if a nonlinear or more general mathematical technique is utilized in model building, the modeler has at his fingertips not this one procedure but the entire gamut of statistical and numerical analysis procedures available for empirical development of functional representations from data. He is able to combine both an inductive and a deductive approach: deductive in that he hypothesizes relationships on the basis of literature and his intuition into the system, and inductive in that he "fits" the mathematical formulation resulting from his deductive logic to available data. Of course, if the modeler is not acquainted with the use of such techniques, then the linear system can be an advantage in terms of its more straightforward method of deriving coefficient values.

C. Scientific Respectability

This subject deals with whether or not the mathematical mechanisms of a model reflect at a given resolution the totality of scientific knowledge available about a particular process contributing to fluctuations in state variables. Failure to develop scientifically respectable models has resulted in a great deal of criticism of mathematical modeling, and justly so. Scientists go to a great deal of trouble to analyze and gain insight into the detailed processes that contribute to observed variability in natural and man-manipulated ecosystems. For a modeler simply to refuse to acknowledge any of this material by using a rigidly structured modeling procedure based on mathematical rather than ecological considerations is somewhat of a scientific heresy. There is no reason, even in terms of expediency as explained above, not to have available the full range and generality of algebraic and other mathematical expressions for the translation of ecological mechanisms into mathematics. For an adept modeler it is no faster to utilize the linear techniques than to utilize general mathematical techniques, whether linear or nonlinear. This same argument could be used to criticize modeling systems other than sets of linear differential equations. Any mathematical framework without sufficient flexibility to accommodate past scientific findings is subject to the above criticism.

One of the problems of ecosystem modeling is the wide range of traditionally trained ecological scientists who do not understand mathematical modeling techniques and are skeptical about their efficacy. These scientists frequently have a great deal of insight into the operation of the systems with which they work. Sometimes this is subjective, sometimes it can be translated into easily understandable objective terms, but it is rarely mathematical. For a modeler to refuse to acknowledge and utilize this information has the effect of increasing rather than decreasing the gap in understanding between ecosystem modelers and traditionally educated and oriented ecologists. On the other hand, a willingness to translate the ideas of these scientists into mathematical terms builds upon past knowledge in ecology and promotes the unity of the science, resulting generally in greater progress rather than a division of ideas and a dichotomy among workers in the field.

D. Insight into the Ecosystem

This subject has to do with model efficacy as a means of gaining understanding of processes which occur in the ecosystem. It deals with the subject of experiment–model feedback in that understanding of the system via analysis of model predictions requires field testing and promotes the building of hypotheses which are subject to experimental tests. There are two directions in which such understanding might go: toward knowledge of detailed processes of the system as they affect system macroproperties, and toward formulation of generalizations about the macroproperties of the system. In terms of understanding detailed processes, it has been pointed out above that the utilization of linear models can contribute little here because the mathematical functions are not analogous, except in rare instances, to what is actually going on in the system. At the other end of the spectrum, the formulation of generalities about the macroproperties of the system might be aided by a linear modeling approach if the predictions made by the linear model were in some sense accurate and realistic. As has been shown above, this will rarely be the case, except for the simplest ecosystems: for example, artificial microcosms constructed in the laboratory. Nonlinear systems have the capability of being as realistic as one desires to build them, given sufficient time and resources.

E. Mathematical Difficulties in Model Solution and Analysis

This subject has to do not with the technical mathematical ability of the modeler, but rather with the theoretical basis for the work that the modeler is doing. It has been the claim of advocates of linear modeling that a sufficient mathematical theory has not been developed to enable nonlinear model building. This has come about because of the fact that mathematicians have not developed analytical techniques (as opposed to numerical techniques) for the solution of general systems of nonlinear differential equations, and the development is not likely to occur soon, if ever. On the other hand, numerical techniques do allow treatment of nonlinear systems with very nearly the same generality as linear systems. The point of view that a nonlinear system of ordinary, first-order differential equations can be considered as simply a linear system whose coefficients are functions of the state variables of the model, rather than simply the driving functions, makes it apparent that numerical solution techniques for linear models should also be appropriate to nonlinear models so long as the functions of the state variables involved in the differential equation system are "well-behaved" in a mathematical sense.

A second area where mathematical difficulties might be involved is in technical construction of the model. This subject has been discussed above in some detail. The point here is that only a rudimentary mathematical ability is necessary to utilize a number of mathematical techniques in nonlinear ecosystem model building. If the modeler does not have such techniques available he is likely to have difficulty in building linear models as well. This is with the possible exception of linear, constant coefficient models in which the simple substitution of values into a basic formula is all that is required.

A third and final area where mathematical difficulties might be encountered is in terms of analysis of the model utilizing such techniques as application of stability criteria, or frequency and sensitivity analysis.

The stability criteria developed by Liapunov apply equally well to nonlinear as to linear models. In fact, they were developed primarily for nonlinear models. Unfortunately, the mathematician's idea of system stability has little in common with the concepts of ecosystem stability as elaborated by ecologists. For example, a mathematician would consider a system exhibiting a constant asymptote of zero for all state variables as a stable one. However, if an ecologist observes that populations of an ecosystem go extinct or to zero, then he considers the system unstable. Nevertheless, the particular mathematical techniques used to determine stability for a mathematical model of an ecosystem apply equally well to nonlinear as to linear systems. The difficulties in application of techniques to nonlinear systems have to do with the "misbehavior" of the system functions in a mathematical sense. Such misbehavior includes the occurrence of singularities, poles, and discontinuities in the equations. Numerical techniques handle simple discontinuities, of the type encountered in ecological systems, rather well. The effect is simply to subdivide the state space into subdomains in which the system is continuous. The more pathological discontinuities of mathematics do not occur in ecological modeling. Biological functions do not have singularities since it is biologically unreal for a state variable of an ecosystem to have a value of infinity, although the concept might find some utility in abstract analysis.

Frequency and sensitivity analysis are generally applied by numerical techniques even to linear models if they are of sufficient size. Analytic techniques of frequency and sensitivity analysis are only practical for application to small systems because of the exorbitant amount of algebra involved. The aid of a computer is required, for example, to determine the eigenvalues and eigenvectors of a constant coefficient matrix in a linear system. As long as it is necessary to use a computer rather than simply paper and pencil, the computer is sufficiently powerful to utilize numerical procedures to develop analogous techniques with nonlinear models. Unfortunately, many analysis procedures for linear models do not apply in the same manner to nonlinear models because of the more complex behavior exhibited by nonlinear systems. For example, the frequency response of a state variable to a given input or driving function is constant at all operating points of a linear system; this is not so for a nonlinear system. The state variables affect frequency response and thus, when one moves from one operating point to another, the frequency response changes as a variable of the state space. The object lesson here is simply that nonlinear systems require more sophisticated concepts for their thorough analysis than do linear systems. However, the purpose of the analysis is to gain ecological insight into the system and, since the ecological system is basically nonlinear, such sophisticated techniques would be required in any event.

VIII. Conclusion

To conclude we might consider that the appropriate mathematical modeling technique for ecosystem analysis depends upon the object to which the model is to be applied. A convenient categorization of model purposes includes the following three areas. A model might be developed for a teaching purpose, that is, for training persons to build and use ecological models. A second purpose might be management, that is, the construction of a model to enable an ecosystem manager to determine consequences of alternative management activities. Such a model will have very specific handles (parameters and driving functions) dictated by the options available to the manager. A third use to which a model might be put could be classified under a general research objective, that is, a model for gaining general insight into the structure and functioning of an ecosystem. A model in this category is very difficult to define thoroughly since one does not know exactly what state variables of the system should be included. The modeler can exercise a great deal of discretion in specification of state variables. Specification of concrete objectives for such a modeling project is extremely difficult, in contrast to specification of objectives for a management-oriented model.

The general conclusion from the above considerations is that linear modeling techniques may be appropriate for a model whose basic purpose is to teach persons how to construct ecological models. Such persons would have frequently a minimum amount of mathematical ability and they may find a linear framework easier to use. At the same time, in teaching a person to use mathematical modeling, it is important to emphasize the shortcomings of the linear approach and the fact that one must eventually master more sophisticated techniques. It is also important to emphasize that models built as a teaching exercise frequently cannot be extended to research or management. The shortcuts used in building such models preclude extensions by limiting the scientific foundation upon which the model is based. For either the research or management purpose, a strictly linear model is to be avoided at all costs.

One must not be hamstrung by arbitrary mathematical constraints in ecological model building for legitimate scientific purposes. The real constraints imposed by the complexity of the ecosystem are enough difficulty with which to cope. The major exception to this rule is the case of systems where the dynamics and deductive considerations do seem to indicate that linear mechanisms are important (e.g., Patten and Witkamp, 1968). Even as an initial venture, such as low-resolution models for later modification, there is no reason to restrict oneself to a linear formulation. As pointed out above, with the minimal mathematical capability necessary for model building, one should simply develop a model of low complexity without regard to whether the mechanisms are linear or nonlinear. Nothing said above should be taken to imply that the occasional or even frequent use of a linear term in an otherwise general nonlinear system of differential equations is invalid or undesirable; the thesis is that it is generally incorrect to restrict a model wholly to linear equations.

Table I summarizes the conclusions of this author relative to the choice of linear or more mechanistic model formulations for each of the three basic model purposes. An L in the table indicates that consideration goes in favor of linear

TABLE I

Comparison of Linear and Nonlinear Modeling[a]

	Teaching	Management	Research
Ease of construction	L	D	D
Data availability	L	N	N
Scientific respectability	N	N	N
Insight into ecosystem	N	N	N
Mathematical difficulties	—	—	—
Model construction	L	N	N
Numerical solution	D	D	D
Model analysis	D	D	D

[a] Explanation in text. Score: linear models, 3; nonlinear models, 10; draw, 8.

models, an N indicates that it goes in favor of nonlinear models, and a D indicates a draw or equal choice of linear versus nonlinear. The final conclusion regarding linear versus nonlinear models for ecosystem description is summarized below the table.*

* Bledsoe's objective arguments are correct as far as they go. Ecosystem *mechanisms* often seem nonlinear. Their nominal and small perturbation *dynamics* appear, however, to be linear, or at least they can be simulated by linear, time-varying models.

The real ecological question concerns the paradox: linear systems from nonlinear components. This problem is not trivial, and some points which bear on it have been assembled in the article by Patten (1975). Results generated by linear and nonlinear approaches will eventually resolve the issue.

EDITOR

(continued)

REFERENCES

Bergner, P. E. (1965). *Isotopes Radiat. Technol.* **3**, 245.

Bledsoe, L. J., Francis, R. C., Swartzman, G. L., and Gustafson, J. D. (1971). PWNEE: A Grassland Ecosystem Model, USIBP Grasslands Biome Tech. Rep. 64. Natur. Resource Ecol. Lab., Colorado State Univ., Ft. Collins, Colorado.

Olson, J. S. (1964). *Health Phys.* **11**, 1385.

Patten, B. C. (1972). A Linear State Space Model for Grassland, unpublished manuscript.

Patten, B. C. (1975). *Amer. Natur.* **109**, 529.

Patten, B. C., and Witkamp, M. (1968). *Ecology* **48**, 813.

While small perturbation dynamics may be linear, the question of what constitutes nominal behavior or a small perturbation in an ecosystem is far from objectively answerable. When, for example, is counterintuitive behavior in a model a realistic projection, and when is it the result of shortcomings in the necessary approximations made in a model? I argue for minimizing these approximations whenever practicable.

The point about results is well-taken. I look forward to more productive and interesting insights into the behavior of ecosystems, rather than further concern with differences in technique.

LEWIS J. BLEDSOE

11

The Aggregation Problem

BERNARD P. ZEIGLER

DEPARTMENT OF APPLIED MATHEMATICS
WEIZMANN INSTITUTE OF SCIENCE, REHOVOT, ISRAEL

I. Introduction

Ecological systems, almost by definition, comprise a multitude of components bound up in a richly interactive relationship. Computer simulation models of such systems, though they are not as restricted in scope as analytic models, are nevertheless subject to severe limitations in descriptive power, considering the wealth of detail known to exist.

Aggregation of variables has long been the modeler's guiding principle in response to this problem. But the principle has not been well-understood and it has yet to achieve full status in the modeler's kit of construction, test, and analytic tools. For example, rarely will a modeler feel compelled to offer an explication of the particular aggregation employed or an a priori justification of its appropriateness. Patten (1970) has pointed out the difficulties often encountered in relating model variables to real world counterparts.

While this state of affairs is due in part to technical difficulties, it is, more fundamentally, a consequence of the absence of a clear conceptual framework in which aggregational issues can be placed in perspective.

In this chapter, recent efforts to provide such a framework are reviewed in an ecological context and an example of ecosystem application is given. Implications of such thinking for future ecological modeling programs are then outlined. The presentation is geared toward the ecologist more interested in the potential applications of recent developments in mathematical systems theory than in the mathematical niceties. Accordingly, most mathematical concepts are not presented in their fullest generality, and proofs of basic results are not given. The interested reader may, however, find the requisite mathematical support in prior work (Zeigler, 1971, 1972a, b) and the references cited therein. Examples of concepts introduced are presented in the appendix. A recently published book collects these concepts in unified form (Zeigler, 1976).

II. Base Model and Experimental Frames

A. Formalism

We imagine the ecosystem to be characterized by a mathematical structure called the *base model*. The base model specifies (i) the set of all possible input segments which may stimulate the system, and (ii) its response in terms of the change in state brought about in the processing of such a segment. To use a more formal and precise means of expression, we say that the base model is a structure $M_B = \langle \Omega_X, Q, \delta \rangle$ where:

1. X is the *input space*. Here we assume X consists of m-tuples of real numbers, that is, $X = R^m$, where m can be thought of as the number of independent sources of stimuli which will be considered to influence the ecosystem but which themselves are not influenced by it.
2. Ω_X is the set of *input segments* or time functions. We assume here that Ω_X is the set of piecewise continuous functions whose domains are finite intervals of the real line (time) and whose range is X. A typical element of Ω_X is of the form ω: $[t_0, t_1] \to X$, where $t_0 \leqslant t_1$ and t_0 and t_1 are in R. This means that the input stimulus is to be considered to start at t_0 and to last until t_1. The

piecewise continuity means that discontinuous jumps in the stimulus sequence occur only finitely often within the application interval $[t_0, t_1]$.

3. Q is the *state space*. Here we assume $Q = R^n$, where n can be thought of as the number of variables deemed to be relevant to the description of the ecosystem at the finest level of resolution of interest.

4. δ is the *transition function*, which characterizes the mechanism or collection of processes that determine the dynamic behavior of the ecosystem. Formally, δ is a mapping from $Q \times \Omega_X$ into Q which is interpreted as follows: Suppose that at time t_0 an input segment ω of duration $[t_0, t_1]$ is applied. Let q be the state of the ecosystem at time t_0; then $\delta(q, \omega)$ is the state of the ecosystem at time t_1. Thus, the transition function determines what state the ecosystem will end up in at the end of the input segment, given the state it was found in at the beginning of the segment and the description of the segment itself.

It is convenient (and some might argue philosophically correct) to assume that δ is *time-invariant*, which means, in essence, that it depends only on the shape of the input segment and not on the absolute time t_0 at which it is initiated. This forces all time-varying effects to be represented explicitly in the state space Q.

B. Base Model

The base model concept is introduced as an attempt to capture what is often meant by the "real system." It serves to formalize the modeler's thoughts on what he would like to find out about the particular ecosystem, recognizing that he is operating within a framework determined largely by the background of general scientific conceptions of the day. Thus, one ecologist may consider a set of variables at the species level to constitute his X and Q spaces. These become for him, the microvariables in the sense of economic theory, the finest resolution of reality of interest. His object of study is then the progressive uncovering of the transition function δ, given initially his a priori conceptions about the general laws governing interactions of the microvariables derived from physics, chemistry, and general biology. Being more reductionist and ambitious, another modeler may consider the proper description to lie at the molecular level or yet at still more refined levels. What I am claiming is that, in any case, the modeler has selected a base model (really a class of base models) which characterizes the level of resolution of interest to him (the choice of X and Q) and what he would like to find out about how the system works (δ).

C. Experimental Frames

The base model is what the ecologist formulates for himself independently of complexity and utilitarian considerations. The possibility of actually building a useful ecosystem model would be hopeless were base models to be the only ones possible. Fortunately, one is not interested in, or may not need to know, all aspects of system behavior at all times. Thus, given a restricted scope of enquiry, it may be possible to construct a model which is valid relative to that scope and which is at the same time much simpler than the base model. We call such a model a *lumped model*, and proceed to outline the relationship between base and lumped models.

A restricted scope of enquiry is characterized formally by a structure called an *experimental frame, E.* Such a frame specifies:

1. a restriction on the set of input segments which will be fed to the ecosystem (base model);
2. a specification of the observational time base, that is, the instants at which ecosystem behavior will be sampled;
3. the initial states in which the base model will be considered to start at the beginning of an experiment;
4. a characterization of the instrumentation employed in making observations, or, in other words, a statement of those aspects of the state behavior which are of interest in this scope of enquiry.

The additional specifications carried by experimental frame E convert the base model into a full-fledged system, S_E. In general, a *system* is a structure $S = \langle \Omega_X, Q, Y, \delta, \lambda \rangle$, where Ω_X, Q, δ are understood as above, Y is an output space, and λ is a mapping from Q to Y. If we consider only the effect of restriction 4 above of the experimental frame (which is enough to demonstrate the general idea), then S_E differs from the base model $\langle \Omega_X, Q, \delta \rangle$ only in the addition of an output space Y and output function λ.

Thus, we can consider each experimental frame E to supply a pair (Y_E, λ_E) which characterizes what aspects of the system one is looking at, or is concerned with, in this frame. For each E, the base model $\langle \Omega_X, Q, \delta \rangle$ is transformed into a full-fledged system, $S_E = \langle \Omega_X, Q, Y, \delta, \lambda \rangle$.

III. Homomorphism and Behavioral Equivalence

Relative to S_E, the state space Q need not be minimal; that is, it may be possible to reduce the state space by lumping together many equivalent states, while still preserving the same input–output behavior. In other words, it may be possible to construct a lumped model S_E' which is much simpler than the base model S_E but is, nevertheless, valid within the frame E.

To give an idea how this may come about we consider systems having the same input specification Ω_X and output specification Y but perhaps different state spaces, transition and output functions Q, δ, λ. Let S be such a system.

With each state q in Q, we associate an *input–output* function β_q mapping Ω_X to Ω_Y. The idea is as follows: Starting in state q and applying the input segment ω, we observe the output segment $\beta_q(\omega)$ (determined as follows: the domain of $\beta_q(\omega)$ is the same as that of ω, say $[t_0, t_1]$; its range is Y and $\beta_q(\omega)(t) = \lambda[\delta(q, \omega_t)]$, where ω_t is to the part of ω starting at t_0 and ending at t; this holds for each t in $[t_0, t_1]$).

We call the collection $\{\beta_q\}$ of input–output functions the *behavior* of S and denote it by $B(S)$. This concept is distinct from the *union* of all the β_q, which we call the *input–output relation* $IO(S)$.

Two systems, S and S', are said to be *behaviorally equivalent* if $B(S) = B(S')$. Again, this is distinct from a weaker equivalence in which $IO(S) = IO(S')$. It is known that the first implies the second but the converse is not true.

The important concept, which this is leading up to, is that of state space *homomorphism*. This is the mechanism by which the state space of a base model can be collapsed into that of a lumped model.

For systems S and S' a mapping h from Q onto Q' is called a *homomorphism* if:

(1) the transition function is preserved,

$$h(\delta(q, \omega)) = \delta'[h(q), \omega]; \tag{1}$$

(2) the output function is preserved,

$$\lambda(q) = \lambda'[h(q)]. \tag{2}$$

S' is said to be a *homomorphic image* of S if there exists such a homomorphism.

Now, the important connections between homomorphism and input–output behavior are contained in the following facts:

Fact 1. (Sufficiency of homomorphism)

If S' is a homomorphic image of S, then S and S' are behaviorally equivalent.

This statement asserts the sufficiency of the homomorphism concept in the sense that if we can ascertain that a homomorphism actually exists from S to S' then we are assured that the input–output behavior of S' is identical to that of S. Moreover, by definition of the map h, the state space of S' is no larger than that of S, and may in fact be much smaller.

Thus S' may be said to be a *valid model* of S in the sense that the input–output behavior is the same for both.

The converse of Fact 1 is not necessarily true except in the important special case when S' cannot be modeled by a smaller system. We say S' is *reduced* when every pair of distinct states q_1, q_1 has distinct input–output functions β_{q1} and β_{q_2}. In other words, in a reduced system no two distinct states "behave," from an external viewpoint, in exactly the same way for all inputs.

Fact 2. (Necessity of homomorphism)

If S and S' are behaviorally equivalent and S' is reduced, then S' is a homomorphic image of S.

Thus, not only does a homomorphism guarantee behavioral equivalence, but it is the only way such equivalence can be obtained when S' is the smallest equivalent system possible.

Facts 1 and 2 can be used in a number of ways. For the present, we discuss the construction of hypothetical base and lumped model pairs in which the lumped model is a homomorphic image of the base model. From previous discussion, we know that, given an experimental frame E, it is logically possible that a base model S_E has a much reduced lumped model behaviorally equivalent to it. Moreover, Facts 1 and 2 assure us that the only way to construct the smallest lumped model equivalent is through the homomorphism concept.

IV. Construction of Base–Lumped Model Pairs

A. Discrete Time Systems

Construction of base–lumped model pairs is facilitated by going over to discrete time systems. A discrete time system S may be specified by a structure $M = \langle X, Q, Y, \delta, \lambda \rangle$, where now δ maps $Q \times X$ into Q and is interpreted as a single time step transition function. This function dictates how a state q at time t and an input x at time t are transformed into a state $\delta(q, x)$ at time $t + 1$. (Every continuous time system S has a family of discrete time systems equivalent to it in the sense that, for each $h > 0$, the behavior of S at instants . . ., $-2h$, $-h$, 0, h, $2h$, . . . is replicated exactly by an associated discrete time system M_h. Usually, however, the input space X, which now consists of function segment names, is too large to deal with and approximations must be made via numerical analysis.)

B. Construction of Output Free Base–Lumped Model Pairs

Our goal now is to demonstrate the construction of a class of base–lumped model pairs of possible ecological significance. The approach is first to consider systems without input and output and build pairs which satisfy the first condition for homomorphism—the preservation of transition structure, that is, Eq. (1). Then we return to consider the crucial role of input–output properties in determining the existence of valid lumped models. This involves consideration of Eq. (2), that is, the requirements for output function preservation.

1. *Base Model Definition*

The base model $M = \langle Q, \delta \rangle$ is specified as follows:

1. The ecosystem is considered to be a collection of n *species*. The species are named either by greek letters $\alpha, \beta, \gamma, \ldots$, or by subscripted greek letters $\alpha_1, \alpha_2, \alpha_3, \ldots$, as convenient.
2. Each species α has an associated real-valued state variable referred to as q_α. This can be interpreted as the biomass of the species. Thus the state space $Q = \{(q_{\alpha_1}, q_{\alpha_2}, \ldots, q_{\alpha_n}) | q_{a_i} \in R\} = R^n$.
3. The species are divided into a number n' of *compartments*. The compartments are named by greek letters primed $\alpha', \beta', \gamma', \ldots$, or $\alpha_1', \alpha_2', \alpha_3', \ldots$. The set of species in a compartment α' is denoted by (α').
4. Each species α is governed by a first-order equation,

$$q_\alpha(t+1) = a_\alpha q_\alpha(t) + i_\alpha(t), \tag{3}$$

where a_α is a time constant (related to the continuous model turnover rate), and i_α will account for the influences of other species.

5. The time constants of all species in a compartment are the same. Thus, we can set $a_\alpha = a_{(\alpha')}$ for each species α in compartment (α').
6. The input i_α to a species α is sensitive only to the total biomass of any compartment that influences it. Thus, the individual species of a compartment exert their influences only in an aggregate way. Therefore, i_α can be expressed as

$$i_\alpha(t) = \sum_{i=1}^{n'} C_{\alpha,\beta_i'} \sum_{\beta \in (\beta_i')} q_\beta(t), \tag{4}$$

where $C_{\alpha,\beta'}$ is the coefficient of influence of compartment β' on species α, and $\Sigma_{\beta \in (\beta')} q_\beta(t)$ is the total biomass in compartment (β').

Note that $C_{\alpha,\beta'}$ may be different for different species α, even if they are in the same compartment. Also, $C_{\alpha,\beta'} = 0$ if and only if species α is not influenced by compartment β'.

7. From 4–6 above we may write

$$q_\alpha(t+1) = a_{(\alpha')} q_\alpha(t) + \sum_{i=1}^{n'} c_{\alpha,\beta_i'} \sum_{\beta \in (\beta_i')} q_\beta(t). \tag{5}$$

Then the transition function $\delta: Q \to Q$ has been defined since

$$\delta[q_{\alpha_1}(t), q_{\alpha_2}(t), \ldots, q_{\alpha_n}(t)] = [q_{\alpha_1}(t+1), q_{\alpha_2}(t+1), \ldots, q_{\alpha_n}(t+1)]. \tag{6}$$

2. *Lumped Model Definition*

The lumped model $M' = \langle Q', \delta' \rangle$ is specified as follows:

1. Each compartment is considered to be an indivisible unit. Thus, the coordinates of the state space are the names of the compartments, α', β', γ', and so on. With each coordinate α' we associate a real-valued state variable q_{α}' (which we will come to associate with the total biomass in compartment α'). Thus, the state space

$$Q' = \{(q_{\alpha_1'}, q_{\alpha_2'}, \ldots, q_{\alpha_{n'}})|q_{\alpha_i'} \in R\} = R^{n'}. \tag{7}$$

2. The dynamics of each compartment α' are very simple:

$$q_{\alpha'}(t+1) = a_{(\alpha')}q_{\alpha'}(t) + \sum_{i=1}^{n'} k_{\alpha', \beta_i'}\, q_{\beta_i'}(t), \tag{8}$$

where $a_{(\alpha')}$ is the same coefficient as previously defined in IV.B.1.5 above, with the new interpretation of being the turnover rate for compartment α'. We set

$$k_{\alpha',\beta'} = \sum_{\alpha \in (\alpha')} c_{\alpha',\beta'}, \tag{9}$$

and $k_{\alpha',\beta'}$ is the coefficient of influence of compartment β' on α'. As before, $\delta'\colon Q \to Q'$ is now defined since

$$\delta'[q_{\alpha_1'}(t), \ldots, q_{\alpha_{n'}'}(t)] = [q_{\alpha_1'}(t+1), \ldots, q_{\alpha_{n'}'}(t+1)]. \tag{10}$$

3. *Verification of Transition Function Preservation*

Now we can show that M' preserves the transition structure of M. This involves finding a mapping h from Q onto Q' such that Eq. (1) holds. Since there is no input, we have to show that

$$h[\delta(q)] = \delta'[h(q)]. \tag{11}$$

Another way of saying this is that when $q(t)$ goes to $q(t+1)$ under the action of δ it should be the case that the corresponding state $q'(t)$ goes to the corresponding state $q'(t+1)$ under the action of δ', the correspondence being that dictated by h, namely $q' = h(q)$. Less formally stated: the way that the individual species change states must be reflected in the way that the compartments as individual entities change states.

The appropriate mapping is that which associates with each compartment α' the sum of the state variables represented by that compartment. Thus, in the biomass interpretation, the state variable associated with α' is the total biomass of the species it contains. More formally,

$$h(q_{\alpha_1}, q_{\alpha_2}, \ldots, q_{\alpha_n}) = (q_{\alpha_1'}, q_{\alpha_2'}, \ldots, q_{\alpha_{n'}'}), \tag{12}$$

where

$$q_{\alpha'} = \sum_{\alpha \in (\alpha')} q_{\alpha}. \tag{13}$$

Let us check that Eq. (11) is satisfied with the map of Eq. (12). In the base model, we have

$$\sum_{\alpha\in(\alpha')} q_\alpha(t+1) = \sum_{\alpha\in(\alpha')} \Big[a_{(\alpha')} q_\alpha(t) + \sum_{i=1}^{n} c_{\alpha,\beta_i'} \sum_{\beta\in(\beta_i')} q_\beta(t)\Big]$$

[using Eq. (5)]

$$= a_{(\alpha')} \sum_{\alpha\in(\alpha')} q_\alpha(t) + \sum_{i=1}^{n'} \sum_{\alpha\in(\alpha')} c_{\alpha,\beta_i'} \sum_{\beta\in(\beta_i')} q_\beta(t)$$

$$= a_{(\alpha')} \sum_{\alpha\in(\alpha')} q_\alpha(t) + \sum_{i=1}^{n'} k_{\alpha',\beta_i'} \sum_{\beta\in(\beta_i')} q_\beta(t) \tag{14}$$

[using Eq. (9)].

In the lumped model we have, from Eq. (8),

$$q_{\alpha'}(t+1) = a_{(\alpha')} q_{\alpha'}(t) + \sum_{i=1}^{n'} k_{\alpha',\beta_i'} q_{\beta_i'}(t). \tag{15}$$

Equations (14) and (15) become identical when we set $q_{\beta'}(t) = \Sigma_{\beta\in(\beta')} q_\beta(t)$ and $q_{\alpha'}(t+1) = \Sigma_{\alpha\in(\alpha')} q_\alpha(t+1)$, as required under the mapping h. Thus Eq. (11) holds for each compartment and so the homomorphism has been validated.

C. The Role of Experimental Frames

So far we have considered models without output and we have found that M' preserves the transition structure of M. Whether M and M' can be extended to a base–lumped model pair now depends on the experimental frame chosen for M. Briefly stated, M' cannot serve as a valid lumped model for M in any frame which allows observation of M at the species level, but can serve as a valid model in any frame which enables observation no finer than the compartment level.

To see this, let $\{Y, \lambda\}$ be specified by a frame E so that the base model in E becomes $M = \langle Q, Y, \delta, \lambda\rangle$. Now $M' = \langle Q', Y, \delta', \lambda'\rangle$ is a homomorphic image of M if and only if the second equation, enforcing output preservation [Eq. (2)]

$$\lambda(q) = \lambda'[h(q)],$$

is satisfied (since Eq. (1) has already been satisfied). Thus, it follows that a λ' exists if and only if λ does not vary over states that have the same image under the mapping h.

Thus, for example, if we want to track the behavior of a particular species α_1, we specify

$$\lambda(q_{\alpha_1}, q_{\alpha_2}, \ldots, q_{\alpha_n}) = q_{\alpha_1}. \tag{16}$$

But M' cannot be extended to a valid lumped model in this frame since, for example, if α_1 and α_2 are in the same compartment ($\alpha_1{}'$), then

$$\lambda(0, 1, q_{\alpha_3}, \ldots, q_{\alpha_n}) = 0 \neq \lambda(1, 0, q_{\alpha_3}, \ldots, q_{\alpha_n}),$$

but

$$h(0, 1, q_{\alpha_3}, \ldots, q_{\alpha_n}) = h(1, 0, q_{\alpha_3}, \ldots, q_{\alpha_n}).$$

However, if we want to account for the growth behavior of the ecosystem populations as a whole, then we set

$$\lambda(q_{\alpha_1}, q_{\alpha_2}, \ldots, q_{\alpha_n}) = \sum_{i=1}^{n} q_{\alpha_i}, \tag{17}$$

and there is an extended lumped model valid for this frame, namely

$$\lambda'(q_{\alpha_1'}, q_{\alpha_2'}, \ldots, q_{\alpha_{n'}'}) = \sum_{i=1}^{n'} q_{\alpha_i'}. \tag{18}$$

Thus we see here illustrated the crucial role of the experimental frame in determining the existence of valid lumped models for a given base model.

The technique illustrated just now is often called "aggregation of variables." By using the homomorphism concept, Zeigler and Weinberg (1970) were able to give necessary and sufficient conditions under which aggregation of variables can yield valid models. The ability of aggregation to preserve component functional relationships and connectivity is explored by Zeigler (1972a). Application of these ideas to nerve net modeling is demonstrated in Zeigler (1972b). Formal statement of the concepts and their application to a variety of fundamental problems in modeling and simulation appears in Zeigler (1976).

V. Some Implications for Ecological Modeling

Like all large-scale systems, the complexity of an ecosystem is likely to necessitate a plurality of models, appropriate for various purposes, rather than a single model adequate for all purposes. Thus, it seems that a multitude of models will have to be constructed, validated, and organized into a coherent whole. The concepts outlined in this chapter may be suggestive for dealing with the following problems associated with this task:

1. *Formulation of Ecosystem Research*

Attempting to formulate the underlying base model and the experimental frames of interest could help to encourage program designers to delineate the ultimate level of description sought, the levels which could serve for various restricted purposes, and the general postulates and laws to be taken for granted in the construction of models.

2. *Construction of Models*

Such concepts as homomorphism and aggregation may be used to aid in the construction of simulateable lumped models adequate for limited scopes of application and to place models in perspective relative to an underlying base model.

3. *Validation of Models and Structural Inference*

The concepts outlined here can be extended to discuss the empirical validation of models and the justification for inferring base model unknown structure from known lumped model structure.

4. *Model Correlation and Organization*

The base model concept can serve as a background against which otherwise disparate models may be placed in perspective. Using the inference concepts just mentioned, information validated via the lumped models may be stored as constraints on structural aspects of the base model. An organization might be imposed by having models linked together and to the base model via homomorphism and like concepts.

5. *Model Retrieval and Application*

Having a good model organization may facilitate the determination of which lumped model is most appropriate for a given applications problem. Complexity considerations, level of validation, and similarity of experimental frame to that demanded by the problem would be some of the factors entering the calculation.

Needless to say, the theoretical problems entailed surpass, by far, current capabilities. However, as papers presented in this volume already suggest, the problems of ecosystem modeling may necessitate, and profit from, an integrated approach such as here suggested.

Appendix. Example of Concepts in the Case of Linear Systems

A. Base Model and Restriction to Experimental Frame

1. *Base Model*

$$M_B = \langle \Omega_X, Q, \delta \rangle.$$

where

$$X = R^m,$$

$$\Omega_X = \{\omega | \omega:[t_0, t_1] \to X; t_0 \leqslant t_1; t_0, t_1 \in R,\ \omega \text{ is piecewise continous}\},$$

$$Q = R^n,$$

$\delta: Q \times \Omega_X \to Q$ is given by

$$\delta(q, \omega_{[t_0, t_1]}) = e^{At} q + \int_{t_0}^{t_1} \exp [A(t_1 - t')B\ \omega_{[t_0, t_1]}(t')\, dt',$$

where A is an $n \times n$ matrix and B is an $m \times n$ matrix. Thus, the system is derived from the set of n first-order differential equations denoted $\dot{q} = Aq + Bx$.

2. *Base Model Restricted to Experimental Frame, E*

$$S_E = \langle \Omega_X, Q, Y, \delta, \lambda \rangle,$$

where $Y = R^p$, and λ: $Q \to Y$ is given by $p \times n$ matrix C, so that $\lambda(q) = Cq$.
The pair (Y, λ) is determined by E.

B. Input-Output Function

For $\beta_q : \Omega_X \to \Omega_Y$, the input–output function is given by $\beta_q(\omega) = \mu$, where if ω: $[t_0, t_1] \to X$, then μ: $[t_0, t_1] \to X$ such that

$$\mu(t) = Ce^{At}q + \int_{t_0}^{t} C \exp [A(t - t')]\ B\omega(t')\, dt'.$$

C. Input-Output Relation

$$IO(S) = \bigcup_{q \in Q} \beta_q = \{(\omega, \mu) | \quad \text{there exists} \quad q \in Q \quad \text{such that} \quad \beta_q(\omega) = \mu\}.$$

D. Systems S and S' Such That S' Is a Homomorphic Image of S

Let S and S' be specified by matrices $\{A, B, C\}$ and $\{A', B', C'\}$, respectively. Then, if there exists a matrix P such that

$$A'P = PA, \qquad B' = PB, \qquad \text{and} \qquad C = CP,$$

then P (as linear transformation) is a homomorphism from S to S' and S' is a homomorphic image of S.

E. Reduced System

$$\beta_q = \beta_{q'} \Leftrightarrow Ce^{At}q + \int_{t_0}^{t} C\exp[A(t-t')]\,B\omega(t')\,dt'$$

$$= Ce^{At}q' + \int_{t_0}^{t} C\exp[A(t-t')]\,B\omega(t')\,dt'$$

for all t and ω,

$$\Leftrightarrow Ce^{At}(q-q') = 0 \text{ for all } t,$$

$$\Leftrightarrow [C, CA, CA^2, \ldots, CA^{n-1}] \text{ has rank } p,$$

where C is $n \times p$.

F. Discrete Time Systems

A *linear discrete* time system $M = \langle X, Q, Y, \delta, \lambda\rangle$ is specified by matrices $\{A, B, C\}$ such that

$$\delta(q, x) = Aq + Bx,$$
$$\lambda(q) = Cq.$$

This can be interpreted as

$$q(t+1) = \delta[q(t), x(t)] = Aq(t) + Bx(t),$$
$$y(t) = \lambda[q(t)] = Cq(t).$$

REFERENCES

Patten, B. C. (1970). *Simulation* **15**, 264.

Zeigler, B. P. (1971). *In* "Computers and Automata" (J. Fox, ed.). Wiley, New York.

Zeigler, B. P. (1972a). *J. Ass. Comput. Mach.* **19**, 742.

Zeigler, B. P. (1972b). *In* "Theory and Applications of Variable Structure Systems" (R. R. Mohler and A. Ruberti, eds.), p. 69. Academic Press, New York.

Zeigler, B. P. (1976). "Theory of Modelling And Simulation." Wilev, New York.

Zeigler, B. P., and Weinberg, R. (1970). *J. theor. Biol.* **29**, 35.

12

The Validation Problem

HAL CASWELL*

DEPARTMENT OF ZOOLOGY AND DIVISION OF ENGINEERING RESEARCH
MICHIGAN STATE UNIVERSITY, EAST LANSING, MICHIGAN

I. Introduction

In this chapter I want to present some thoughts on model "validation," particularly as it confronts the ecologist constructing a large, complicated ecosystem model. Little of what I will offer would appear new to someone trained in the philosophy of science. My justification for presenting the material is its complete absence from the systems ecology literature. I will not give any dramatic solutions or new techniques; I will be satisfied to direct some attention to a facet of modeling that will become more and more important as more and more ecological modeling projects near completion.

* Present address: Ecology Section U-42, University of Connecticut, Storrs, Connecticut.

II. Normal Concept of Validation

What *is* "validation," as it is used in systems ecology? My dictionary sent me from "validate" to "verify," which was defined as "to establish the truth, accuracy, or reality of." This definition already hints at the roots of the ambiguity in our use of the word. As I will show later, the distinction between truth or reality and accuracy is an important one.

The use of the word in modeling acknowledges the basic scientific practice of comparing one's ideas with reality to see how they measure up. Although there are signs in this book that the situation is changing, the process in systems ecology currently seems to go something like this: A model, consisting of a set of differential or difference equations, is constructed, and the parameters estimated from field data or literature. The initial conditions of the model are set to correspond with the initial point of some sequence of data (hopefully different data from that used to evaluate the parameters), and the resultant solution of the model generated by simulation or analysis. This solution is compared with the data sequence: if the agreement is good, the model is validated; if not, the usual procedure is to modify the model—perhaps by changes in parameter values, perhaps by more basic structural changes—until agreement is obtained.

III. A Model Paradox

I want to discuss briefly an example that highlights one of the problems with this approach to validation. It is taken from an interesting series of papers by von Foerster *et al.* (1960, 1961a,b, 1962) and von Foerster (1966). The problem at hand is to model the growth of the human population of the entire world.

Let $N(t)$ represent the total world population at time t. Then we can write

$$dN(t)/dt = f(N)N(t), \tag{1}$$

where the growth rate function $f(N)$ expresses the effect of density on population growth. This much is ecological old hat; one of the classic derivations of the logistic equation follows by letting $f(N)$ be a decreasing linear function of N. This is a sensible first approximation for a group of organisms whose only interaction is to interfere with one another in a scramble for a limited supply of resources. But, reasons von Foerster, this is not the case with man at all. The logistic model describes a population whose members take part in a multiperson, zero-sum game among themselves for resources. Man's ability to construct societies, civilizations, and technologies turns the situation into a two-person game, man versus nature, and man's success varies directly with population size. Under this hypothesis $f(N)$ is not a decreasing but an increasing function of N, for which von Foerster chooses

$$f(N) = aN^{1/K}, \qquad K \simeq 1, \tag{2}$$

where a is a constant and K is slightly less than one. This is still ecologically reasonable; in fact, von Foerster is not alone in proposing $f(N)$ functions with at least some regions of positive slope (e.g., Slobodkin, 1953; Odum and Allee, 1954; Rosenzweig and MacArthur, 1963; Kilmer, 1972). The resulting equation is

$$dN(t)/dt = aN(t)^{1+1/K}, \tag{3}$$

which has the solution

$$N(t) = N(t_0)\ [(t^* - t_0)/(t^* - t)]^K, \tag{4}$$

where t^* is a constant whose importance will become apparent shortly.

Very well then, here we have a model of an ecological system. A much simpler model, to be sure, than any ecosystem model, but one which should be amenable to the same validation procedure. Von Foerster and his colleagues estimate the parameters in the model from 24 estimates of world population size, carefully screened for mutual independence, over the last 2000 years or so. Figure 1 shows the agreement between the model and the data. Considering the magnitude of the sampling error which must be inherent in this data, the agreement is outstanding. More importantly, there is no sign of any systematic deviation from the straightline relation predicted by the model.

Table I shows another indication of the accuracy of the model. The United Nations' median projection of world population size for the year 2000 A.D. is compared with the prediction based on Eq. (4). This projection is based on far more detailed demographic data than is Eq. (4). Yet, as the U.N. accumulates more and more information, its projections seem to be converging to the prediction made by von Foerster's model.*

TABLE I

U.N. estimate of population at 2000 A.D., estimate made in: 1950	1957	1958	1959	1970	Estimate from von Foerster's model, made in 1960
3.20	5.00	5.70	6.20	6.49	6.91 (billions)

**Note added in proof:* As of 1975, there is additional evidence of the accuracy of Eq. (3). The Population Reference Bureau's estimate for world population in mid-1975 is 3.97 billion; von Foerster's prediction, made 15 years earlier, is 3.65 billion. We are actually ahead of schedule on the way to Doomsday. See J. Serrin (1975).

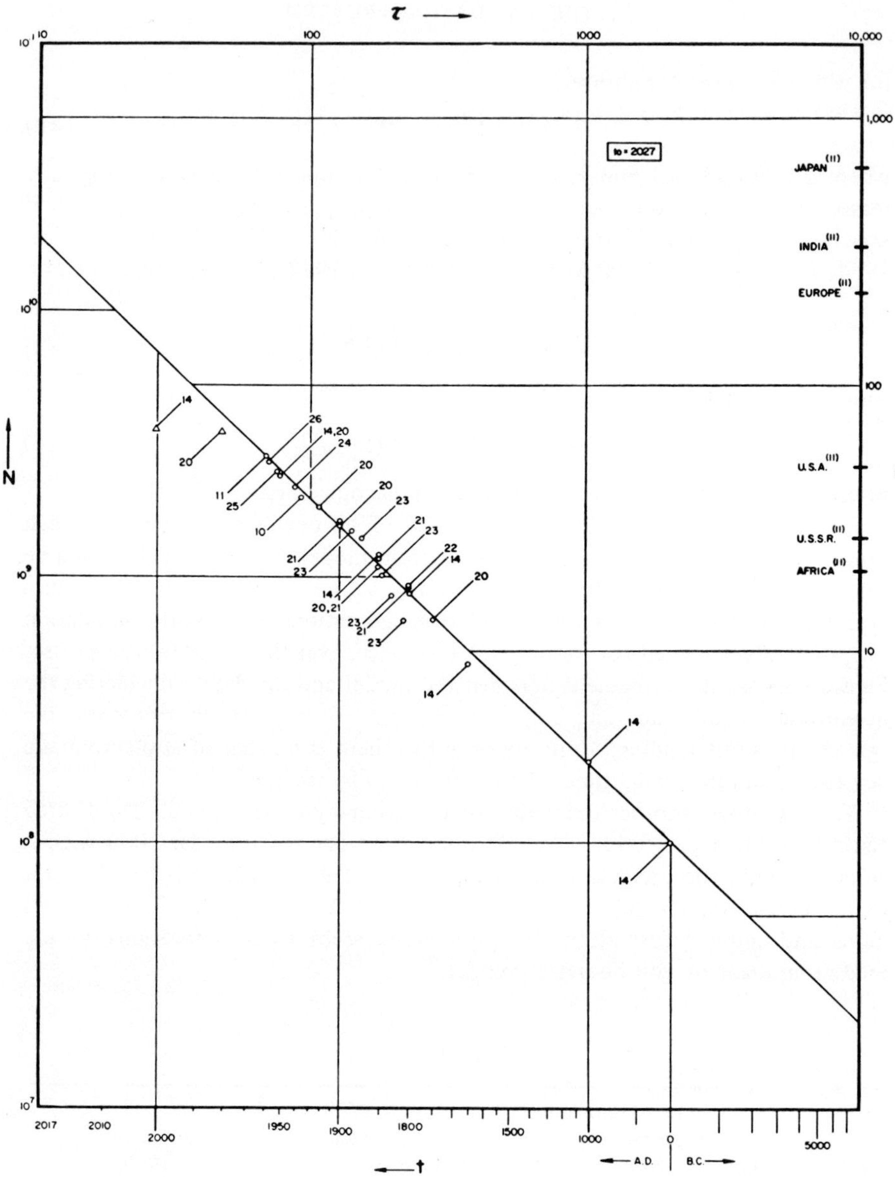

FIG. 1. A log–log plot of world population size N against time t (note the reversed scale for t), compared to the straight line predicted by Eq. (4). The upper abscissa measures time in units of $\tau = t^* - t$; the right-hand ordinate expresses population density (individuals per square mile) n, with density estimates for various countries in 1958 indicated for comparison. The triangular points on the graph are not data points, but projections made by various authors. The numbers indicate references in von Foerster's original paper. Reproduced with permission from von Foerster *et al.* (1960). Copyright 1960 by the American Association for the Advancement of Science.

So far, so good. The agreement of the model and available data is satisfactory; certainly it has passed the usual validation procedure. But so far we have examined only its accuracy; what can we say about its truth or reality? A glance at the denominator in Eq. (4) reveals that this model is a member of a class of models which reach, not approach but *reach,* infinite values in finite time. When $t = t^*$, $N(t) = \infty$. The parameter values obtained by von Foerster *et al.* (1960) allow t^*, "Doomsday" as they call it, to be estimated. Their estimate places it at 2026.87 A.D. ± 5.5 years. If their model is correct, the number of people on earth will become infinite in 50 years.

So we are faced with a quandary. This model gives an excellent fit to data, but generates projections that seem unbelievable. (Systems ecology is not devoid of similar examples; consider a donor-controlled compartment model of a food chain which produces solutions that agree with the data, but which also claims that the system is completely insensitive to variations in the top trophic level.) Has the model of the world population been validated or not?

IV. Two Purposes of Modeling

The problem, which lies at the root of this quandary, is one of purpose. The purposes for which a model is constructed have much to say about how important accurate predictions and a good fit to data are in the process of validation.

Models of systems are systems themselves, the interacting components of which are mathematical variables and expressions. They are, moreover, man-made systems, "artificial" systems in the terminology of Simon (1969). The construction of such artificial systems is a design problem, and the process of design is in essence a search for agreement between properties of the artificial system and a set of demands placed on it by the designer (Simon, 1969; Alexander, 1964; Chermayeff and Alexander, 1963). It is impossible to evaluate the success or failure of a design attempt without specification of these demands, the task environment in which the artificial system is to operate. Validation of a model is precisely such an evaluation, so we need to consider the purposes for which models are constructed.

In particular I will distinguish between (i) models that are constructed primarily to provide accurate *prediction* of the behavior of a system, and (ii) models that, as scientific theories, are attempts to gain *insight* into how the system operates. Both the means and the ends of what we call "validation" are different for these two types of models.

A. Models for Understanding

The goal of scientific hypotheses or theories is to increase our ability to explain how nature works. "Theories," says Popper (1959), "are nets cast to catch what we call 'the world'. . . . We endeavor to make the mesh ever finer and finer." In logical form, theories are universal statements, claims that something is always and/or everywhere true. Philosophers of science are fond of the example, "All ravens are black."

Figure 2 outlines the statements involved in a systems theoretical model. Statements are made concerning the identity and behavior of objects in the system and describing the interactions between objects. These statements are embodied in mathematical representations as "free-body models" and "constraint equations," respectively (Caswell *et al.*, 1972). Note that the mathematical representation of a particular set of statements is not necessarily unique. The free-body models and constraint equations are combined to give a mathematical representation of the behavior of the entire system. Conclusions about the system are derived from this representation by deductive logic in the form of mathematical analysis or simulation. As in the example above, these

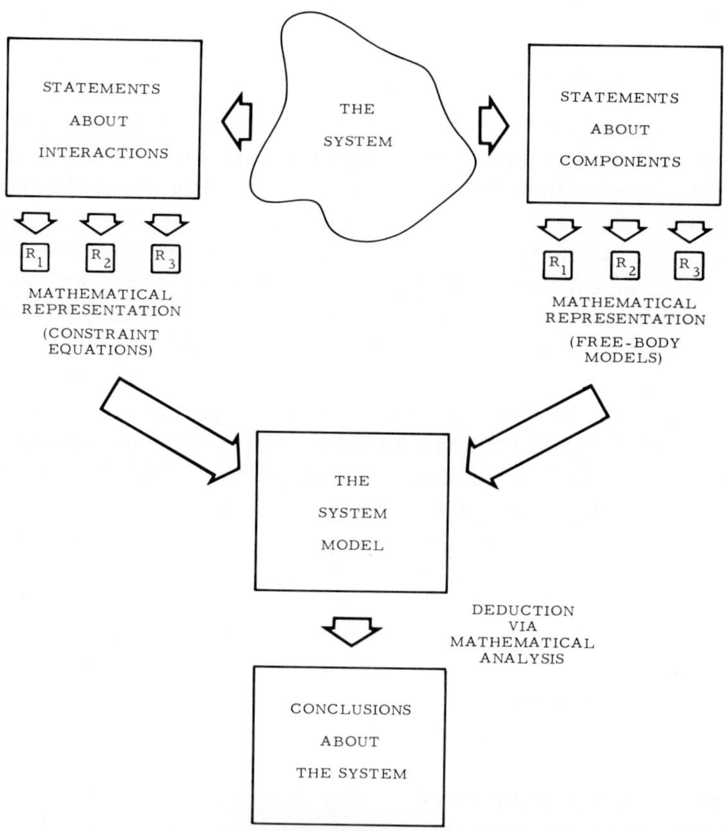

FIG. 2. The theoretical statements involved in the construction of a system model.

conclusions include not only solutions of the model but more qualitative aspects dealing with limiting behavior, sensitivity, etc.

For these theoretical models the problem of validation is simply the old philosophical problem of induction—how can one logically infer the truth of universal statements from observations of particular phenomena? Popper (1959, 1963) has studied this problem in depth, and concluded that in fact there is *no* way to show a theory or any universal statement to be true via induction. No matter how many black ravens we see, the next one may be an albino. More importantly, he makes a convincing case that the procedure of science should be quite the opposite; the scientist's energies should be devoted to attempts to *invalidate* his theories. This invalidation is logically possible because one can deduce, from the universal statements of the theory and certain singular statements defining the details of the situation ("initial conditions"), sets of conclusions which can disagree with particular observations. If this happens, the universal statements from which the conclusions were derived have been invalidated. Science, Popper maintains, is a sequence of "conjectures and refutations," hypotheses or theories put forward by the imagination of the scientist and tested until they are refuted. A theory inherently incapable of refutation belongs not to science, but to metaphysics.

B. Models for Prediction

The process of evaluating models by attempting to refute them needs to be looked at in more detail, but first I want to examine the validation problem as it applies to another class of models. These I refer to as *predictive* models, models designed primarily to provide accurate quantitative predictions of the behavior of the system.

Predictive models crop up regularly in situations where one's livelihood depends on the ability to make accurate predictions—in applied sciences, in resource management, in engineering. The important point is that the truth or reality of the model is never at issue. You can navigate on the basis of a model of the universe that ignores relativity, survey as if the world were flat, and build a telescope without worrying about whether light is a set of waves, a stream of particles, or something more mystifying. Perhaps the canonical form of purely predictive model building is stepwise multiple regression, where variables are included in or rejected from the model solely on the basis of their contribution to its accuracy. Figure 3 is an outline of the construction of a purely predictive regression model, written with industrial application in mind. Nowhere is there any hint of testing the truth or reality of the model. In fact, as Popper (1963) points out, more often than not such testing is redundant, since the model is known in advance to be false. Instead, our concern is to evaluate its accuracy and the range of conditions over which it is useful.

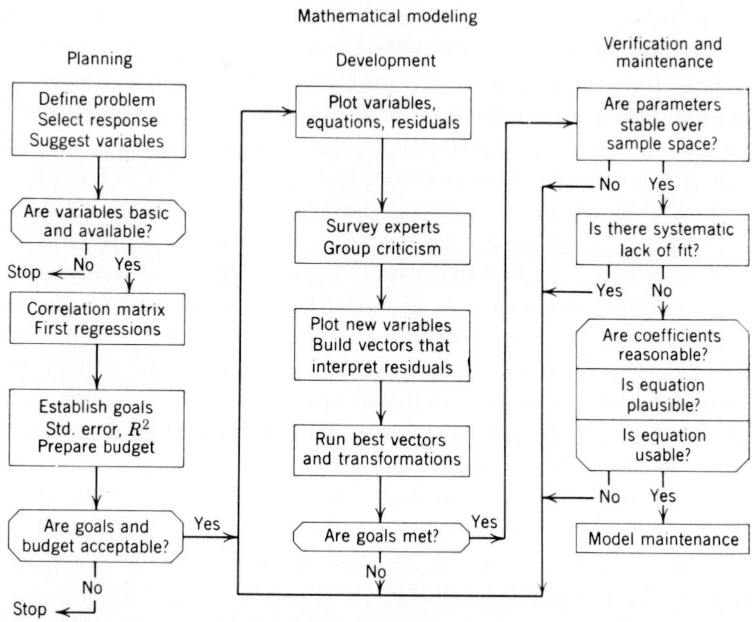

FIG. 3. A flow chart outlining steps in construction of a predictive model by multiple regression analysis. From Draper and Smith (1966) with permission.

V. Corroboration versus Validation

"Validation" of a predictive model, then, means something very different from "validation" of a scientific theory. In the first case the goal is to determine when and where the model is a useful predictor. In the second case the goal is to attempt to refute the model, and our confidence in its statements increases as it passes more and more severe tests. To distinguish these two approaches I will henceforth use Popper's term "corroboration" to refer to the theoretical case:

> So long as a theory withstands detailed and severe tests and is not superseded by another theory in the course of scientific progress, we may say that it has proved its mettle or that it is *corroborated* (Popper, 1963, p. 33).

I will reserve "validation" for the case of the predictive models.

What then of von Foerster's world population model? If it is interpreted as a predictive model, it seems to be doing reasonably well; it is at least more highly validated than either the exponential or logistic models. If it is a theoretical model, it may be considered refuted, unless we are willing to refute instead some highly corroborated theories, which maintain that there is only a finite amount of matter in the universe. To the extent that it mixes those purposes we are left in confusion. Many ecological modeling efforts do in fact mix the two purposes. This is particularly true of the International Biological Program (IBP) modeling activities, for the IBP contains the mixture in its basic philosophy. This is not necessarily bad, and at any rate it is unavoidable. The distinction should be kept in mind though; allowances should be made for the existence of models (or segments of models) that have been refuted but validated, or corroborated but invalidated.

There is a very important practical difference between corroboration of a theoretical model and validation of a predictive model. As Fig. 3 shows, it is possible to establish, a priori, criteria that allow one to stop and say that a predictive model is satisfactorily validated. These criteria can be stated in terms of the error of prediction, the proportion of the variance explained, the cost of improving the model versus the cost of a given level of error, etc. There is no way this can be done for a theoretical model. The process of testing must continue indefinitely, since no matter how many tests the model has passed, it may fail the next one.

VI. Testing Theoretical Models

I want to return now to the process of testing theoretical models. The usual procedure of "validation" described earlier can be a highly inefficient way of testing a model. Most ecological models, and in particular ecosystem models, are very complex, at least in terms of the number of variables and parameters they contain. This makes it difficult to know when to reject a model by comparing output time series with observed data. Statistical criteria of rejection are usually not applicable, and without such criteria it is difficult to decide just *how much* disagreement between model and data is required for rejection. But the corroboration of a theoretical model increases with the severity of the tests it passes. If we want highly corroborated models, statements about ecosystems in which we can have a high degree of confidence, then we need approaches that will reduce to a minimum the uncertainty about rejection of a false model. Some suggestions follow.

First, don't test entire models; test them in pieces (the component models *and* the interaction constraints). This not only reduces the complexity of the models being tested, it has other advantages that will become apparent shortly. Secondly, theories are not to be tested in a vacuum, but are to be compared to alternative theories (Popper, 1959). Platt (1964) has given the systematic application of this technique the name "strong inference." He outlines the procedure as follows:

1. devising alternative hypotheses,
2. devising a crucial experiment, the outcome of which will exclude at least one of the hypotheses,
3. carrying out the experiment, and
4. returning to 1, refining the problem further and further.

Platt claims that the different rates of progress in different branches of science are due in large measure to the frequency with which their scientists practice strong inference. While it may often be difficult to decide if the behavior of a model is aberrant enough to warrant rejection, it is much easier to compare the results of two and decide which is better. The almost total neglect of the construction of alternative models in systems ecology is no doubt a result of our training. Most ecologists encounter the idea of alternative hypotheses only in statistics. In statistics, the construction of alternative hypotheses demands little or no ingenuity beyond the decision between a one-tailed or two-tailed test. When dealing with theories on the order of complexity of ecological models, developing alternative models demands as much ingenuity as constructing the initial model, but it is effort well-spent.

A third point—don't overlook the role of designed experiments in testing theoretical models. The conclusions of a theory that provide the most severe tests, and hence the highest degree of corroboration, are those that are the most surprising; that are expected on the basis of the theory in question, but *not* on the basis of existing or alternative theories (Popper, 1959, 1963). Such unexpected conclusions are not likely to involve time series of model output. Any pair of alternative models will both generate output, and if the modeling process has proceeded very far the outputs can be expected to be quite similar. The surprising consequences of a model may often be qualitative, or at least testable in qualitative fashion. Conclusions concerning the number, position, and stability properties of equilibria, or the insensitivity of certain variables to others (cf., Blalock, 1971) would seem to be prime candidates for this type of testing. For example, a test comparing a donor-controlled compartment model with a Volterra-style model (using cross-product terms) might be carried out much more rigorously by comparing the sensitivity of the herbivore level to changes in predator abundance than by comparing output time series from the two models with data.

The corroboration of a theoretical model demands that it be subjected to as severe a set of tests as possible. Toward this end we can test component and interaction models rather than entire system models, and devise explicit alternative hypotheses and crucial distinguishing experiments. But after this, what? The process of scientific development outlined by Popper and Platt is cyclical; each refutation must be followed by a new set of conjectures. It is not enough to reject a model and throw it away. This would be a waste of the insight that went into the model in the first place, and would quickly drain the imagination of any scientist. Instead we need to be able to pin the blame, at least tentatively, on some portion of the model. By changing this portion and continuing the testing procedure we can incorporate what was learned from the experiments or observations that led to rejection.

In the case of von Foerster's world population model the blame can be fixed relatively easily. We have enough experience with alternative models of similar form to guess that the growth rate function, $f(N)$, is responsible for the conclusion of infinite population size. We would propose then, as an alternative model, that at some critical value of N, $f(N)$ will change from $aN^{1/K}$ to some other function. This is in fact von Foerster's interpretation. He points out (von Foerster *et al.*, 1961a) that, in physical systems described by such equations, the values do not actually become infinite. Before that happens the system undergoes a radical change—a change of phase, ionization, breakdown, an explosion—in other words, the system breaks, so that it no longer exhibits the same behavior.

The ease with which we can tentatively attribute the refutation of von Foerster's model to a particular statement is due to its small size. It would be impossible to do the same with an ecosystem model in which a behavior is the result of the interaction of literally hundreds of variables, parameters, and functions. This is another reason to conduct tests on portions of the model rather than the whole thing. This may be the strongest reason, for without the ability to recycle our testing procedure we have gained nothing by refuting a model. The use of explicit alternative models is also helpful. If two models differ in only some of their statements, differences in their behavior can be at least tentatively attributed to the differing statements.

I have encountered the suggestion that sensitivity analysis be applied to this problem. The idea is that the parameters to which the misbehaving variable is most sensitive should be prime candidates for adjustment to correct the model. This makes some sense if the model is purely predictive, but none at all in the case of a theoretical model. In the latter situation such alterations represent ad hoc modifications of the statements of the model, and moreover assume that only a small subset (those dealing with parameter values) of the statements are open to correction.

It might be objected that if the corroboration process I have outlined was followed rigorously, systems ecology would be brought to a grinding halt. No one, it might be claimed, can build a model of an ecosystem that cannot be refuted almost immediately. In one sense this is true, but in another important sense it is not. I alluded earlier to the possibility that a model might be invalidated as a predictor but *not* refuted as a theory. The conclusions of a theoretical model that are usually tested are those deduced from one of the mathematical representations of the theoretical statements (see Fig. 2). This may force us to deal with far more detail than our statements intend. Suppose the statement "the variable X varies in the same direction as the variable Y" is embodied in the mathematical representation $X = a + bY$. This representation might be a poor predictor of X, and a test applied to the mathematical representation might result in refutation of the model. But a test of the *statement* itself might corroborate it. Efforts could then be directed to incorporating more detail in the statement. The point is that we must be careful to keep in mind just what we are testing when we attempt to refute a theoretical model.

There is another side to this coin which is worth mentioning in passing. The conclusions from a model might be a result of the particular mathematical representation chosen (see, e.g., May, 1973) rather than of the statements. Because he deals with relatively simple mathematical representations of rather subtle ideas, Levins (1968) has been particularly concerned with this problem. He refers to results which can be obtained from several different mathematical representations as "robust theorems." This is an internal check on the model, not an attempt to refute it by comparison with the real world; hence it is not really a form of corroboration.

One often gets the impression that "validation" is something to be done to a model after it is completed. The outline I have presented suggests that generating alternative models, devising tests, and revising models should be carried out, component by component and interaction by interaction, while the model is being constructed. It also speaks strongly for constructing separate models of the components and interactions that make up the system, a point which has been made before for other reasons (Caswell *et al.*, 1972).

VII. Conclusion

I hope that this discussion of the validation and corroboration of models will be useful to systems ecologists engaged in actual modeling efforts. We need models that do more than accurately describe ecological systems; we need models that represent steps toward a theory of ecosystem dynamics. We won't get them until we begin treating our models like theories and subjecting them to rigorous tests. I would like to see alternative models and the results of comparative tests set out explicitly in modeling publications. At this point we have as much to learn from our failures as our successes.

Acknowledgments

This research was supported by the National Science Foundation through Grant GI-20 and a Graduate Fellowship.

REFERENCES

Alexander, C. M. (1964). "Notes on the Synthesis of Form." Harvard Univ. Press, Cambridge, Massachusetts.
Blalock, H. M. (1971). "Causal Models in the Social Sciences." Aldine-Atherton, Chicago, Illinois.
Caswell, H., Koenig, H., Resh, J., and Ross, Q. (1972). *In* "Systems Analysis and Simulation in Ecology" (B. C. Patten ed.), Vol. II, p. 3. Academic Press, New York.
Chermayeff, S., and Alexander, C. M. (1963). "Community and Privacy." Doubleday, Garden City, New York.
Draper, N. R., and Smith H. (1966). "Applied Regression Analysis." Wiley, New York.
Kilmer, W. L. (1972). *J. Theoret. Biol.* **36,** 9.
Levins, R. (1968). "Evolution in Changing Environments." Princeton Univ. Press, Princeton, New Jersey.
May, R. M. (1973). *Amer. Natur.* **107,** 46.
Odum, H. T., and Allee W. C. (1954). *Ecology* **35,** 95.
Platt, J. R. (1964). *Science* **146,** 347.
Popper, K. (1959). "The Logic of Scientific Discovery." Harper, New York. (Originally published in 1935 as *Logik der Forschung.*)
Popper, K. (1963). "Conjectures and Refutations: The Growth of Scientific Knowledge." Harper, New York.
Rosenzweig, M., and MacArthur, R. (1963). *Amer. Natur.* **97,** 209.
Serrin, J. (1975). *Science* **189,** 86.
Simon, H. A. (1969). "The Sciences of the Artificial." M.I.T. Press, Cambridge, Massachusetts.
Slobodkin, L. B. (1953). *Ecology* **34,** 430.
von Foerster, H. (1966). The Numbers of Man, Rep. #13.0, Biological Comput. Lab., Univ. of Illinois, Urbana, Illinois.
von Foerster, H., Mora, P., and Amiot, L. W. (1960). *Science* **132,** 1291.
von Foerster, H., Mora, P., and Amiot, L. W. (1961a). *Science* **133,** 936.
von Foerster, H., Mora, P., and Amiot, L. W. (1961b). *Science* **133,** 1931.
von Foerster, H., Mora, P., and Amiot, L. W. (1962). *Science* **136,** 173.

PART V

THEORY IN ECOSYSTEM ANALYSIS

This final section of the book presents chapters concerned with theoretical aspects of ecosystems.

Chapter 13 by Drs. Waide and Webster explores the applicability of engineering systems analysis procedures to ecosystem study. Time domain, frequency response, stability, and sensitivity analyses are examined. Time domain analysis investigates timing and magnitude characteristics of simple standard signals (singularity functions) propagating through linear systems. Frequency response analysis characterizes a system's response change in magnitude, phase, or angular frequency compared to these properties of sinusoidal inputs. Stability analysis investigates resistance (stability against displacement) and resilience (recovery following displacement). For ecologically realistic patterns of connectivity, increased connectance and system size both increase stability. Sensitivity analysis is a method of investigating relationships of system state variables to parameters. All systems possess a linear sensitivity substructure which mediates propagation of small parameter deviations to state variables. The chapter concludes that engineering and ecological systems analysis are fundamentally different. Engineers seek to design and build systems according to well-specified operating criteria. Ecologists must analyze and manipulate systems that already exist as culminations of a long evolutionary history. Some engineering techniques may apply, others may not, but more analysis is needed to augment the heavy commitment to simulation which so far has marked systems ecology.

Chapter 14 by Drs. Mulholland and Sims illustrates the applicability of optimal control theory to problems in ecosystem regulation. A method is presented for controlling large systems with a relatively small number of parameters or inputs. The philosophy is to reduce a problem to comprehensible proportions by mapping key indicators of larger systems into state variables of smaller system models. To regulate an ecosystem a desired performance (e.g., insect pest reduction) must be specified. Then a control policy (predator introduction) can be developed and implemented. The problem is complicated by multiple objectives (minimize pests while maximizing crop production), and tradeoffs are difficult to decide when systems are large. General solution of the problem is provided, and extensions to linear and nonlinear examples given. In conclusion, modern systems analysis and control theory concepts together with ecological theory made operational in models are seen as providing the technical basis for an ecosystem engineering of the future.

Chapter 15 by Drs. Astor, Patten, and Estberg further elaborates the system sensitivity substructure concept as introduced in Chapter 13. Developments are illustrated by examples utilizing simple population models familiar in elementary ecology. The mathematical context is differential equations. Sensitivity is the difference between perturbed system behavior and some unperturbed (nominal) reference dynamics. A first-order approximation to this difference is developed, then formulated as the solution of a set of linear sensitivity equations. Relationships of sensitivity to concepts of stability are investigated. A system is asymptotically stable with respect to a perturbation if its sensitivity in relation to nominal dynamics vanishes with large time. This vanishing sensitivity criterion is a necessary but not sufficient condition for stability. Difficulties of specifying sufficient conditions lead to a new approach to stability under parameter perturbations. A theorem concerning sensitivity in a finite time interval for a class of time-varying, nonhomogeneous, linear systems is proved. Regarding applications to ecosystems, sensitivity analysis can assess effects of measurement errors and perturbations, aid in evaluating stability, and help set research priorities by identifying most significant data. Heuristically, sensitivity accounts for the propagation of small perturbations through the causal network of ecosystems. It is this propagation network that is referred to as "sensitivity substructure."

In *Chapter 16* Dr. Conrad is concerned with patterns of biological control in ecosystems whose properties are determined by reproductive success of component populations. Ecosystem disturbances arise both externally and internally. Control mechanisms to minimize dysfunction include physical and organizational reliability, which combat endogeneous disturbance, and selective indifference and adaptability, which cope with environmental uncertainty. Examples of each are provided. Entropy and conditional entropy measures of state transition schemes, representing system and environment behavioral uncertainties, are formulated. Systems are more efficient or fit if their behavioral uncertainty is not higher, their anticipation of environment not better, and their tolerance for injury and selective independence not greater than necessary. Control mechanisms are manifested at different levels of organization. Adaptability of a level is formulated according to a general law of biological compensation: Restricted adaptability of a level must be compensated either by increased adaptability of others or by increased indifference to the environment. If adaptability is large a level is regulating; if small it is regulated. The different control patterns within populations interlock with each other and with control processes at higher levels to regulate the ecosystem as a whole. The answer to the question, "How is behavioral uncertainty related to environmental unpredictability?" is, "Adaptability plus potential indifference to environment must exceed environmental uncertainty, and tends in evolution in the direction of equalling it."

Chapter 17, by B. C. Patten, R. W. Bosserman, J. T. Finn, and W. G. Cale, concludes the book with a causal model of the organism/environment relation as basis for a two-environment ecosystem theory. The causal bond (cause → effect) is modeled as an abstract, input-output object, called a *holon*. The input side of the holon is recognized as a distinct object, *creaon*, and the output side as another object, *genon*. Each defines a unique and specific *environment*. A holon is determinate if its output time segment corresponding to an input segment is unique. The role of state in holon determinacy is described. A holon models the causal bond if it is both determinate and nonanticipatory. Causal networks, such as ecosystem food webs or biogeochemical cycles, are decomposable into causal sequences—serial cascades of causal bonds. The holon is thus the building block of causal sequences and networks. Causal interaction of two system variables is formulated as direct dependence, and different kinds of system variables are distinguished. Parameters originate causality and state variables propagate it to outputs. Several matrix representations of direct dependence are shown. Serial propagation in causal chains is expressed as sequential dependence. Then inputs, state variables, and outputs in causal chains are distinguished by a set of sequential dependency theorems. Mutual dependence of system variables is defined, and a conjecture made that ecosystems have all state variables mutually dependent and thus subject to the same set of causes. Sequential dependency matrices are derived by multiplication of direct dependency matrices. A transitive closure matrix is one that represents all possible sequential dependency paths of any length in a system. Mason's theorem expressing propagation from one point to another in a network is developed, interpreted, and illustrated. Then, input-output flow analysis is used to quantify direct and indirect causation by matrix methods. Creaon inflow analysis determines causality from different sources required to maintain specified outputs, and genon outflow analysis determines the propagated effects generated from specified inputs. These analyses depend on existence of associated transitive closure matrices, which is assured only when input or output contact with environment occurs. This is likened to Goedel's theorem on consistency of deductive systems. Flow analysis measures of network causation are derived as indexes (path length, cycling efficiency, and cycling index) and illustrated with several simple ecosystem models. The theory concludes with discussion of the reciprocal, mutually defining nature of a holon and its two distinct environments, and ecosystem as the ultimate unit of coevolution. Holon/environment coevolution represents a new generalization about the organism/environment relation—a point of departure for a more ecologically consistent concept of organic evolution in the ecosystem context.

13

Engineering Systems Analysis: Applicability to Ecosystems

JACK B. WAIDE† AND JACKSON R. WEBSTER*

INSTITUTE OF ECOLOGY AND DEPARTMENT OF ZOOLOGY
UNIVERSITY OF GEORGIA, ATHENS, GEORGIA

University of Georgia, *Contributions in Systems Ecology*, No. 29.

* Present address: Department of Biology, Virginia Polytechnic Institute and State University, Blacksburg, Virginia.

† Present address: Department of Zoology, Clemson University, Clemson, South Carolina.

I. Introduction

The systems approach is finding ever-increasing application in ecology, generally as a natural outgrowth of the ecosystem viewpoint, and more specifically as a result of the emphasis given to this approach within the International Biological Program. Dale (1970) has compared the systems approach to problem solving, identifying four phases in each: (i) choice of system components (lexical), (ii) definition of relationships among components (parsing), (iii) functional specification of these relationships (modeling), and (iv) validation of the model and investigation of its properties (analysis). In this chapter we examine only this final phase, investigation of the properties of the model, for which we would reserve the term "systems analysis."

To date systems ecology has largely ignored model validation and systems analysis. Indeed, most systems approaches in ecology which have reached the open literature have been somewhat modest in results if not in goals, perhaps prompting Gifford's (1971) comment that application of the systems approach to ecology has so far yielded only complex numerical statements of the obvious. Yet, if experience in other fields is any indication, the inclusion of analysis techniques in systems ecology offers to make this approach eventually a powerful one for the elucidation of ecosystem paradigms.

This chapter seeks to illustrate the potential benefits of systems analysis to ecology. Our concern is with four general classes of analysis commonly applied in other disciplines. Time domain analysis involves investigation of the timing and magnitude characteristics of simple signals fluxing through systems. We examine specifically system responses to unit singularity functions. A second broad class of analysis, frequency response analysis, concentrates on the ability of a system to attenuate (or amplify) and phase shift sinusoidal inputs. Also, investigation of systems in the frequency domain allows evaluation of feedback effects on system performance. The third category, stability analysis, seems very relevant to ecosystem theory. Stability considerations in ecology have often been plagued by vague and nonequivalent definitions of what it means for a system to be stable. We have recognized two ecological notions of stability and identified these with classical engineering definitions. Further, the relationships both of a "donor-controlled" versus a "recipient-controlled" modeling approach and of the connectedness of a system to stability are considered. Finally, techniques of sensitivity analysis are examined in some detail. Reasons for performing sensitivity analyses are considered in relation to the information that can be gained. We also discuss computational methods for performing such analyses, and present a theoretical interpretation of sensitivity results in both ecological and evolutionary terms.

One of the current controversies in systems ecology is the question of linear versus nonlinear models (Bledsoe, this volume). Our discussion of engineering techniques focuses on methods developed primarily for linear systems, and our examples are restricted to linear models. Since the previous chapter had little to say in favor of linear models, we feel obliged to present briefly our rationale for perhaps overemphasizing the importance of linear models for systems ecology.

No ecosystem model can ever be exact. Bellman (1968, pp. 6–9) stated, "... we must face the fact that every equation we employ to describe the physical world is approximate." This point is just as true, if not more so, when modeling biological systems. The question is, then, how exact do we want a model to be? And, within what operating range do we want this accuracy? Answers to these questions will be quite different for different people and for different modeling goals. In any modeling study, a variety of approaches is possible. Given the exact specifications of the study objectives and the questions that are being addressed, the investigator should be able to choose a modeling approach best suited to the purposes at hand. Large-scale ecosystem modeling to date has suffered more from incomplete specification of modeling objectives than from incorrect application of modeling techniques. In any case, for some study objectives the linear approach will be quite useful; in other cases it will be completely inappropriate. In still other situations linear techniques may provide initial, approximate analyses which may then be refined in subsequent studies. But it seems premature at best to abandon completely the vast body of linear systems theory that has proved so useful in diverse disciplines. All such techniques must be continually evaluated for the ecological insights they provide, but their utility for addressing certain classes of ecological questions has been confirmed in numerous applications. Other justifications for using linear modeling techniques in ecology have been given elsewhere (e.g., Waide *et al.*, 1974; Patten, 1975; Patten *et al.*, 1975; Webster *et al.*, 1975).

For those situations where linear approaches are appropriate, the discussions that follow should provide helpful background and basic references. We have indicated which techniques are applicable only to linear models and those that have a greater range of applicability.

II. Time Domain Analysis

A. Underlying Rationale

Systems engineers have classically studied systems in two separate domains. In the time domain input signals are simple and nonperiodic, and outputs are functions of time. In the frequency domain input signals are sinusoidal, and outputs are functions of the frequency and magnitude of input. In both cases, the underlying rationale is the same, namely to characterize the response of a system to certain test signals. While ecologists generally apply realistic input signals to their models, the application of simple signals may yield information about the functioning of the system, as well as provide information on the system response to other, more complex inputs. We discuss time domain approaches below, and techniques in the frequency domain in the following section.

In the time domain, the elementary components into which input signals are resolved are identical in shape, but are applied at different instants in time. Thus, time domain components always involve time as a parameter (Schwarz and Friedland, 1965).

The elementary functions into which inputs are resolved are termed unit singularity functions. A whole family of such functions exists, including the following: unit doublet, $u_{-1}(t)$; unit impulse, $u_0(t)$; unit step, $u_1(t)$; and unit ramp, $u_2(t)$. These functions are related according to the equation (Schwarz and Friedland, 1965)

$$u_{i+1}(t) = \int_{-\infty}^{t} u_i(\xi)\,d\xi, \qquad i = \ldots, -2, -1, 0, 1, 2, \ldots. \tag{1}$$

We shall concentrate on methodologies associated with the unit impulse and unit step responses. Such approaches are not applicable to the analysis of nonlinear models.

B. Unit Singularity Responses

1. *Unit Impulse Function* $[u_0(t)]$

The unit impulse is a function of infinite height (at $t = t_0$) and zero width. If $p_\tau(t)$ is a pulse of height $1/\tau$ and width τ, that is,

$$p_\tau(t) = \begin{cases} 1/\tau & 0 \leqslant t \leqslant \tau \\ 0 & \text{otherwise,} \end{cases} \tag{2}$$

then the impulse function may be defined as

$$u_0(t) = \lim_{\tau \to 0} p_\tau(t). \tag{3}$$

The unit impulse is termed the (Dirac) delta function $\delta(t)$ by some authors (e.g., DiStefano *et al.*, 1967).

The unit impulse response is the response of a linear system with zero initial conditions when the input is the unit impulse function. In linear, first-order systems such a response shows an initial rapid rise from zero to a peak value, followed by an exponential decay to zero. Besides being interested in the general qualitative nature of an impulse response, we may also define certain specific time domain specifications relative to it. For example, Waide *et al.* (1974) defined three such measures: (i) t_0, the time required to reach the peak value, (ii) t_1, the time required for 0.632 $(1 - 1/e)$ of decay from the peak to occur, and (iii) k, the peak value of the impulse response.

One final property of the impulse response is of interest. Once the unit impulse response of a linear system is known, it is possible to calculate the response to any input signal according to the convolution integral

$$y(t) = \int_0^t z(\xi)\, h(t-\xi)\, d\xi, \tag{4}$$

where y is the response to arbitrary input z, and h is the unit impulse response.

2. *Unit Step Function* $[u_1(t)]$

The unit step function is defined according to the equation

$$u_1(t-t_0) = \begin{cases} 1, & t > t_0 \\ 0, & t \leqslant t_0. \end{cases} \tag{5}$$

The unit step response is the output of a system when the input signal is the unit step function and the initial conditions are zero. For a linear system the step response may show an asymptotic rise to its steady-state value, or it may overshoot this steady-state value and then oscillate about it with the amplitude of oscillation decreasing exponentially to zero as time increases.

Several time domain specifications are typically defined relative to the unit step response (DiStefano *et al.*, 1967):

1. steady-state error—the difference between the input signal and the output signal at steady state;
2. overshoot—the difference between the input signal and the maximum value of the output signal;
3. delay time—the time required for the step response to reach 50% of its final value;
4. rise time—the time required for the step response to rise from 10% to 90% of its final value;
5. settling time—the time required for the step response to come within some specified percentage, usually 2% or 5%, of its final value;
6. predominant time constant—the time required for the step response (or its exponential envelope) to reach 63.2% of its final value; this is equivalent to the turnover time.

The last five measures are all transient response specifications, of which no. 2 is a measure of relative stability and nos. 3–6 are measures of the speed of response.

The system response (of a linear system) may also be obtained from knowledge of the step response, according to the convolution integral

$$y(t) = \int_0^t \dot{z}(\xi)\, a(t-\xi)\, d\xi, \tag{6}$$

where y is the response to an input z, with time derivative $\dot{z}$, and a is the unit step response (DiStefano *et al.*, 1967).

C. Ecological Applications of Time Domain Analysis

Time domain approaches have, to date, found little application in ecosystem modeling and analysis. However, such approaches should yield valuable information about systems under study, as well as allowing comparisons between various ecosystems. Several possible ecological applications of time domain analysis are considered in the remainder of this section.

1. *Model Development*

The actual construction of ecosystem models is an important step in the overall analysis. However, ecologists have no general precepts or laws which tell them how to compartmentalize or to formulate system equations. Observing the response of the tentative model to unit singularity functions, especially the unit impulse, should help to fomulate the model, or to choose among different approaches.

2. *Research Applications*

At the present time, little direct connection exists between field research and modeling. Experimental utilization of singularity techniques may help to bridge this gap. For example, application of radionuclides to ecosystems or their components can be done in the form of a step or pulse excitation, with the monitored response yielding information on the functioning of the system. Such responses can be helpful in formulating the model (e.g., Goldstein and Elwood, 1971). Another example might be use of the impulse function in conceptual development of hydrological models (e.g., Curlin, 1970).

3. *Ecosystem Analysis*

Time domain approaches also have obvious value in analyzing the system model once constructed. Under the rationale discussed above, one should be able to predict a model's response to any input once the singularity response is characterized. This should be beneficial not only in developing ecological theory, but also as a resource management technique.

Such an approach also will help in understanding model behavior in terms of examining (i) the behavior of input signals as they move through the system, (ii) the effect of these signals on the system components, and (iii) the effect of system components on the magnitude and timing of signals fluxing through the system. For example, Waide *et al.* (1974) employed time domain analysis as one portion of their analysis of a model of the calcium cycle in a forest ecosystem. They determined unit impulse and step responses for each compartment decoupled from the system, coupled within the system, and coupled within the system with the major feedback loop eliminated. Also, total system response was determined for both the open and closed loop cases. Finally, singularity functions were separately applied to each individual system input and compartment as well as total system responses monitored. Such techniques allow analysis of the dynamic system's behavior in the time domain in a fairly comprehensive fashion.

III. Frequency Response Analysis

A. The Basic Approach

As noted above the basic philosophy behind frequency response analysis is identical with that underlying time domain approaches: namely, to characterize the response of a system to known test signals. In the frequency domain, as opposed to the time domain case, these signals are applied at the same instant of time but vary in shape (i.e., frequency). Thus, angular frequency ω is the basic parameter in the frequency domain (Schwarz and Friedland, 1965).

The elementary signal into which inputs in the frequency domain are resolved is the sine wave. The term "frequency response" refers to the steady-state response of a system subjected to sinusoidal inputs of fixed amplitude but with varying frequencies.

The sine function is a good signal for analytical purposes (Child and Shugart, 1972). It is a well-behaved, manageable function in the mathematical sense, whose properties are well-known. Also, many natural signals impinging upon ecosystems are sinusoidal, or very nearly so. Further, many irregular, periodic curve forms can be decomposed into sine waves of various frequencies by Fourier analysis.

The basic frequency response techniques to be discussed below were designed primarily for linear systems, especially those containing feedback. However, they also may be applied to nonlinear models. By using a describing function analysis on a nonlinear representation (e.g., Shinners, 1972), one essentially produces a linearized transfer function representation which is amenable to analysis by frequency response techniques.

B. Frequency Response Specifications

The response of a linear system to a sinusoidal excitation of amplitude A is also a sine wave of the same frequency, but the output may have a different amplitude, and it may be shifted out of phase from the input. That is, an input of $A \sin \omega t$ applied to a linear system will produce an output $B \sin(\omega t + \phi)$. There are two parameters of this response. The magnitude ratio (MR) is the ratio of output to input amplitude ($\mathrm{MR} = B/A$). A linear system may attenuate ($\mathrm{MR} < 1$) or amplify ($\mathrm{MR} > 1$) input signals. The second parameter is the phase shift (ϕ), given in degrees. A linear system may lead ($\phi > 0^{\circ}$) or lag ($\phi < 0^{\circ}$) sinusoidal input signals.

The dependency of these two parameters on the input frequency is generally depicted in Bode plots, which include graphs of MR vs log ω (the magnitude portion) and ϕ vs log ω (the phase portion). In the magnitude plot, the magnitude ratio generally is expressed in decibels (db), where

$$\mathrm{db} = 20 \log_{10} \mathrm{MR}. \tag{7}$$

Such plots provide a convenient way to examine changes in sinusoidal signals as they flux through systems.

The characteristics of Bode plots for linear systems in which signals cascade through serially connected compartments can be described in general. In the magnitude portion of the plot the system has little influence on the input signal at low frequencies. As ω increases, the system begins to attenuate or amplify the input, with the amount of influence increasing with increasing ω. If the output being examined is from the kth compartment in the cascade, in which only the first compartment receives an input signal, then the MR plot asymptotically approaches a line of slope $\pm k$. The sign of k is negative if the compartment is attenuating and positive if it is amplifying the input signal. For a first-order compartment, the frequency at which the magnitude ratio is ± 3 db defines the turnover rate of that compartment.

The phase portion of the Bode plot similarly exhibits several distinct regions. Again, at low frequencies the system has little effect on the input. As input frequency increases, the system shifts the input phase increasingly with increasing ω, up to a maximum rate at which the phase graph shows a point of inflection. Past this inflection point the rate of increase in ϕ decreases with ω, until ϕ reaches a constant value. For a kth-order cascade, the inflection point occurs at $k \cdot (\pm 45^{\circ})$, and the high-frequency asymptote is the line given by $\phi = k \cdot (\pm 90^{\circ})$. These signs are positive for leading systems and negative for lagging ones. In a first-order system the turnover rate is defined by the frequency at which the point of inflection occurs.

The responses described above for linear, cascade-type systems generally are not obtained from systems containing multiple inputs or feedback paths.

Responses of such systems are more complex. However, for linear systems the principle of superposition holds, so that knowledge of the qualitative behavior of responses to sinusoidal inputs for linear cascades provides information about what to expect from sinusoidal excitations of more complex systems.

Regardless of the complexity of the system under study, the response to sinusoidal forcings of each individual compartment within the system, as well as of the total system, is of interest. Basically, two main types of systems exist, open-loop and closed-loop. Because the closed-loop system is not necessarily stable over the range of all possible input frequencies its frequency response cannot always be determined directly, but must occasionally proceed graphically from Nichols chart analysis (DiStefano *et al.*, 1967; Shinners, 1972). A Nichols chart is a plot of ϕ vs MR, with both polar and rectangular coordinates. One first breaks the major feedback loop in a closed-loop system, determines the frequency response of the open-loop system, and then plots this open-loop response on the rectangular coordinates of the Nichols chart. The closed-loop response can then be read off the polar coordinates of the same graph.

A number of techniques have been developed to determine frequency response characteristics of systems and their components. One digital computer approach which is applicable to linear ecosystem models is the SFR program written by Kerlin and Lucius (1966).

C. Ecological Applications of Frequency Response Analysis

As stated above for time domain analysis, frequency response approaches have so far been little used in ecosystem modeling efforts. Yet, these techniques also seem to have much potential benefit for ecologists concerned with understanding the properties of complex systems. Much similarity exists between approaches in the time and the frequency domains. However, each type of analysis will yield different kinds of information. Two ecological applications of frequency response techniques are examined briefly below.

1. *Model Development*

Frequency response characteristics can aid in the formulation of state equations governing the behavior of a system or its components (Child and Shugart, 1972). If the order of response of a linear system or compartment is known, and its response to one high-frequency input can be determined, the magnitude portion of the Bode plot can then be developed. This information allows the Laplace transform of the state equation to be written, from which the actual differential equation can be derived by operator inversion.

2. *Ecosystem Analysis*

Frequency response techniques should be especially beneficial in analyzing the properties of total ecosystem models. Child and Shugart (1972), in an analysis of a model of potassium cycling in a tropical forest ecosystem, illustrated three ecological applications of frequency response techniques. First, a comparison of responses of individual compartments both coupled within and decoupled from the system allowed determination of compartment effects on input signals, and vice versa. Out-of-system responses characterize the intrinsic properties of the compartments, while in-system behavior shows what effects being coupled within the system and receiving signals from other compartments have on these properties. These authors demonstrated the importance of the time constant, or turnover time, to such considerations. Second, Nichols chart analysis allowed determination of the frequency response characteristics of the total system. Third, a comparison of the total system closed-loop response with that of the system when the major feedback loop had been broken afforded an analysis of the control action exerted by this feedback. Waide *et al.* (1974) applied similar techniques to the calcium cycle in a temperate forest ecosystem, providing some interesting comparisons with the tropical model. Shugart *et al.* (1976) also applied frequency response techniques to assess the control action afforded by various monitoring points in a model of the calcium cycle in a *Liriodendron* forest. System state variables were examined individually in an attempt to maintain the calcium cycle in this forest at some nominal state by monitoring a given system component, and then using this information to regulate the input to the system, for example, of some toxic metal by-product of a nearby industry.

Additional frequency response techniques exist for analyzing stability of ecosystem models. These approaches will be discussed in Section IV. D.

IV. Stability Analysis

A. Introduction

Margalef (1969) stated that a system is stable if, when changed from a steady state, it develops forces that tend to restore it to its original condition. Since we today find ourselves pushing natural systems farther and farther from their natural steady states, it is imperative that we learn about the capacity of these systems for self-restoration.

From a modeling viewpoint stability analysis represents a connection between the mathematical model and the real-world dynamic system (Tomović, 1963). In concluding his discussion of the stability concept Cunningham (1963) said, "... the concept of stability offers one a moderately simple way of learning about the operation of a dynamic system." We can, then, identify two reasons for performing stability analysis: first, to relate a model to the real system, and second, to learn something about the real system's operation, the primary objective.

B. Ecological Notions of Stability

A review of ecological literature reveals two major stability concepts which can be identified with two classically defined types of stability (Margalef, 1968; May, 1971b). The first concerns persistence of an ecosystem or population, which corresponds to "neighborhood stability" (Lewontin, 1969). The second has to do with exogenous parameter variations, that is, those caused by agents outside the system (Margalef, 1968). This may be related to the notion of "structural stability" (Lewontin, 1969; Tomović, 1963; Levin, 1970). (Many authors formally restrict the concept of structural stability to a consideration of the effects of variations in the structure of system equations on system behavior—e.g., of variations in the pattern of connection among system state variables. Other authors broaden the concept to include consideration of variations in values of system parameters. It is this latter, broader sense in which we are using the term structural stability.)

Neighborhood stability has been the subject of many ecological studies, particularly in population genetics and population ecology (e.g., Wright, 1935; Rosenzweig and MacArthur, 1963). This type of stability concerns a system's ability to return to an equilibrium point after a state disturbance. That is, for a given set of starting values will the system grow without bound, cease to exist (go to zero), or reach some steady state (stability)? Methods of determining neighborhood stability have been extensively developed in engineering and mathematical literature, and have been applied not only to simple population systems but also to more complex ones. Examples are Verhoff and Smith (1971), Levin (1970), May (1971a, b, 1973a, b), and Gallopín (1971b).

The ecological application of the structural stability concept can be attributed to MacArthur (1955). This type of stability exists in an ecosystem if the abundance of species tends to remain quite constant in the face of parameter variation. Structural stability in the ecological sense assumes neighborhood stability and then asks questions about the frequency and magnitude of fluctuations in the history of each species in the ecosystem (Leigh, 1965). As such, structural stability is a measure of relative stability. Structural stability may be the more important concept for ecologists to investigate. Besides MacArthur, several other authors have discussed this subject. Dunbar (1960) compared the nonoscillatory character of tropical communities with the highly oscillatory nature of arctic communities, giving rise to Patten's (1961, 1962) field methods for measuring stability. Odum (1969) considered the increase in stability, defined as resistance to external perturbations, through succession. Watt (1969) examined three fundamental patterns of fluctuations and what they mean in relation to large environmental changes. Leigh (1965) used a statistical measure of the frequency of fluctuations. Holling (1972) presented the notion of a "domain of stability." His idea was that equilibrium states are seldom, if ever, reached exactly, but that there is a range around the equilibria over which the system components vary. Outside this "domain of stability" radical changes can occur. This notion is very similar to Lewontin's (1969) concept of statistical ensembles of stability points.

Several other stability concepts can be found in ecological literature, though they are not entirely distinct from the two ideas expressed above. A system is said to be "globally stable" if it returns to the same equilibrium point from all points in its state space. Lewontin (1969) presented a fairly complete discussion of this notion, as did Tomović (1963) under the heading "stability in the large." Relative stability is considered in several recent ecological papers. Patten and Witkamp (1967) and Jordan, Kline, and Sasscer (1972) considered the magnitude of system fluctuation due to perturbations of linear models that were inherently stable in the neighborhood sense. They used a stability measure that combined both the magnitude and the duration of response to perturbation. Child and Shugart (1972) and Waide *et al.* (1974) used frequency response techniques to derive separate measures for the two complementary aspects of perturbation response (magnitude, duration). Webster *et al.* (1975) provided a more thorough discussion of these relative stability concepts and related them to specific attributes of ecosystem elemental cycles. These stability concepts were also used by Waide and Swank (1975) to analyze potential impacts of intensive timber management on the nitrogen cycle in two important southeastern forest ecosystem types.

C. Engineering Definitions and Methods

1. *Neighborhood Stability*

Neighborhood stability relates directly to the engineer's concept of stability attributed to Liapunov (1892) and Poincaré (1892). A system is stable if its unit impulse response approaches zero as time approaches infinity or, alternatively, a system is stable if any bounded input produces a bounded output (DiStefano *et al.*, 1967). For a system formulated in the state variable manner with the origin moved to an equilibrium (singularity) point, the system is said to be stable in the sense of Liapunov if, for any given $\epsilon > 0$, there exists a $\delta > 0$ such that

$$\|\mathbf{x}(0)\| < \delta \qquad \text{implies} \qquad \|\mathbf{x}(t)\| < \epsilon \qquad \text{for all} \quad t > 0.$$

It can be shown (e.g., Bellman, 1967) that a linear system is stable in the sense of Liapunov if the characteristic roots of the system matrix have negative real parts (Theorem of Liapunov and Poincaré). A number of methods have been developed for determining the nature of the roots without their actual calculation. The Routh criterion (Routh, 1892), Hurwitz criterion (Hurwitz, 1892), Nyquist criterion (Nyquist, 1932), root locus method (Evans, 1948), and Bode plots (Bode, 1945) are some of the most often encountered methods, and are treated in most textbooks on linear systems (e.g., Shinners, 1972; DeRusso *et al.,* 1967; Schwarz and Friedland, 1965).

Two methods, both attributed to Liapunov, have been developed for determining the stability of nonlinear systems. Liapunov's first method involves linearization by Taylor series expansion and truncation, after which linear methods can be applied. Liapunov's second method requires finding a function of the state variables which is positive definite. If such a function can be found which has a derivative that is negative definite, then the function is stable at the equilibrium point. Liapunov's methods are discussed in many books (e.g., Tomović, 1963; Sánchez, 1968).

2. *Structural Stability*

Structural stability has not been extensively discussed or analyzed in engineering and mathematical literature. Tomović (1963) denotes structural stability or "parameter insensitivity" as the relationship of the system to disturbances in parametric space. This idea is quite closely related to the notion of parameter sensitivity discussed in the last section of this chapter and is treated briefly there. Tomović notes that structural stability is very difficult to study in a purely analytical fashion, and must therefore be studied with computers. With digital computers systems of equations can be solved very rapidly for various parameter values. With analog computers the system may be simulated directly without describing it by a system of equations. This is, then, still a trial and error technique. As Holling (1972) says, "In this role the computer is used largely as a guide to intuition, as the ecologist explores a multidimensional response surface. . . ."

Tomović also cited several Russian references on structural stability. Additionally, Thom (1969) discussed the concept of structural stability in relation to the biological problem of morphogenesis.

3. *Relative Stability*

Relative stability is often expressed in terms of allowable ranges of parameter values. Other more precise methods have been developed using frequency response approaches (DiStefano *et al.,* 1967). Several examples of these techniques are presented in the next section.

D. Stability Analysis in the Frequency Domain

We now present two examples of stability analysis as applied to ecology. The first considers frequency domain specifications in a comparison of stability in tropical and temperate forest ecosystems. The second examines the relationship between system stability, connectedness, and the importance of donor-controlled relationships.

In addition to the frequency response methods discussed earlier, several frequency domain techniques exist for analyzing the stability characteristics of dynamic systems. We would like to consider briefly some of these methods, while contrasting the stability of tropical and temperate forest ecosystems. Child and Shugart (1972) employed these techniques in analyzing a model of the magnesium cycle in a tropical forest, while Waide *et al.* (1974) applied them to the calcium cycle in a temperate forest.

Systems engineers have found it convenient to approximate many systems with second-order representations (e.g., second-order differential equations). Such approximations often work quite well, so that extensive methodologies developed for second-order systems can be applied to a variety of systems of higher order (DiStefano *et al.,* 1967). Both studies cited above utilized such a second-order approximation. The reader is referred to Child and Shugart (1972) for a more complete discussion of the approach.

In general the phase portion of a second-order frequency response should show a high-frequency asymptote at $-180°$, and a point of inflection at $-90°$ which corresponds to the natural frequency ω_n. This latter quantity is determined from the gain portion of the response curve, as that frequency at which the high-frequency asymptote (slope = -2) intersects the abscissa. For the tropical and temperate forest models (Table I), the phase shift at the natural frequency was $-87°$ and $-93°$, respectively, indicating that the second-order approximation was a good one in each case.

A second parameter of interest for second-order systems is the damping factor, ζ. This may be calculated by a variety of analytical or graphical techniques (e.g., Saucedo and Schiring, 1968; Savant, 1964).

These two parameters, ω_n and ζ, relate to the response of a system following perturbation away from equilibrium. As such, they may be used as measures of two of the ecological notions of relative stability discussed above. The natural frequency ω_n is related to how hard a system is to perturb from equilibrium, while the damping factor ζ indicates the length of time required for the system, once perturbed, to return to steady state. Table I gives values for each of these parameters for the two systems. These values reveal that the calcium dynamics of the temperate forest system are more stable than the magnesium dynamics of the tropical one in that it is harder to perturb the former from steady state [ω_n (temperate) $< \omega_n$ (tropical)], and also more stable in the sense that it

TABLE I

Comparison of Selected Frequency Response Stability Measures for a Tropical[a] and a Temperate[b] Forest Ecosystem[c]

	Tropical	Temperate
ϕ at $\omega = \omega_n$	$-87°$	$-93°$
ω_n	4.0×10^{-1}	2.6×10^{-2}
ζ	1.44	1.21
ϕ_m	$-170°$	$-175°$
GM	+95 db	+125 db
Frequencies of maximum instability	$3.6 - 7.3 \times 10^{-2}$	$4.0 - 7.0 \times 10^{-2}$

[a] Child and Shugart (1972).
[b] Waide *et al.* (1974).
[c] Tropical values, originally given in radians per day, have been converted to radians per year. See text for meaning of symbols.

takes the temperate system less time to return to steady state once perturbed [ζ (temperate) $<$ ζ (tropical)].

As noted above, ecologists are interested not only in whether a system is stable or not, but also how stable it is, that is, in relative stability. Frequency domain analysis also allows this concept to be quantified in terms of gain and phase margins. The gain margin (GM) is defined as the value of the magnitude ratio (db) where the total system phase shift is $\pm 180°$, while the phase margin (ϕ_m) is $\pm 180°$ minus the phase shift at the 0-db point of the total system magnitude ratio (DiStefano *et al.,* 1967). These two measures indicate how much gain or phase shift must be added to the system response to produce a maximally unstable system. Values for ϕ_m and GM (Table I) suggest that the temperate and tropical systems are about equally stable in phase shift terms, but that the temperate forest is relatively more stable in its signal attenuation characteristics.

Finally, one may use frequency response techniques to determine frequencies leading to unstable system behaviors. Maximal instability in system behavior occurs in closed-loop systems such as ecosystems when an input signal is not attenuated, but is shifted out of phase by $\pm 180°$. In this case, the feedback signal augments the input, leading to a signal which grows without bound. Frequencies leading to such unstable behavior can be determined from a Nichols chart, which is a plot of the phase shift versus the magnitude ratio (DiStefano *et al.,* 1967). Both systems under consideration were completely stable in this sense. However, frequencies which take the system closest to the point of maximum instability may also be determined. Table I shows that this range of potentially destabilizing frequencies is almost identical for the two systems. As noted by Child and Shugart (1972), this analysis is a necessary condition for system stability, but not sufficient unless the model is isomorphic or homomorphic to the real system. Thus, to the extent that these two models accurately represent the steady-state characteristics of the real systems, the above results suggest that the same range of frequencies may lead to unstable behaviors in the two systems. The reader is referred to Waide *et al.* (1974) for a complete discussion of these stability comparisons.

Webster *et al.* (1975) refined and extended these relative stability ideas. These authors argued that ecosystem relative stability is largely determined by the availability, storage, and recycling of essential elemental resources. Two complementary aspects of relative stability were defined. Resistance is the ability of a system to resist displacement from its nominal dynamics, while resilience is the rate with which it recovers to an initial state once displaced from it. A second-order frequency response approach was again employed to quantify these concepts, resistance being measured by the natural frequency and resilience by the damping ratio. Random matrices, subject to mass conservation limitations, and hypothetical ecosystem models, constructed according to the characterization of alternative properties of elemental cycling, were analyzed to determine relationships between the two aspects of relative stability and specific properties of elemental cycles.

Resistance was shown to be related to large storage pools and to long turnover times, while resilience was related to rapid turnover and recycling rates. Thus, resistance and resilience are somewhat inverse concepts. Perhaps most importantly, terrestrial and aquatic systems were shown to be quite different in the relative balance between resistance and resilience they exhibited, with most hypothetical terrestrial systems examined being more resistant than most aquatic systems, which were more resilient. This finding suggests a continuum of ecosystem adaptations for persistence, ranging at one end from planktonic systems which appear to persist by recovering rapidly following disturbance (resilience), to temperate forests which seem to persist by deviating little in response to perturbation (resistance). The degree and frequency of fluctuation of the physical environment and abiotic limitations placed on biological uptake of essential resources are among the factors that govern the degree to which a given ecosystem exhibits adaptations for resistance or resilience. Finally, these authors examined the contribution of turnover rates of ecosystem components to the balance between resistance and resilience, and concluded that (1) this balance is determined by the relative proportions of *r*- and *K*-selected populations in the ecosystem; (2) early successional species largely determine the rate of recovery of an ecosystem from disturbance and thus contribute to the resilience of the system; and (3) high levels of utilization of primary production by first-order biophages leads to rapid regeneration of assimilated resources, and hence to rapid turnover and ecosystem resilience (Webster *et al.*, 1975).

These ideas have also been considered by Waide and Swank (1975), who applied relative stability concepts to an analysis of potential disturbance by intensive logging practices of the nitrogen cycle in southeastern loblolly pine and mixed hardwood forests. The relative stability approach was shown to predict the susceptibility of the nitrogen cycle in these systems to disturbance by intensive timber utilization and indicated that such forestry practices must be thoroughly evaluated before being generally applied.

E. Neighborhood Stability, Connectedness, and Donor Control

In this section attention is limited to neighborhood stability. We feel it intuitive that undisturbed ecosystems remain near equilibrium under normal conditions (Waide *et al.,* 1974; Patten, 1974; Webster *et al.,* 1975). Like Slobodkin (1964), we "... accept as an initial axiom (as did Darwin and Wallace) that, barring drastic physical changes or the introduction of new species, the number and kind of animals found in any region in the world tend to remain essentially constant from year to year." Therefore, it can be stated that natural ecosystems possess neighborhood stability unless under extreme stress. As Holling (1972) noted, radical changes can occur beyond the domain of stability. The question we would ask then, is, what characteristics do ecosystems possess which enable them to be stable?

We would like to consider system stability in relation to the connectedness of the system and to the control of flow through the system. The degree to which system components are interconnected, and the effect of this connectedness on system stability, is important in such large, complex systems such as ecosystems. This analysis is limited to linear models.

Gardner and Ashby (1970) defined the connectance or connectedness (C) of a linear system as the percentage of nonzero values in the system matrix. Using the Hurwitz stability criterion, they examined the stability of matrices in which the probability of an off-diagonal element being nonzero was equal to C. Nonzero values were drawn from a uniform distribution between -1 and $+1$. The elements of the main diagonal were all negative and distributed evenly between -1 and -0.1. Hence, the authors assured that each component of the system was inherently stable by itself.

In general, they found that increased connectedness decreased the probability of stability. As they increased the number of system components from $n = 4$ to $n = 10$ they showed that the probability of stability dropped suddenly from near 100% to near 0% as the connectedness was increased above a critical value of 13%.

May (1972) reexamined the conclusion of critical connectedness from an analytical standpoint. He considered only large matrices with nonzero elements drawn from a distribution with a mean of zero and a mean square value of α. He found that the system was almost certain to be stable if $\alpha < (nC)^{-1/2}$, and almost certainly unstable if $\alpha > (nC)^{-1/2}$. These analytical conclusions agreed quite well with Gardner and Ashby's computer results. May suggested two ecosystem applications. First, a conclusion very similar to one reached by Margalef (1968): species that react with many other species do so weakly, and those that interact strongly do so with but a few species. Secondly, he suggested that ecosystems might be organized into small interacting units with fairly high connectivity, a feature which he noted may be observed in many ecosystems.

The conclusion reached by both Gardner and Ashby, and May, that stability (or the probability of stability) decreases as the connectedness of a system increases, is not in accord with the often quoted statement of MacArthur (1955) that stability increases as the number of links in the trophic web increases, or with the many ramifications of the stability–diversity argument in general. Perhaps the problem lies in the definition of stability, as discussed above, or perhaps variety is not ". . . a necessity for the long life of the ecosystem . . ." (Odum, 1969). Certainly this is an area which has received much verbal treatment by ecologists and yet is still without solid foundation. May (1972, 1973b) warned against any simple belief that increased stability is a mathematical consequence of increasing trophic web complexity.

Other authors have also examined this relationship between ecosystem stability, complexity (i.e., the number of state variables), and the degree of connectedness. Somorjai and Goswami (1972) showed that the notion of a critical connectedness, below which the system is almost certainly stable and above which it is unstable, may not have meaning for nonlinear systems. Šiljak (1974) verified the existence of a sharp transition between stability and instability in a wide variety of linear and nonlinear systems. However, Daniels and Mackay (1974), by actually computing eigenvalues of random matrices, questioned the existence of a sharp boundary between stability and instability, arguing that such a result was an artifact produced by Gardner and Ashby's (1970) use of the Hurwitz stability criterion. Roberts (1974) introduced the notion of a "feasible random ecosystem," wherein all populations are limited to having equilibrium values greater than zero. When ecologically unreasonable cases of negative population sizes were dropped from the analyses, Roberts showed an increase in stability with increasing system size. However, Gilpin (1975) showed this result to be an artifact of the assumptions of Roberts' model, and showed a decrease in stability with increasing number of system components. Other aspects of these relationships have been explored by May (1973b) and Maynard Smith (1974). In all the above analyses, positive and negative signs of interaction terms in the system matrix were considered to be equally likely.

Our approach to the problem was to examine stability based on the characteristic roots of the system matrix. Our modeling rationale has been based on two rather stringent restrictions: (i) the change in any system component is given by income minus loss (mass balance), and (ii) any flow from one system component to another is proportional to the contents of the donor compartment. Thus, in state variable notation, the system can be represented by a set of linear, first-order differential equations which in matrix representation appear as

$$\dot{\mathbf{x}} = A\mathbf{x} + \mathbf{z}, \tag{8}$$

where A is the system matrix, $\mathbf{x}$ is the vector of system variables, and $\mathbf{z}$ is the vector of system inputs. Since any off-diagonal element a_{ij}, $i \neq j$, of A represents the proportionality constant for an input into the ith component, these values are inherently nonnegative. Elements of the main diagonal, on the other hand, represent summed output flows and are inherently negative. That is,

$$a_{ii} \leqslant 0, \qquad a_{ij} \geqslant 0, \quad i \neq j. \tag{9}$$

Further, the sum of losses from a component must be equal (though opposite in sign) to the inputs to other system components from that component, minus any losses from the system via that compartment. That is,

$$a_{jj} = -\sum_{i=1, i\neq j}^{n} a_{ij} - a_{0j}, \tag{10}$$

where a_{0j} represents losses from the system. This can also be stated as

$$\left|a_{jj}\right| \geqslant \left|\sum_{i=1, i\neq j}^{n} a_{ij}\right|. \tag{11}$$

Hearon (1963) describes systems in which Eq. (11) is true as having a dominant main diagonal (dominant with respect to columns, in this case).

These restrictions, taken together, are quite far-reaching. Of major consequence is the result that the real parts of the characteristic roots of such a matrix are nonpositive (Hearon, 1953, 1963). Thus, any model formulated in this manner will be inherently stable.

The greatest restriction (i.e., the one with the most apparent lack of ecological validity) is that of donor control. Smith (1972) noted that, although passive flow (donor control) promotes stability, the essence of being alive includes effort to obtain resources (recipient control). Thus, we wish to consider the effect of including both recipient and donor control, in varying proportions, on the stability of linear systems. Recipient control inherently leads to instability in single population models; however, we will show that some amount of recipient control can be tolerated in ecosystems.

For a flow from the jth to the ith compartment which is proportional to the contents of the recipient compartment, the nonzero off diagonal element is a_{ji}, not a_{ij} as in donor control. Moreover, this term now represents an output from

x_j, so that this off-diagonal element is inherently negative. Equation (10) still holds, that is, the main diagonal is the negative sum of elements in the same column, minus any coefficients for losses from the system via that component. However, Eq. (11) may no longer hold. Specifically, if the recipient-controlled flows into a compartment are greater than its donor-controlled losses, the main diagonal will no longer be dominant. To summarize, the differences between donor control and recipient control are as follows:

Donor control	Recipient control
(1) $a_{ij} \geqslant 0, \quad i \neq j$	(1) $a_{ij} \leqslant 0, \quad i \neq j$
(2) $a_{jj} = -\sum_{i=1, i\neq j}^{n} a_{ij} - a_{0j}$	(2) $a_{jj} = -\sum_{i=1, i\neq j}^{n} a_{ij} - a_{0j}$
(3) $a_{jj} \leqslant 0$	(3) a_{jj} may be $= 0, <0$, or >0, depending on relative magnitudes of $\sum_{i=1, i\neq j}^{n} a_{ij}$ and a_{0j}.

These conclusions may become somewhat clearer by examining a two-compartment model: (1) with no interaction,

$$\dot{x}_1 = -a_{01}x_1 + z_1, \qquad \dot{x}_2 = -a_{02}x_2 + z_2, \qquad A = \begin{bmatrix} a_{11} & 0 \\ 0 & a_{22} \end{bmatrix}$$

where

$$a_{11} = -a_{01}, \qquad a_{22} = -a_{02};$$

(2) with a donor-controlled flow from x_1 to x_2,

$$\dot{x}_1 = (-a_{21} - a_{01})x_1 + z_1, \qquad \dot{x}_2 = a_{21}x_1 - a_{02}x_2 + z_2, \qquad A = \begin{bmatrix} a_{11} & 0 \\ a_{21} & a_{22} \end{bmatrix}$$

where

$$a_{11} = -(a_{21} + a_{01}), \qquad a_{22} = -a_{02};$$

(3) with a recipient-controlled flow from x_1 to x_2,

$$\dot{x}_1 = -a_{01}x_1 - a_{12}x_2 + z_1, \qquad \dot{x}_2 = a_{12}x_2 - a_{02}x_2 + z_2, \qquad A = \begin{bmatrix} a_{11} & -a_{12} \\ 0 & a_{22} \end{bmatrix}$$

where

$$a_{11} = -a_{01}, \qquad a_{22} = a_{12} - a_{02},$$

or, if we note that $a_{12} < 0$,

$$A = \begin{bmatrix} a_{11} & a_{12} \\ 0 & a_{22} \end{bmatrix}$$

where

$$a_{11} = -a_{01}, \qquad a_{22} = -(a_{12} + a_{02}).$$

To examine the relationship between percent connectedness (C), percent donor control (D), and stability, we defined percent donor control as the probability that a nonzero off-diagonal element was positive. We then constructed random matrices in a fashion similar to Gardner and Ashby and calculated characteristic roots to determine stability. Off-diagonal elements were chosen to be nonzero with a probability equal to C. The actual value given to these nonzero off-diagonal elements was chosen at random from a uniform distribution between 0 and 1. The sign was changed to negative with a probability of $1 - D$. All main diagonal elements were given an initial value selected at random from a uniform distribution between 0 and -1 to correspond to an output from the system. Then the sum of elements in the same column was subtracted. The resulting matrix was subjected to an eigenvalue routine (Westley and Watts, 1970), and the characteristic roots examined for positive real parts.

The results are shown in Fig. 1, for $n = 5$, 10, and 15. Each data point for $n = 5$ and 10 represents at least 50 trials. Data points for $n = 15$ represent 30 trials. For $D = 50\%$ (positive and negative off-diagonals equally probable) the graphs generally resemble those of Gardner and Ashby, showing the critical connectedness value of about 13% for larger systems ($n \geqslant 10$). (These results actually correspond more closely to the gradual transitions seen by Daniels and Mackay (1974) than they do to the sharp transitions of Gardner and Ashby (1970).) Note that some amount of recipient-controlled interactions can be tolerated in stable systems. Above the critical connectedness value this level appears to be independent of the connectedness. These results show that fairly high levels of connectedness can be maintained by fairly high levels of donor control. The most obvious conclusion is that donor control leads to a greater probability of stability.

Assuming that ecosystem stability has selective value for the populations in the ecosystem (Dunbar, 1960; Darnell, 1970), we would like to consider the role of donor control in ecosystem evolution and regulation. It was stated above that donor control apparently has little ecological meaning, but in light of our computer results we would like to reconsider this point. We suggest that donor control is nothing more than population regulation by means of resource limitation. As such, our concept of donor control is related to the "balance of nature" arguments (Hairston *et al.,* 1960; Murdock, 1966; Ehrlich and Birch,

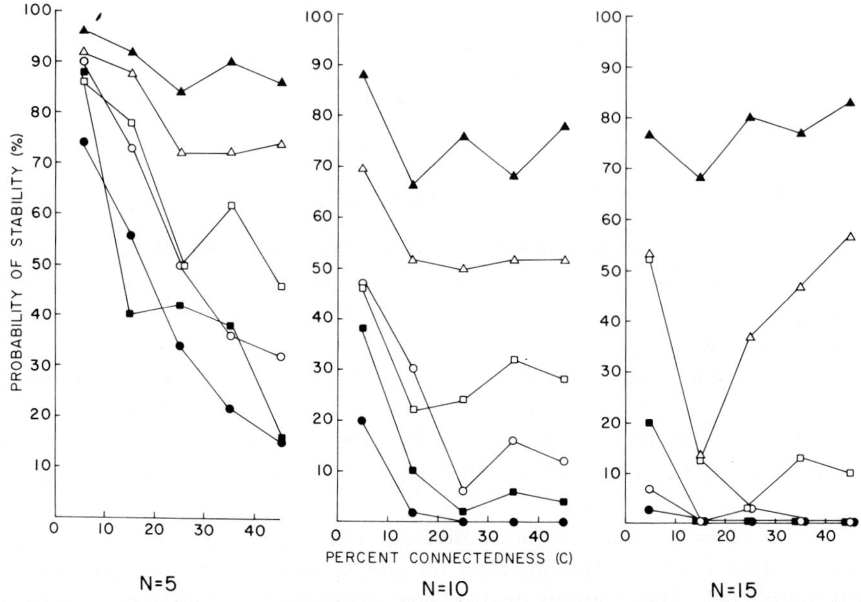

FIG. 1. Probability of stability as a function of connectedness (*C*) and percent donor control (*D*) for linear systems of 5, 10, and 15 components. Percent donor control: ●, 40%; ■, 50%; ○, 60%; □, 70%; △, 80%; ▲, 90%.

1967; Slobodkin *et al.*, 1967), especially as restated by Wiegert and Owen (1971).

Wiegert and Owen suggested that, in general, all populations are ultimately regulated by the resource base available to them. However, in some circumstances, resource limitation and direct competition for resources among populations within a trophic level may not be operative. If a population, through more efficient resource utilization, is able significantly to alter its rate of resource supply such that extreme fluctuations and extinction become a distinct possibility, then that population is in general not limited by resource supply. Regulation generally emanates from the next higher trophic level (recipient control). However, those populations that do not affect their rate of resource supply are usually limited by resources (i.e., donor-controlled). Even in those systems such as forests, where consumer populations may not be regulated in a strictly donor-controlled fashion, some amount of donor control may be operative. For example, the many reported cases of allelochemical and structural adaptations of forest trees to reduce grazing pressure are mechanisms of donor control.

Further, Wiegert and Owen point out that in a successional sequence the relative importance of donor control increases through successional time. This argument can be related to *r* and *K* selection, in the sense that succession generally leads to the increase of equilibrium species (*K*-selected, resource-limited, density-dependent, donor-controlled) at the expense of opportunistic species (*r*-selected, not resource-limited, density-independent, recipient-controlled) (e.g., Pianka, 1970, 1972; Gadgil and Solbrig, 1972).

Finally, Murdoch's (1969) consideration of predator–prey relationships can be related to donor control. He concluded that switching by predators among prey species, based on the densities of the prey species involved, could stabilize such relationships. This conclusion obviously suggests donor control.

Webster and Waide (in prep.) extended these preliminary results. The same basic methods of matrix construction as described above were used. Five system sizes were examined: $n = 5$, 10, 15, 20, and 30. For each of the first three system sizes, 200 matrices were examined for each combination of donor control D and connectedness C. For $n = 20$ and 30, 100 matrices were examined for each pair of D and C. Again, the higher the level of donor control, the greater the probability of stability. The resulting curves were much smoother in these analyses than in Fig. 1, due to the larger numbers of matrices examined. Several new results were also obtained. First, for $n = 10$, the probability of stability as a function of connectedness decreased to the value of $C = 30\%$, for all values of D, but increased for connectedness above 30%. For system sizes of 15 or above, stability was an increasing function of connectedness for values of D at or above 70%. The rise in probability of stability with increasing connectedness was especially dramatic in the $n = 30$ case. Second, for high values of connectedness ($C \geqslant 30\%$) and of donor control ($D = 90\%$), the probability of stability increased with system size. Finally, the community matrix type of system model was also subjected to these analyses; in this case, the mass conservation limitation of Eq. (10) was not present. For such matrices stability decreased with increases of positive, off-diagonal interaction terms above 50%, the reverse of the results summarized above and shown in Fig. 1. But the basic result of this section, that donor control enhances system stability in systems subject to mass conservation restraints, was confirmed by these more extensive analyses. DeAngelis (1975) has also confirmed such a relation between stability and donor control.

In conclusion, donor control is seen as an important mechanism operative in ecosystems, increasing the stability of these systems in the face of extremely diverse and complex patterns of relationships (i.e., high connectedness) found in such systems. The donor control referred to, however, is that interpreted in several more subtle ways than just regarding donors as forcing energy or material into passive receivers, which is ecologically untenable.

V. Sensitivity Analysis

A. The Concept of Sensitivity

A final class of questions typically posed in systems analysis lies in the general area of sensitivity analysis. Basically, sensitivity analysis refers to any attempt to elucidate the effect on the mathematical model of a system of change or uncertainty in the system state description. Mankin and Brooks (1971) suggest several reasons for performing a sensitivity analysis. In the first place, the mathematical model of the real system under consideration is only an abstraction of certain specific properties of that system; the model is not an exact mimic of the real world. Second, even if the model completely replicated the natural system, some discrepancy between assumed values of the model parameters and their true values is bound to exist due to errors in measurement or estimation. Additionally, even if the above two considerations were not valid, the parameters are quite likely to change with changes in the system's environment. In essence, then, sensitivity analysis is one way of approaching issues of uncertainty when employing deterministic modeling techniques, the conclusions of which are based on point estimates of parameters rather than on stochastic distributions (de Neufville and Stafford, 1971).

Another important reason for performing a sensitivity analysis is to gain insight into the structure and internal working of the modeled system. Sensitivity analysis is a method of investigating the relationships of system state variables to system parameters. The following discussions hopefully should provide some general direction to the application of sensitivity techniques to ecosystem analysis.

1. *The Basic Ideas of Sensitivity**

Most of the uses of sensitivity analysis rest on the original work of Miller and Murray (1953). These authors presented an elegant and rigorous theory underlying the error analysis of differential analyzers. They were specifically concerned with the effect of errors on the machine solution of systems of differential equations. Of special interest to these discussions is their belief that, besides the importance of knowing the effect of errors on machine solutions, another very important reason for performing such an error (i.e., sensitivity) analysis is that it gives one an insight into the mathematical or theoretical structure of the system of differential equations being considered.

The work of Tomović (1963) has done much to extend the theory of sensitivity analysis. This author identified two basic problems in dynamic systems analysis: (i) the solution of the differential equations characterizing the system, and (ii) the stability of motion of these solutions. Viewing sensitivity as the analysis of stability of motion of dynamic systems within a finite time interval, Tomović defined sensitivity with respect to perturbations in the model parameters. Following a parameter perturbation of Δp, the system

* Some of the original symbols used by authors cited in this section have been changed for consistency with notations to be developed later. These changes have in no way altered the meanings of the equations or ideas to be reviewed.—Ed.

$$F(\dot{x}, x, t, p_0) = 0, \qquad x = x(t, p_0), \tag{12}$$

is given by

$$F(\dot{x}, x, t, p_0 + \Delta p) = 0. \tag{13}$$

If we let

$$\Delta s = [x(t, p_0 + \Delta p) - x(t, p_0)] / \Delta p, \tag{14}$$

then the sensitivity coefficient of the dynamic system, denoted $s(p_0, t)$, is defined by

$$\lim_{\Delta p \to 0} \Delta s = \frac{\partial x(t, p_0)}{\partial p_0} = s(p_0, t). \tag{15}$$

Three kinds of sensitivity coefficients were defined by Tomović: (i) for steady-state systems, the static sensitivity coefficient $s_s = s(p_0)$; (ii) for nonsteady-state systems, the dynamic sensitivity coefficient $s_t = s(p_0, t)$; and (iii) for systems in which each parameter varies over some finite region of parameter space, the parameter sensitivity coefficient $s_p = s(p_1, \ldots, p_m, t)$. Tomović presented general discussions on the calculation and interpretation of sensitivity coefficients, and of their relationship to stability considerations. We shall return to some of his ideas later.

One of the more common implementations of sensitivity analysis has been the use of parameter influence (sensitivity) coefficients in parameter optimization and identification problems (Meissinger, 1964a; Meissinger and Bekey, 1966; Harvey *et al.*, 1966), as well as in substitution methods for the verification of analog computer solutions of systems of differential equations (Meissinger, 1964b). As an example, the parameter optimization or identification problem may be formulated as follows (Meissinger and Bekey, 1966): Let the dynamic system under investigation be defined by the equation

$$\dot{\mathbf{X}} = \mathbf{F}(\mathbf{X}, t, \alpha), \qquad \mathbf{X}(0) = \mathbf{X}_0, \tag{16}$$

where $\mathbf{X}$ is an $n \times 1$ state vector and $\boldsymbol{\alpha}$ is an $m \times 1$ unknown parameter vector. The form of the system equations, given by the vector-valued vector function $\mathbf{F}$, is assumed to be known. Corresponding to the system (16) is a mathematical model, given by

$$\dot{\mathbf{x}} = \mathbf{F}(\mathbf{x}, t, \mathbf{p}), \qquad \mathbf{x}(0) = \mathbf{x}_0, \tag{17}$$

where $\mathbf{p}$ represents the m adjustable model parameters. The objective of parameter optimization is to adjust $\mathbf{p}$ such that some appropriate criterion function is minimized. Most such functions are based on the output error,

$$\mathbf{e} = \mathbf{x} - \mathbf{X},$$

of system and model outputs resulting from identical input perturbations of the system and its model. The parameter adjustment usually is considered successful if the output error criterion attains some steady-state minimum at the correct parameter values, that is,

$$\lim_{t \to \infty} \mathbf{p}(t) = \alpha(t),$$

or some close approximation thereto. In this approach, parameter influence or sensitivity coefficients, obtained from the solution of a set of sensitivity equations, are used as feedback for a gradient method which yields the optimal parameter values.

The sensitivity equations for this procedure arise from differentiation of the model equation (17) with respect to p_i to yield

$$\frac{\partial}{\partial p_i}\left(\frac{d\mathbf{x}}{dt}\right) = \left(\frac{\partial \mathbf{F}}{\partial \mathbf{x}}\right)\left(\frac{\partial \mathbf{x}}{\partial p_i}\right) + \frac{\partial \mathbf{F}}{\partial p_i}. \tag{18}$$

If $\mathbf{x}$ is piecewise continuous and differentiable in both t and p_i, we may reverse the order of differentiation to obtain

$$\frac{d}{dt}\left(\frac{\partial \mathbf{x}}{\partial p_i}\right) = \left(\frac{\partial \mathbf{F}}{\partial \mathbf{x}}\right)\left(\frac{\partial \mathbf{x}}{\partial p_i}\right) + \frac{\partial \mathbf{F}}{\partial p_i}, \tag{19}$$

or

$$\dot{\mathbf{s}}(p_i) = (\partial \mathbf{F}/\partial \mathbf{x})\, \mathbf{s}(p_i) + \partial \mathbf{F}/\partial p_i, \tag{20}$$

where $\mathbf{s}(p_i) = \partial \mathbf{x}/\partial p_i$. Equation (20) is then the sensitivity equation (Meissinger and Bekey, 1966).

Recent work by Pridgen (1970) has added somewhat of a new interpretation to sensitivity analysis. This author examined probabilistic and deterministic, as well as single- and multiparameter sensitivities, while providing a design tool which has as output most of the types of sensitivity measures previously defined. Starting from a state space model, Pridgen formulated a state sensitivity model in the same fashion as Meissinger and Bekey (1966). Differentiating the state transition equation,

$$\dot{\mathbf{x}} = A\mathbf{x} + B\mathbf{z}, \tag{21}$$

with respect to the parameter p yields

$$\partial \dot{\mathbf{x}}/\partial p = A\ \partial \mathbf{x}/\partial p + A'\mathbf{x} + B\mathbf{z}' + B'\mathbf{z}, \tag{22}$$

where the prime indicates differentiation with respect to p. Reversing the order of differentiation yields the equation

$$\frac{d}{dt}\left(\frac{\partial \mathbf{x}}{\partial p}\right) = A\frac{\partial \mathbf{x}}{\partial p} + A'\mathbf{x} + B\mathbf{z}' + B'\mathbf{z}, \tag{23}$$

or

$$\dot{\mathbf{s}}(p, t) = A\mathbf{s}(p, t) + \mathbf{c}, \tag{24}$$

where the state sensitivity operator $s_i(p, t)$ is defined as

$$s_i(p, t) = \partial x_i(t)/\partial p \tag{25}$$

and

$$c = A'\mathbf{x} + B\mathbf{z}' + B'\mathbf{z}.$$

Thus, Eq. (24) leads to a state sensitivity model.

Given the above, it is possible to write an equation for the state of the system following a perturbation in the parameter p of $\Delta p = p_0 - p_{\text{nom}}$ such that (Pridgen, 1970)

$$\mathbf{x}(t, p_0) \approx \mathbf{x}(t) + \mathbf{s}(p, t)\,(\Delta p). \tag{26}$$

Hence, under this interpretation, sensitivity is viewed as a state operator which moves the system through parameter space in response to changes or perturbations in system parameters.

A wealth of other references on sensitivity analysis could be cited. However, the above contain the basic ideas underlying sensitivity techniques, especially those to which we wish to refer. The recent work by Tomović and Vukobratović (1972) has presented a general theory of sensitivity analysis. It remains to be seen to what extent such methodologies are applicable to ecosystem models.

2. *Applications in the Biological Sciences*

Sensitivity techniques have been used in biology in areas other than ecology. Savageau (1971) implemented parameter sensitivity as a means of assessing the effectiveness of a biochemical control system. In particular, he suggested the criterion of parameter sensitivity as relevant to investigating "the ability of such control systems to 'buffer' the end product concentration against changes internal to the system itself." He used a measure of relative parameter sensitivity, normalized with respect to the amounts or values of the variables and parameters present at steady state.

The concept of sensitivity has rarely been applied explicitly to ecological problems. Kaye and Ball (1969) used the SFR-3 program (Kerlin and Lucius, 1966) to compute transient and frequency response sensitivities for a model of radionuclide movement in a tropical ecosystem. Smith (1970) employed sensitivity analysis of a simple model of a freshwater ecosystem as an illustration of the application of systems analysis techniques to ecosystem analysis. He calculated relative sensitivities, normalized relative to the smallest sensitivity value obtained, by successive simulation trials using different parameter values. In his analysis of bass energetics, Glass (1971) also computed relative sensitivities, defined as percent change in a variable of interest in response to a certain percent change in some parameter. A somewhat more comprehensive look at sensitivity was provided by Patten (1969) and Brylinsky (1972) in an examination of the sensitivity structure of a marine ecosystem. These authors considered steady-state sensitivities of state variables both to other variables and to parameters. Both relative and absolute sensitivities were analyzed. Of particular importance was the distinction in these articles between direct sensitivity effects and indirect ones as mediated through the pattern of system connectivity. Gist (1972) utilized analog computer approaches to investigate sensitivities of state variables to parameters and forcings in his analysis of calcium and potassium fluxes in a cryptozoan food web. Finally, Gallopín (1971a) employed sensitivity methods to assess the effects of parameter changes on steady-state solutions of a model of a resource–population system.

General discussions, though brief, of the possible ecological uses of sensitivity analysis may be found in Kowal (1971), Mankin and Brooks (1971), and Funderlic and Heath (1971).

B. Calculation of the Sensitivity Operators

In this section we present a straightforward method, readily implementable on analog and digital computers, for calculating sensitivity operators of linear ecosystem models. The method is based on Kerlin (1967). We thus assume either a linear system, or a nonlinear one which has been linearized in the region of an operating point. However, the technique is general and can be extended to nonlinear representations.

We start with a state variable representation of an ecosystem, given by

$$\dot{\mathbf{x}} = A\mathbf{x} + \mathbf{z}, \tag{27}$$

where $\mathbf{x}$ is an $n \times 1$ vector of system state variables, with $\dot{\mathbf{x}}$ its time derivative; A is an $n \times n$ matrix of system parameters (transfer coefficients); and $\mathbf{z}$ is an $n \times 1$ vector of inputs, where z_i is the input to x_i.

The sensitivity of the state vector $\mathbf{x}$ can be obtained by differentiating Eq. (27) with respect to a_{ij}, yielding

$$\partial\dot{\mathbf{x}}/\partial a_{ij} = A\, \partial\mathbf{x}/\partial a_{ij} + (\partial A/\partial a_{ij})\mathbf{x}. \tag{28}$$

Reversing the order of differentiation, and defining

$$\mathbf{s}(a_{ij}, t) = \partial\mathbf{x}/\partial a_{ij}$$

to be the sensitivity of $\mathbf{x}$ to a_{ij}, we have

$$\dot{\mathbf{s}}(a_{ij}, t) = A\mathbf{s}(a_{ij}, t) + (\partial A/\partial a_{ij})\mathbf{x} = A\mathbf{s}(a_{ij}, t) + \xi_{ij}\mathbf{x}, \tag{29}$$

where ξ_{ij} is the elementary matrix with 1 in the (i, j) position and zeros elsewhere.

Equation (29) is similar in form to Eq. (27). Hence, the method used to find solutions to Eq. (27) can also be used to solve Eq. (29). As an example, MATEXP (Ball and Adams, 1967; Patten, 1971), a program which can be used to solve differential equations of the form (27), has a subroutine which calculates sensitivities to elements of the A matrix using Eq. (29) and the matrix exponential method of integration. However, this program requires some modification before it can be used correctly to calculate sensitivities for such state variable models. For such models, the diagonal elements a_{ii} ($i = 1, \ldots, n$) of the matrix A generally represent turnover rates of the variables x_i, and are calculated from the equation

$$a_{ii} = -\sum_{k=1,\, k\neq i}^{n} a_{ki} - b_{0i}, \tag{30}$$

where b_{0i} is the loss rate from x_i to the system's environment. With this constraint on A, we have that

$$\partial A/\partial a_{ij} = \xi_{ij} - \xi_{jj}. \tag{31}$$

Thus, Eq. (29) becomes

$$\dot{\mathbf{s}}(a_{ij}, t) = A\mathbf{s}(a_{ij}, t) + (\xi_{ij} - \xi_{jj})\mathbf{x}, \tag{32}$$

or, for any specific x_k,

$$\dot{s}_k(a_{ij}, t) = \mathbf{a}_{k\cdot}\, \mathbf{s}(a_{ij}, t) + \mathbf{e}_{k\cdot}\mathbf{x}, \tag{33}$$

where $s_k(a_{ij}, t)$ is the sensitivity of x_k to a_{ij}, $\mathbf{a}_{k\cdot}$ is the kth row of A, and $\mathbf{e}_{k\cdot}$ is the kth row of $(\xi_{ij} - \xi_{jj})$. More specifically, we may express Eq. (33) as

$$\dot{s}_k(a_{ij}, t) = \begin{cases} \mathbf{a}_{k\cdot}\, \mathbf{s}(a_{ij}, t) & \text{if } k \neq i, \quad k \neq j \\ \mathbf{a}_{k\cdot}\, \mathbf{s}(a_{ij}, t) + x_j & \text{if } k = i \\ \mathbf{a}_{k\cdot}\, \mathbf{s}(a_{ij}, t) - x_j & \text{if } k = j, \quad j \neq i. \end{cases}$$

In the derivation of Eq. (32), we did not require that the elements of A be constant. In particular, they may be time varying.

This development has concerned sensitivity to the elements of A only (i.e., parameter sensitivity). Other measures could be devised as well. Kerlin (1967) also considered sensitivity to the initial conditions of the state variables, sensitivity in the frequency domain, and pole sensitivity. Tomović (1963) suggested the extension of sensitivity theory to state variables and inputs. Patten (1972) recognized three categories of sensitivity as important for ecological studies: state sensitivity, input sensitivity, and parameter sensitivity. In this discussion we limit ourselves to parameter sensitivity in the time domain.

C. The State Sensitivity Model

Solution of Eq. (32) will yield a great number of values for the sensitivity of x_k to a_{ij}, for all x_k and a_{ij}. Additionally, however, it leads to a state sensitivity model $\dot{s}(A, x, t)$ for the system under investigation, as noted by Pridgen (1970). Two features of this model should be emphasized.

First, for linear state variable models, the state and sensitivity models are of the same general form (Meissinger and Bekey, 1966; Tomović, 1963). Thus, as noted above, both models can be solved by the same approach. However, the two models are not independent, since some of the sensitivity equations include state variables as forcings. Therefore, the two models must be solved concurrently. Further, their homogeneous portions are identical, so that the two systems will have the same characteristic equation. Hence, system stability can be investigated using the sensitivity equations as well as the state equations. This latter point, regarding the importance of the sensitivity model to stability considerations, also applies to nonlinear representations. In many cases, linear as well as nonlinear, the sensitivity model is a more efficient method for investigating stability than is the state model (Tomović, 1963).

Second, no matter whether the state model is linear or nonlinear, the sensitivity model is always linear (Tomović, 1963; Mankin and Brooks, 1971). This has led to considerations concerning when a nonlinear system may be represented by its linear sensitivity eqûations. Tomović (1963) provides a general discussion of the problem. More importantly, perhaps, this fact suggests that all systems possess a linear sensitivity substructure embedded within the total system structure. We seek to clarify this notion in the remainder of this section.

D. Ecology and the Concept of Sensitivity

1. *Reasons for Performing a Sensitivity Analysis*

As evidenced by the paucity of citations above, sensitivity analysis has not found much application in ecosystem modeling to date. We might at this point consider, then, what kinds of information the ecologist can derive from sensitivity analysis. Five broad categories, necessarily overlapping to a great extent, suggest themselves:

a. *Model Construction.* Sensitivity analysis may aid in model formulation. Such an approach could, for example, assist in selecting among various formulations, or in choosing appropriate parameter values for the model, as in the parameter optimization problem illustrated above. Plinston (1971) discusses such an approach for the choice of parameters to be used in hydrological models.

b. *Error Analysis.* This use of sensitivity techniques refers to the attempt to elucidate effects of noise or errors in model parameters on conclusions to be drawn from the model.

c. *Resource Management.* Questions appropriately posed in sensitivity terms can be of immense value in the management of natural resources. Such questions are of the "What if we do this?" sort. This use of sensitivity is somewhat analogous to successive simulation trials with differing parameter values, yet yields a great deal more information for less effort.

d. *Research Effort Allocation.* Identification of parameters to which a specific system is most sensitive will aid in directing research effort to maximize the information gained (at the total system level) in any ecosystem investigation. As an example, Meyer (1972) illustrates such a use of sensitivity analysis to guide data gathering in the construction of a hydrologic model.

e. *Understanding Ecosystems.* Sensitivity analysis should prove to be an invaluable tool in analyzing the structure and function of ecosystems.

Although distinctions among the above five categories are more artificial than real, the discussions to follow will focus on the fifth category, that of understanding ecosystems. The particular proposition is that ecosystems (and any system in general) manifest an inherent sensitivity substructure, a structure which is a fundamental property of the specific system being considered (see also Astor *et al.*, this volume).

2. *Sensitivity and the Pattern of System Connectivity*

What is the nature of this system sensitivity substructure? Three aspects will be emphasized: (i) in the remainder of this section, the relation between the sensitivity structure and the pattern of system component interactions; (ii) in the next section, the relation of this system substructure to the dynamic motion of the system; and (iii) in the section following that, the dynamic nature of these sensitivity relationships.

Components of ecosystems may interact via the flow of energy and materials, and through transfers of information, however one wishes to define this latter concept. Many of these system components are directly and causally connected. Further, many parts of the system are linked to others both by direct relationships and also by indirect ones mediated through a series of intervening interactions. Additionally, many components which are not directly connected are causally but indirectly related through the patterns of component interactions. Both the relatively high degree of connectivity typically found within ecosystems and the possibility for both direct and indirect relationships leads to a highly diverse network of signal transmission in such systems.

The influence of system components on one another is also propagated along this system network. In this sense, the sensitivity substructure of an ecosystem is related to the propagation of influence within the system. That is, this sensitivity structure is composed of the causal relationships among the various parts of the ecosystem, relative to the pattern of component interactions within the system.

Based on the above, one reasonable approach to the sensitivity analysis of total ecosystem models would be the elucidation of pattern. Besides being interested in the magnitude of sensitivity of some state variable to a parameter, also of interest are the pathways over which this influence is transmitted. For example, several paths may exist between any two points in a system. By judiciously zeroing appropriate intervening parameters along the several paths, and then performing sensitivity analyses as indicated by Eq. (32), paths could be ranked according to the relative magnitude of the influence they transmitted, providing valuable insight into the functioning of the system being investigated.

Hence, to reiterate, one essential part of the concept of an ecosystem sensitivity structure is the notion of a pattern of influence propagation along the network of component interactions.

3. *Sensitivity as a State Operator*

Consider a system state variable x_i which is causally related to the parameter p. Assume that at an instantaneous point in time t_0, p is perturbed by an amount Δp from its original value p_0 to a new value $p_0 + \Delta p$. Associated with the specific interaction, direct or indirect, between x_i and p is the relationship $s_i(p, t)$ which expresses the sensitivity of x_i to p. Because x_i and p are causally related, the perturbation Δp can be used to compute the change in value of x_i, as well as its new value, in relation to the applied perturbation in p, according to the equations

$$\Delta x_i(\Delta p, t) = s_i(p_0, t)(\Delta p) + R, \tag{34a}$$

and

$$x_i(p_0 + \Delta p, t) = x_i(p_0, t) + s_i(p_0, t)(\Delta p) + R. \tag{34b}$$

In the above, R represents the higher-order terms in the Taylor series expansion. In this sense, it represents the error of the linear approximation. It is assumed that the linear approximation is good for a small perturbation Δp (i.e., $R \approx 0$). A similar analysis by Pridgen (1970) was illustrated above [Eq. (26)].

Equations (34) above imply that sensitivity operators may be used to calculate perturbed states of the system in response to a small parameter perturbation. However, in terms of the unperturbed dynamic behavior of ecosystems, these equations suggest something more. In this spirit, and analogously to the concept of state, the system sensitivity structure is viewed as providing sufficient information to move the system through time from its present state to a new state in response to time dynamics of the system parameters.

Specifically, time can be viewed as a natural force acting to perturb the value of system parameters. Although stationary models of ecosystems can be constructed, it is certainly true that most real-world parameters change with time-induced changes in the system's environment. Following Astor *et al.* (this volume), the time behavior of parameter p may be expressed by the equation

$$p(t) = p(t_0) + kq(t), \tag{35}$$

where $q(t)$ is a function of time (possibly constant) and k is a scalar relating q to p, with

$$q(t_0) = 0.$$

Over some small, finite time interval $(t_0, t_0 + \Delta t]$, p will change in value according to the equation

$$p(t_0 + \Delta t) = p(t_0) + \Delta p. \tag{36}$$

To write an equation for the change in state of x_i caused by this time-induced parameter variation, it is desirable to eliminate time from the equations and work strictly within the parameter domain. As shown by Tomović (1963), this is possible for a finite time interval. In such a case, the difference between the actual state change Δx_i and the linear approximation to it [e.g., Eq. (34a)] is always finite. Thus, for a sufficiently small time-induced change Δp around some point p_0 in parameter space, we may always represent disturbances of the real-world system by linear approximations. This result obtains from the theory of differential equations as it relates to the analytical dependence of solutions on parameters. The only conditions which must be satisfied are those guaranteeing uniqueness and existence of solutions. Thus, time may be eliminated from the equations (Tomović, 1963).

Influenced by the natural variation in the parameter p indicated by Eq. (35), as it is transmitted through the sensitivity substructure, state variable x_i assumes a new value according to the equations

$$\Delta x_i(\Delta p) = s_i(p_0)\,(\Delta p) + R, \tag{37a}$$

and

$$x_i(p) = x_i(p_0) + s_i(p_0)\,(\Delta p) + R. \tag{37b}$$

Again, these equations produce a first-order approximation to the state change causally influenced by the parameter change Δp. Hence, under this interpretation, the sensitivity operator acts as a state operator, moving the system through state space in response to time dynamics of the system parameters.

Equations (34) and (37) are valid since only a single parameter p causally influencing x_i was perturbed [Eqs. (34)], or varied with time [Eqs. (37)]. However, in any natural ecosystem, each state variable x_i is causally influenced by a number of parameters p_j, each of which is subject to time-induced changes Δp_j [Eq. (36)]. Thus, the resultant effect on the motion of x_i will depend on the sum of the individual changes $\Delta x_i(\Delta p_j)$ induced by time variation in each of the p_j ($j = 1, \ldots, m$). Therefore, the change in x_i is given by the equations

$$\Delta x_i(\Delta \mathbf{p}) = \sum_{j=1}^{m} s_i(p_{j_0})(\Delta p_j) + R, \tag{38a}$$

and

$$x_i(\mathbf{p}) = x_i(\mathbf{p}_0) + \sum_{j=1}^{m} s_i(p_{j_0})(\Delta p_j) + R. \tag{38b}$$

Thus, in any case, the sensitivity substructure is viewed as that portion of the system which relates dynamics in parametric space to dynamics in state space, where the former act through this sensitivity structure to regulate the latter (see also Astor *et al.*, this volume).

It should be emphasized that this treatment of sensitivity is different from the classical approach, wherein sensitivity relates to errors or applied pertubations in parameters. Here, we have presented a theoretical treatment of sensitivity relationships in ecosystems, as they relate to natural time variation in system parameters.

4. *Sensitivity Dynamics*

In this section, we consider the dynamic nature of sensitivity relationships in ecosystems, in both the time and the parameter domain.

a. *Time Dynamics of the Sensitivity Operators.* In the preceding section, we considered the relation between the sensitivity structure of a system and the dynamic motion of that system, without mentioning anything about dynamics in the sensitivity operators. However, the sensitivity relations of a natural ecosystem also change with time.

Let us rewrite Eq. (37a) as

$$\Delta x_i(\Delta p) \approx s_i(p_0)(\Delta p). \tag{39}$$

This equation expresses a direct relationship between the changes in x_i and p and the sensitivity of x_i to p. In particular, it tells us something about the dynamics of the sensitivity relationship $s_i(p_0)$.

Any system state variable x_i must vary within certain finite bounds if that component is to persist through time. That is, $x_i(t)$ will vary an amount Δx_i in relation to the natural variation of Δp_j ($j = 1, \ldots, m$) in the parameters $\mathbf{p}$ [Eq. (38a)]. But the change in x_i may not grow without bound. This implies that, as the parameters $\mathbf{p}$ vary in time according to Eq. (35), the sensitivity relationships $s_i(p_{j_0})$ will change so that the variability in x_i remains within certain limits. Thus, we see that the sensitivity structure of an ecosystem exhibits dynamic behavior in response to natural dynamics in the system parameters.

b. *Sensitivity Relations in Parametric Space.* So far in this discussion, nothing has been said about the dependency of sensitivity on the exact parameter values being considered. However, the sensitivity relationship between any state variable and the parameters to which it is related depends quite heavily on the location in the space of system parameters of the specific parameter set being studied.

Tomović (1963) examined this question in detail. He introduced the notion of bifurcation set(s) of parameters, in the neighborhood of which dramatic changes in the characteristics of the system trajectories occur (e.g., change from a stable to an unstable system). Tomović's exact formulation of the problem is not important here. What is important is the realization that, within certain regions of parametric space, the nature of the system behavior undergoes important shifts. Put in other terms, within these regions, the system is extremely sensitive to changes in the system's parameters.

This realization is important for the design of sensitivity analyses of ecosystem models. It should be quite instructive to determine such bifurcation sets of system parameters. For example, such a set of parameters might correspond to a shift in the mode of system operation at a certain point in the annual cycle. Or, it might represent, in a successional sequence in southeastern old fields, a switch from the forb to the *Andropogon* stage of succession. Determination of such values is of considerable importance to the ecologist, for both management and theoretical reasons.

E. Sensitivity, Stability, and the Evolution of Sensitivity Relationships

In the preceding sections, we considered the relation between the linear sensitivity structure of an ecosystem and the dynamic behavior of that system, as well as the sensitivity dynamics in both the time and parameter domains. In this section, we wish to examine sensitivity and stability on an evolutionary time scale. The above discussions pertaining to state and sensitivity time dynamics should still be valid on this expanded time scale.

1. *Parameter Sensitivity and Structural Stability*

Much of stability theory has rested on the methods of Liapunov, and has been concerned with stability with respect to small perturbations in the initial states of the system (Section IV. C. 1). However, it is not at all clear that such an approach has much potential benefit for ecosystem analysis (e.g., Lewontin, 1969). We suggest that the concept of structural stability, and its relation to parameter sensitivity, may be of more help in understanding ecosystem structure and function. Structural stability refers to the stability of a system with respect to variations or perturbations in parameter space. Some authors refer to this notion as parameter insensitivity (Tomović, 1963). As discussed below, system stability may be related to the evolution of insensitivity to parameters of the system. Model-oriented sensitivity analyses should be of great value in such considerations.

2. *The Evolution of Sensitivity Relationships*

Let us rewrite Eq. (37a) as

$$\Delta x_i(\Delta p) - R = s_i(p_0)\,(\Delta p). \tag{40}$$

In this formulation we interpret R to be the maximum variability in x_i for it to persist through time. In other words, we are equating the stability of system component x_i with its persistence in time.

Equation (40) shows that, under this interpretation, only two factors contribute to variation in x_i, namely $s_i(p_0)$ and Δp. If either or both of s_i and Δp are quite large, then x_i has a good chance of exceeding the bounds on its permissible variation and becoming unstable (i.e., going extinct). Hence, if Δp is large, $s_i(p_0)$ must be correspondingly small to maintain x_i within its defined limits of variability. In other words, Eq. (40) implies that natural selection will act to decrease the sensitivity of a system component to a highly variable parameter.

The obvious question at this point is the method by which selection may act to decrease sensitivity. In answering this question, it is convenient to use expressions such as the sensitivity of one system state variable to another, a meaningless expression in terms of parameter sensitivity. However, such an expression will be understood to mean the sensitivity of one system component to those factors (parameters) controlling the second component. For example, the sensitivity of a primary consumer to a producer might refer to the sensitivity of that consumer to the producer's gross productivity. Nor is it at all clear that this latter expression is not what we as ecologists are really interested in.

We suggest four mechanisms by which sensitivities may be reduced. These arguments may be seen to follow many of the diversity–stability arguments to some extent.

1. A system component may become less sensitive to one state variable if it becomes more sensitive to another one. Consider a consumer population with two resources, one of which is highly variable while the other is reasonably stable. The consumer is more likely to persist over time if it is less sensitive to the variable resource and more sensitive to the stable one.

2. A component may become less sensitive to one parameter if the number of parameters to which it is sensitive is increased. That is, all other factors held constant, a population with a broad resource base and a relatively small sensitivity to any one of these resources is more stable (potentially) than one with a narrow resource base with relatively greater sensitivities to the individual resources. This point is quite similar to one emphasized by May (1972) and Margalef (1968) relating both to the strength of interaction among species and to the number of interacting species.

3. One system component may become less sensitive to a parameter if its distance from the parameter is increased. This gets back to the first notion of a sensitivity substructure, that of the pattern of system connectivity. Thus, a primary consumer may be more sensitive to a producer than would be a secondary consumer. Efficiency of energy utilization, of course, places a limit on the length of food chains, and hence on the ability of such a force as this to reduce sensitivity.

4. Negative feedback mechanisms tend to reduce sensitivity to system parameters (Shinners, 1972). Hence, one would expect selection to act to produce negative feedback control mechanisms in ecosystems.

Thus, based on a stability criterion, persistence through time, we have considered the evolution of sensitivity relationships. The first mechanism discussed above relates to decreased sensitivity to a highly variable parameter, while the last three arguments relate to reduction of the magnitude of sensitivity operators in general.

F. Linearity of the System Sensitivity Substructure

We have alluded to linearity of the system sensitivity substructure at several points in the above. In Section VI. C on the state sensitivity model it was stated that, regardless of the nature of the state space model, the state sensitivity model is always linear. Further, in all the equations relating to the various ramifications of the sensitivity concept, only linear terms in the Taylor series expansions were considered. Rather than viewing this linearity of the sensitivity relationship as being a fortuitous result with mathematical consequences only, we feel that it has important theoretical bearing on the functioning of natural systems. While the full import of this idea is not appreciated, we do wish to consider it breifly.

Tomović (1963) showed that, for nonlinear electrical networks, computation of the sensitivity operators often is equivalent to measurement of the current in a linear subnetwork of the system. Hence, for such physical systems the linear sensitivity structure has an exact physical interpretation. It is not at all certain whether such an exact interpretation of the sensitivity substructure of ecosystems is possible.

Sensitivity essentially is related to change in the system parameter set within parametric space. In the normal operating behavior of ecosystems, such changes are probably small over a given time span, with dynamic motion of the system resulting from the total variation in parameters over the entire time interval of interest. Thus, the linear sensitivity structure would seem to move the system through time in response to the normal, sequentially small variations in values of system parameters. However, this substructure is not sufficient to account for system behavior in response to major shifts in the system's environment. Such large changes require factors outside the linear sensitivity structure to maintain invariant the system operating characteristics.

Within this normal mode of system operation, parametric sensitivity is related to homeostatic control of system behavior, as discussed by Tomović (1963). In the first place, sensitivity relationships can be both positive and negative. That is, a certain system component can increase or decrease in relation to variations in the parameters to which it is related. Thus, when variations in one or more parameters would cause a component to increase or decrease drastically, the sensitivity to other parameters may act in a compensatory fashion to stabilize the behavior of this component. Further, as noted earlier (Section V.D.4), the sensitivity substructure also exhibits dynamics in response to natural variations in the system parameters. Thus, decreasing magnitudes of sensitivity may help compensate for a potentially destabilizing variation in parameters, again stabilizing the trajectories of system components. Hence, the linear sensitivity structure is related to maintaining invariant the characteristics of system behavior. Conrad (1972) also pointed out the importance of such compensatory behaviors toward maintenance of ecosystem stability. Waide *et al.* (1974) demonstrated the existence of such compensatory behaviors in a model of the calcium cycle in a deciduous forest ecosystem.

VI. Conclusions

In the above discussions we examined four general classes of systems analysis techniques. We attempted to present the basic approaches in each case, as well as to examine the applicability of these techniques to ecosystem modeling efforts. In some cases we have done no more than essentially review available techniques and note how they have been or may be applied to ecological problems. In other cases we have presented new ideas or interpretations. In the stability analysis section, thoughts concerning the relationship between neighborhood stability, system connectedness, and donor control were examined. However, as was brought out in the discussion following Lewontin's (1969) contribution to the Brookhaven symposium on diversity and stability, and reiterated above, it is not at all clear that the neighborhood stability approach is appropriate for many ecological considerations, especially those involving natural ecosystems. In the sensitivity portion of this paper, a theoretical interpretation of sensitivity relationships in ecosystems, as they relate to time dynamics of the system parameters, was presented. This is a different approach to sensitivity questions than the usual one of examining the effect of parameter perturbations or errors on system performance.

We certainly did not seek to be exhaustive in our discussions. Many other aspects of these techniques, as well as numerous other analytical approaches, could have been examined. For example, we did not consider control theory and its applicability to the homeostatic regulation of ecosystems. Some of the techniques which were considered center around the classical transfer function representation of systems. Such representations assume that the system under study is both observable and controllable, in the control theory sense. However, it is not at all clear that ecosystems are either completely controllable or completely observable. Investigations of these concepts may be of considerable importance to ecosystem theory and management. Thus, control theory approaches may ultimately prove to be of more benefit to ecosystem analysis than the transfer function approach (see also Mulholland and Sims, this volume).

Several of the points raised above, as well as other considerations, have led us to a general conclusion. That is, given the current state of systems ecology, we feel that engineering and ecological systems analysis have fundamental differences. The essence of the systems engineering approach is the design of specific systems, subject to certain generally well-defined constraints and criteria. Also, the behavior of these systems and their parts, as well as the environments in which they operate, usually are known or knowable with fairly good accuracy. However, in ecology we attempt to analyze systems already in existence, which are not only difficult to measure but also difficult even to define. We know that natural ecosystems, when not stressed greatly, are stable. Yet ecologists have not agreed on general principles governing ecosystem organization or regulation. So, what we are really doing in systems ecology is attempting a backdoor, after-the-fact approach to the analysis of the structure and function of ecosystems. Thus, many of the systems engineering methodologies either may not be applicable to ecosystem analysis, or may require modification before they can be applied. At the very least, all such analytical approaches which are applicable will require interpretation in ecological terms. This does not imply that the extensive techniques and experience of systems engineering should be ignored, but that they should be examined critically before being applied.

We would argue emphatically that the analysis phase of the total systems approach is a powerful one for the elucidation of ecosystem properties. Analysis is every bit as important a problem as modeling or model validation, for it is only in analysis, whether by computer exploration or by strictly analytical techniques, that we can arrive at explanations, rather than just descriptions, of ecosystem structure and function. But to a large extent, systems ecology, inasmuch as such a discipline exists, has been mostly a matter of description and bookkeeping. Simulation and analysis should complement one another in a true systems approach to the study of ecosystems. We therefore suggest that more attention should be paid to this analysis phase in systems ecology, to the modification of engineering techniques for ecological application if necessary, as well as to the actual application and interpretation of such techniques.

Acknowledgments

We gratefully acknowledge the direction and motivation given to this study by Dr. Bernard C. Patten. A number of techniques discussed were first brought to our attention in a class given at the University of Georgia by Dr. Leslie G. Callahan, Jr., and by Dr. Seicchi Fujita of the School of Industrial and Systems Engineering, Georgia Institute of Technology. Dr. J. B. Mankin, Oak Ridge National Laboratory, read and criticized an early form of the sensitivity section. Mr. John Leffler made several helpful criticisms in later stages. Mrs. Nancy Payne typed the manuscript.

The sensitivity section represents collective thinking within our group over a period of several years. The original stimulus to pursue sensitivity analysis, as well as the germ of several ideas developed herein, was provided by an unpublished draft manuscript by Dr. Patten describing a state space model for grassland. Subsequent seminar discussions with the group, whose membership also included Drs. Richard G. Wiegert, Jeffrey J. Lee, Clayton S. Gist, Donald C. Inman, and William G. Cale, led to earlier results presented in Patten (1972). The specific notion (Section V. D. 3) of an implicit sensitivity substructure in every system that mediates nominal (i.e., unperturbed) state dynamics in response to the natural time dynamics of parameters is in part due to Dr. Patten, as elaborated in his chapter with Astor *et al.* (this volume). Extensions, clarifications, and interpretations of this basic idea, contained in Sections V. D. 3–4 and V. E–F, are our own.

Computer funds for the stability analysis portion of the work as well as support for JRW were provided by the Coweeta Site of the Eastern Deciduous Forest Biome, U.S. IBP. JBW was supported by a NSF Predoctoral Fellowship during the course of the study.

REFERENCES

Ball, S. J., and Adams, R. K. (1967). MATEXP, A General Purpose Digital Computer Program for Solving Ordinary Differential Equations by the Matrix Exponential Method. Oak Ridge Nat. Lab., Tech. Memo. 1933.

Bellman, R. E. (1967). "Introduction to the Mathematical Theory of Control Processes," Vol. I. Academic Press, New York.

Bellman, R. E. (1968). "Some Vistas of Modern Mathematics: Dynamic Programming, Invariant Imbedding and the Mathematical Biosciences." Univ. of Kentucky Press, Lexington Kentucky

Bode, H. W. (1945). "Network Analysis and Feedback Amplifier Design." Van Nostrand-Reinhold, Princeton, New Jersey.

Brylinsky, M. (1972). *In* "Systems Analysis and Simulation in Ecology" (B. C. Patten, ed.), Vol. II, pp. 81–101. Academic Press, New York.

Child, G. I., and Shugart, H. H. (1972). *In* "Systems Analysis and Simulation in Ecology" (B. C. Patten ed.), Vol. II, pp. 103–135. Academic Press, New York.

Conrad, M. (1972). *J. Theoret. Biol.* **34**, 325.

Cunningham, W. J. (1963). *Amer. Sci.* **51**, 425.

Curlin J. W. (1970). *In* "Analysis of Temperate Forest Ecosystems" (D. E. Reichle, ed.), pp. 268–285. Springer-Verlag, Berlin and New York.

Dale, M. B. (1970). *Ecology* **51**, 2.

Daniels, J., and Mackay, A. L. (1974). *Nature (London)* **251**, 49.

Darnell, R. M. (1970). *Amer. Zool.* **10**, 9.

De Angelis, D. L. (1975). *Ecology* **56**, 238.

deNeufville, R., and Stafford, J. H. (1971). "Systems Analysis for Engineers and Managers." McGraw-Hill, New York.

DeRusso, P. M., Roy, R. J., and Close, C. M. (1967). "State Variables for Engineers." Wiley, New York.

DiStefano, J. J., Stubberud, A. R., and Williams, I. J. (1967). "Feedback and Control Systems." McGraw-Hill, New York.

Dunbar, M. J. (1960). *Amer. Natur.* **94**, 129.

Ehrlich, P. R., and Birch L. C. (1967). *Amer. Natur.* **101**, 97.

Evans, W. R. (1948). *Trans. AIEE* **67**, 547.

Funderlic, R. F., and Heath, M. T. (1971). Linear Compartmental Analysis of Ecosystems. Oak Ridge Nat. Lab.-IBP-71-4.

Gadgil, M., and Solbrig, O. T. (1972). *Amer. Natur.* **106**, 14.

Gallopín, G. C. (1971a). *Oecologia* **7**, 382.

Gallopín, G. C. (1971b). *Oecologia* **7**, 414.

Gardner, M. R., and Ashby W. R. (1970). *Nature (London)* **228**, 784.

Gifford, D. (1971). *Bull. Brit. Ecol. Soc.* **2**, 2.

Gilpin, M. E. (1975). *Nature (London)* **254**, 137.

Gist, C. S. (1972). Analysis of Mineral Pathways in a Cryptozoan Foodweb. Ph.D. Dissertation, Univ. of Georgia, Athens, Georgia.

Glass, N. R. (1971). *In* "Systems Analysis and Simulation in Ecology" (B. C. Patten, ed.), Vol. I, pp. 325–363. Academic Press, New York.

Goldstein, R. A., and Elwood, J. W. (1971). *Ecology* **52**, 935.

Hairston, N. G., Smith, F. E., and Slobodkin, L. B. (1960). *Amer. Natur.* **94**, 421.

Harvey, R. A., Taylor, G. R., and Benham, R. D. (1966). *Simulation* **6**, 181.

Hearon, J. Z. (1953). *Bull. Math. Biophys.* **15**, 121.

Hearon, J. Z. (1963). *Ann. N.Y. Acad. Sci.* **108**, 36.

Holling, C. S. (1972). *Proc. Int. Symp. Modelling Tech. Water Resources Syst.* **1**, 3.

Hurwitz, H. (1895). *Math. Ann.* **46**, 273.

Jordan, C. F., Kline, J. R., and Sasser, D. S. (1972). *Amer. Natur.* **106**, 237.

Kaye, S. V., and Ball, S. J. (1969). *In* "Symposium on Radioecology" (D. J. Nelson and F. C. Evans, eds.), pp. 731–739. USAEC Rep. CONF-670503.

Kerlin, T. W. (1967). *Simulation* **8**, 337.

Kerlin, T. W., and Lucius, J. L. (1966). The SFR-3 Code—A Fortran Program for Calculating the Frequency Response of a Multivariable System and Its Sensitivity to Parameter Changes. Oak Ridge Nat. Lab., Tech. Memo. 1575.

Kowal, N. E. (1971). *In* "Systems Analysis and Simulation in Ecology" (B. C. Patten, ed.), Vol. I, pp. 123–194. Academic Press, New York.

Leigh, E. (1965). *Proc. Nat. Acad. Sci. U.S.* **53,** 777.

Levin, S. A. (1970). *Amer. Natur.* **104,** 413.

Lewontin, R. C. (1969). *Brookhaven Symp. Biol.* **22,** 13.

Liapunov, M. A. (1892). "Problème Général de la Stabilité du Mouvement." Kharhov (reprinted as *Annals Math. Study* No. 17. Princeton Univ. Press, Princeton, New Jersey, 1949).

MacArthur, R. H. (1955). *Ecology* **36,** 533.

Mankin, J. B., and Brooks, A. A. (1971). Numerical Methods for Ecosystems Analysis. Oak Ridge Nat. Lab.-IBP-71-1.

Margalef, R. (1968). "Perspectives in Ecological Theory." Univ. of Chicago Press, Chicago, Illinois.

Margalef, R. (1969). *Brookhaven Symp. Biol.* **22,** 25.

May, R. M. (1971a). *Math. Biosci.* **12,** 59.

May, R. M. (1971b). *Proc. Ecol. Soc. Aust.* **6,** 17.

May, R. M. (1972). *Nature (London)* **238,** 413.

May, R. M. (1973a). *Amer. Natur.* **107,** 46.

May, R. M. (1973b). "Stability and Complexity in Model Ecosystems. " Princeton Univ. Press, Princeton, New Jersey.

Maynard Smith, J. (1974). "Models in Ecology." Cambridge Univ. Press, London and New York.

Meissinger, H. F. (1964a). *Simulation* **3,** 52.

Meissinger, H. F. (1964b). *Simulation* **3,** 55.

Meissinger, H. F., and Bekey, E. A. (1966). *Simulation* **6,** 94.

Meyer, C. F. (1972). *Water Resources Res.* **8,** 212.

Miller, K. S., and Murray F. J. (1953). *J. Math. Phys.* **32,** 136.

Murdoch, W. W. (1966). *Amer. Natur.* **100,** 219.

Murdoch, W. W. (1969). *Ecol. Monogr.* **39,** 335.

Nyquist, H. (1932). *Bell Syst. Tech. J.* **11,** 126.

Odum, E. P. (1969). *Science* **164,** 262.

Patten, B. C. (1961). *Science* **134,** 1010.

Patten, B. C. (1962). *Limnol. Oceanogr.* **7,** 266.

Patten, B. C. (1969). *Trans. Amer. Fisheries Soc.* **98,** 570.

Patten, B. C. (1971). *In* "Systems Analysis and Simulation in Ecology" (B. C. Patten, ed.), Vol. I, pp. 3–121. Academic Press, New York.

Patten, B. C. (1972). *In* "Theory and Applications of Variable Structure Systems" (R. R. Mohler and A. Ruberti, eds.), pp. 213–232. Academic Press, New York.

Patten, B. C. (1974). *In* "Structure, Functioning, and Management of Ecosystems," Proc. 1st Internat. Congr. Ecology (supplement). Centre for Agricultural Publishing and Documentation, Wageningen, Netherlands.

Patten, B. C. (1975). *Amer. Natur.* **190,** 529.

Patten, B. C., and Witkamp, M. (1967). *Ecology* **48,** 813.

Patten, B. C., Egloff, D. A., Richardson, T. H., *et al.* (1975). *In* "Systems Analysis and Simulation in Ecology" (B. C. Patten, ed.), Vol. III, pp. 205–421. Academic Press, New York.

Pianka, E. R. (1970). *Amer. Natur.* **104,** 592.

Pianka, E. R. (1972). *Amer. Natur.* **106,** 581.

Plinston, D. T. (1971). *In* "Mathematical Models in Ecology" (J. N. R. Jeffers, ed.), pp. 237–247. Oxford Univ. Press (Blackwell), London and New York.

Poincaré, H. (1892). "Les Méthodes Nouvelles de la Mécanique Céleste." Gauthier-Villars, Paris (Reprinted by Dover, New York, 1957).

Pridgen, J. H. (1970). Computation of Sensitivity Measures. Ph.D. Dissertation, Oklahoma State Univ., Stillwater, Oklahoma.

Roberts, A. (1974). *Nature (London)* **251**, 607.

Rosenzweig, M. L., and MacArthur, R. H. (1963). *Amer. Natur.* **97**, 209.

Routh, E. J. (1892). "Dynamics of System Rigid Bodies." Macmillan, New York.

Sánchez, D. A. (1968). "Ordinary Differential Equations and Stability Theory: An Introduction." Freeman, San Francisco, California.

Saucedo, R., and Schiring, E. E. (1968). "Introduction to Continuous and Digital Control Systems." Macmillan, New York.

Savageau, M. A. (1971). *Nature (London)* **229**, 542.

Savant, C. J., Jr. (1964). "Control System Design." McGraw-Hill, New York.

Schwarz, R. J., and Friedland, B. (1965). "Linear Systems." McGraw-Hill, New York.

Shinners, S. M. (1972). "Modern Control System Theory and Application." Addison Wesley, Reading, Massachusetts.

Shugart, H. H., Jr., Reichle, D. E., Edwards, N. T., and Kercher, J. R. (1976). *Ecology* **57**, 99.

Šiljak, D. D. (1974). *Nature (London)* **249**, 280.

Slobodkin L. B. (1964). *Amer. Sci.* **52**, 342.

Slobodkin, L. B., Smith, F. E., and Hairston, N. G. (1967). *Amer. Natur.* **101**, 109.

Smith, F. E. (1970). *In* "Analysis of Temperate Forest Ecosystems" (D. E. Reichle, ed.), pp. 7–18. Springer-Verlag, Berlin and New York.

Smith, F. E. (1972). *Trans. Conn. Acad. Arts Sci.* **44**, 309.

Somorjai, R. L., and Goswami, D. N. (1972). *Nature (London)* **236**, 466.

Thom, R. (1969). *Topology* **8**, 313.

Tomović, R. (1963). "Sensitivity Analysis of Dynamic Systems." MacGraw-Hill, New York.

Tomović, R., and Vukobratović, M. (1972). "General Sensitivity Theory." American Elsevier, New York.

Verhoff, F. H., and Smith, F. E. (1971). *J. Theor. Biol.* **33**, 131.

Waide, J. B., and Swank, W. T. (1975). *Proc. Soc. Amer. For.* (in press).

Waide, J. B., Krebs, J. E., Clarkson, S. P., and Setzler, E. M. (1974). *In* "Progress in Theoretical Biology" (R. Rosen and F. M. Snell, eds.), Vol. III, pp. 261–345. Academic Press, New York.

Watt, K. E. F. (1969). *Brookhaven Symp. Biol.* **22**, 142.

Webster, J. R., and Waide J. B. (in prep). Complexity, resource limitation, and the stability of ecosystems.

Webster, J. R., Waide, J. B., and Patten, B. C. (1975). *In* "Mineral Cycling in Southeastern Ecosystems" (F. G. Howell, J. B. Gentry, and M. H. Smith, eds.), pp. 1–27. ERDA Symp. Series, CONF–740513.

Westley, G. W. and Watts, J. A. (eds.) (1970). The Computing Technology Center Numerical Analysis Library. Computing Technol. Center, Oak Ridge Nat. Lab., CTC-39.

Wiegert, R. G., and Owen, D. F. (1971). *J. Theoret. Biol.* **30**, 69.

Wright, S. (1935). *J. Genet.* **30**, 257.

14

Control Theory and the Regulation of Ecosystems

ROBERT J. MULHOLLAND AND CRAIG S. SIMS

CENTER FOR SYSTEMS SCIENCE AND SCHOOL OF ELECTRICAL ENGINEERING
OKLAHOMA STATE UNIVERSITY, STILLWATER, OKLAHOMA

I. Introduction

Most quantitative studies of ecosystems begin with the construction of a model. For an ecological energetics study this may amount to constructing an energy flow diagram, which may then lead to a compartment model with its associated dynamics modeled mathematically by a set of differential equations. The mathematical model, which in some sense describes the operation of an ecosystem, depends upon ecological properties as well as the objectives and desired results of the mathematical modeling effort. The ecological properties include the definition of system variables, the assumptions made regarding individual system components, and the characterization of interactions which may exist between these components. The objectives of the modeling effort influence the choice of mathematical techniques employed, whether they are algebraic or dynamic, qualitative, analytical, or computational. The appropriateness of any mathematical model is judged by how close it reproduces and predicts the observed behavior of the ecosystem. The modeling problem is in general a difficult task, which usually requires problem-orientated insight before appropriate mathematical models can be constructed.

This paper considers mathematical models of ecosystems which are described by ordinary differential equations. Both linear and nonlinear models are considered which are of sufficient generality to be applicable to ecosystem analysis. The objective is to present a methodology which enables regulation of the operation of large-scale ecosystems by the adjustment of free parameters or inputs. The purpose is to improve the operation of ecosystems which, for a seminatural, engineered waste disposal process, might take the form of an increase in ecosystem carrying capacity. Or, in a pond managed for sport fishing, the objective may be to maximize bass population.

For many years engineers have been concerned with the analysis and design of automatic control systems. The techniques of control theory now available enable the design and construction of automatic controls for complex systems ranging from automation in the production and preparation of food products to the Viking spacecraft guidance system. Indeed, the control of large-scale systems is essential to the complex industrial base of the economy that exists today in the United States.

The general objective of optimal control theory is to make a system operate in a more desirable way, which may imply that the controlled system is required to operate more efficiently, more economically, or perhaps more reliably. Until quite recently the design of control systems was accomplished by trial and error methods combined with engineering intuition. A mathematical theory of optimal control has developed during the past two decades, primarily stemming from efforts of Bellman (1957), Kalman (1963), and Pontryagin (1962). This mathematical theory attempts to take the design of controlled systems from an art to a science.

Application of the mathematical theory of optimal control applies universally to a large area of the sciences, including electrical, mechanical and aerospace, economic, and biological systems. For example, in the medical sciences there exist automatic control systems to regulate flow of oxygen to infant incubators, thereby meeting the generalized medical goal of prolonging life and reducing pain.

The modern high-speed digital computer has made possible applications of optimal control theory to practical problems. It is the engineers' interest in creating "better" systems that constantly reinforces the development of these two technologies. However, the mere existence of computational ability is not sufficient to deal with large-scale systems. The systems designer must come to grips with that which constitutes desired system performance and how it can be implemented using an optimal control policy.

There is a definite need for adequate methodologies for the treatment of large-scale systems. This fact is particularly apparent in those areas of systems theory which are beginning to receive increasing attention, such as socioeconomic and environmental systems. Methods of treating large-scale systems problems include the techniques of aggregation, and hierarchical control policies (Chidambara and Schainker, 1971; Aoki, 1968; Takahara and Mesarovic, 1969). The technique which is suggested here is certainly related to these topics, and is proposed as a simple but useful control technique for systems whose state vector contains many elements. The basic philosophy of the design procedure suggested here is related to that of model-reference or model-following control systems (Tyler, 1964; Motyka, 1972).

The problem of controlling large-scale systems has many aspects. Of particular importance is the fact that one does not know how to think about higher-order state variables, their couplings, and their associated desired behavior. It is apparent that the mind is limited as to the number of factors which it can consider in a reasonable way. This fundamental problem is definitely not solved by application of more and larger computers, as the difficulty occurs in the early conceptual stages, indeed in the very formulation of problems. No amount of computational ability will be of any use until the problem is understood and formulated correctly.

The philosophy proposed here is that one should control the large-scale system in a way which is specifically designed to allow one to consider a much smaller system whose variables are key indicators of the behavior of the larger system. The reduced system is to have a dimension which is small enough so that the designer will be able to postulate reasonably that which constitutes desirable performance.

II. Ecosystem Modeling

Compartment models have been used to model biological systems for many years. Atkins (1969) traces the origins of these techniques back to the work of Hevesey (1923), who employed radioactive lead in order to demonstrate the uptake and loss process of lead ions by the roots of plants. The first applications of compartmental analysis to animals were also made by Hevesey, who modeled the metabolism of radioactive bismuth in rabbits (Christiansen *et al.,* 1924). These early applications were primarily confined to tracer studies of physiological systems. The term *compartment,* first used by Sheppard (1948), is defined as a quantity of substance which is assumed to have uniform and distinguishable kinetics of transformation or transport.

A great deal of the literature on compartment models deals with steady-state analysis and the evaluation of kinetic rate constants. Other studies are concerned with the structural features of multicompartment systems, including whether they are open or closed with respect to the flow of substances, whether the system is strongly or weakly connected with respect to the flows, or whether the system has flow cycles and hence the possibility of oscillations (Thron, 1972). Several general treatments of these structural properties have been published, including Sheppard (1948), Sheppard and Householder (1951), Hearon (1953), and Berman and Schoenfeld (1956). The complete mathematical verification of compartmental tracer analysis, with respect to the stability and asymptotic properties of general models, was finally published by Hearon (1963).

Although the compartment analysis approach to ecological systems is sometimes traced back to the early work of Kostitzin (1935), it has matured of late into an area of biological modeling that in some sense is quite separate from that of tracer analysis or physiological kinetics. Ecosystem compartment models have been developed for both linear and nonlinear analysis with detail approaching that of tracer studies (Patten, 1971). However, these ecosystem models have unique characteristics in which environmental effects, such as temperature or light, are coupled into the model equations through periodically varying time functions.

A. Linear Models

Consider an ecosystem in which n functional groups of species have been identified. These n species are then associated with compartments in which their biomass, or equivalent energy, is lumped and represented by the variables $x_1, x_2, \ldots, x_n$. The possible flow of biomass, or energy, from compartment j to compartment i is indicated by F_{ij}, so that the mass balance equation for each compartment is

$$\dot{x}_i = \frac{dx_i}{dt} = \sum_{j=0}^{n} F_{ij} - \sum_{j=0}^{n} F_{ji} \qquad (i = 1, 2, \ldots, n), \tag{1}$$

where the first summation gives the inflow and the second sum the outflow, and F_{i0} and F_{0i} represent interactions with the environment. The assumption that flows are linear and donor compartment controlled requires that

$$F_{ij} = a_{ij} x_j \tag{2}$$

for all $j \neq 0$ and i indicated, where a_{ij} is the rate coefficient of the transfer from compartment j to compartment i. The flows F_{i0} are regarded as inputs which drive the ecosystem into steady state. Thus, Eqs. (1) and (2) become

$$\dot{x}_i = F_{i0} + \sum_{j=1}^{n} a_{ij}x_j - \left(a_{0i} + \sum_{j=1}^{n} a_{ji}\right)x_i \tag{3}$$

for each $i = 1, 2, \ldots, n$. In matrix form, this is compactly written as

$$\dot{\mathbf{x}} = A\mathbf{x} + \mathbf{b} \tag{4}$$

where $\mathbf{x}$ and $\mathbf{b}$ are the column vectors of components x_i and F_{0i}, respectively, and A is the $n \times n$ matrix composed of the rate coefficients of Eq. (3). This class of models is common to the current ecological literature, and its ecological validity rests upon representing small perturbations experienced by ecosystems which are observed to be in steady state.

The average rate coefficients which appear as entries in A are computed from average flows F_{ij}^0 and the average standing crops $x_j{}^0$ as

$$a_{ij}^0 = F_{ij}^0 / x_j{}^0 . \tag{5}$$

This computation assumes that the ecosystem is in steady state, which for the mathematical model requires $\dot{\mathbf{x}} = 0$ on the average.

Ecosystems and of course all of the flows which describe their structure, are temperature dependent. The temperature dependence of interactions between compartments in an ecosystem model is defined through the use of a temperature coefficient which is usually designated by the symbol Q_{10}. This Q_{10} factor describes the ratio of the flow F_{ij} measured at two temperatures differing by 10° C. In general, the van't Hoff equation (Giese, 1957, p. 63) gives

$$Q_{10} = [F_{ij}(T_2)/F_{ij}(T_1)]^{10/(T_2 - T_1)}, \tag{6}$$

where $F_{ij}(T_1)$ is the flow measured at temperature T_1 and $F_{ij}(T_2)$ is the flow at T_2. The temperature of an ecosystem can be determined as a causal function of time: $T = T(t)$, a sometimes useful approximation being

$$T(t) = T_{av} + T_{var} \sin 2\pi t/t_0, \tag{7}$$

where T_{av} is the average temperature, T_{var} is the variance, and t_0 is the period of the temperature cycle.

It is clear that the flows of the compartment model for an ecosystem are time varying. Equation (5) gives only the average flow. The inclusion of temperature effects gives

$$F_{ij} = q_{ij}a_{ij}^0 x_j, \tag{8}$$

where

$$q_{ij} = Q_{10}^r, \qquad \text{and} \qquad r = [T(t) - T_{av}]/10.$$

Note that these equations allow for different Q_{10} factors for each flow. Thus, the flows have been represented as donor controlled with periodic rate coefficients of period t_0, where $a_{ij}(t + t_0) = a_{ij}(t)$ for all t.

The ecosystem inputs, representing interaction of the ecosystem with its environment, are dependent upon, for example, temperature and light and their causal relationships to time. Thus, the components of the vector **b** are t_0-periodic functions of time.

B. Nonlinear Models

In many cases, the linear hypothesis is not sufficiently well justified to be employed for a successful model. Several nonlinear modeling efforts have been put forth in recent years. Most of these models are based upon the Volterra-Lotka hypothesis with recent theoretical and phenomenological interpretations given by Ulanowicz (1972) and Huang and Morowitz (1972). The Hairston–Smith–Slobodkin (1960) hypothesis provides an excellent example of a general rule for constructing nonlinear ecosystem models. This rule prescribes the flows from autotrophs to herbivores to be recipent controlled, so that

$$F_{ij} = \phi_{ij} x_i. \tag{9}$$

The ecological interpretation of this assumption is that competition for a limited primary production does not control the herbivores. Rather, the herbivores are controlled at the next higher trophic level, by the carnivores. The heterotrophic feeding flows between herbivores and carnivores are both donor and recipient controlled:

$$F_{ij} = \phi_{ij} x_i x_j. \tag{10}$$

Thus, the herbivores are controlled by the carnivores which in turn are controlled by competition for their herbivore food supply. Under certain conditions, it has been found that self-limiting terms are required in order to stabilize the heterotrophic flows (Garfinkel, 1962, 1967; Garfinkel and Sack, 1964). Appropriately modified, Eq. (10) becomes

$$F_{ij} = \alpha_{ij} x_i x_j - \beta_{ij} x_i^2 - \alpha_{ij} x_j^2. \tag{11}$$

The general mathematical form for nonlinear models is

$$\dot{\mathbf{x}}(t) = \mathbf{f}(\mathbf{x}(t), \mathbf{v}(t), t), \tag{12}$$

where $\mathbf{x}(t)$ is the state and $\mathbf{v}(t)$ represents a vector of only those inputs which are controllable, or free for adjustment by the system designer. The indicated time dependence of **f** reflects the environmental inputs, such as solar radiation, temperature, precipitation, etc. A number of ecosystems have been modeled using these techniques and assumptions, and their behavior has been analyzed in quantitative terms using large-scale computer simulations.

However, it is desirable to analyze the properties of systems of the form of Eq. (12) independently of their size. For example, are models constructed in this way, using ecological principles, always mathematically stable? That is, regardless of the initial state $\mathbf{x}(0)$, do solutions of the model equations converge to a steady state? Another important question is related to the control aspects of large systems. How does one specify a strategy for controlling the solutions of Eq. (12) by choosing the inputs $\mathbf{v}(t)$ so that the system performs in some optimal way? In general, the behavior of large-scale systems is difficult to describe and hence hard to control.

Before proceeding with the control aspects of ecosystems, it is useful to note that some elegant solutions exist regarding the behavior of large-scale systems. It has been proven by Hearon (1963) that the averaged form of Eq. (3), regardless of size, yields solutions which converge to the average standing crop values $\mathbf{x}^0$ as a steady state. The method of proof is based upon the diagonally dominant nature of the compartment matrix A, in which all column sums yield strictly negative numbers. This is a direct result of the rate coefficients a_{0i} which give the interaction of the ecosystem with its environment. When energy is involved, these coefficients are primarily determined by the compartmental heat loss, or respiration. This concept is equivalent to the well-known idea based upon steady-state thermodynamics, in which the maintenance or stability of life is viewed as a continual process of pumping out disorder in the form of heat loss through respiration to provide internal order within each compartment (Odum, 1971, p. 38).

III. Control Problem Formulation

A successful compartment model will replicate and predict the observed ecosystem behavior. The next step in systems analysis is manipulation of the ecosystem behavior using the mathematical model. In order to control or regulate the behavior of an ecosystem, a desired performance must be obtained. A control policy then results. For example, the reduction of pest population levels in an agricultural or natural ecosystem may represent a desirable objective. The control policy may involve the introduction of a predator to accomplish this reduction in pest populations. It is required that the optimum predator population level be computed given the level of the pests.

The control policy must take into account the total concept of desired ecosystem performance. In the agricultural pest problem example, minimization of the pest population level is not sufficient as this minimum must take place under the constraint of maximum crop production. Clearly, the control policy results from tradeoffs common to the behavior of the ecosystem under study. When systems are large, it becomes increasingly difficult, if not impossible, to include all behavioral effects on system performance accurately within a mathematical expression of control policy.

A. Optimal Control Laws

Optimal control theory seeks to prescribe control policies in mathematical terms, thus placing the control problem within the mathematical domain of the calculus of variations. The result of this analysis leads to an optimal control law:

$$\mathbf{v}(t) = \mathbf{g}(\mathbf{x}(t), t), \tag{13}$$

which is derived from the model Eq. (12) and a mathematical statement of performance. The mathematical structure of this control law is of prime importance to the ecosystem designer. If the function $\mathbf{g}$ is not explicitly dependent upon the state vector $\mathbf{x}(t)$, then the control $\mathbf{v}(t)$ is a prescribed function of time. In this case, the control law is called *open loop.* This type of law is often undesirable as it is clear that any external disturbances or perturbations of the system state, which may occur after the open-loop regulation of the system is begun, are not accounted for by the control signal $\mathbf{v}(t)$. When the control $\mathbf{v}(t)$ depends upon the system state $\mathbf{x}(t)$, the control law is called *closed loop* or it is said to be in feedback form. In this case, the control of the system responds directly to changes in the behavior of the system. This concept in the natural control exerted by biological systems is well known as homeostasis.

B. Performance Measures

It is assumed in optimal control theory that the system designer can describe mathematically the desired system performance. The general form of this expression, called the performance index, may be written as a scalar functional ϕ given by

$$\mathrm{PI} = \phi(\mathbf{x}, \mathbf{v}, t). \tag{14}$$

This measure of performance, depending upon the system state, the control, and time, varies in accordance with the problem in question. For example, in an agricultural ecosystem, the performance index would combine into a scalar measure the effects of maximizing crop production while penalizing excessive applications of fertilizer and insecticides. For economic-based systems, the factors of interest could include economic growth, inflation, unemployment, and excessive expenditure of resources.

Suppose that in an ecosystem the desired behavior can be described as a time series: $\mathbf{x}_d(t)$, which could represent a steady-state. The error between the desired behavior and that predicted by the model is $\mathbf{x}_d(t) - \mathbf{x}(t)$, which is to be minimized. However, this error minimization must take place within a meaningful context in which excessive control input energy is penalized. Thus, a typical performance measure employed for this class of problems is

$$\mathrm{PI} = \int_{t_i}^{t_f} [(\mathbf{x}_d(t) - \mathbf{x}(t))^2 + \mathbf{v}^2(t)]\, dt, \tag{15}$$

where t_i is the time at which control is applied and t_f is the final time. In Eq. (15), the integral between t_i and t_f is employed to assess the cumulative effect of error and control, which have been made positive by the use of quadratic terms.

When the model equations for the ecosystem are linear, as in Eq. (4), the quadratic performance measure of Eq. (15) leads to a linear control law in feedback form (Kalman, 1960):

$$\mathbf{v}(t) = K(t)\, \mathbf{x}(t), \tag{16}$$

where $K(t)$ is an $m \times n$ matrix. If the model equations are nonlinear or the performance measure is nonquadratic then the optimal control does not exhibit the feedback structure of Eq. (16). In such cases, the general problem solution leads to an open-loop control law derived from the iterative solution of a two-point boundary value problem (Leondes, 1966).

The main difficulty encountered with the solution of optimal control problems usually deals with that of computation. But, this is not the only difficulty. When large-scale systems are analyzed, it is extremely difficult to prescribe mathematically the desired performance of the system, let alone prescribe a control policy which results in this behavior. A new scheme for the regulation of large-scale systems is presented in the next section.

IV. Control of Large-Scale Systems

The following analysis presents a new methodology for the control of large-scale systems. This technique is based upon the formulation of a small-scale dynamic optimization problem derived from certain key indicators of behavior which are related to the large-scale system. A definite computational advantage results from this methodology, since only an algebraic problem need be solved to obtain the control policy for the large-scale system in terms of the optimum performance as prescribed by the indicator functions.

A. Problem Statement

It is assumed that a mathematical model is available for the large-scale ecosystem under consideration:

$$\dot{\mathbf{x}} = \mathbf{f}(\mathbf{x}, \mathbf{v}), \tag{17}$$

where $\mathbf{x}$ is the state vector and $\mathbf{v}$ the primary control vector. A new vector is defined which is of smaller dimension than $\mathbf{x}$, and presumably reflects the designer's opinion regarding what variables, related to $\mathbf{x}$, should be considered as indicators of overall system behavior. The vector $\mathbf{y}$ is functionally related to $\mathbf{x}$ through the equation

$$\mathbf{y} = \boldsymbol{\rho}(\mathbf{x}), \tag{18}$$

where $\boldsymbol{\rho}$ is a vector-valued function. The designer may be able to postulate in a reasonable way how he would expect $\mathbf{y}$ to behave so that good performance is achieved. For example, he may have arrived at a description of $\mathbf{y}$ in which the rate of change of $\mathbf{y}$ is functionally related to $\mathbf{y}$:

$$\dot{\mathbf{y}} = \mathbf{g}(\mathbf{y}), \tag{19}$$

which may occur in a natural way from solving an optimization problem involving $\mathbf{y}$ and $\dot{\mathbf{y}}$, using the calculus of variations. An example of this situation occurs in inner product control theory (Mulholland and Rhoten, 1972a, b; Leeper and Mulholland, 1972). For increased generality, and to allow for the possibility of a hierarchical structure, a more general description of $\mathbf{y}$ is adopted:

$$\dot{\mathbf{y}} = \mathbf{g}(\mathbf{y}, \mathbf{u}), \tag{20}$$

where $\mathbf{u}$ is an auxiliary control vector, perhaps chosen to minimize a performance measure:

$$J = \int_{t_i}^{t_f} L(\mathbf{y}, \mathbf{u})\, dt, \tag{21}$$

where L represents mathematically the desired performance of Eq. (20). Obviously if a feedback solution is available to the small-scale control problem specified by Eqs. (20) and (21), then Eq. (19) is of sufficient generality. The system described by Eq. (20) is a conceptual aid in the sense that it is easier to understand a small-scale dynamical system than a large-scale system such as Eq. (17). It is of interest then to determine a control law for Eq. (17), that is, to select $\mathbf{v}$ so that the variables of principal interest as defined by Eq. (18) satisfy Eq. (20).

B. Problem Solution

The problem solution is by means of algebra, and is rather similar to the technique of differential approximation (Sage, 1969), although no approximations are necessary as the problem is stated here. The resulting algebraic equations which must be solved are of the same form as those obtained by Motyka (1972); however, they arise in a different context here. The objective here is not to make the observed state of the system and state of a model behave in the same way, but rather to relate the state of a large-scale system to a smaller-dimensioned set of variables which indicate desired behavior. From

Eqs. (17) and (18) it is clear that

$$\dot{\mathbf{y}} = \mathbf{h}(\mathbf{x}, \mathbf{v}), \tag{22}$$

where **h** is defined as

$$\mathbf{h} = (d\rho/d\mathbf{x})\,\mathbf{f}(\mathbf{x}, \mathbf{v}). \tag{23}$$

Using Eqs. (18) and (20), one can also write $\dot{\mathbf{y}}$ in terms of **x** and **u** as

$$\dot{\mathbf{y}} = \mathbf{g}[\rho(\mathbf{x}), \mathbf{u}] = \psi[\mathbf{x}, \mathbf{u}]. \tag{24}$$

Equating Eqs. (22) and (24), it is true that **v** must be selected to satisfy the requirement that

$$\mathbf{h}(\mathbf{x}, \mathbf{v}) = \psi[\mathbf{x}, \mathbf{u}]. \tag{25}$$

Solving Eq. (25) for **v** in terms of **x** and **u** may not always be possible, and, if it is possible, it may not be easy. The important point is that it is an algebraic problem rather than a large-scale dynamical optimization problem. Thus, a significant reduction in complexity can be achieved. Except for the linear case, it is difficult to say much in general about the solution to Eq. (25), and no generalized results are suggested at this time. Some aspects of the general problem are illustrated by considering the linear case and a class of nonlinear problems of interest.

C. The Linear Case

For many systems, Eq. (17) can be written in the following linear form:

$$\dot{\mathbf{x}} = A\mathbf{x} + B\mathbf{v}, \tag{26}$$

where A is an $n \times n$ matrix and B is an $n \times m$ matrix. The indicators of overall system behavior also are assumed to satisfy a linear relationship:

$$\mathbf{y} = M\mathbf{x}, \tag{27}$$

where M is an $r \times n$ matrix. Remaining consistent, Eqs. (26) and (27) then give rise to a linear dynamical reduced system for **y**:

$$\dot{\mathbf{y}} = F\mathbf{y} + G\mathbf{u}, \tag{28}$$

where F is an $r \times r$ matrix and G is an $r \times s$ matrix. The solutions of Eq. (28) which correspond to those of Eq. (26) are prescribed by $\mathbf{y}(0)$ as obtained from Eq. (27).

The linear assumption of Eq. (27) is not unreasonable in the theory of aggregation, at least as it is applied to ecological systems (O'Neill, 1971). The collective biomass, or the equivalent energy, of each biological species determines the state of an ecological system. However, the total number of species in an ecosystem may be unknown and, if known, may number in the hundreds or even thousands. Therefore, species aggregation in ecological systems usually takes the form of lumping species linearly in functional groups, that is, groups of organisms which have the same or similar function within the ecosystem. The first level of aggregation thus requires the use of ecological species in place of taxonomic species. A more severe aggregation is sometimes employed in ecosystems about which little is known, in which species are linearly grouped according to their feeding habits, i.e., primary producers, herbivores, carnivores, omnivores, etc. This gives rise to reduced ecological system models with state dimensions of order ten or less. The aggregation process then reduces, for example, a large-scale ecosystem model based upon functional groups of species as in Eq. (26) to a smaller model given by Eq. (28), in terms of trophic interactions, which is easier for the designer to comprehend in terms of total system performance.

In order to develop a control policy for Eq. (26), consider the reduced system Eq. (28) and substitute from Eq. (27) to give

$$F\mathbf{y} + G\mathbf{u} = M(A\mathbf{x} + B\mathbf{v}), \tag{29}$$

or

$$MB\mathbf{v} = (FM - MA)\mathbf{x} + G\mathbf{u}. \tag{30}$$

The question now of determining $\mathbf{v}$ in terms of $\mathbf{x}$ and $\mathbf{u}$ has been reduced to one of linear algebra. Clearly a solution of Eq. (30) exists if and only if

$$\operatorname{rank}[MB] = \operatorname{rank}[MB; (FM - MA); G]. \tag{31}$$

Furthermore, the solution of Eq. (30) is unique if and only if MB is of full rank r.

For example, consider the single input case with $B = (1, 0, \ldots, 0)^T$, and the linear sum aggregation $M = (1, 1, \ldots, 1)$. Then $MB = 1$, and the solution of Eq. (30) is prescribed uniquely given the designer's choice of system performance. Indeed, MB has an inverse for all $M = (C_1, C_2, \ldots, C_n)$ as long as $C_1 \neq 0$, which clearly indicates that the controlled variables must enter into the aggregation scheme employed. This is obvious when stated in ecological terms; for example, in order to control ecosystem production through the adjustment of nutrient levels, the controlled primary producer compartments must appear in the aggregated system.

D. Nonlinear Example

To illustrate the application of the indicated design philosophy, the following nonlinear system is considered

$$\dot{\mathbf{x}} = A(\mathbf{x}) + B(\mathbf{x})\mathbf{v}. \tag{32}$$

A vector of indicator functions is defined as having elements

$$y_i = \tfrac{1}{2}\alpha_i^T(\mathbf{x})\,\alpha_i(\mathbf{x}), \qquad i = 1, 2, \ldots, r, \tag{33}$$

where $\boldsymbol{\alpha}_i$ is a vector-valued function of $\mathbf{x}$. Thus one can write

$$\mathbf{y} = \boldsymbol{\rho}(\mathbf{x}), \tag{34}$$

with the nonnegative elements of $\boldsymbol{\rho}$ defined according to Eq. (33). In terms of the indicator functions, the performance measure of interst is assumed to be

$$J = \int_0^\infty \left[\sum_{i=1}^{r} (\beta_i^2 y_i^2 + \dot{y}_i^2)\right] dt, \tag{35}$$

whereupon the Euler–Lagrange equations in first integral form for Eq. (35) are

$$\dot{y}_i = -\beta_i y_i, \qquad i = 1, 2, \ldots, r, \tag{36}$$

or

$$\dot{\mathbf{y}} = -\beta\mathbf{y} = -\beta\rho(\mathbf{x}), \tag{37}$$

where β is a diagonal matrix with elements β_i. Using Eqs. (22) and (23), one can obtain another expression for $\dot{\mathbf{y}}$:

$$\dot{\mathbf{y}} = \Gamma(\mathbf{x})\,[A(\mathbf{x}) + B(\mathbf{x})\mathbf{v}], \tag{38}$$

where $\boldsymbol{\Gamma}(\mathbf{x})$ has its ith row defined as

$$\Gamma_i(\mathbf{x}) = \alpha_i^T(\mathbf{x})\,\partial\alpha_i/\partial\mathbf{x}. \tag{39}$$

Equatirg Eqs. (37) and (38), it is seen that $\mathbf{v}$ must be chosen to satisfy

$$\Gamma(\mathbf{x})\,[A(\mathbf{x}) + B(\mathbf{x})\mathbf{v}] = -\beta\rho(\mathbf{x}). \tag{40}$$

If, for every $\mathbf{x}$ on the optimum trajectory, the matrix $\boldsymbol{\Gamma}(\mathbf{x})\,B(\mathbf{x})$ has an inverse, then an optimum feedback control may be found by solving Eq. (40), resulting in the expression

$$\mathbf{u}(\mathbf{x}) = -[\Gamma(\mathbf{x})\,B(\mathbf{x})]^{-1}\,[\Gamma(\mathbf{x})\,A(\mathbf{x}) + \beta\rho(\mathbf{x})]. \tag{41}$$

A special case of the above problem will be considered to demonstrate that Eq. (40) may have a solution even when $\boldsymbol{\Gamma}(\mathbf{x})\,B(\mathbf{x})$ is not even a square matrix. Let the system be linear, of the form

$$\dot{\mathbf{x}} = A\mathbf{x} + \mathbf{v}, \tag{42}$$

and consider a scalar indicator function which is a measure of the distance of the state from the origin

$$y = \tfrac{1}{2}\mathbf{x}^T\mathbf{x}. \tag{43}$$

Further, assume that this distance is to be minimized with a smoothness constraint placed upon the norm of the state trajectories (Mulholland and

Rhoten, 1972a), so that the performance measure is of the form indicated by Eq. (35),

$$J = \int_0^\infty [y^2 + \dot{y}^2] \, dt. \tag{44}$$

Equation (37) is then the scalar equation

$$\dot{y} = -y.$$

Thus Eq. (40) becomes

$$\mathbf{x}^T A \mathbf{x} + \mathbf{x}^T \mathbf{v} = -\tfrac{1}{2} \mathbf{x}^T \mathbf{x}, \tag{45}$$

and

$$\mathbf{v} = -[\tfrac{1}{2}\mathbf{x} + A\mathbf{x} + S\mathbf{x}] \tag{46}$$

provides the required solution, where S is any $n \times n$ skew-symmetric matrix. It is interesting to note that optimal feedback solutions of this problem are nonunique whenever $\mathbf{x}$ is of dimension greater than 1.

V. Summary and Conclusions

Perhaps the mathematical modeling of ecosystems is approaching a level of realism sufficient to warrant a need for a contribution such as this one on control theory. If this is not the case, then this paper serves to point out the difference between the way a systems scientist, or engineer, views large-scale systems compared to that of an ecologist, or ecosystem modeler. In short, the ecologist is concerned with observing, analyzing, and understanding how ecosystems operate, while the engineer is primarily interested in manipulating and controlling their behavior to achieve some goal.

The history of mankind contains a series of technological achievements that have removed man from a strict dependence upon natural ecological systems for his subsistence. Man today, however, is still critically dependent upon the assimilative capacity of natural ecosystems for his waste disposal, which is a primary reason for the present so-called ecological crisis (Koenig *et al.*, 1972).

Innovative solutions employing recycling process combinations, both natural and artificial, are needed to solve waste control problems now and in the future. Perhaps the most important technological problem facing society today involves the engineering of ecological systems capable of degrading industrial waste products. The modern concepts of systems analysis and control theory coupled with ecological theory and research clearly provide the basic mathematical, scientific, and engineering framework for the description and solution of problems of waste disposal from the total ecosystem approach. The waste problem solution requires an in-depth understanding of the control of large-scale ecosystems.

A methodology for controlling large-scale dynamical systems has been considered in this chapter. It is clear that one could extend the proposed technique to the hierarchical case by considering Eq. (20) to be representative of an intermediate dynamical system and choosing **u** in such a way as to relate y to an even smaller-dimensioned system, thus repeating the procedure described herein. The process could be repeated as many times as necessary, subject to the constraint that the required algebra be solvable. It is possible that the algebraic constraints may even offer a clue as to how one should set up a hierarchy, although such a remark is purely conjectural at this point. In general, it appears difficult to draw any strong conclusions about the proposed method, regarding whether or not it is an appropriate approach. It is the authors' viewpoint that the idea of selecting the primary control vector specifically to simplify the problem formulation, so that one may consider a smaller more tractable problem, has considerable intuitive appeal for many large-scale ecosystem problems.

Acknowledgments

This research was supported in part by NSF Departmental Science Grant GU-3160 to the Center for Systems Science and by NSF Research Grant GK-34084 to the first-named author. The authors are grateful to Dr. B. C. Patten of the University of Georgia for helpful comments regarding the content and scope of the paper. We also wish to acknowledge the many helpful conversations with Dr. Kent W. Thornton of the Center for Systems Science and Department of Zoology at Oklahoma State University.

REFERENCES

Aoki, M. (1968). *IEEE Trans. Automat. Contr.* **AC-13,** 246.

Atkins, G. L. (1969). "Multicompartment Models for Biological Systems." Methuen, London.

Bellman, R. E. (1957). "Dynamic Programming." Princeton Univ. Press, Princeton, New Jersey.

Berman, M., and Schoenfeld, R. (1956). *J. Appl. Phys.* **27,** 1361.

Chidambara, M. R., and Schainker, R. B. (1971). *IEEE Trans. Automat. Contr.* **AC-16,** 175.

Christiansen, I. A., Hevesey G., and Lomholt, S. (1924). *C. R. Hebd. Séanc. Acad. Sci. Paris* **178,** 1324.

Garfinkel, D. (1962). *Nature (London)* **194,** 856.

Garfinkel, D. (1967). *J. Theor. Biol.* **14,** 46.

Garfinkel, D., and Sack, R. (1964). *Ecology* **45,** 502.

Geise, A. C. (1957). "Cell Physiology." Saunders, Philadelphia, Pennsylvania.

Hairston, H. G., Smith F. E., and Slobodkin, L. B. (1960). *Amer. Natur.* **94,** 421.

Hearon, J. Z. (1953). *Bull. Math. Biophys.* **15,** 121.

Hearon, J. Z. (1963). *Ann. N.Y. Acad. Sci.* **108**, 36.
Hevesey, G. (1923). *Biochem. J.* **17**, 439.
Huang, H. -W., and Morowitz, H. J. (1972). *J. Theor. Biol.* **35**, 489.
Kalman, R. E. (1960). *Bol. Soc. Mat. Mex.* **5**, 102.
Kalman, R. E. (1963). *In* "Mathematical Optimization Techniques" (R. E. Bellman, ed.), Chapter 16. California Univ. Press, Berkeley, California.
Koenig, H. E., Cooper, W. E., and Falvey, J. M. (1972). *IEEE Trans Syst. Man Cybernet.* **SMC-2**, 319.
Kostitzin, V. A. (1935). "Evolution de l'Atmosphère." Hermann, Paris.
Leeper, J. L., and Mulholland, R. J. (1972). *IEEE Trans. Automat. Contr.* **AC-17**, 401.
Leondes, C. T. (1966). "Advances in Control Systems: Theory and Applications," Vol. 3. Academic Press, New York.
Motyka, P. R. (1972). *IEEE Trans. Automat. Contr.* **AC-17**, 809.
Mulholland, R. J., and Rhoten, R. P. (1972a). *J. Franklin Inst.* **294**, 73.
Mulholland, R. J., and Rhoten, R. P. (1972b). *IEEE Trans. Automat. Contr.* **AC-17**, 95.
Odum, E. P. (1971). "Fundamentals of Ecology," 3rd ed. Saunders, Philadelphia, Pennsylvania.
O'Neill, R. V. (1971). *In* "Systems Analysis and Simulation in Ecology" (B. C. Patten, ed.), Vol. I, pp. 441–477. Academic Press, New York.
Patten, B. C. (1971). *In* "Systems Analysis and Simulation in Ecology" (B. C. Patten, ed.), Vol. I, pp. 3–121. Academic Press, New York.
Pontryagin, L. S., Boltyanskii, V. G., Gamkrelidze, R. V., and Mishchenko, E. F. (1962). *In* "The Mathematical Theory of Optimal Processes" (K. N. Trirogoff, translator, L. W. Neustadt, ed.). Wiley (Interscience), New York.
Sage, A. P. (1969). "Optimal Systems Control." Prentice-Hall, Englewood Cliffs, New Jersey.
Sheppard, C. W. (1948). *J. Appl. Phys.* **19**, 70.
Sheppard, C. W., and Householder, A. S. (1951). *J. Appl. Phys.* **22**, 510.
Takahara, Y., and Mesarovic, M. D. (1969). *IEEE Trans. Automat. Contr.* **AC-14**, 688.
Thron, C. D. (1972). *Bull. Math. Biophys.* **34**, 277.
Tyler, J. S. (1964). *IEEE Trans. Automat. Contr.* **AC-9**, 485.
Ulanowicz, R. E. (1972). *J. Theor. Biol.* **34**, 239.

15

The Sensitivity Substructure of Ecosystems

PETER H. ASTOR

ENVIROSPHERE COMPANY: A DIVISION OF EBASCO SERVICES
NEW YORK, NEW YORK

BERNARD C. PATTEN

DEPARTMENT OF ZOOLOGY, UNIVERSITY OF GEORGIA, ATHENS, GEORGIA

AND

GERALD N. ESTBERG

DEPARTMENT OF PHYSICS
UNIVERSITY OF SAN DIEGO, ALCALA PARK, SAN DIEGO, CALIFORNIA

University of Georgia, *Contributions in Systems Ecology,* No. 27.

I. Introduction

The dynamic motion of total or partial ecosystems can be simulated by state models consisting of sets of coupled differential equations. The state variables $x_1(t), \ldots, x_n(t)$ may represent population sizes or densities, or biomass, energy, or nutrient contents of various functional aggregates (compartments). The parameters $p_1(t), \ldots, p_m(t)$ may include (i) exogenous input (driving, forcing) variables, such as light, temperature, precipitation, immigration, or import; (ii) endogenous process variables, representing birth, death, respiration, feeding, excretion, or export rates; and (iii) physical or biological constants. Parameters are distinguished from state variables in being independent of the states of the defined system.

The state equations for a general linear or nonlinear system can be written in vector form as

$$d\mathbf{x}(t)/dt = \mathbf{f}[\mathbf{x}(t), \mathbf{p}(t)] \qquad t \in [t_0, t_1], \tag{1}$$

where $\mathbf{x}(t) \in R^n$ is the state vector, $\mathbf{p}(t) \in R^m$ is a parameter vector, and $\mathbf{f}$ is the vector of n functions describing rates of change of $x_1(t), \ldots, x_n(t)$. R^n is the state space of the model and $[t_0, t_1]$ is the closed time interval over which the system dynamics is to be generated. For nonautonomous systems one of the elements of the parameter vector is t itself. The time development of the state vector is based upon the initial time and state of the system (e.g., Zadeh and Desoer, 1963),

$$\mathbf{x}(t_0) = \mathbf{x}_0, \tag{2}$$

which is usually obtained from observation, measurement, experimentation, or conjecture. The unique solution $\mathbf{x}(t)$ of Eq. (1) subject to Eq. (2) represents the nominal dynamics in state space of the modeled system over the selected interval $[t_0, t_1]$, and serves as a basis for simulation and prediction.

The accuracy of simulation clearly depends upon accurate determination of the functions $f_1, \ldots, f_n$, the parameters $p_1(t), \ldots, p_m(t)$, and the initial states $x_1(t_0), \ldots, x_n(t_0)$. Given that $\mathbf{f}$ is correct, implicit in the system structure as defined by Eq. (1) is a substructure representing the deviation from $\mathbf{x}(t)$ caused by error in measurement or perturbation of one or more of the parameters or initial states. This will be referred to as the *sensitivity substructure* of the system.

In this chapter aspects of this substructure are developed for several classical population models preliminary to further investigations into the sensitivity analysis of ecosystems. The ecosystem emphasis is intended to be central, and the population models serve mainly to illustrate the discussion. The study draws upon the original sensitivity investigations of Miller and Murray (1953), subsequent extensions by Tomović (1963), Meissinger and Bekey (1966), and

Pridgen (1970), and the recent general theory provided by Tomović and Vukobratović (1972). A recent collection of papers in Cruz (1973) also serves as useful background.

II. Fixed versus Variable Structure Systems

Equation (1) defines a fixed structure system if the functions $f_1, \ldots, f_n$ remain the same on $[t_0, t_1]$. If they change, however, then Eq. (1) represents a variable structure system (Mohler and Ruberti, 1972). In ecosystems, when processes are disrupted the resulting disturbances propagated through the system may or may not cause structural alterations that would necessitate modifying $\mathbf{f}$ in a corresponding model. In this chapter only disturbances that do not cause structural changes will be considered. With $\mathbf{f}$ fixed, only parameter and state perturbations are allowed, and the chapter thus concerns the sensitivity substructure of fixed structure systems, illustrated by fixed structure population models. Attention will be confined to relatively small perturbations, as defined in the next section.

III. Parameter Sensitivity

A. The General Parameter Perturbation

Suppose the initial value problem (1) and (2) yields a unique solution. Then, for a given value of $t \in [t_0, t_1]$ there is a unique corresponding point $\mathbf{x}(t)$ in state space. If the parameter vector $\mathbf{p}(t)$ is changed to $\mathbf{p}^*(t)$, then the unique solution to

$$d\mathbf{x}^*(t)/dt = \mathbf{f}[\mathbf{x}^*(t), \mathbf{p}^*(t)], \qquad \mathbf{x}^*(t_0) = \mathbf{x}_0,$$

evaluated at the same t, yields a point $\mathbf{x}^*(t)$ in the same state space as $\mathbf{x}(t)$. In general, $\mathbf{x}^*(t)$ is different from $\mathbf{x}(t)$.

With this viewpoint in mind, the state vector can be regarded as a function of the parameter vector $\mathbf{p}$, and written

$$\mathbf{x} = \mathbf{x}(t, \mathbf{p}), \qquad \mathbf{p} \in P,$$

where P is the parameter space of admissible vector-valued functions. By an admissible vector-valued function is meant one which induces a unique solution to Eqs. (1) and (2). Such a space P would, for example, contain all uniformly bounded, integrable vector functions on $[t_0, t_1]$ (Tomović and Vukobratović, 1972). By considering parameter sensitivity, then, it is desired to show how variations in the space P affect the system dynamics in state space.

In the space P of parameters, if $\mathbf{p}(t)$ is the nominal parameter dynamics and

$\mathbf{p}^*(t)$ a perturbed or error trajectory, variations $\mathbf{p}^*(t) - \mathbf{p}(t)$ defined by

$$\mathbf{p}^*(t) = \mathbf{p}(t) + \epsilon\mathbf{v}(t), \qquad \epsilon \in R, \quad \mathbf{v}(t) \in P, \tag{3}$$

will be considered. The variational function $\mathbf{v}(t)$ takes on values in R^m and is an arbitrary function in P. Once chosen, however, all possible parameter variations are simply multiples of it.

If

$$\mathbf{v}(t) = \begin{cases} \mathbf{0} & \text{when} \quad t_0 \leqslant t < t_k \\ \mathbf{v}(t) & \text{when} \quad t_k \leqslant t \leqslant t_1, \end{cases} \qquad t_k \in [t_0, t_1], \tag{4}$$

then Eq. (4) represents a parameter change due to some event at $t = t_k$ that is unaccounted for in the nominal model. Radical natural phenomena such as earthquakes, tidal waves, volcanic eruptions, or serious interventions into an ecosystem by some species such as man can be represented by Eq. (4) to the extent that they do not alter the fixed structure $\mathbf{f}$. Milder disturbances that qualify as parameter perturbations only can always be cast in this general form.

With $\mathbf{x}(t)$ the nominal dynamics of state corresponding to $\mathbf{p}(t)$, let $\mathbf{x}^*(\epsilon, t)$ represent the perturbed state dynamics corresponding to $\mathbf{p}^*(t)$. Then, $\mathbf{x}(t)$ satisfies Eqs. (1) and (2) on the subinterval $[t_0, t_k]$, and on the remaining interval the state behavior is $\mathbf{x}^*(\epsilon, t)$, satisfying

$$d\mathbf{x}^*(\epsilon, t)/dt = \mathbf{f}[\mathbf{x}^*(\epsilon, t), \mathbf{p}^*(t)], \qquad t \in [t_k, t_1], \tag{5}$$

subject to

$$\mathbf{x}^*(\epsilon, t_k) = \mathbf{x}(t_k). \tag{6}$$

The latter is true since $\mathbf{x}^*$ follows the unperturbed time development until $t = t_k$. Parameter sensitivity concerns the variation $\mathbf{x}^*(\epsilon, t) - \mathbf{x}(t)$ induced in state dynamics by the variation $\mathbf{p}^*(t) - \mathbf{p}(t)$ of system coefficients or inputs.

B. Definition of the Senstivity Measure

Intuitively, a sensitivity measure is expected to represent the difference $\mathbf{x}^* - \mathbf{x}$ between a perturbed and nominal solution. The following argument provides a first-order approximation to this difference.

The perturbed solution $\mathbf{x}^*(\epsilon, t)$ is a function of the real variables ϵ and t. Therefore, for fixed t, $\mathbf{x}^*(\epsilon, t)$ may be expanded in a Taylor series around $\epsilon = 0$:

$$\mathbf{x}^*(\epsilon, t) = \sum_{k=0}^{\infty} \frac{(\epsilon - 0)^k}{k!} \frac{\partial^k x^*(\epsilon, t)}{\partial \epsilon^k}\bigg|_{\epsilon=0} \tag{7}$$

$$= \mathbf{x}^*(0, t) + \epsilon \frac{\partial \mathbf{x}^*}{\partial \epsilon}\bigg|_{\epsilon=0} + o(\epsilon^2).$$

The term $\mathbf{o}(\epsilon^2)$ represents a vector function, all of whose components have factors of ϵ^2 or higher. When $\epsilon = 0$, however, then $\mathbf{x}^*(0, t) = \mathbf{x}(t)$, the nominal trajectory. Thus, for small values of ϵ, Eq. (7) becomes

$$\mathbf{x}^*(\epsilon, t) - \mathbf{x}(t) \approx \epsilon \left.\frac{\partial \mathbf{x}^*}{\partial \epsilon}\right|_{\epsilon=0}, \tag{8}$$

and sensitivity can be defined (Tomović and Vukobratović, 1972) as

$$\mathbf{s}_{\mathbf{v}}(t) \triangleq \left.\frac{\partial \mathbf{x}^*}{\partial \epsilon}\right|_{\epsilon=0} = \lim_{\epsilon \to 0} \frac{\mathbf{x}^*(\epsilon, t) - \mathbf{x}(t)}{\epsilon}. \tag{9}$$

The subscript $\mathbf{v}$ in this definition emphasizes that sensitivity depends upon the particular parameter variation chosen [Eq. (3)]. For notational convenience, however, the subscript will be omitted in subsequent discussions.

1. *Example 1. Exponential Growth Model with Time-Invariant Parameter Perturbation*

The simplest population growth model (Malthus, 1798) with number or density of individuals $N(t)$ as the state variable is given by

$$dN(t)/dt = r(t)N(t), \qquad N(t) \in R, \quad r(t) \in R, \quad t \in [0, t_1], \tag{10}$$

with initial conditions

$$N(0) = N_0. \tag{11}$$

The parameter $r(t)$ is the biotic potential (birth rate minus death rate). The solution of Eq. (10) subject to Eq. (11) is

$$N(t) = N_0 \exp\left\{\int_0^t r(\xi)\, d\xi\right\}. \tag{12}$$

Let $r(t)$ be varied throughout $[0, t_1]$ by a constant amount ϵ:

$$r^*(t) = r(t) + \epsilon, \tag{13}$$

where $v(t) = 1$ is taken for $t > 0$. Equation (10) becomes

$$dN^*(\epsilon, t)/dt = [r(t) + \epsilon]\, N^*(\epsilon, t), \qquad t \in [0, t_1]. \tag{14}$$

With Eq. (11) as the initial state, the solution of Eq. (14) is

$$\begin{aligned} N^*(\epsilon, t) &= N_0 \exp\left\{\int_0^t [r(\xi) + \epsilon]\, d\xi\right\} \\ &= N_0\left[\exp\left\{\int_0^t r(\xi)\, d\xi\right\} \exp\left\{\int_0^t \epsilon\, d\xi\right\}\right] \\ &= N_0\, e^{\epsilon t} \exp\left\{\int_0^t r(\xi)\, d\xi\right\}. \end{aligned} \tag{15}$$

The sensitivity of $N(t)$ with respect to $r(t)$ is

$$s(t) = \lim_{\epsilon \to 0} \frac{N^*(\epsilon, t) - N(t)}{\epsilon}$$

$$= \lim_{\epsilon \to 0} \frac{N_0 e^{\epsilon t} \exp\left\{\int_0^t r(\xi)\, d\xi\right\} - N_0 \exp\left\{\int_0^t r(\xi)\, d\xi\right\}}{\epsilon}$$

$$= \lim_{\epsilon \to 0} \frac{N_0 \exp\left\{\int_0^t r(\xi)\, d\xi\right\} (e^{\epsilon t} - 1)}{\epsilon}$$

$$= \lim_{\epsilon \to 0} N(t) \left[\frac{e^{\epsilon t} - 1}{\epsilon}\right],$$

or, applying L'Hôpital's rule,

$$= \lim_{\epsilon \to 0} N(t)\, [t e^{\epsilon t}].$$

Thus,

$$s(t) = tN(t). \tag{16}$$

2. *Accuracy of the Sensitivity Measure*

The degree to which $\epsilon s(t)$, that is, $\epsilon \cdot \partial \mathbf{x}^*/\partial \epsilon$ [Eq. (8)], approximates the true state perturbation $\mathbf{x}^*(\epsilon, t) - \mathbf{x}(t)$ can be illustrated for Example 1. The nominal time development of the population [Eq. (12)] is perturbed to Eq. (15) when the biotic potential is altered according to Eq. (14). The actual perturbation is the difference between Eqs. (15) and (12):

$$N^*(\epsilon, t) - N(t) = N_0 \exp\left\{\int_0^t r(\xi)\, d\xi\right\} (e^{\epsilon t} - 1) = N(t)\, [e^{\epsilon t} - 1]. \tag{17}$$

From the sensitivity equation (16), the approximate perturbation is

$$N^*(\epsilon, t) - N(t) \approx \epsilon\, \partial N^*/\partial \epsilon = \epsilon t N(t). \tag{18}$$

Equations (17) and (18) are nearly equal for $\epsilon t \ll 1$ since then, in Eq. (17),

$$e^{\epsilon t} - 1 = [(1 + \epsilon t + \cdots) - 1] \approx \epsilon t.$$

If the interval $[0, t_1]$ over which the model operates is large, then $\epsilon t \ll 1$ will not longer be valid. Thus, the sensitivity measure is accurate for reasonable intervals of time, but does not give a realistic approximation of the perturbation effect for indefinitely large times.

3. *Single-Parameter Perturbations*

Some attention has been given (Tomović and Vukobratović, 1972) to parameter sensitivity when only one parameter, $p_j(t)$ say, has been perturbed. That is, $p_j{}^*(t) = p_j(t) + \epsilon v_j(t)$, where $\mathbf{v}(t)$ in Eq. (3) is

$$\mathbf{v}(t) = (0, \ldots, 0, v_j(t), 0, \ldots, 0).$$

Because only one parameter is perturbed, this sensitivity is termed a one-dimensional sensitivity, and notated as $\mathbf{s}_{v_j}$:

$$\mathbf{s}_{v_j} = \lim_{\epsilon \to 0} \epsilon^{-1} [\mathbf{x}(p_1, \ldots, p_j + \epsilon v_j, \ldots, p_m) - \mathbf{x}(p_1, \ldots, p_m)], \qquad j = 1, \ldots, m. \tag{19}$$

C. Implementation of the Sensitivity Definition

Implementing definition (9) involves solving two systems of differential equations for $\mathbf{x}(t)$ and $\mathbf{x}^*(\epsilon, t)$, both possibly nonlinear, and then taking a limit. An alternate and simpler method is contained in the following formal derivation.

Equation (5) is differentiated with respect to ϵ so that the ith component of the state vector satisfies

$$\frac{\partial}{\partial \epsilon}\left(\frac{dx_i{}^*}{dt}\right) = \sum_{j=1}^{n} \frac{\partial f_i}{\partial x_j{}^*} \frac{\partial x_j{}^*}{\partial \epsilon} + \sum_{j=1}^{m} \frac{\partial f_i}{\partial p_j{}^*} \frac{\partial p_j{}^*}{\partial \epsilon}, \qquad i = 1, \ldots, n.$$

Assuming that each $x_i{}^*(\epsilon, t)$ is continuously differentiable in both ϵ and t, the order of differentiation can be reversed (Meissinger and Bekey, 1966); that is,

$$\frac{d}{dt}\left(\frac{\partial x_i{}^*}{\partial \epsilon}\right) = \frac{\partial}{\partial \epsilon}\left(\frac{dx_i{}^*}{dt}\right).$$

From Eq. (3), $\partial p_j{}^*/\partial \epsilon = v_j(t)$, so that, setting $\epsilon = 0$, the result in matrix-vector form is

$$d\mathbf{s}(t)/dt = J(t)\,\mathbf{s}(t) + K(t)\,\mathbf{v}(t), \tag{20}$$

where

$$J(t) = [\partial f_i/\partial x_j], \qquad i, j = 1, \ldots, n,$$

and

$$K(t) = [\partial f_i/\partial p_j], \qquad i = 1, \ldots, n, \quad j = 1, \ldots, m,$$

are Jacobian matrices. Equation (20) comprises a system of linear differential equations for the sensitivity vector $\mathbf{s}(t)$. Further, if the perturbation $\mathbf{v}(t)$ is assumed zero for $t < t_k$, where $t_k \in [t_0, t_1]$, then $\mathbf{x}^*(\epsilon, t_k) = \mathbf{x}(t_k)$ and $\mathbf{x}^*(\epsilon, t_k) - \mathbf{x}(t_k) \approx \epsilon\,\mathbf{s}(t_k)$ [Eq. (8)], implying that $\mathbf{s}(t_k) = \mathbf{0}$. Equation (7) thus implies that

$$s(t_k) = 0 \tag{21}$$

since all the coefficients in the Taylor series must vanish at $t = t_k$. The unique solution to Eqs. (20) and (21) is the desired sensitivity, to within a first-order approximation.

Example 2. Exponential Growth Model with Time-Invariant Parameter Perturbation, Second Method

Before proceeding with the general time-varying case, the applicability of Eqs. (20) and (21) can be demonstrated for Example 1 under the previously considered time-invariant perturbation of the biotic potential. Differentiating Eq. (14) with respect to ϵ,

$$\partial(dN/dt)/\partial\epsilon = r(t)\,\partial N/\partial\epsilon + \epsilon\,\partial N/\partial\epsilon + N.$$

Setting $\epsilon = 0$, and reversing the order of differentiation, yields

$$d(\partial N/\partial\epsilon)/dt = ds/dt = r(t)s + N, \tag{22}$$

subject to

$$s(0) = 0. \tag{23}$$

Equation (22) has the following correspondences to Eq. (20):

$$J(t) = \partial f/\partial N = r(t), \qquad K(t) = \partial f/\partial r = N(t), \qquad v(t) = \partial r^*/\partial\epsilon = 1.$$

Equation (22) can be rewritten

$$ds/dt - r(t)s = N(t) = N_0 \exp\left\{\int_0^t r(\xi)\,d\xi\right\}.$$

Applying the integrating factor

$$\exp\left[-\int_0^t r(\xi)\,d\xi\right]:$$

$$\frac{ds}{dt}\exp\left[-\int_0^t r(\xi)\,d\xi\right] - sr(t)\exp\left[-\int_0^t r(\xi)\,d\xi\right] = N_0.$$

Recognizing that

$$-r(t)\exp\left[-\int_0^t r(\xi)\,d\xi\right] = \frac{d}{dt}\left(\exp\left[-\int_0^t r(\xi)\,d\xi\right]\right),$$

the left side of the last equation can be rewritten and the equation becomes

$$\frac{d}{dt}\left(s(t)\exp\left[-\int_0^t r(\xi)\,d\xi\right]\right)=N_0.$$

Integrating both sides with respect to time and applying the initial condition (23) yields

$$s(t)\exp\left[-\int_0^t r(\xi)\,d\xi\right]=tN_0.$$

Solving for $s(t)$ and substituting $N(t)$ for

$$N_0\exp\left[\int_0^t r(\xi)\,d\xi\right],$$

the sensitivity of $N(t)$ to $r(t)$ is produced:

$$s(t)=tN(t). \tag{24}$$

This result is the same as before [Eq. (16)], illustrating that $s(t)$ obtained by solving a differential equation (22) is identical to $s(t)$ computed directly as a limit.

D. Formal Derivation of the Sensitivity Equations

The derivation of Eq. (20) can be made more rigorous if proper smoothness in the dependence of $\mathbf{x}^*(\epsilon, t)$ on the real variable ϵ is assumed. That is, suppose $\mathbf{x}^*$ may be written in a Taylor series around $\epsilon = 0$, as in Eq. (7),

$$\mathbf{x}^*(\epsilon,t)=\mathbf{x}^*(0,t)+\epsilon\mathbf{s}(t)+\cdots,$$

and that the function $\mathbf{f}[\mathbf{x}^*, \mathbf{p}^*]$ in Eq. (5) also has that property. That is,

$$\mathbf{f}[\mathbf{x}^*,\mathbf{p}+\epsilon\mathbf{v}]=\mathbf{f}[\mathbf{x},\mathbf{p}]+\epsilon[J(t)\mathbf{s}+K(t)\mathbf{v}]+\cdots.$$

Then, since $\mathbf{x}^*(0, t) = \mathbf{x}(t)$,

$$\mathbf{x}+\epsilon\dot{\mathbf{s}}+\cdots=\dot{\mathbf{x}}^*=\mathbf{f}[\mathbf{x},\mathbf{p}]+\epsilon[J(t)\mathbf{s}+F(t)\mathbf{v}]+\cdots,$$

and comparing coefficients of ϵ yields Eq. (20). Further, $\mathbf{x}^*(\epsilon, t_k) = \mathbf{x}(t_k)$ so that

$$\mathbf{0}=\epsilon\mathbf{s}(t_k)+\cdots,$$

implying $\mathbf{s}(t_k) = \mathbf{0}$ [Eq. (21)].

For any given system the sensitivity equations (20) and (21) are easily derived and their solution $\mathbf{s}(t) = [s_1(t), \ldots, s_n(t)]^T$ represents the time-varying sensitivity of state dynamics on $[t_0, t_1]$ due to perturbations or error in the parameters $p_1(t), \ldots, p_m(t)$. Here, $s_i(t)$ represents the sensitivity of $x_i(t)$ $(i = 1, \ldots, n)$ due to all the given parameter perturbations. The fact that the sensitivity equations of any dynamical system (linear or nonlinear) are linear has significant implications both computationally and theoretically, as discussed later.

1. *Example 3. Exponential Growth Model with Time-Varying Parameter Perturbation*

The exponential growth model can now be examined under time-varying perturbations. Let

$$r^*(t) = r(t) + \epsilon v(t), \qquad t \in [0, t_1], \tag{25}$$

implying

$$dN^*/dt = [r(t) + \epsilon v(t)]N^*, \qquad t \in [0, t_1], \tag{26}$$

with initial condition

$$N^*(\epsilon, 0) = N_0. \tag{27}$$

Differentiating Eqs. (26) and (27) with respect to ϵ and setting $\epsilon = 0$ gives

$$ds(t)/dt = r(t)\, s(t) + N(t)\, v(t), \tag{28}$$

subject to

$$s(0) = 0. \tag{29}$$

The correspondence of Eq. (28) to Eq. (20) is obvious.

The solution of Eq. (28) is

$$s(t) = \exp\left[\int_0^t r(\xi)\, d\xi\right] \int_0^t \exp\left[-\int_0^\xi r(\eta)\, d\eta\right] N(\xi)\, v(\xi)\, d\xi,$$

and, substituting for $N(t)$ [Eq. (12)],

$$s(t) = \exp\left[\int_0^t r(\xi)\, d\xi\right] \int_0^t \exp\left[-\int_0^\xi r(\eta)\, d\eta\right] N_0 \exp\left[\int_0^\xi r(\eta)\, d\eta\right] v(\xi)\, d\xi$$

$$= N_0 \exp\left[\int_0^t r(\xi)\, d\xi\right] \int_0^t v(\xi)\, d\xi.$$

Using the solution for $N(t)$ again, the final result is

$$s(t) = N(t) \int_0^t v(\xi)\, d\xi, \qquad t \in [0, t_1]. \tag{30}$$

Note that, if $v(t) = 1$, $s(t) = tN(t)$ as before [Eqs. (16) and (24)].

2. *Example* 4. *Logistic Growth Model*

The Pearl–Verhulst logistic model of populations (Verhulst, 1844; Pearl, 1924) modifies the exponential model to account for self-inhibition of growth with increasing population size. The logistic equation is

$$dN/dt = r(t)\,N[1-\alpha(t)N],$$
$$N(t)\in R, \quad r(t)\in R, \quad \alpha(t)\in R, \quad t\in[0,t_1], \tag{31}$$

with initial condition

$$N(0) = N_0. \tag{32}$$

The differential equation is nonlinear, and $\alpha(t)$ is the self-inhibition coefficient.

The two parameters may be perturbed as follows:

$$r^*(t) = r(t) + \epsilon v_1(t), \qquad v_1(t)\in R, \tag{33}$$

$$\alpha^*(t) = \alpha(t) + \epsilon v_2(t), \qquad v_2(t)\in R. \tag{34}$$

The perturbed state equation is then

$$dN^*/dt = [r(t) + \epsilon v_1(t)]N^*\{1-[\alpha(t)+\epsilon v_2(t)]N^*\}, \tag{35}$$

with initial state

$$N^*(\epsilon, 0) = N_0. \tag{36}$$

Differentiating Eq. (35) with respect to ϵ gives

$$\frac{\partial}{\partial\epsilon}\left(\frac{dN}{dt}\right) = r\frac{\partial N}{\partial\epsilon} - 2r\alpha N\frac{\partial N}{\partial\epsilon} + \epsilon v_1\frac{\partial N}{\partial\epsilon} - 2r\epsilon v_2 N\frac{\partial N}{\partial\epsilon} - 2\alpha\epsilon v_1 N\frac{\partial N}{\partial\epsilon}$$
$$-2\epsilon^2 v_1 v_2 N\frac{\partial N}{\partial\epsilon} + Nv_1 - rv_2N^2 - \alpha v_1 N^2 - 2\epsilon v_1 v_2 N^2.$$

Reversing the order of differentiation, setting $\epsilon = 0$, and rearranging the remaining terms,

$$ds/dt = [r(1-2\alpha N)]s + [N(1-\alpha N)]v_1 + [-rN^2]v_2, \tag{37}$$

with initial condition

$$s(0) = 0. \tag{38}$$

Equation (37) can be put in the standard form of Eq. (20) as follows:

$$\frac{ds}{dt} = [r(t)\{1-2\alpha(t)N\}]s + [N\{1-\alpha(t)N\}, -r(t)N^2]\begin{bmatrix} v_1(t)\\ v_2(t)\end{bmatrix}. \tag{39}$$

In this example $J(t)$ is a 1×1 matrix, $K(t)$ is 1×2, and $\mathbf{v}(t)$ is 2×1. Solution of Eq. (39) subject to Eq. (38) gives the sensitivity of logistic population dynamics to either or both of the parameters $r(t)$ and $\alpha(t)$ [see Example 12, Eq. (93)]. Note that the sensitivity substructure (39) is linear even though the system state structure is nonlinear [Eq. (31)].

3. *Example 5. Lotka–Volterra Competition Model*

The well-known Lotka–Volterra equations in two dimensions (Lotka, 1925; Volterra, 1927) form a model of two competing logistic species:

$$\begin{aligned} dN_1/dt &= r_1(t)\,N_1\,[1-\alpha_1(t)N_1-\beta_1(t)N_2] \\ dN_2/dt &= r_2(t)\,N_2\,[1-\beta_2(t)N_1-\alpha_2(t)N_2], \end{aligned} \tag{40}$$

$$r_i(t)\in R,\quad \alpha_i(t)\in R,\quad \beta_i(t)\in R,\quad t\in[0,t_1],\qquad i=1,2,$$

with initial conditions

$$N_1(t_0)=N_1(0),\qquad N_2(t_0)=N_2(0). \tag{41}$$

The α coefficients represents self-inhibition, and the β coefficients denote competitive inhibition of growth.

Consider the following vector $\mathbf{p}^*$ of perturbed parameters:

$$\begin{bmatrix} r_1^*(t) \\ \alpha_1^*(t) \\ \beta_1^*(t) \\ r_2^*(t) \\ \beta_2^*(t) \\ \alpha_2^*(t) \end{bmatrix} = \begin{bmatrix} r_1(t) \\ \alpha_1(t) \\ \beta_1(t) \\ r_2(t) \\ \beta_2(t) \\ \alpha_2(t) \end{bmatrix} + \epsilon \begin{bmatrix} v_1(t) \\ v_2(t) \\ v_3(t) \\ v_4(t) \\ v_5(t) \\ v_6(t) \end{bmatrix}. \tag{42}$$

The perturbed state equations are then

$$\begin{aligned} dN_1^*/dt &= (r_1+\epsilon v_1)N_1^* - (r_1+\epsilon v_1)(\alpha_1+\epsilon v_2)N_1^{*2} - (r_1+\epsilon v_1)(\beta_1+\epsilon v_3)N_1^*N_2^* \\ dN_2^*/dt &= (r_2+\epsilon v_4)N_2^* - (r_2+\epsilon v_4)(\beta_2+\epsilon v_5)N_1^*N_2^* - (r_2+\epsilon v_4)(\alpha_2+\epsilon v_6)N_2^{*2}. \end{aligned} \tag{43}$$

Differentiating with respect to ϵ, eliminating terms containing ϵ (ϵ is set equal to zero), and reversing the order of differentiation reveals the following sensitivity substructure:

$$\begin{aligned} ds_1/dt &= r_1(1-2\alpha_1N_1-\beta_1N_2)s_1 - r_1\beta_1N_1s_2 + N_1(1-\alpha_1N_1-\beta_1N_2)v_1 \\ &\quad - r_1N_1^2v_2 - r_1N_1N_2v_3 \\ ds_2/dt &= -r_2\beta_2N_2s_1 + r_2(1-\beta_2N_1-2\alpha_2N_2)s_2 + N_2(1-\beta_2N_1-\alpha_2N_2)v_4 \\ &\quad - r_2N_1N_2v_5 - r_2N_2^2v_6. \end{aligned} \tag{44}$$

The initial conditions are

$$s_1(0) = s_2(0) = 0. \tag{45}$$

In the standard form of Eq. (20) the sensitivity equations (44) become

$$\begin{bmatrix} ds_1/dt \\ ds_2/dt \end{bmatrix} = \begin{bmatrix} r_1(1-2\alpha_1 N_1 - \beta_1 N_2) & -r_1\beta_1 N_1 \\ -r_2\beta_2 N_2 & r_2(1-\beta_2 N_1 - 2\alpha_2 N_2) \end{bmatrix} \begin{bmatrix} s_1 \\ s_2 \end{bmatrix}$$

$$+ \begin{bmatrix} N_1(1-\alpha_1 N_1 - \beta_1 N_2) & -r_1 N_1^2 & -r_1 N_1 N_2 & 0 & 0 & 0 \\ 0 & 0 & 0 & N_2(1-\beta_2 N_1 - \alpha_2 N_2) & -r_2 N_1 N_2 & -r_2 N_2^2 \end{bmatrix} \begin{bmatrix} v_1(t) \\ v_2(t) \\ v_3(t) \\ v_4(t) \\ v_5(t) \\ v_6(t) \end{bmatrix} \tag{46}$$

These equations are readily solved, at least numerically [see, e.g., Pridgen (1970) for algorithms and computer programs]. Note once again that the sensitivity equations (46) are linear even though the state structure of the system (40) is nonlinear.

IV. State Sensitivity

A. Perturbations of State

State theory (e.g., Zadeh and Desoer, 1963) asserts that the current state of a system summarizes the entire history of the system to that point in time. Future state dynamics unfold from the current condition in response to present and future inputs, with no recourse to the past. Spontaneous state perturbations, that is, those not caused by parameter deviations from nominal, may be treated as a problem in initial conditions sensitivity, with the time of perturbation t_k being interpreted as initial time.

A state-perturbed system then satisfies Eqs. (1) and (2) on $[t_0, t_k]$, leaving the system in state $\mathbf{x}(t_k) = \mathbf{x}_k$ at time t_k. Subsequently, the system's motion is given by

$$d\mathbf{x}^*/dt = \mathbf{f}[\mathbf{x}^*(\epsilon, t), \mathbf{p}(t)], \qquad t \in (t_k, t_1], \tag{47}$$

with initial conditions

$$\mathbf{x}^*(\epsilon, t_k) = \mathbf{x}_k^*, \tag{48}$$

where

$$\mathbf{x}_k^* = \mathbf{x}_k + \epsilon \mathbf{v}_k, \qquad \mathbf{v}_k(t) \in R^n. \tag{49}$$

B. Derivation of the State Sensitivity Measure

In an analysis equivalent to that of parameter sensitivity, both sides of Eqs. (47) and (48) are differentiated with respect to ϵ:

$$\partial(d\mathbf{x}^*/dt)/\partial\epsilon = J(t)\,\partial\mathbf{x}^*/\partial\epsilon + K(t)\,\partial\mathbf{p}/\partial\epsilon, \tag{50}$$

$$\frac{\partial\mathbf{x}^*}{\partial\epsilon}(t_k) = \mathbf{v}_k(t_k). \tag{51}$$

Since $\mathbf{p}(t)$ is unperturbed, $\partial\mathbf{p}/\partial\epsilon = \mathbf{0}$. Further, interchanging the order of differentiation in Eq. (50), setting $\epsilon = 0$, and letting

$$\mathbf{s}(t) = \partial\mathbf{x}^*/\partial\epsilon\Big|_{\epsilon=0},$$

$$d\mathbf{s}/dt = J(t)\mathbf{s}, \qquad t \in (t_k, t_1). \tag{52}$$

The initial conditions become

$$\mathbf{s}(t_k) = \mathbf{v}_k(t_k). \tag{53}$$

$J(t)$ is again the Jacobian matrix $[\partial f_i/\partial x_j]$, $i, j = 1, \ldots, n$.

The only differences in comparison to parameter sensitivity are: (i) the initial state vector rather than the parameter vector is considered a function of ϵ, (ii) the resulting sensitivity equations (52) are homogeneous rather than nonhomogeneous (20), and (iii) the initial sensitivity is nonzero [Eq. (53) versus Eq. (21)]. The sensitivity substructure is still linear irrespective of the linearity or nonlinearity of the state dynamics.

1. *Example 6. State Perturbation of Exponential Growth Model*

For the exponential population model [Eqs. (10) and (11)] with state perturbed or measured with error at time $t_k = 0$, Eq. (11) becomes

$$N^*(0) = N_0 + \epsilon v_0, \tag{54}$$

and the solution is now

$$N^*(\epsilon, t) = (N_0 + \epsilon v_0)\exp\left[\int_0^t r(\xi)\,d\xi\right], \qquad t \in [0, t_1]. \tag{55}$$

Differentiating Eq. (55) with respect to ϵ and letting $\epsilon = 0$,

$$s(t) = v_0 \exp\left[\int_0^t r(\xi)\,d\xi\right].$$

Since

$$N(t) = N_0 \exp\left[\int_0^t r(\xi)\,d\xi\right],$$

the right-hand side represents $v_0 N(t)/N_0$, so that the sensitivity to an initial state change is

$$s_{v_0} = (v_0/N_0)\,N(t), \qquad N_0 \neq 0. \tag{56}$$

Alternatively, the system is described by

$$dN^*/dt = f[N^*(\epsilon,t), r(t)] = r(t)\,N^*(\epsilon,t), \qquad t \in [0,t_1], \tag{57}$$

with

$$N^*(\epsilon,0) = N_0^* = N_0 + \epsilon v_0. \tag{58}$$

Applying Eq. (52), with $J(t) = \partial f/\partial N = r(t)$ from Eq. (57),

$$ds/dt = r(t)s, \tag{59}$$

with initial condition

$$s(0) = v_0. \tag{60}$$

The solution of Eq. (59) subject to Eq. (60) is

$$s_{v_0} = v_0 \exp\left[\int_0^t r(\xi)\,d\xi\right] = (v_0/N_0)\,N(t), \qquad N_0 \neq 0,$$

as in Eq. (56).

2. *Example 7. State Perturbation of the Logistic Model*

Let the logistic model [Eqs. (31) and (32)] be state perturbed at time $t_k = 0$. Then,

$$\begin{aligned} dN^*/dt &= f[N^*(\epsilon,t), r(t), \alpha(t)] \\ &= r(t)\,N^*(\epsilon,t)\,[1-\alpha(t)\,N^*(\epsilon,t)], \qquad t \in [0,t_1], \end{aligned} \tag{61}$$

with initial condition

$$N^*(\epsilon,0) = N_0^* = N_0 + \epsilon v_0. \tag{62}$$

Applying Eq. (52) with $J(t)$ as given in Eq. (39),

$$ds/dt = [r(t)\{1-2\alpha(t)\,N(t)\}]\,s, \tag{63}$$

subject to

$$s(0) = v_0. \tag{64}$$

The solution is

$$s(t) = v_0 \exp\left(\int_0^t [r(\xi)\{1-2\alpha(\xi)\,N(\xi)\}]\;d\xi\right), \tag{65}$$

where $N(t)$ satisfies Eqs. (31) and (32).

3. *Example 8. State Perturbation of the Lotka–Volterra System*

Deviations in the initial state of the Lotka–Volterra model [Eqs. (40) and (41)] at $t_k = 0$ result in

$$\begin{aligned} dN_1^*/dt &= r_1(t)N_1^*[1-\alpha_1(t)N_1^* - \beta_1(t)N_2^*] \\ dN_2^*/dt &= r_2(t)N_2^*[1-\beta_2(t)N_1^* - \alpha_2(t)N_2^*], \qquad t \in [0, t_1], \end{aligned} \tag{66}$$

subject to

$$\begin{aligned} N_1^*(\epsilon, 0) &= N_1(0) + \epsilon v_1 \\ N_2^*(\epsilon, 0) &= N_2(0) + \epsilon v_2. \end{aligned} \tag{67}$$

The sensitivity equations are the homogeneous part of Eq. (47):

$$\begin{bmatrix} ds_1/dt \\ ds_2/dt \end{bmatrix} = \begin{bmatrix} r_1[1-2\alpha_1 N_1 - \beta_1 N_2] & -r_1\beta_1 N_1 \\ -r_2\beta_2 N_2 & r_2[1-\beta_2 N_1 - 2\alpha_2 N_2] \end{bmatrix} \begin{bmatrix} s_1 \\ s_2 \end{bmatrix}, \tag{68}$$

with initial conditions

$$s_1(0) = v_1, \qquad s_2(0) = v_2. \tag{69}$$

These equations can be solved by numerical methods.

In conclusion, state sensitivity analysis can be useful as a first approximation to error propagation in the time development of a system due to inaccurate measurement of the initial state. However, state sensitivity can also be employed to assess the propagated change when the system state is intentionally altered. Ecosystem manipulations that might be most conveniently treated as state perturbations include, for example, adding pollutants to a lake or stream, stocking fish in a pond, or cutting away a portion of a cattle herd. Such actions clearly provide sudden state changes that need not necessarily be modeled as parameter alterations.

V. Sensitivity and Stability

A. Basic Stability Concepts

The subject of stability has always been implicit in ecology through historical association with problems of population fluctuations, extinctions, and epidemic outbreaks. In recent times, MacArthur (1955) gave impetus to ecosystem stability considerations by suggesting a relationship between ecosystem stability and species diversity. The Brookhaven diversity–stability symposium some years ago (Woodwell and Smith, 1969) summarized the state of ecologists' thinking on the subject at that time, with generally inconclusive results. The literature on ecological stability is currently being expanded as ecologists investigate the applicability of stability theory as developed in the physical sciences. There is an often expressed opinion that formal stability theory may not be appropriate for ecological problems. This remains to be seen. Modifications and extensions will almost certainly be needed, and the tentative treatment of the subject provided below in the context of sensitivity analysis is offered in a spirit of exploration.

About the fact that stability theory is a complex subject there can be no doubt. The concept of stability itself is difficult to define. Although definition and implementation are simple enough for linear time-invariant systems, extension to nonlinear or time-varying systems becomes quite complicated.

1. *Homogeneous, Time-Invariant Linear Systems*

One concept of stability is that a system is stable if it does not blow up, that is, if its behavior is bounded. For unforced, linear, time-invariant systems,

$$d\mathbf{x}(t)/dt = A\mathbf{x}(t), \tag{70}$$

where A is a matrix of constant coefficients, stability in this bounded sense is assured if all the eigenvalues of A lie in the left half of the complex plane or on the imaginary axis. This is investigated either by calculating exact roots of the characteristic equation (directly, or by the root-locus method), or by determining the region of parameter space where roots are guaranteed to have the requisite properties for stability (Routh–Hurwitz criterion, or Nyquist criterion), namely nonpositive real parts.

Another concept of stability is based upon the dissipative tendency of many systems. In this sense, systems of the form of Eq. (70) would be considered stable only if they tended to zero at equilibrium (i.e., $\mathbf{x}(t) \to \mathbf{0}$ as $t \to \infty$). The ecologist would question if this fits his concept of stability, but in most cases without input he would probably agree to the zero state as a reasonable endpoint. In this instance, for the system to be stable, the eigenvalues of A must all have negative real parts, and thus lie in the left half of the complex plane but not on the imaginary axis.

The concept of zero endpoint has more applicability than is first apparent. For example, suppose $\mathbf{x}(t)$ is to satisfy $d\mathbf{x}/dt = \mathbf{f}(\mathbf{x}, \mathbf{p})$, and suppose $\mathbf{x}(t) = \mathbf{x}^0$ is a constant solution (i.e., $\mathbf{f}(\mathbf{x}^0, \mathbf{p}) = \mathbf{0}$ for all t). The stability question here is that if $\mathbf{x}(t) \neq \mathbf{x}^0$ originally, does it eventually approach this value for large t? The question can be rephrased as follows: If $\Delta\mathbf{x}(t) = \mathbf{x}(t) - \mathbf{x}^0$, does $\Delta\mathbf{x}(t) \to \mathbf{0}$ as $t \to \infty$? Here the zero endpoint for $\Delta\mathbf{x}(t)$ is both reasonable and germane. Note that Patten *et al.* (1975) have argued that ecosystem states $\mathbf{x}(t)$ represent deviations $\Delta\mathbf{x}(t)$ above the zero state, $\mathbf{x}^0 = \mathbf{0}$, making the zero condition particularly significant in connection with ecosystems.

In general, it is true that stability of linear, homogeneous, time-invariant systems is determined by coefficients of the model, and is independent of initial state (Bellman, 1953). Furthermore, such systems are usually either unambiguously stable or unstable, depending upon the definition of stability used.

2. *Example 9. Stability of the Time-Invariant Exponential Population Model*

The homogeneous, linear, time-invariant population model formed from Eq. (10) when $r(t) = r$ (constant) has the characteristic equation

$$\lambda - r = 0.$$

The eigenvalue λ is thus

$$\lambda = r, \tag{71}$$

and the system is consequently stable in the bounded sense if $r \leqslant 0$, stable in the zero state trending sense if $r < 0$, and unstable otherwise.

3. *Homogeneous, Time-Varying Linear Systems*

For a time-varying exponential population, Eq. (12) shows that the time integral of the biotic potential must merely exist for stability according to the boundedness concept. That is,

$$\int_0^\infty r(\xi)\, d\xi < \infty;$$

$r(t)$ must be integrable over the infinite interval $[0, \infty]$. On the other hand, for stability by the return-to-zero concept, the specific condition

$$\int_0^\infty r(\xi)\, d\xi = -\infty$$

is both necessary and sufficient. These two conditions, in relation to the two different definitions of stability to which they correspond, illustrate the subtlety that often accompanies stability analysis of systems with time-varying endogenous parameters.

The general homogeneous, time-varying linear system is given by

$$d\mathbf{x}(t)/dt = A(t)\,\mathbf{x}(t), \tag{72}$$

where $A(t)$ is nonconstant. The unique solution of Eq. (72) produced when $\mathbf{x}(t_0) = \mathbf{0}$ is

$$\mathbf{x}^0(t) = \mathbf{0}, \tag{73}$$

the zero state. Let $\|\mathbf{x}(t)\|$ be a norm of the state vector,

$$\|\mathbf{x}(t)\| = \sum_{i=1}^{n} |x_i(t)|, \tag{74}$$

measuring distance from the origin to $\mathbf{x}(t)$ in state space. A small region of the state space, $\|\mathbf{x}(t)\| < \sigma$, can be defined to see if there exists a small region of nonzero perturbations around the initial state $\mathbf{x}(t_0) = \mathbf{0}$ that produce trajectories which remain within $\|\mathbf{x}(t)\| < \sigma$. If this is true for all $\sigma > 0$, no matter how small, then the boundedness concept of stability can be expressed as:

Definition 1

The equilibrium state $\mathbf{x}^0(t) = \mathbf{0}$ is *stable in the sense of Liapunov* (stable, i.s.L.) if, for every t_0 and every $\sigma > 0$, there exists some $\delta > 0$, no matter how small, such that if $\|\mathbf{x}(t_0)\| < \delta$ (representing a perturbation from equilibrium), then $\|\mathbf{x}(t)\| < \sigma$ for all $t > t_0$.

This is the canonical notion of system stability, and it applies also to nonlinear systems with solution $\mathbf{x}^0(t) = \mathbf{0}$ (Wiberg, 1971).

A stronger stability requirement than boundedness of $\|\mathbf{x}(t)\|$ following an equilibrium state displacement is return to the steady state. This concept is embodied in the notion of asymptotic stability:

Definition 2

The equilibrium state $\mathbf{x}^0(t) = \mathbf{0}$ is *asymptotically stable* if (i) it is stable i.s.L., and (ii) for any t_0 and any $\mathbf{x}(t_0)$ sufficiently close to $\mathbf{0}$, $\mathbf{x}(t) \to \mathbf{0}$ as $t \to \infty$.

This definition also holds for nonlinear systems. There are systems where $\mathbf{x}(t) \to \mathbf{0}$ as $t \to \infty$ that are not stable i.s.L., and these are regarded as unstable. Time-invariant linear systems [Eq. (70)] are Liapunov stable or asymptotically stable if all eigenvalues of A have nonpositive or negative real parts, respectively. For time-varying linear systems [Eq. (72)] simple criteria for stability do not exist. However, results for certain special cases have been established (e.g., Zadeh and Desoer, 1968; Bellman, 1959). In these cases the criteria for stability depend upon the integral of the eigenvalues, as in Examples 12 and 13 presented below.

4. *Nonhomogeneous Linear Systems*

Inputs (exogenous parameters) also complicate the problem. The general nonhomogeneous linear system may be represented by

$$d\mathbf{x}(t)/dt = A(t)\,\mathbf{x}(t) + B(t)\,\mathbf{z}(t), \tag{75}$$

$$\mathbf{y}(t) = C(t)\,\mathbf{x}(t), \tag{76}$$

where $\mathbf{z}(t)$ is an input vector, $A(t), B(t)$, and $C(t)$ are time-varying matrices of endogenous parameters, and $\mathbf{y}(t)$ is the system's output (response) vector. If every bounded input (b.i.) produces a bounded output (b.o.), the system is externally stable, or stable, b.i.b.o. That is, if $A(t)$ and $B(t)$ are bounded and continuous, then:

Definition 3

A system with input (75) is *externally stable* (stable b.i.b.o.) if, for any initial state $\mathbf{x}(t_0) = \mathbf{x}_0$ corresponding to any t_0, and for any $\mathbf{z}(t)$ such that $\|\mathbf{z}(t)\| < \kappa_1$, there exists a constant κ_2 dependent only on $t_0, \mathbf{x}_0$ and κ_1 such that $\|\mathbf{y}(t)\| \leqslant \kappa_2$ for all $t \geqslant t_0$.

In general, b.i.b.o. stability does not relate to the zero input stability concepts outlined above because it concerns the time behavior of outputs (rather than states). However, with $C(t) = I$ (identity matrix) and $B(t)$ nonsingular (and its inverse bounded), system (75) is externally stable if and only if its homogeneous portion is asymptotically stable for all t_0 (Wiberg, 1971). For $B(t)$ to be nonsingular, $\mathbf{z}(t)$ must be an n vector.

5. *Nonlinear Systems*

Stability of nonlinear systems may depend on inputs and initial states as well as coefficients. Further, stability may not always be clearly defined as it is for linear systems. Liapunov (1892) initiated the basic definitions and methods pertinent to nonlinear stability. His *first method* concerns the class of nonlinear systems for which the state equation can be solved. The equation is linearized about some singular point, called an operating point. This point is most often an equilibrium state $\mathbf{x}^0(t)$. The linearized equations are then analyzed for stability according to Definitions 1–3 by methods appropriate to the particular equations. Stability of the linearized equations implies *local stability* of the nonlinear system, that is, stability in the neighborhood of the operating point. Liapunov's *second method* requires no solution to the state equation. A positive function called a *Liapunov function* is formed and differentiated with respect to time. If the derivative of this function is negative along the trajectories of the system in state space, the system is asymptotically stable.

B. Stability of Linearized Systems

Stability analysis by linearization of a nonlinear system produces a deviation substructure similar to the state sensitivity substructure. Thus, if Eq. (1) represents a nonlinear system, its solution behavior $\mathbf{x}(t)$ can be locally approximated as linear deviation dynamics $\Delta\mathbf{x}(t)$ about an operating point trajectory $\mathbf{x}^0(t)$ that satisfies

$$d\mathbf{x}^0(t)/dt = \mathbf{f}[\mathbf{x}^0(t), \mathbf{p}(t)]. \tag{77}$$

That is, in Eq. (1),

$$\mathbf{x}(t) = \mathbf{x}^0(t) + \Delta\mathbf{x}(t), \tag{78}$$

where $\mathbf{x}^0(t)$ denotes reference (operating point) values of the states. Equation (1) can then be rewritten:

$$\frac{d}{dt}[\mathbf{x}^0(t) + \Delta\mathbf{x}(t)] = \frac{d}{dt}[\mathbf{x}^0(t)] + \frac{d}{dt}[\Delta\mathbf{x}(t)] = \mathbf{f}[\mathbf{x}^0(t) + \Delta\mathbf{x}(t), \mathbf{p}(t)]. \tag{79}$$

If the deviations are small, the ith component of Eq. (79) can be expanded in a Taylor series around the reference trajectory:

$$\frac{d}{dt}(x_i^0) + \frac{d}{dt}(\Delta x_i) \approx f_i(\mathbf{x}^0, \mathbf{p}) + (\partial f_i/\partial x_1)^0 \Delta x_1 + \cdots + (\partial f_i/\partial x_n)^0 \Delta x_n, \quad i = 1, \ldots, n. \tag{80}$$

If the reference behavior satisfies Eq. (77), the first terms of both sides of Eq. (80) cancel, and the deviation dynamics can be written

$$\frac{d}{dt}(\Delta x_i) \approx (\partial f_i/\partial x_1)^0 \Delta x_1 + \cdots + (\partial f_i/\partial x_n)^0 \Delta x_n, \qquad i = 1, \ldots, n. \tag{81}$$

In matrix notation this becomes

$$\frac{d}{dt}[\Delta\mathbf{x}(t)] \approx J^0(t)\, \Delta\mathbf{x}(t), \tag{82}$$

where

$$J^0(t) = [(\partial f_i/\partial x_j)^0], \qquad i, j = 1, \ldots, n.$$

Equation (82) specifies that small deviation dynamics $\Delta\mathbf{x}(t)$ about an operating point $\mathbf{x}^0(t)$ are approximately linear.

This classical linearization procedure can be generalized to determine the deviation dynamics $\Delta\mathbf{x}(t)$ about an operating point trajectory and an operating point parameter simultaneously. Let $\mathbf{x}^0(t)$ satisfy Eq. (77) while $\mathbf{x}(t) = \mathbf{x}^0(t) + \Delta\mathbf{x}(t)$ satisfies

$$d\mathbf{x}(t)/dt = \mathbf{f}[\mathbf{x}(t), \mathbf{p}(t) + \Delta\mathbf{p}(t)].$$

By proceeding as before, the equation for the deviation dynamics corresponding to Eq. (82) becomes

$$\frac{d(\Delta x_i)}{dt} \approx \left(\frac{\partial f_i}{\partial x_1}\right)^0 \Delta x_1 + \cdots + \left(\frac{\partial f_i}{\partial x_n}\right)^0 \Delta x_n + \left(\frac{\partial f_i}{\partial p_1}\right)^0 \Delta p_1 + \cdots + \left(\frac{\partial f_i}{\partial p_m}\right)^0 \Delta p_m, \quad i = 1, \ldots n.$$

In matrix notation the differential equation is

$$d[\Delta\mathbf{x}(t)]/dt = J^0(t)\, \Delta\mathbf{x}(t) + K^0(t)\, \Delta\mathbf{p}(t), \tag{83}$$

where $J^0(t)$ is defined as before, and

$$K^0(t) = (\partial f_i/\partial p_j)^0, \qquad i = 1, \ldots, n, \quad j = 1, \ldots, m.$$

If only state sensitivity is to be considered a nonzero solution of Eq. (83) when $\Delta\mathbf{p}(t) = \mathbf{0}$ is sought. Conversely, the solution of Eq. (83) with zero initial condition but nonzero parameter deviation $\Delta\mathbf{p}$ is the previously discussed parameter sensitivity. Since Eq. (83) is linear, its general solution combined with a nonzero initial condition leads to a sum of these two sensitivities.

Example 10. *Stability of the Logistic Model*

Asymptotic stability of the logistic model [Eq. (31)] may be investigated by linearizing it about an equilibrium trajectory. May (1973) used this approach for the special case of constant $r(t)$ and $\alpha(t)$. If $\alpha(t) = \alpha$, a constant, then $N^0 = 1/\alpha$ is an equilibrium point since both sides of Eq. (77) vanish. Using the generalized linearization technique, $N^0(t) = 1/\alpha$ is the operating trajectory satisfying Eq. (77), $\mathbf{p}(t) = [r(t), \alpha]^T$ are the operating parameters, $\Delta\mathbf{p}(t) = [\Delta r(t), \Delta\alpha]^T$ are the parameter deviations, and ΔN_0 is the initial condition. This gives

$$J^0(t) = (\partial f/\partial N)^0 = r(t) - 2r(t)\alpha N^0(t) = -r(t),$$

$$K^0(t) = [(\partial f/\partial r)^0, (\partial f/\partial \alpha)^0] = [N^0(t) - \alpha N^0(t)^2, -r(t)N^0(t)^2]$$
$$= [0, -r(t)/\alpha^2],$$

and thus the equation

$$d\Delta N(t)/dt = -r(t)\,\Delta N - [r(t)/\alpha^2]\,\Delta\alpha(t). \tag{84}$$

The solution of this initial value problem is

$$\Delta N(t) = \Delta N_0 \exp\left[-\int_0^t r(\xi)\,d\xi\right]$$
$$- \frac{1}{\alpha^2} \exp\left[-\int_0^t r(\xi)\,d\xi\right] \int_0^t \exp\left[\int_0^\xi r(\eta)\,d\eta\right] r(\xi)\,\Delta\alpha(\xi)\,d\xi. \tag{85}$$

Note that the solution does not depend upon $\Delta r(t)$, and that it is the sum of the state and parameter sensitivities for the logistic equation [Eqs. (38), (39), and (65)].

Let us consider the special case involving Liapunov stability. If in Eq. (84) $\Delta\alpha = 0$, then

$$\Delta N(t) = \Delta N_0 \exp\left[-\int_0^t r(\xi)\,d\xi\right]. \tag{86}$$

In the special case where $r(t) = r$, a constant,

$$\Delta N(t) = \Delta N_0\, e^{-rt}.$$

This solution is bounded and the system stable i.s.L. about $N^0(t) = 1/\alpha$ if $r = 0$; $\Delta N(t) \to 0$ as $t \to \infty$ and the system is asymptotically stable about the equilibrium point if $r > 0$, as concluded by May (1973). In the case of $r(t)$ nonconstant, if $r(t) \geqslant 0$ and

$$\int_0^\infty r(\xi)\, d\xi = \infty,$$

then the disturbance dies away asymptotically. This local stability analysis holds for small values of ΔN_0 and is no guarantee that the trajectory will approach the operating point after large disturbances. In this case, however, there exists a Liapunov function (a positive definite function whose time derivative is never positive), namely,

$$L(N) = (1-\alpha N)^2. \tag{87}$$

Clearly $L(N) > 0$ for $N \neq 1/\alpha$, while

$$dL/dt = -2\alpha(1-\alpha N)\, dN/dt,$$

which becomes

$$dL/dt = -2\alpha\, r(t)\, N(1-\alpha N)^2 \tag{88}$$

after using Eq. (31). The right-hand side of Eq. (88) is never positive, implying $L(N)$ is indeed a Liapunov function for the logistic model. By Liapunov's criteria, then, global stability is legitimately characterized by the above local analyses.

C. Sensitivity Definition of Stability

The ease of formulating sensitivity dynamics of a system, the fact that a sensitivity substructure is always linear, and the fact that one method of analyzing stability of nonlinear systems is through a linear deviation substructure similar to state sensitivity, motivates a connection between stability and sensitivity, and a generalization based upon parameter sensitivity. Tomović (1963) examined the relation of sensitivity to stability with conclusions (pp. 87–88) embodied in the following definition;

Definition 4

A system defined by Eqs. (1) and (2) is asymptotically stable with respect to a perturbation if its perturbed response $\mathbf{x}^*(\epsilon, t)$ converges to its nominal dynamics $\mathbf{x}(t)$ for large time; that is,

$$\lim_{t\to\infty} [\mathbf{x}^*(\epsilon, t) - \mathbf{x}(t)] = \mathbf{0} \tag{89}$$

for all $\epsilon > 0$. The system can be said to have *state asymptotic stability* if the perturbation is of the type depicted in Eq. (49), and *parameter asymptotic stability* if the perturbation is as represented in Eq. (3). Each of these stabilities will be considered local if the limit holds for small ϵ [i.e., if there exists a $\delta > 0$ such that Eq. (89) is true for all $\epsilon < \delta$].

From the definition of sensitivity [Eq. (9)], Eq. (89) implies vanishing sensitivity:

$$\lim_{t \to \infty} \mathbf{s}(t) = \mathbf{0}. \tag{90}$$

Unfortunately, the converse may not be true, and Eq. (90) therefore gives necessary but not sufficient conditions for parameter or state asymptotic stability. The contrapositive of Definition 4 provides a useful test, however: if $\mathbf{s}(t)$ does *not* tend to zero as time gets large, then the system is *not* asymptotically stable with respect to the perturbation.

The next examples illustrate Definition 4 in terms of the population models. Then a more formal treatment will be presented in Section V. D, wherein a modified converse statement is shown to hold.

1. *Example* 11. *Sensitivity and Stability of the Exponential Model*

The exponential population model [Eq. (10)] has for its parameter sensitivity Eq. (30):

$$s(t) = N(t) \int_0^t v(\xi)\, d\xi, \tag{30}$$

where $v(t)$ is the perturbation. Assuming $v(t)$ is positive and integrable,

$$0 < \int_0^t v(\xi)\, d\xi < \infty,$$

then a necessary condition for parameter asymptotic stability of $N(t)$ with respect to $v(t)$ is that $N(t) \to 0$ as $t \to \infty$, implying [Eq. (30)] that $s(t) \to 0$ as $t \to \infty$ also.

When $r(t) = r$, a constant, $N(t) \to 0$ and hence $s(t) \to 0$ as $t \to \infty$ only if $r < 0$, which is the condition previously noted (Example 9) for asymptotic stability of the time-invariant model. For the time-varying case, as indicated in the discussion following Example 9, the necessary condition for parameter asymptotic stability is

$$\int_0^\infty r(\xi) d\xi = -\infty.$$

State sensitivity for the exponential model is given by Eq. (56):

$$s(t) = (v_0/N_0)\, N(t), \qquad N_0 \neq 0, \tag{56}$$

where v_0 is the initial state perturbation. For state asymptotic stability with respect to this perturbation, it is necessary that $N(t) \rightarrow 0$ [and hence $s(t) \rightarrow 0$] as $t \rightarrow \infty$. That $s(t) \rightarrow 0$ is not sufficient for parameter asymptotic stability can be illustrated for the exponential growth model by letting

$$r(t) = -2t/(t^2 + 1).$$

The solution to Eqs. (10) and (11) is then

$$N(t) = N_0/(t^2 + 1), \qquad t \geqslant 0.$$

By perturbing the biotic potential to

$$r^*(t) = r(t) + \epsilon,$$

the new trajectory satisfies Eq. (14) and is given by

$$N^*(\epsilon, t) = [N_0/(t^2 + 1)]\, e^{\epsilon t}. \tag{91}$$

The sensitivity, as given by Eq. (16), is

$$s(t) = tN(t) = N_0 [t/(t^2 + 1)]. \tag{92}$$

Equation (92) shows that $s(t) \rightarrow 0$ as $t \rightarrow \infty$ but, by using L'Hospital's rule, Eq. (91) shows that $N^*(\epsilon, t) \rightarrow \infty$ for all $\epsilon > 0$. Vanishing sensitivity is therefore not sufficient for parameter asymptotic stability, only necessary.

It should become clear at this point that local state asymptotic stability is identical to classical asymptotic stability as posed by considerations of deviation dynamics; both ask whether small displacements in initial conditions will produce convergent solutions. In particular, if $\mathbf{x}(t) = \mathbf{x}^0(t)$ is an operating point satisfying Eq. (77) [i.e., $\mathbf{x}^0(t)$ is an equilibrium trajectory], then the deviation dynamics for $\Delta\mathbf{x}(t)$ described approximately by Eq. (82) and an initial condition are equivalent to the state sensitivity equations [Eqs. (52) and (53)], where

$$\Delta\mathbf{x}(t_k) = \epsilon \mathbf{v}_k.$$

2. *Example* 12. *Parameter Sensitivity and Stability of the Logistic Model*

Example 10 indicated conditions for state asymptotic stability of the time-invariant logistic model in the vicinity of its equilibrium state $N^0(t) = 1/\alpha$. These conditions were obtained from the explicit solution [Eq. (86)] to the deviation equation.

For parameter asymptotic stability Eq. (39) can be solved subject to Eq. (38) through use of an integrating factor in the same manner as employed in Example 2:

$$s(t) = \exp\left[\int_0^t \lambda(\xi)\, d\xi\right] \int_0^t \exp\left[-\int_0^\xi \lambda(\eta)\, d\eta\right] [k_1(\xi) v_1(\xi) + k_2(\xi) v_2(\xi)]\, d\xi,$$

$$t \in [0, t_1]. \tag{93}$$

Here, $\lambda(t)$ is the eigenvalue, and $k_1(t)$ and $k_2(t)$ correspond to elements of the $K(t)$ matrix in Eq. (39):

$$\lambda(t) = r(t)\,[1-2\,\alpha(t)\,N(t)], \tag{94}$$

$$k_1(t) = N(t)\,[1-\alpha(t)\,N(t)], \tag{95}$$

$$k_2(t) = -r(t)N(t)^2. \tag{96}$$

For parameter asymptotic stability it is necessary that $s(t) \to 0$ as $t \to \infty$. The nominal case where $r(t)=r$ and $\alpha(t)=\alpha$ are both positive constants will be considered. There are two situations, either $N(0) = 1/\alpha$ or $N(0) \neq 1/\alpha$.

In the first case the system starts at equilibrium and remains there. The unique nominal solution is $N(t) = 1/\alpha$. Substituting this into Eqs. (94)–(96) gives $\lambda(t) = -r,\ k_1(t) = 0$, and $k_2(t) = -r/\alpha^2$. Therefore, Eq. (93) becomes

$$s(t) = (r/\alpha^2)\, e^{-rt} \int_0^t e^{r\xi}\, v_2(\xi)\, d\xi, \tag{97}$$

which must approach zero as $t \to \infty$ for asymptotic stability, and which does so, for example, whenever $r > 0$ and $|v_2(t)| \leqslant e^{-bt}$, $b > 0$. It is interesting to note that the question of stability in this case is completely independent of the function $v_1(t) = (r^* - r)/\epsilon$. This means experimentally that the accepted value of r is not critical, whereas the accepted value of α is, as indicated by the presence of $v_2(t) = (\alpha^* - \alpha)/\epsilon$ in Eq. (97).

The second situation mentioned, $N(0) \neq 1/\alpha$, leads to a nominal solution $N(t)$ which approaches the steady state $1/\alpha$ as $t \to \infty$ (Example 10). Let deviation from $1/\alpha$ be represented as a time-invariant perturbation, for example, Eq. (13). Then, for any small $\epsilon > 0$ there is a τ such that, whenever $t > \tau$,

$$\alpha^{-1} - \epsilon < N(t) < \alpha^{-1} + \epsilon. \tag{98}$$

The solution for sensitivity is still given by Eq. (93), and it is now desired to find conditions on the perturbations $v_1(t)$ and $v_2(t)$ which force $s(t) \to 0$ as $t \to \infty$.

Equation (93) can be partitioned as follows:

$$s(t) = \sigma_1(t) + \sigma_2(t),$$

where

$$\sigma_1(t) = \exp\left[\int_\tau^t \lambda(\xi)\, d\xi\right] \left\{ \exp\left[\int_0^\tau \lambda(\xi)\, d\xi\right] \right.$$

$$\left. \times \int_0^\tau \exp\left[-\int_0^\xi \lambda(\eta)\, d\eta\right] [k_1(\xi)\, v_1(\xi) + k_2(\xi)\, v_2(\xi)]\, d\xi \right\}, \tag{99}$$

$$\sigma_2(t) = \exp\left[\int_\tau^t \lambda(\xi)\,d\xi\right] \int_\tau^t \exp\left[-\int_\tau^\xi \lambda(\eta)\,d\eta\right] [k_1(\xi)\,v_1(\xi) + k_2(\xi)\,v_2(\xi)]\,d\xi. \tag{100}$$

In the first expression the only term containing t is

$$\exp\left[\int_\tau^t \lambda(\xi)\,d\xi\right],$$

and for vanishing sensitivity it will be sufficient to show that $\sigma_1(t)$ and hence this exponential must vanish as $t \to \infty$. The integral of the eigenvalue,

$$\int_\tau^t \lambda(\xi)\,d\xi,$$

must therefore approach $-\infty$ as $t \to \infty$. Since Eq. (98) provides bounds on $N(t)$, the following bounds on $\lambda(t)$, $k_1(t)$, and $k_2(t)$ are determined by substituting the former into Eqs. (94)–(96). All of the following inequalities hold for $t > \tau$:

$$r(-1-2\alpha\epsilon) < \lambda(t) < r(-1+2\alpha\epsilon),$$

$$r(-1-2\alpha\epsilon)(t-\tau) < \int_\tau^t \lambda(\xi)\,d\xi < r(-1+2\alpha\epsilon)(t-\tau), \tag{101}$$

$$|k_1(t)| < \alpha\epsilon(\alpha^{-1}+\epsilon) = \hat{k}_1, \qquad |k_2(t)| < r(\alpha^{-1}+\epsilon)^2 = \hat{k}_2.$$

Since $r > 0$, then by the second inequality in Eq. (101) a choice of $\epsilon < 1/2\alpha$ forces

$$\int_\tau^t \lambda(\xi)\,d\xi \to -\infty,$$

and thereby $\sigma_1(t) \to 0$, as $t \to \infty$.

The second expression (100) is more difficult as t is the upper limit of the convolution integral. However, Eq. (101) provides bounds for $\sigma_2(t)$, as follows:

$$|\sigma_2(t)| < \exp[r(-1+2\alpha\epsilon)(t-\tau)] \int_\tau^t \exp[-r(-1-2\alpha\epsilon)(\xi-\tau)]\,(\hat{k}_1|v_1| + \hat{k}_2|v_2|)\,d\xi. \tag{102}$$

By considering only those factors in Eq. (102) which contain the variable t, the condition on v_1, v_2 becomes, for ϵ small enough,

$$\lim_{t\to\infty} \exp[-r(1-2\alpha\epsilon)t] \int_\tau^t \exp[r(1+2\alpha\epsilon)\xi]\,|v_i(\xi)|\,d\xi = 0, \qquad i = 1, 2, \tag{103}$$

implying $\sigma_2(t) \to 0$ as $t \to \infty$. Thus, both $\sigma_1(t)$ and $\sigma_2(t)$, and hence $s(t)$, vanish with large time, as is necessary for parameter asymptotic stability.

Expression (103) may seem difficult to work with, but a simple example will illustrate its value. Let $v_i(t) = a\,e^{-bt}$, or any combination of such functions where $a, b > 0$. Then, Eq. (103) becomes

$$a \exp[-r(1-2\alpha\epsilon)t] \int_{\tau}^{t} \exp[(r+2r\alpha\epsilon-b)\xi]\,d\xi. \tag{104}$$

By letting $m = r + 2r\alpha\epsilon - b$, Eq. (104) becomes

$$(a/m)[e^{-bt} - e^{m\tau} \exp(-r(1-2\alpha\epsilon)t)]$$

if $m \neq 0$, and

$$a\,e^{-bt}(t-\tau)$$

if $m = 0$. If $\epsilon < 1/2\alpha$, then both of these vanish as $t \to \infty$.

Example 12 illustrates a point about the relationship of eigenvalues to stability when eigenvalues are time varying [e.g., Eq. (94)]. In this case stability may not follow from the fact that the real parts of the characteristic roots are instantaneously nonpositive or negative. The analysis performed in the example indicates that it is not the instantaneous values, but rather the integrals, of nonstationary eigenvalues which determine stability. This is further illustrated in the next example.

3. *Example* 13. *State and Parameter Asymptotic Stability of the Lotka–Volterra Model*

The dynamics of the two-dimensional Lotka–Volterra system are governed by Eqs. (40) and (41). The biotic potentials r_1, r_2, self-inhibiting rates α_1, α_2, and competition rates β_1, β_2 are all taken to be positive and constant. Nonzero equilibrium trajectories are given by solutions to the simultaneous equations

$$1-\alpha_1\bar{N}_1-\beta_1\bar{N}_2 = 0, \qquad 1-\beta_2\bar{N}_1-\alpha_2\bar{N}_2 = 0, \tag{105}$$

which result upon setting $\dot{N}_1 = \dot{N}_2 = 0$ in Eqs. (40) and (41). These equations are solved to give

$$\bar{N}_1 = (\alpha_2-\beta_1)/D, \qquad \bar{N}_2 = (\alpha_1-\beta_2)/D, \tag{106}$$

where

$$D = \alpha_1\alpha_2-\beta_1\beta_2. \tag{107}$$

Since, at equilibrium, $\bar{N}_1, \bar{N}_2$ are to be positive, Eq. (106) implies either

(1) $\alpha_2-\beta_1, \alpha_1-\beta_2$, and D are all positive, or
(2) $\alpha_2-\beta_1, \alpha_1-\beta_2$, and D are all negative.

As established below, only condition (1) guarantees state asymptotic stability and, for suitable conditions on the perturbations of α_1 and β_1, parameter asymptotic stability as well.

State asymptotic stability can be shown by considering the classical deviation dynamics around the equilibrium points $\bar{N}_1$, $\bar{N}_2$ given by Eq. (106). Recalling that neighborhood dynamics around equilibrium points are given by Eq. (82),

$$J_0 = \left(\frac{\partial f_i}{\partial x_j}\right)\bigg|_{\bar{N}_1, \bar{N}_2} = \begin{pmatrix} -r_1 \alpha_1 \bar{N}_1 & -r_1 \beta_1 \bar{N}_1 \\ -r_2 \beta_2 \bar{N}_2 & -r_2 \alpha_2 \bar{N}_2 \end{pmatrix},$$

which has for a determinant

$$|J_0| = r_1 r_2 \bar{N}_1 \bar{N}_2 D.$$

The characteristic equation is given by $\lambda^2 + b\lambda + c = 0$, where

$$b = r_1 \alpha_1 \bar{N}_1 + r_2 \alpha_2 \bar{N}_2, \qquad c = |J_0|.$$

The solutions to this quadratic are the eigenvalues, and are given by

$$\lambda_1 = \tfrac{1}{2}[-b-(b^2-4c)^{1/2}], \qquad \lambda_2 = \tfrac{1}{2}[-b+(b^2-4c)^{1/2}]. \tag{108}$$

The key to stability lies in the investigation of these eigenvalues.

Note that the discriminant

$$b^2 - 4c = (r_1 \alpha_1 \bar{N}_1 + r_2 \alpha_2 \bar{N}_2)^2 - 4 r_1 r_2 \bar{N}_1 \bar{N}_2 D,$$

can be rewritten as

$$b^2 - 4c = (r_1 \alpha_1 \bar{N}_1 - r_2 \alpha_2 \bar{N}_2)^2 + 4 r_1 r_2 \bar{N}_1 \bar{N}_2 \beta_1 \beta_2.$$

Hence, for all permissible values of r_1, r_2, and β_1, β_2, the discriminant is positive. Therefore, both eigenvalues are real. Clearly, $\lambda_1 < 0$ and it remains to find the sign of λ_2. Invoking condition (1) (i.e., $D > 0$), then

$$c = r_1 r_2 \bar{N}_1 \bar{N}_2 D > 0$$

so that

$$b^2 - 4c < b^2,$$

implying

$$\lambda_2 = \tfrac{1}{2}b + \tfrac{1}{2}(b^2 - 4c)^{1/2} < -\tfrac{1}{2}b + \tfrac{1}{2}b = 0.$$

Incidentally, condition (1) may be interpreted by saying that state asymptotic stability exists if and only if each species' self-inhibiting rates α_1, α_2 exceed its competition rates β_1, β_2.

Using the information derived above, sufficient conditions can be specified for vanishing of the parameter sensitivity vector in a special case, namely that when the nominal dynamics are

$$N_1(t) = \bar{N}_1, \qquad N_2(t) = \bar{N}_2. \tag{109}$$

Conditions (if any) on the perturbations of $r_1, r_2, \alpha_1, \alpha_2, \beta_1$, and β_2 that result in parameter asymptotic stability are to be investigated.

The sensitivity equations are given in Eqs. (44) and (45) and, using Eq. (109),

$$ds(t)/dt = J_0 s(t) + B(t), \tag{110}$$

where

$$s(t) = \begin{bmatrix} s_1(t) \\ s_2(t) \end{bmatrix}, \qquad s(0) = \mathbf{0}, \qquad B(t) = -\begin{pmatrix} r_1 \bar{N}_1{}^2 v_2 + r_1 \bar{N}_1 \bar{N}_2 v_3 \\ r_2 \bar{N}_1 \bar{N}_2 v_5 + r_2 \bar{N}_2{}^2 v_6 \end{pmatrix},$$

and J_0 is as given before.

The solution to Eq. (110) is

$$s(t) = e^{J_0 t} \int_0^t e^{-J_0 \xi} B(\xi)\, d\xi, \tag{111}$$

where

$$e^{J_0 t} = \frac{e^{\lambda_2 t} - e^{\lambda_1 t}}{\lambda_2 - \lambda_1} (J_0 - I) = k(t)\,(J_0 - I),$$

where I is the 2×2 identity matrix, and λ_1, λ_2 are as given in Eq. (108). Equation (111) implies

$$\begin{aligned} s(t) &= \left(\frac{e^{\lambda_2 t} - e^{\lambda_1 t}}{\lambda_2 - \lambda_1}\right)(J_0 - I) \int_0^t \left(\frac{e^{-\lambda_2 \xi} - e^{-\lambda_1 \xi}}{\lambda_2 - \lambda_1}\right)(J_0 - I)\, B(\xi)\, d\xi \\ &= k(t)\,(J_0 - I)^2 \int_0^t k(-\xi)\, B(\xi)\, d\xi \\ &= -(J_0 - I)^2\, k(t) \int_0^t \begin{bmatrix} r_1 \bar{N}_1{}^2\, k(-\xi)\, v_2(\xi) + r_1 \bar{N}_1 \bar{N}_2\, k(-\xi)\, v_3(\xi) \\ r_2 \bar{N}_1 \bar{N}_2\, k(-\xi)\, v_5(\xi) + r_2 \bar{N}_2{}^2\, k(-\xi)\, v_6(\xi) \end{bmatrix} d\xi \\ &= -(J_0 - I)^2 \begin{bmatrix} r_1 \bar{N}_1{}^2\, k(t) \int_0^t k(-\xi)\, v_2(\xi)\, d\xi + r_1 \bar{N}_1 \bar{N}_2\, k(t) \int_0^t k(-\xi)\, v_3(\xi)\, d\xi \\ r_2 \bar{N}_1 \bar{N}_2\, k(t) \int_0^t k(-\xi)\, v_5(\xi)\, d\xi + r_2 \bar{N}_2{}^2\, k(t) \int_0^t k(-\xi)\, v_6(\xi)\, d\xi \end{bmatrix}. \end{aligned}$$

Thus, $s(t) \to \mathbf{0}$ as $t \to \infty$ if and only if each component tends to zero as t increases. Therefore, we want

$$k(t) \int_0^t k(-\xi)\, v_j(\xi)\, d\xi \to 0 \qquad \text{as} \quad t \to \infty, \quad j = 2, 3, 5, 6.$$

This becomes

$$\frac{e^{\lambda_2 t} - e^{\lambda_1 t}}{(\lambda_2 - \lambda_1)^2} \int_0^t (e^{-\lambda_2 \xi} - e^{-\lambda_1 \xi})\, v_j(\xi)\, d\xi,$$

or

$$\frac{1}{(\lambda_2-\lambda_1)^2}\left[e^{\lambda_2 t}\left(\int_0^t e^{-\lambda_2\xi}\,v_j(\xi)\,d\xi-\int_0^t e^{-\lambda_1\xi}\,v_j(\xi)\,d\xi\right)\right.$$

$$\left.-e^{\lambda_1 t}\left(\int_0^t e^{-\lambda_2\xi}\,v_j(\xi)\,d\xi-\int_0^t e^{-\lambda_1\xi}\,v_j(\xi)\,d\xi\right)\right],$$

which goes to zero as $t\to\infty$ if

$$e^{\lambda_q t}\int_0^t e^{-\lambda_r\xi}\,v_j(\xi)\,d\xi\to 0,\qquad q,r=1,2,\quad j=2,3,5,6.$$

All these conditions will be met if, for example, we require

$$v_j(t)\leqslant e^{-\gamma t},\qquad \text{where}\quad \gamma>(b^2-4c)^{1/2}>0.$$

Again, these are necessary but not sufficient conditions for asymptotic stability of the time-invariant Lotka–Volterra system to parameter perturbations.

The difficulties, illustrated by these last few examples, of determining stability when parameters are perturbed motivates the theoretical treatment undertaken in the next section.

D. Stability under Parameter Perturbations

As usual, let $\mathbf{x}(t)$ be the nominal state structure of a system given by Eqs. (1) and (2), and $\mathbf{x}^*(\epsilon, t)$ be the perturbed structure given by Eqs. (5) and (6). The following is asserted as a new starting point from which to consider system stability under parameter perturbations:

Definition 5

A system is stable under property P if, whenever $\mathbf{x}(t)$ has property P, then so does $\mathbf{x}^*(\epsilon, t)$ for all ϵ.

This definition is more general than Definitions 1 and 2 in not being a priori limited to operating point neighborhoods. At the same time, it provides an underlying rationale for Definition 4. Under Definition 5, a number of useful theorems can be proved regarding the relationship of $\mathbf{x}^*(\epsilon, t)$ to $\mathbf{x}(t)$. They can be found, for example, in Bellman (1953), and will be stated here without proof.

1. *Perturbation Bounds*

Suppose, in Eq. (1),

$$\mathbf{f}[\mathbf{x}(t), \mathbf{p}(t)] = A(t)\,\mathbf{x}(t); \tag{112}$$

the entries $a_{ij}(t)$ of the $n\times n$ time-varying matrix $A(t)$ comprise the set of parameters. Thus, $m = n^2$. Any or all of these parameters may be perturbed by simply adding another $n \times n$ matrix $B(t)$ to the original one:

$$p_{ij}^*(t) = a_{ij}(t) + \epsilon b_{ij}(t), \qquad i, j = 1, \ldots, n. \tag{113}$$

The perturbed system is

$$d\mathbf{x}^*(\epsilon, t)/dt = [A(t) + \epsilon B(t)]\, \mathbf{x}(\epsilon, t), \tag{114}$$

where $B(t)$ replaces the perturbation vector $\mathbf{v}(t)$ so that additional restrictions may be placed upon it. One such restriction is that the perturbations $b_{ij}(t)$ become small as t gets large. Otherwise, $\mathbf{x}^*(\epsilon, t)$ could not be expected to behave anything like $\mathbf{x}(t)$ for large t, in violation of Definition 5.

The size of the entries in $B(t)$ can be measured by a matrix norm such as

$$\|B(t)\| = \sum_{i,j=1}^{n} |b_{ij}(t)|. \tag{115}$$

Similarly for the parameter matrix:

$$\|A(t)\| = \sum_{i,j=1}^{n} |a_{ij}(t)|. \tag{116}$$

This particular norm has the useful properties

$$\|AB\| \leqslant \|A\|\, \|B\|, \tag{117}$$

and

$$\|\textstyle\int A\, dt\| \leqslant \int \|A\|\, dt. \tag{118}$$

Here, $\int A\, dt$ is the matrix whose ijth element is $\int a_{ij}\, dt$.

Bellman (1953, p. 34, Theorem 1) proved the following theorem: If all solutions of $d\mathbf{x}/dt = A\mathbf{x}$, where A is a constant matrix, are bounded as $t \to \infty$, then solutions of

$$d\mathbf{x}^*/dt = [A + \epsilon B(t)]\, \mathbf{x}^* \tag{119}$$

are bounded also for all ϵ provided that the integral of the norm of $B(t)$ is bounded:

$$\int_{t_0}^{\infty} \|B(\xi)\|\, d\xi < \infty.$$

This restriction on $B(t)$ is more severe than simply asking that $B(t) \to 0$ as $t \to \infty$. For example, in the one-dimensional case $B(t) = 1/(t + 1)$,

$$\lim_{t \to \infty} B(t) = 0,$$

but

$$\int_{t_0}^{\infty} d\xi/(\xi+1)$$

does not exist.

On the other hand, if the constant matrix A in the nominal situation is such that all solutions of $d\mathbf{x}/dt = A\mathbf{x}$ approach $\mathbf{0}$ as $t \to \infty$, then much less is required of the perturbing matrix $B(t)$ to guarantee that $\mathbf{x}^* \to \mathbf{0}$ also as $t \to \infty$, as is necessary for stability, i.s.L. Specifically, Bellman (1953, p. 36, Theorem 2) proves that all that is required is to have $\|B(t)\| \leqslant c$ for $t \geqslant t_0$, where c is a constant which depends upon the eigenvalues of A.

At this point it is tempting to extend these two theorems to the time-varying case, where $A(t)$ is nonconstant. However, the above conditions are not enough (Bellman, 1953, p. 42, Theorem 5) to guarantee stability in the sense of Definitions 1 and 5. The new conditions required depend upon the nominal parameters $A(t)$, and the many special cases are not germane to the present discussion. For this reason, generalization to the time-varying case will not be considered further.

2. *Sensitivity in Finite Time Intervals*

It is relevant, however, to consider the behavior of $\mathbf{x}^*(\epsilon, t)$ with respect to that of $\mathbf{x}(t)$ *in any finite interval,* $[0, t_1]$. Two problems concerning finite intervals have previously been identified: (i) error in the approximation of sensitivity to the deviation $\mathbf{x}^* - \mathbf{x}$ can grow with large time (Section III. B. 2), so t_1 should be kept small, but (ii) the sensitivity measure of asymptotic stability (Definition 4) cannot readily be implemented on a finite interval. The following result is established below on the interval $[0, t_1]$ for a class of time-varying, nonhomogeneous, linear systems. This class is defined by the equations

$$d\mathbf{x}(t)/dt = A(t)\,\mathbf{x}(t) + \mathbf{z}[\mathbf{p}(t)], \qquad \mathbf{x}(0) = \mathbf{x}_0, \tag{120}$$

where the vector function $\mathbf{z}$ is linear in $\mathbf{p}(t)$ [i.e., $\mathbf{z}(\alpha\mathbf{p}_1 + \beta\mathbf{p}_2) = \alpha\mathbf{z}(\mathbf{p}_1) + \beta\mathbf{z}(\mathbf{p}_2)$ for all vectors $\mathbf{p}_1$, $\mathbf{p}_2$ and real numbers α, β]. Here, the parameters in $A(t)$ are considered as endogenous rates, and those in $\mathbf{z}(\mathbf{p})$ represent exogenous inputs.

Theorem

For ϵ sufficiently small $\mathbf{x}^*(\epsilon, t)$, in norm, cannot deviate from $\mathbf{x}(t)$ on a finite intercal $[0, t_1]$ by more than a constant times the maximum of the sensitivity vector; more precisely,

$$\|\mathbf{x}^*(\epsilon, t) - \mathbf{x}(t)\| \leqslant \alpha \max_{t \in [0, t_1]} \|\mathbf{s}(t)\|, \tag{121}$$

where $\alpha = \epsilon \exp(\epsilon c t_1)$, and c is a constant dependent on the perturbation.

Proof:

A general time-varying perturbation of Eq. (120) is

$$A^*(\epsilon, t) = A(t) + \epsilon B(t), \qquad \mathbf{p}^*(\epsilon, t) = \mathbf{p}(t) + \epsilon \mathbf{v}(t). \tag{122}$$

The perturbed solution of Eq. (120) will then satisfy

$$d\mathbf{x}^*(\epsilon, t)/dt = [A(t) + \epsilon B(t)]\ \mathbf{x}^*(\epsilon, t) + \mathbf{z}[\mathbf{p}(t) + \epsilon \mathbf{v}(t)], \tag{123}$$

subject to

$$\mathbf{x}^*(\epsilon, 0) = \mathbf{x}_0.$$

It is well-known (e.g., Coddington and Levinson, 1955) that the solution $\mathbf{x}^*(\epsilon, t)$ is analytic in ϵ for each fixed t in the common domain of continuity of $A(t)$, $B(t)$, $\mathbf{p}(t)$, and $\mathbf{v}(t)$. Thus,

$$\mathbf{x}^*(\epsilon, t) = \mathbf{x}(t) + \epsilon \mathbf{s}(t) + \epsilon^2 \mathbf{u}_2(t) + \epsilon^3 \mathbf{u}_3(t) + \cdots, \tag{124}$$

where

$$\mathbf{x}(0) = \mathbf{x}_0, \qquad \text{and} \qquad \mathbf{s}(0) = \mathbf{u}_2(0) = \mathbf{u}_3(0) = \cdots = \mathbf{0} \tag{125}$$

are the initial conditions.

Differentiating Eq. (124) with respect to t and substituting in Eq. (123) yields

$$\begin{aligned}\dot{\mathbf{x}} + \epsilon\dot{\mathbf{s}} + \epsilon^2 \dot{\mathbf{u}}_2 + \epsilon^3 \dot{\mathbf{u}}_3 + \cdots = {} & [A\mathbf{x} + \mathbf{z}(\mathbf{p})] + \epsilon[A\mathbf{s} + B\mathbf{x} + \mathbf{z}(\mathbf{v})] \\ & + \epsilon^2(A\mathbf{u}_2 + B\mathbf{s}) + \epsilon^3(A\mathbf{u}_3 + B\mathbf{u}_2) + \cdots,\end{aligned}$$

or the system of equations

$$\begin{aligned}\dot{\mathbf{x}} &= A\mathbf{x} + \mathbf{z}(\mathbf{p}) \\ \dot{\mathbf{s}} &= A\mathbf{s} + B\mathbf{x} + \mathbf{z}(\mathbf{v}) \\ \dot{\mathbf{u}}_2 &= A\mathbf{u}_2 + B\mathbf{s} \\ \dot{\mathbf{u}}_3 &= A\mathbf{u}_3 + B\mathbf{u}_2 \\ &\ \vdots\end{aligned} \tag{126}$$

subject to the initial conditions (125). To obtain these equations, the fact that $\mathbf{z}(\mathbf{p})$ is linear in $\mathbf{p}$ was used. Equations (126) may now be solved iteratively by the method outlined below.

Let $\Phi(t, t_0)$ be the transition (or fundamental) matrix (e.g., Wiberg, 1971, p. 99), representing the solution to

$$\partial\Phi(t, t_0)/\partial t = A(t)\ \Phi(t, t_0),$$

subject to the initial condition $\Phi(t_0, t_0) = I$, the $n \times n$ identity matrix. Then, the solution for $\mathbf{u}_2(t)$, for example, can be written as

$$\mathbf{u}_2(t) = \int_0^t \Phi(t, \xi)\, B(\xi)\, \mathbf{s}(\xi)\, d\xi. \tag{127}$$

Likewise, the solutions for $\mathbf{u}_{n+1}(t)$ can be written

$$\mathbf{u}_{n+1}(t) = \int_0^t \Phi(t, \xi)\, B(\xi)\, \mathbf{u}_n(\xi)\, d\xi, \qquad n = 2, 3, \ldots. \tag{128}$$

By defining

$$\mathcal{L}\mathbf{u} = \int_0^t \Phi(t, \xi)\, B(\xi)\, \mathbf{u}(\xi)\, d\xi,$$

and

$$\mathcal{L}^n \mathbf{u} = \mathcal{L}[\mathcal{L}^{n-1}\mathbf{u}], \qquad \mathcal{L}^0 \mathbf{u} = \mathbf{u},$$

Eqs. (127) and (128) may be written in the notational form

$$\begin{aligned} \mathbf{u}_2 &= \mathcal{L}\mathbf{s} \\ \mathbf{u}_3 &= \mathcal{L}\mathbf{u}_2 = \mathcal{L}^2 \mathbf{s} \\ \mathbf{u}_4 &= \mathcal{L}\mathbf{u}_3 = \mathcal{L}^3 \mathbf{s} \\ &\vdots \end{aligned}$$

Therefore, $\mathbf{x}^*(\epsilon, t)$, as defined in Eq. (124), can be written as

$$\mathbf{x}^* - \mathbf{x} = \epsilon \sum_{n=0}^{\infty} \epsilon^n \mathcal{L}^n \mathbf{s},$$

so that

$$\|\mathbf{x}^* - \mathbf{x}\| \leqslant |\epsilon| \sum_{n=0}^{\infty} |\epsilon|^n \|\mathcal{L}^n \mathbf{s}\|. \tag{129}$$

By properties (117) and (118),

$$\|\mathcal{L}\mathbf{s}\| \leqslant \int_0^t \|\Phi(t, \xi)\, B(\xi)\| \cdot \|\mathbf{s}(\xi)\|\, d\xi. \tag{130}$$

Considering some fixed interval $[0, t_1]$, where $t_1 > 0$, and assuming that $A(t)$ and $B(t)$ are continuous on $[0, t_1]$, then

$$\|\Phi(t, \xi)\, B(\xi)\| \leqslant c,$$

some constant which can be found, and

$$\|\mathbf{s}(\xi)\| \leqslant M$$

on $[0, t_1]$. Hence, by replacing the norms by these maximums, Eq. (130) becomes

$$\|\mathcal{L}\mathbf{s}\| \leqslant cM \int_0^t d\xi = cMt$$

on $[0, t_1]$. Likewise, since on $[0, t_1]$

$$\mathcal{L}^2\mathbf{s} = \int_0^t \Phi(t, \xi)\, B(\xi)\, \mathcal{L}\mathbf{s}(\xi)\, d\xi$$

$$\mathcal{L}^3\mathbf{s} = \int_0^t \Phi(t, \xi)\, B(\xi)\, \mathcal{L}^2\mathbf{s}(\xi)\, d\xi$$

$$\vdots$$

we have

$$\|\mathcal{L}^2\mathbf{s}\| \leqslant c \int_0^t \|\mathcal{L}\mathbf{s}\|\, d\xi \leqslant c(cM) \int_0^t \xi\, d\xi = (c^2 t^2/2)M$$

$$\|\mathcal{L}^3\mathbf{s}\| \leqslant c \int_0^t \|\mathcal{L}^2\mathbf{s}\|\, d\xi \leqslant (c^2/2)M \int_0^t \xi^2\, d\xi = (c^3 t^3/3!)M$$

$$\vdots$$

and in general

$$\|\mathcal{L}^n\mathbf{s}\| \leqslant [(ct)^n/n!]M, \qquad n = 1, 2, 3, \ldots. \tag{131}$$

But, $0 \leqslant t \leqslant t_1$ so that Eq. (131) becomes

$$\|\mathcal{L}^n\mathbf{s}\| \leqslant [(ct_1)^n/n!]M. \tag{132}$$

The final result, expression (121), can now be shown. From Eqs. (128) and (131),

$$\|\mathbf{x}^* - \mathbf{x}\| \leqslant |\epsilon| \sum_{n=0}^{\infty} |\epsilon|^n [(ct_1)^n/n!]M \leqslant |\epsilon| \sum_{n=0}^{\infty} [(|\epsilon| ct_1)^n/n!]M = \epsilon\, e^{\epsilon c t_1} M.$$

Letting $\alpha = \epsilon \exp(\epsilon c t_1)$,

$$\|\mathbf{x}^* - \mathbf{x}\| \leqslant \alpha M = \alpha \max_{t \in [0, t_1]} \|\mathbf{s}(t)\| \tag{121}$$

for all $t \in [0, t_1]$.

Admittedly, Eq. (121) is only an approximation of the disparity between $\mathbf{x}(t)$ and $\mathbf{x}^*(\epsilon, t)$ for $t \in [0, t_1]$, but it does indicate that sensitivity can provide a measure for this disparity, and thus justifies Definition 4 for evaluating stability. It should be emphasized, however, that this inequality does not hold over an infinite interval since c is not defined and α increases without bound as $t \to \infty$. This can result in no bound, as was illustrated in the last part of Example 11 where, with $r(t) = -2t/t^2 + 1$, the nominal behavior of $N(t)$ vanished as $t \to \infty$ whereas the perturbed solution $N^*(\epsilon, t)$ became unbounded, even though sensitivity also vanished with large time.

VI. Ecosystem Applications

Sensitivity analysis provides a ready means for assessing effects of measurement errors or perturbations on system dynamics. It also exhibits potential to be developed into a useful stability measure. Therefore, it would appear to be of immediate use in ecosystem modeling and related applied problems.

Large-scale ecosystem models always require more data than are available from associated empirical studies. Gaps must be filled in from the literature and even educated guesses. Sensitivity analysis allows the determination of where a model will suffer most from data deficiencies, and thus can aid in setting research priorities. Once a model is validated, the effects of state or parameter perturbations can be systematically evaluated and expressed as sensitivity dynamics—a powerful form of systems analysis.

The sensitivity equations, being linear, are easy to solve, and uniqueness is guaranteed. Computations are particularly convenient to perform in the case of linear state models, and such models (nonhomogeneous and time varying) have been shown to be appropriate for large-scale ecosystem representations (e.g., Patten, 1972a; Patten *et al.*, 1975). These models are constructed around the form

$$d\mathbf{x}(t)/dt = A(t)\,\mathbf{x}(t) + B(t)\,\mathbf{z}(t), \tag{75}$$

where $\mathbf{x}(t)$ is the state vector, $\mathbf{z}(t)$ is a vector of inputs, and $A(t)$ and $B(t)$ are coefficient matrices. With the latter three quantities interpreted as parameters, then $\partial f_i/\partial p_k$ for the kth parameter will be either states $\mathbf{x}(t)$ or elements of $B(t)$ and $\mathbf{z}(t)$. These become the elements of $K(t)$ in the sensitivity equation (20). Similarly, $\partial f_i/\partial x_j = a_{ij}(t) \in A(t)$, and $J(t)$ is identical to $A(t)$. The sensitivity substructure of Eq. (75) is therefore simply

$$d\mathbf{s}(t)/dt = A(t)\,\mathbf{s}(t) + K(t)\,\mathbf{v}(t), \tag{133}$$

where $\mathbf{v}(t)$ is the perturbation vector, and where the entries in $K(t)$ are drawn from $\mathbf{x}(t)$, $B(t)$, and $\mathbf{z}(t)$. This equation, containing so many components from the state model (75), is exceedingly easy to implement together with the latter in a computer program.

VII. Sensitivity and Causality

Aside from the practical aspects of sensitivity analysis, the value of sensitivity as a heuristic concept for understanding ecosystems is also to be considered. In this section the relationship of sensitivity to causality in the fundamental organization of dynamic systems is examined to indicate how the state transition function may be interpreted as sensitivity to nominal parameter dynamics.

Hubbell (1973) has recently commented on the slow growth of a systems perspective in ecology, despite the fact that in the ecosystem the ecologist confronts the ultimate complex system. Part of the reason he attributes to causality being viewed traditionally in single factor terms. Populations are construed as autonomous units whose abundance is regulated by this or that factor. They are not treated in the perspective of being surrounded by, and enmeshed within, a total ecosystem, receiving energy, material, and information inputs from a multiplicity of direct and indirect sources, and generating state dynamics and outputs in response. Further, although Hutchinson (1948) pointed it out a generation ago, it is not generally appreciated that ecosystem states tend to be mutually causal because of the closed-loop interactive structure conferred by matter recycling. The ecosystem is, in effect, nested within these grand recycle loops, and the systems significance of this is that the dynamics of each state and parameter variable affects that of every state variable through a complex feedback network of influence propagation (Patten, 1972b). Cause and effect originating anywhere in the ecosystem structure are transmitted, ultimately, everywhere.

That sensitivity and causality are related is implicit in the ability of sensitivity to estimate the effect on any state variable of both state and parameter perturbations anywhere in a system. A perturbation is propagated through the interconnection causing deviations from nominal dynamics all along the way. These effects become subsequent causes in an interactive sequence that, in ecosystems, is endless because of the outer loop structure. The sensitivity substructure [Eqs. (20)–(21) and (52)–(53)] mediates this causation through the interaction matrices $J(t)$ and $K(t)$.

However, since sensitivity is defined [Eq. (9)] as a deviation referenced to a nominal trajectory, in unperturbed systems there is technically no sensitivity since there are no deviations from nominal. Still, an intuitive sense of sensitivity remains in the causality that continues to organize the unperturbed time development: (i) the system is "sensitive" to, and responds to, nominal influences, and (ii) the system retains a capability to react to perturbations, a "sensitivity" that is present and changing whether a disturbance occurs or not. It is these properties that are not actually represented in the sensitivity measure $s(t)$, and yet have intuitive appeal.

The properties are, of course, formally embodied in the concept of state (e.g.,

Zadeh and Desoer, 1963), and it is the rate of state change of a system, that is, the state transition function, for which a sensitivity interpretation is now suggested. Three facts about system dynamics may be stated, as follows:

(1) The nominal time development is provided by knowledge of the primary state structure, that is, by the present state $\mathbf{x}(t)$ and the state transition function $\mathbf{f}(t)$:

$$\mathbf{x}(t+dt) \approx \mathbf{x}(t) + dt\, \mathbf{f}(t), \qquad t \in [t_0, t_1), \tag{134}$$

for small time increments $dt > 0$.

(2) State deviations from nominal are approximated by a linear substructure, the sensitivity $\mathbf{s}(t)$:

$$\mathbf{x}^*(\epsilon, t) \approx \mathbf{x}(t) + \epsilon\mathbf{s}(t), \qquad t \in [t_0, t_1], \tag{135}$$

for small perturbations $\epsilon > 0$.

(3) The perturbed time development follows from combining Eqs. (134) and (135):

$$\begin{aligned} \mathbf{x}^*(\epsilon, t+dt) &\approx \mathbf{x}^*(\epsilon, t) + dt\, \mathbf{f}(t) \\ &\approx \mathbf{x}(t) + \epsilon\mathbf{s}(t) + dt\, \mathbf{f}(t), \qquad t \in [t_0, t_1], \end{aligned} \tag{136}$$

It is apparent, in comparing Eqs. (134) and (135), that the state transition function $\mathbf{f}(t)$ plays a comparable role in generating time dynamics to that of sensitivity $\mathbf{s}(t)$ in producing deviations from nominal states. That is, the transition function can be interpreted as a special form of sensitivity: namely, "sensitivity" of the state vector to time "perturbations". The time increments $dt > 0$ represent such perturbations, a special case of the general perturbation $\epsilon > 0$. The correspondences of dt and ϵ and $\mathbf{s}(t)$ and $\mathbf{f}(t)$ in Eq. (136) are apparent.

Thus, sensitivity as a concept can be extended to encompass the state response of a dynamic system to all influences, including nominal motion due to unperturbed parameter dynamics. This expanded concept of sensitivity and its relationship to the propagation of cause and effect in system interconnections will be further considered in future research.

VIII. Summary

This chapter explores the idea that implicit in the primary state structure of linear or nonlinear dynamic systems is a linear substructure, sensitivity, that (i) measures effects of disturbances, (ii) determines stability, and (iii) mediates causality.

The sensitivity equations are formulated for constant and time-varying cases of parameter perturbation or measurement error, as well as for perturbations of state. Principal results are illustrated for simple population models (exponential, logistic, and two-dimensional Lotka–Volterra competition), and the known properties of these classical systems are thereby extended.

The relationship of sensitivity to stability is examined, and sensitivity criteria for asymptotic stability defined and qualified. The criteria are applied to certain cases of the population models, and more general conditions for stability of these are stated than have appeared heretofore in the literature. Asymptotic stability is extended as a concept from its classical definition in terms of state perturbations (state asymptotic stability) to the more general case where parameters are perturbed (parameter asymptotic stability).

Finally, the sensitivity concept is expanded to include a set of ideas concerning causality in the nominal dynamics of systems. It is suggested that the state transition function in effect represents system sensitivity to unperturbed parameter dynamics in time.

Acknowledgments

The interactions leading to this paper were begun at an NSF institute, Systems Approach to the Life Sciences, held during summer 1972 at Utah State University. Doctors George F. Sheats (State University College, Plattsburg, New York), Walter J. Hudetz (California State College, Fullerton, California), and Robert H. Foulkes (Youngstown State University, Youngstown, Ohio) participated in the discussions and contributed to aspects of sensitivity analysis not represented here.

In addition, BCP wishes to acknowledge continuing dialog over a period of several years with students and colleagues at the University of Georgia, particularly Jack B. Waide, Jackson R. Webster, William G. Cale, and Dr. Jeffrey J. Lee. These interactions reinforced the expanded interpretation of sensitivity, leading eventually to recognition of the state transition function as a form of sensitivity, as indicated in Section VII.

REFERENCES

Bellman, R. (1953). "Stability Theory of Differential Equations." McGraw-Hill, New York.

Coddington, E. A., and Levinson, N. (1955). "Theory of Ordinary Differential Equations." McGraw-Hill, New York.

Cruz, J. B. (ed.) (1973). "System Sensitivity Analysis." Dowden, Stroudsburg, Pennsylvania.

Hubbell, S. P. (1973). *Amer. Natur.* **107**, 94.

Hutchinson, G. E. (1948). *Ann. N.Y. Acad. Sci.* **50**, 221.

Liapunov, M. A. (1892). Le problème général de la stabilité du mouvement, *Ann. Fac. Sci. Toulouse* **9**, 203–474, 1907 (originally published in Russian in 1892; reprinted as *Annals Math. Stud.* no. 17. Princeton Univ. Press, Princeton, New Jersey).

Lotka, A. J. (1925). "Elements of Physical Biology." Williams and Wilkins, Baltimore, Maryland (Reprinted 1956, "Elements of Mathematical Biology." Dover, New York).

MacArthur, R. H. (1955). *Ecology* **36**, 533.

Malthus, T. R. (1798). "An Essay on the Principle of Population as it Affects the Future Improvement of Society, with Remarks on the Speculations of Mr. Godwin, M. Canducet, and Other Writers." First published 1798.

May, R. M. (1973). *Amer. Natur.* **107**, 46.

Meissinger, H. F., and Bekey, E. A. (1966). *Simulation* **6**, 94.

Miller, K. S., and Murray, F. J. (1953). The mathematical basis for the error analysis of differential analyzers, *MIT J. Math. Phys.* Nos. 2 and 3.

Mohler, R. R., and Ruberti, A. (eds.) (1972). "Theory and Applications of Variable Structure Systems." Academic Press, New York.

Patten, B. C. (1972a). *Simulation* **19**, 177.

Patten, B. C. (1972b). *In* "Theory and Applications of Variable Structure Systems" (R. R. Mohler and A. Ruberti, eds.), pp. 213–232. Academic Press, New York.

Patten, B. C., Egloff, D. A., Richardson, T. H., and 38 Coauthors (1975). *In* "Systems Analysis and Simulation in Ecology" (B. C. Patten, ed.), Vol. III, pp. 205–421. Academic Press, New York.

Pearl, R. (1924). "Studies in Human Biology." Williams and Wilkins, Baltimore, Maryland.

Pridgen, J. H. (1970). Computation of Sensitivity Measures. Ph.D. thesis, Oklahoma State Univ., Stillwater, Oklahoma.

Tomović, R. (1963). "Sensitivity Analysis of Dynamic Systems." McGraw-Hill, New York.

Tomović, R., and Vukobratović, M. (1972). "General Sensitivity Theory." Amer. Elsevier, New York.

Verhulst, P. F. (1844). Recerches mathématiques sur la loi d'accroissement de la population, *Mém. Acad. Roy. Bruxelles,* T XVIII, 1–58.

Volterra, V. (1927). *Scientia* **41**, 85.

Wiberg, D. M. (1971). "State Space and Linear Systems." McGraw-Hill, New York.

Woodwell, G. M., and Smith, H. H. (eds.) (1969). Diversity and Stability in Ecological Systems. Brookhaven Nat. Lab. Symp. in Biol. 22. Upton, New York.

Zadeh, L. A., and Desoer, C. A. (1963). "Linear System Theory. The State Space Approach." McGraw-Hill, New York.

16

Patterns of Biological Control in Ecosystems

MICHAEL CONRAD*

INSTITUTE FOR INFORMATION SCIENCES
UNIVERSITY OF TÜBINGEN, TÜBINGEN, GERMANY

I. Introduction

The ecosystem consists of units which are capable of reproducing themselves. According to the theory of evolution by variation and natural selection, the

* Present address: Department of Computer and Communication Sciences, University of Michigan, Ann Arbor, Michigan.

properties of the ecosystem are determined by the relative reproductive success of these units. This is necessarily so since populations of relatively successful units have higher growth rates and therefore predominate in the course of evolution.

The relative reproductive success of an organism is determined by its interaction with the environment, including its biological environment. In order to reproduce the organism must be adapted to the specific features of its environment. It must also cope with the unpredictability of this environment as well as with noise processes (or inherent tendencies to malfunction).

It is quite difficult to analyze the specific adaptations which make particular organisms more or less successful. These adaptations are not readily amenable to a common characterization just because they are particular and specific. However, this is not true for the more general adaptations which allow systems to cope with noise and environmental uncertainty.

This chapter presents a qualitative discussion of these more general adaptations (or control characteristics). First, the basic mechanisms of control in biological systems are classified and characterized. Then, how these general mechanisms manifest themselves at different levels of biological organization and how the quantity of control is related to the behavior of the environment are described. Finally, the factors which determine how biological systems utilize these different mechanisms and modes of control, that is, the role they play in the ecosystem as a whole, are indicated. These strategies (or patterns) of utilization are important for understanding the sequential change of the ecosystem in succession and evolution. This is inevitably the case since the ability to cope with internal and external disturbances is crucial for determining the relative growth rates of populations and communities.

The formal basis of the discussion which follows will be presented in a separate paper. The general mathematical background derives from the work of Shannon (1948) and also from Ashby (1956) and Kinchin (1957).

II. The Ecosystem in a General Sense

The ecosystem in a general sense is a complete biological system along with its physical environment. A biological system is complete when it is capable of persisting indefinitely without the help of any other biological systems. This persistence is possible because of the system's capacity to evolve by means of imprecise self-reproduction. The self-reproduction process requires a cycle of matter. This is driven by an external energy source.

In order to understand the ecosystem in a general sense it is necessary to consider all the processes which support it. In particular this is true for the control processes. To consider these we must first decide how to describe the ecosystem.

Naturally, the ecosystem may be described in terms of its fundamental components—in terms of all the atoms and molecules in some region of space. However, this kind of complete description is not very useful. In practice the ecosystem is described hierarchically in terms of various units (or compartments) at different levels of organization, for example, populations, organisms, cells, molecular aggregates, and so forth. In this case a partial description of the ecosystem is given in terms of the arrangement of populations, then a partial description of each population in terms of organisms, and so forth, until the bottom level is reached.

The behavior of the ecosystem is its sequence of states (or trajectory). These states may be complete physical states or they may be "many-tuples" of partial states, generated by the hierarchical and compartmental decomposition of the system. The sequence of states may be complex, but nevertheless predictable; or it may be quite simple and unpredictable. As the objective uncertainty of the sequence of states increases we will say that the sequence becomes more uncertain.

A hierarchical approach to ecosystem biology has the advantage that it really reflects fundamental features of biological organization. Indeed, in the absence of this approach it would not be possible to analyze the relation between different types of control processes in biology.

III. Fundamental Sources of Dysfunction

The behavior of the living part of the ecosystem (the biota) is subject to both external and internal disturbances. The external disturbances arise from macroscopic uncertainty of the environment, that is, from statistical irregularity in its behavior. The internal disturbances arise from spontaneous noise processes or from microscopic fluctuations in the environment. Needless to say the distinction between internal and external disturbances is not always sharp.

The behavior of the system is also disturbed by its inability to anticipate the statistically regular, or certain, behavior of the environment. For example, the system's life cycle may not be well correlated to periodic changes in the environment. In this case the system's "model" of the environment is not as good as possible, at least in the sense that it fails to utilize certain environmental regularities.

These regularities may be either temporal or spatial. If the regularities are temporal this means that there are correlations between environmental states at different times, as in the natural periodicities. If the regularities are spatial, this means that there are correlations among environmental states at different places.

This determines the possible ways in which a disturbance can be distributed over a biological system. In particular, the effect of the disturbance is multiplied if different parts of the system respond to it independently.

The system may also be able to anticipate behavior of the environment on the basis of very short-term regularities. This is possible because environmental disturbances are often preceded by preliminary, minor disturbances, which function as signals. In this case the system operates on the basis of two time scales—one associated with the disturbance and the other associated with the signal, which of course cannot be anticipated.

In general, internal noise processes interfere with anticipation. This is because they interfere with information flow from the environment to the system. They interfere with anticipation which is based on intrinsic timing processes as well.

The correlations, particularly temporal correlations in the environment, are partly a biological property. Thus, the environment of a well-organized (well-balanced) ecosystem will be more certain than the environment of an unbalanced ecosystem. Also, what appears as an internal disturbance from the standpoint of the biota as a whole may contribute to environmental uncertainty from the standpoint of an individual population or organism. In this case the organism or population may be uncertain about the behavior of other biological systems, systems which are now included in its environment.

IV. Fundamental Mechanisms of Control

A. The Mechanisms

Biological systems contend with these sources of dysfunction either by physical isolation or by selectively dissipating them into an ensemble of functional states (or trajectories). These states may be macroscopically equivalent, at most involving only temporary changes in system behavior, or they may be macroscopically distinct, involving relatively permanent changes. The possible mechanisms of control grade into one another, but for purposes of discussion it is useful to classify them into the following four categories: (i) physical reliability; (ii) organizational reliability; (iii) selective indifference; (iv) adaptability. Physical and organizational reliability allow the system to combat noise or malfunction of internal origin. Selective indifference and adaptability allow the system to cope with unpredictability of the environment (see Table I).

B. Physical and Organizational Reliability

The physical reliability mechanism allows the system to withstand noise because its components function with greater precision. This precision is

TABLE I

General Mechanisms of Biological Control

Origin of disturbance	Mechanism	Characteristic feature	Example
Internal	Physical reliability	Component precision	Thermal isolation
	Organizational reliability	Component redundancy	Duplication of units
External	Constitutive adaptability	Exigency machinery always present	Constitutive enzyme
	Inducible adaptability	Exigency machinery produced when needed	Inducible enzyme
	Selective adaptability	Population of different types of machines	Genetic diversity, trial and error learning
	Selective indifference	System isolated from exigencies	Protective structures of biological or environmental origin
Internal or external	Failure	Behavior of environment uncorrelated to behavior of system	Subfunctional behavior (injury, disease, death)

determined by their quantum structure and by the number of states which are equivalent from the macroscopic point of view. In particular, energy levels of components may be organized so as to decrease their sensitivity to the thermal environment (cf. Pattee, 1968). Also, the system and its components are less vulnerable to fluctuation if many states are equivalent from the macroscopic point of view.

The system may also withstand noise by virtue of its spatial or temporal organization. This organization must allow for families of functionally equivalent states (or trajectories) correlated to the environment in the same way. These are states (or trajectories) that allow for the same chance of reproducing even though they may be macroscopically distinct. As the ensemble of functionally equivalent states (or trajectories) becomes larger it absorbs more of the noise, but without any adverse effect on the behavior of the system. In order for this mechanism to work the system must maintain the coherence of the family. This (selectively dissipative) maintenance process must itself be reliable. Therefore, the organizational mechanism of reliability ultimately depends on physically reliable components. If these components are expensive it may be advantageous for the system to develop families of functionally equivalent states and make heavy use of a few reliable components to maintain the coherence of these families.

The simplest example of organizational reliability is provided by duplication of units. For example, a cell may have multiple copies of genetic material. If one of these copies is damaged this does not disturb the function of the system; furthermore, some reliable components (certain enzymes) may be used to return the damaged system to its original state. Naturally the duplication of units allows the system to cope with damage of external as well as internal origin. Also, spatial and temporal redundancy can be used for error correction in much more sophisticated ways. This is quite important, in particular, for the transmission and processing of environmental variations which play the role of signals (Shannon, 1948; Winograd and Cowan, 1963).

C. Adaptability and Indifference

The selective indifference mechanism allows the system to withstand macroscopic variations of the environment by dissipating these into an ensemble of macroscopically equivalent states. The variations of the environment induce different system trajectories. However, these differences are mediated by static structures rather than by the dynamics of the system. The differences are relatively impermanent and allow the system to function as well as possible in the new environment.

The simplest example of the selective indifference mechanism is a protective structure (such as a shell). The state of the protective structure may change temporarily in response to an environmental variation, but without any effect on the essential operations of the system.

The final mechanism of control is adaptability. Adaptability allows the system to withstand macroscopic variations of the environment by assuming behavior which is essentially different but which is nevertheless functional in the new environment. In this case, however, the differences are mediated by the dynamics of the system rather than by static structures. If the differences are transient the adaptability is *constitutive.* If they are (or are capable of becoming) relatively permanent the adaptability is *inducible* or *selective.*

In constitutive adaptability the ensemble of trajectories depends on the constitution of the system, that is, on machinery which is always available to it. Variations in the environment induce different trajectories in the system, but these differences converge on what is essentially the same final situation.

The simplest example of constitutive adaptability is provided by a microorganism which is constitutive for some enzyme. In this case the microorganism may be indifferent to the presence or absence of some substrate; that is, differences in the detailed chemical behavior induced by presence or absence of the substrate are of no essential significance. In general a system with few requirements or many possible sources of these requirements is functionally independent in the sense that it is indifferent to the presence or absence of some of these sources or requirements. Naturally constitutive mechanisms may also allow the system to cope with the variability of more general environmental features.

In general, constitutive adaptability is quite similar to selective indifference, except that one is dynamic and the other static. Indeed the two mechanisms can be placed in the same category, as a form of indifference or adaptability, depending on whether it is practical to distinguish and correlate the different trajectories of the system to variations in the environment. This is not true for inducible and selective adaptability. These are quite different from selective indifference.

Inducible adaptability allows the system to cope with disturbances which are associated with relatively long lasting changes in the environment. The environmental change pushes the system into behavior which is appropriate for the new environment. Ordinarily, the change in behavior of the system is relatively permanent, that is, is coordinated to a structural change. Essentially the change in the environment induces latent machinery in the system.

The simplest example of inducible adaptability is provided by a cell which is inducible for some enzyme. As another example, different environments may induce different patterns of growth, in particular among plants.

The final type of adaptability is selective. This is the trial and error mechanism. In this case the environment induces changes in system behavior by a selection process. For example, the system may contain a population of subsystems which are rough replicas of one another. If the environment changes, the distribution of replicas changes. This changes the behavior of the system as a whole. If the environment changes permanently, the system as a whole changes permanently.

Genetic adaptability is the best example of the selective mechanism. Here the species shuffles its genes to form an ensemble of genotypes which are functionally dissimilar, though often only slightly so. This ensures that the population contains a number of types, at least some of which are capable of coping with each of a number of particular environments. The species has various mechanisms (ring chromosomes, periodic mutation, "binomial" distribution of genes, etc.) which it uses to maintain the coherence of this family of genotypes in a statistically stable environment, but which nevertheless allows changes in composition of the family to be induced (by selection) if the environment changes permanently. Essentially the interaction with the environment amplifies the relative amount of pre-existing machinery, in contrast to releasing the construction of latent machinery.

Selective adaptability is similar to organizational reliability in the sense that both mechanisms are based on redundancy. However, the role of noise is different in the two cases. In organizational reliability the redundancy is used to eliminate the detectible effects of noise (i.e., error) as nearly as possible. This elimination cannot be complete in selective (or trial and error) adaptability since the noise helps to generate the ensemble of rough replicas. However, the effect of the noise is reduced if many possible internal disturbances are able to generate a satisfactory ensemble for the environment to select from. This means that the ensemble of possible trajectories of the system as a whole must be large. In general this is necessary if selective adaptability is to be effective, just as it is necessary in order for organizational reliability to be effective (Conrad, 1972a).

In sum, aside from physical isolation, the mechanisms of control are all statistical. If disturbances arise from noise processes, they may be damped by an ensemble of functionally equivalent states (organizational reliability). If the disturbances are of environmental origin they induce new behavior in the system. This may be associated with protective structures (selective indifference) or with permanent internal structures of the system (constitutive adaptability). The disturbances may also induce latent structures (inducible adaptability) or they may select new structures and modes of behavior (selective adaptability).

It is important that these general mechanisms grade into one another. It is also important that not all the changes in the system induced by variations in the environment contribute to adaptability or to any other form of control. Needless to say, an environmental variation may induce states of injury or force the system to assume other abnormal behavior, that is, behavior which is inappropriately correlated to the environment.

V. Statistical Description of the Ecosystem

To consider the relationship between statistical properties of the system and statistical properties of the environment, the ecosystem may be described in terms of the quintuple

$$\langle A_1, A_2, T, \omega_1, \omega_2 \rangle,$$

where A_1 is the set of biota states, A_2 is the set of environmental states, T is a set of times ($t = 1, 2, 3, \ldots$), ω_1 represents the transition scheme of the biota, and ω_2 the transition scheme of the environment. The transition schemes (or sets of transition probabilities) are defined by

$$\omega_1 \equiv \{p[\alpha_1(t+1)|\beta(t)]\}, \qquad \text{and} \qquad \omega_2 \equiv \{p[\alpha_2(t+1)|\beta(t)]\},$$

where α_1 and α_2 are variables which run over A_1 and A_2, $\beta = (\alpha_1, \alpha_2)$, and it is supposed for convenience that the relationship between biota and physical environment is included in α_1.

The entropy of scheme ω_1 is given by

$$H(\omega_1) = -\sum_{\alpha_1, \beta} p[\beta(t)]\, p[\alpha_1(t+1)|\beta(t)] \log p[\alpha_1(t+1)|\beta(t)],$$

and similarly for scheme ω_2. The conditional entropy of scheme ω_1 given the transition in scheme ω_2 is given by

$$H(\omega_1|\omega_2) = -\sum_{\alpha_1, \alpha_2, \beta} p[\beta(t)]\, p[\alpha_1(t+1), \alpha_2(t+1)|\beta(t)]$$
$$\log p[\alpha_1(t+1)|\beta(t), \alpha_2(t+1)],$$

and likewise for scheme ω_2 given scheme ω_1.

The entropies defined above are connected by the following identity

$$H(\omega_1) - H(\omega_1|\omega_2) \equiv H(\omega_2) - H(\omega_2|\omega_1). \tag{1}$$

$H(\omega_1)$ and $H(\omega_2)$ represent the (unconditional) uncertainties in behavior of the system and environment, respectively. The conditional quantity $H(\omega_1|\omega_2)$ describes the uncertainty in the behavior of the system given the contemporaneous behavior of the environment. This will be called the decorrelation entropy since it increases with disturbances of internal origin or inability to anticipate behavior of the environment. The conditional quantity $H(\omega_2|\omega_1)$ describes the uncertainty in behavior of the environment given the contemporaneous behavior of the system, that is, the system's absence of information about the environment. If this absence of information is nonselective (or makes no difference to the system's chances of reproducing) it represents behavior of the environment from which the system is isolated or whose effect on the system decays to essential insignificance. This will be called selective indifference since it is evidently based on the selective indifference mechanism and on selective indifference which it is convenient to associate with constitutive adaptability.

The term on the left-hand side of Eq. (1) represents the information which the behavior of the system provides about the contemporaneous behavior of the environment. The term on the right-hand side represents the information which the behavior of the environment provides about the contemporaneous behavior of the system. These quantities are nonzero only if

$$p[\alpha_1(t+1)|\beta(t), \alpha_2(t+1)] \neq p[\alpha_1(t+1)|\beta(t)].$$

This is possible, in our picture, because the system operates on two time scales. In general the information which the system provides about contemporaneous behavior of the environment on the longer (discrete) time scale equals the information which flows from the environment to the system on the shorter (in reality continuous) time scale. This is realistic since a change in the environment may not affect the system immediately and its major effect may be preceded by small, initial effects. Or, the major effect may last a long time, but only damage the system after some time interval. In either case the states of both system and environment at the end of the time interval are correlated to the state of the environment during the time interval.

VI. Inherent Statistical Properties

Equation (1) is always true—indeed it is a trivial identity of probability theory. However, it does not answer the question about the relationship between statistical properties of the biota and statistical properties of the environment. This is because it is just as true in a certain as an uncertain environment; just as true for a living system as a nonliving system.

In order to answer this question the inherent statistical properties of the system must be characterized independently of the predictability of the particular environment in which they are operating. To do this the system must be stressed, subject to some condition. This will determine certain inherent (or potential) statistical quantities.

The simplest way of stressing the system is to find the most unpredictable environment in which it is still alive. The uncertainty of this environment will be called the functional capacity of the system. The condition for the system's remaining alive indefinitely can thus be written

$$\hat{H}(\omega_1) - \hat{H}(\omega_1|\omega_2) + \hat{H}(\omega_2|\omega_1) \geqslant H(\omega_2), \tag{2}$$

where the carets denote inherent entropies, that is, entropies which characterize the system when it is operating at its functional capacity. $\hat{H}(\omega_1)$ is the potential uncertainty in behavior of the system, $\hat{H}(\omega_1|\omega_2)$ is the tolerance for decorrelation, $\hat{H}(\omega_2|\omega_1)$ is the tolerance for error or selective indifference, and $H(\omega_2)$ is the actual uncertainty of the environment.

The term $\hat{H}(\omega_1) - \hat{H}(\omega_1|\omega_2)$ will be called the (functional) adaptability since it represents the system's potentiality for absorbing environmental disturbances without error or other indifference when it is operating at its functional capacity. According to Eq. (2) the adaptability plus the potential indifference must not be less than the actual uncertainty of the environment, at least if the viability of the system is not to be destroyed with certainty. Naturally the system can fulfill this condition, but nevertheless sustain considerable injury.

The adaptability equals the information which the environment provides about the system under the statistically most unfavorable conditions. The system can also be stressed (again by trying different environmental transition schemes) subject to the condition that this is a maximum. In this case the system will be said to be operating at its regular capacity (since this is the capacity as ordinarily defined in information theory). If the adaptability is smaller than the regular capacity, this means that certain disturbances of the environment reduce the reliability with which the system processes information, but that the system can remain alive under these conditions.

VII. Evolutionary Tendencies

Taking account of certain efficiency considerations. Eq. (2) may be replaced by

$$\hat{H}(\omega_1) - \hat{H}(\omega_1|\omega_2) + \hat{H}(\omega_2|\omega_1) \rightarrow H(\omega_2), \tag{3}$$

where the arrow indicates that the left-hand side approaches the right-hand side. The arrow represents an evolutionary tendency, not any precise or rigorous condition.

Equation (3) follows from the assumption that biological systems have a competitive advantage when each of the individual terms on the left-hand side takes on its lowest possible value [or its highest possible value in the case of $\hat{H}(\omega_1|\omega_2)$]. In other words, systems are more efficient (or fit) if they forego any unused inherent entropy: if their inherent behavioral uncertainty is not higher than necessary; if their capability for anticipating the behavior of the environment is not better than necessary; if their tolerance for injury and selective independence are not greater than necessary. If the equality does not hold the system can reduce at least one of these capabilities.

This assumption, which will be called the least entropy assumption, is plausible on biological grounds. For example, suppose that the potential uncertainty in behavior of a system (consistent with its being alive) is greater than necessary, and that it is therefore more adaptable than it has to be in the given environment. In this case the system may carry unnecessary machinery (for example, superfluous constitutive enzymes or food gathering structures) or unused latent variability (for example, unused inducible enzymes or modes of development). Certainly the system must pay for constructing and maintaining this superfluous machinery or for the potentiality of developing it. The system must also pay for the uncertainty in its behavior associated with trial and error learning (selective adaptability). For example, in a constant environment the organism whose behavior patterns are inherited (instinctive) will be more efficient than the system which must acquire (learn) such patterns. Similarly, populations must pay for excessive genetic variability. In general, genetically different types in a species are not equally fit and therefore the fitness of a population (in a given environment) can ordinarily be increased by eliminating certain of these types.

It is also possible for the potential indifference or tolerance for error to be larger than necessary. This is also expensive. In order for the system to be indifferent to the environment it must synthesize or construct protective structures; or if some of the indifference arises from superfluous machinery the system must pay for this (as above). In any case the system's indifference cannot be too large without becoming error. The tolerance for error is also costly, since the system must pay for the ability to recover from a large number of possible states of injury or disease.

Finally, the decorrelation entropy may be smaller than necessary. This means that the potential uncertainty in the behavior of the system given the behavior of the environment is maintained at a low level. Again the system must pay for this since a low decorrelation entropy means that the system anticipates behavior of the environment with high precision. It is also means that the system must suppress uncertainty of biological origin or other internal sources of noise.

The pressures so far discussed, while they may produce the result described by Eq. (3), do not determine the actual values of the inherent entropies. The tendency for an equality can hold, but nevertheless with the individual entropies assuming large values. This is clear since a single equation could not determine these values uniquely. In fact the expansion can be expected if the tendency is to be satisfied, but with the natural provision that the error be small (or the adaptability large).

The error may arise from internal disturbances as well as from disturbances of environmental origin. These internal disturbances depend on the noise characteristics of the system, characteristics which cannot be changed arbitrarily. In general noise can be decreased only by increasing physical precision (at a cost to the system), or by making heavy use of a few reliable components to combat this noise. However, this error correction mechanism depends on a large ensemble of states and therefore on high values for both $\hat{H}(\omega_1)$ and $\hat{H}(\omega_1|\omega_2)$. In general these two quantities may have to be very large in order for the error to be small. Essentially what is happening is that the decorrelation entropy increases, but the functionally equivalent states allow for more selective decorrelation.

In general, the actual values of the inherent entropies are determined by costs and advantages associated with error, physical precision, organizational reliability, and selective indifference. However, the efficiency considerations leading to Eq. (3) apply regardless of any changes in the relative importance of these costs and advantages. It is also important that these considerations suggest that the adaptability tends to equal the regular capacity in the course of evolution. This is because the system would otherwise have to pay for maintaining potentialities which it does not have occasion to use in its actual environment, but which manifest themselves in a less uncertain environment.

How important are these efficiency considerations in nature? They undoubtedly play a role in determining the directions of evolution, in particular at or below the population level. This is plausible since such considerations follow directly from the fact of competition in evolution.

VIII. Levels of Control

Naturally each of the general control mechanisms (and also abnormal behavior) may manifest themselves at different levels of organization, or in different compartments at the same level. These different manifestations may be quite diverse; nevertheless, they all contribute to Eq. (3).

Some particular examples of basic mechanisms have already been described. Now it is necessary to give a more systematic analysis of the way in which these mechanisms operate at different levels of organization. The present section provides an informal discussion of this question. The next section indicates how level structure is incorporated into Eq. (3).

A. Partial States

So far the ecosystem has been described in terms of a certain set of distinguishable states, or configurations of elementary components. However, in order to analyze control processes at different levels it is necessary to redescribe the ecosystem in terms of partial states, or states associated with compartments at these different levels.

The partial state of a compartment is given by (i) specifying its subcompartments at the next lower level and (ii) specifying its configuration in terms of these subcompartments, that is, without distinguishing differences within their boundaries. This makes it possible to describe uncertainty about the behavior of the ecosystem in terms of certain local uncertainties, or local entropies, associated with particular levels of organization or with particular types of compartments at some level of organization. These local entropies characterize the uncertainty about changes in the partial state (structure and configuration) of the compartment or level, but with the provision that this uncertainty is conditioned on appropriate information about changes at other levels of organization.

It is worth making this type of decomposition more explicit. Suppose the complete state of the ecosystem at time t is known. Then, how uncertain the answers to the following questions are at time $t + 1$ must be determined:

1. How is the biota broken up into populations and what is its configuration in terms of these populations?
2. How are populations broken up into organisms and what are their configurations in terms of these organisms?
3. How are organisms broken up into cells and what are their configurations in terms of cells?
4. How are cells broken up into large molecules and what are their configurations in terms of these large molecules?

5. How are large molecules broken up into components and what are the configurations of these molecules in terms of components?

The uncertainty associated with the physical environment may also be described in this way. For example, the environment may be divided into different regions of space and the question asked, how are these different regions broken up into components and what are their configurations in terms of these components? The configuration (spatial organization) of the ecosystem as a whole in terms of the biota and the physical environment is ordinarily uncertain as well.

The elementary components may also be in different states (i.e., be of different types). In general, it is assumed that the number of components is constant and that the number of each type is constant (conservation of matter).

Each of the above uncertainties characterizes the modifiability manifested by the ecosystem at the level in question (assuming that it is operating at capacity). In general these modifiabilities are not independent; that is, the answers to questions 1–5 are not independent. Later it will be seen how these dependencies are expressed and what their significance is.

The above decomposition could be made more refined. For example, trophic groups (say grazing and detritus populations), organ systems, subcellular organelles (such as the genome), and so forth could be included. The main advantage of considering these is that certain types of control processes concentrate in such subsystems.

It is also possible to ask questions about the levels of any particular compartment, for example, of a single population. In this case the remainder of the biota is regarded as part of the environment of this population.

B. Biological Examples

The significance of modifiability at different levels can be seen by taking examples from different levels of the biota. The behavior of the biota itself may be uncertain in terms of its populations. This may be compartmental uncertainty, associated with the way the biota is broken into populations; or it may be configurational uncertainty, associated with changes in the configuration of the biota in terms of its populations. The compartmental uncertainty is not so important since the number of populations is fairly definite (but the ecosystem may have more or less tolerance for the extinction of certain populations). The configurational uncertainty is more important. This is associated with variability in interspecies social relations (such as communication and symbiosis).

The behavior of populations may also be uncertain in terms of organisms. If this is compartmental uncertainty it is associated with adaptability (or reliability) mediated by redundancy of certain types of organisms or variations in the number of organisms in the population. Such variations may arise from inherent noise (fluctuation) or they may be induced by changes in the environment (culturability). If the uncertainty is configurational it is associated with the movement of organisms or variability of social organization within the population.

Similarly, the behavior of organisms may be uncertain in terms of cells. For example, reliability may be associated with duplication of organs, or physiological (short-term) and developmental (long-term) adaptability associated with variability in cell number or in the contiguity relations of cells. This is also true at the cellular level, in the form of redundant organelles, variability in cell size, or variations in enzyme constitution or biochemical pathways induced by the environment. The behavior of large molecules or organelles may also be uncertain in terms of basic components (or smaller molecules). This is especially important since the reshuffling of genes (by recombination, mutation, crossing over, etc.) appears as configurational and, to a lesser extent, compartmental uncertainty of the genome. This is the basis of genetic adaptability. Other configurational changes at the molecular level also play an important role: for example, allosterism in cellular regulation, or configurational changes in memory or immunity.

Naturally many of the changes at different levels are correlated; others are not, or at least not in any detectible way. For example, changes in cell size are certainly manifested at the organism level. From the practical standpoint, however, it is altogether possible for changes in the size of organisms to be virtually undetectable at the cellular level. Also, it is important that variation in components in the biota can manifest itself in terms of cell size, size of organisms, or number of organisms. Later it will be seen that these types of uncertainty, along with the system's capacity to take advantage of (or anticipate) compensating variations in the environment, play an important role in regulating the ecosystem.

IX. Hierarchies of Indeterminacy

Now it will be indicated how the discussion in the preceding section can be incorporated into Eq. (3). In general, the equations which result from this incorporation are complicated, since the structure of the ecosystem is complicated. For the present purposes, however, it will be sufficient to describe the general features of these equations.

The main problem is to write the transition scheme of the biota and physical environment as products of local, partial schemes which describe the behavior of compartments in terms of subcompartments at the next lower level. In fact, this is not so easy since these compartments are not necessarily present at any given time. This difficulty may be eliminated by introducing auxiliary, empty compartments, which either have disappeared or have some probability of appearing. Also, these compartments (real or virtual) can be aggregated into a grosser, but permanent, compartmental structure. The equations will be described in terms of this permanent structure.

The simplest such permanent structure is given by the level structure of the biota itself. In this case each level is a permanent compartment, with its finer compartmental structure contributing to its state. Naturally this permanent structure can be further refined to include different populations (since these are relatively permanent), trophic groups, different types or organs, cell types, organelles, etc. A typical permanent structure is illustrated in Fig. 1.

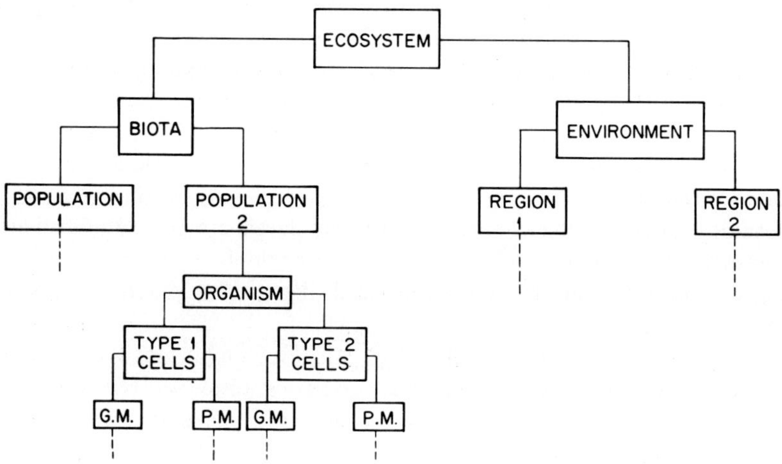

FIG. 1. Compartmental structure of a simplified ecosystem. The biota consists of two populations and the physical environment of two regions of space. The organisms of population 2 consist of two types of cells, each of which is described in terms of certain genotypic molecules (G.M.) and certain phenotypic molecules (P.M.). The substructure of population 1 as well as the types of components (at the bottom level) are not shown. Naturally the compartmental structure of any real ecosystem is much more complicated.

Each state of the permanent structure is a many-tuple whose component states are the states of each permanent compartment. Thus, the state of the biota is given by

$$\alpha_1 = ({}_1\alpha_1{}^n, {}_1\alpha_1^{n-1}, {}_1\alpha_1^{n-2}, {}_1\alpha_2^{n-2}, \ldots, {}_1\alpha_i^j, \ldots, {}_1\alpha_{m-1}^0, {}_1\alpha_m{}^0),$$

where the biota is compartment 1 at level $n-1$, ${}_1\alpha_i^j\,(j \leqslant n-1)$ is the partial state of the ith permanent compartment at the jth level of the biota, and ${}_1\alpha_1{}^n$ is the partial state of the ecosystem (compartment 1 at level n) in terms of the biota and physical environment. This state is here associated with the biota, and it is further assumed that there are m permanent subcompartments at level 0. Each of these states is associated with a partial transition scheme,

$$_1\omega_i^j = \{p[_1\alpha_i^j(t+1)|\beta(t)]\},$$

where the same notation is used for the transition schemes as for the states. Now the complete transition scheme of the biota can be written in terms of its partial schemes:

$$\omega_1 = \prod_{j=0}^{n} {}_1\omega_i^j,$$

where the product is taken over all possible values of the index i, and

$$_1\omega_p{}^q\, {}_1\omega_r{}^s \equiv \{p[_1\alpha_p{}^q(t+1),\, {}_1\alpha_r{}^s(t+1)|\beta(t)]\}.$$

Similarly, the complete transition schemes of the physical environment may be written as a product of local transition schemes, associated with particular regions of space:

$$\omega_2 = \prod_{j=0}^{1} {}_2\omega_i^j,$$

where the physical environment is compartment 2, the compartments at level 1 represent the regions of space, and the higher-level compartments are regarded as either degenerate or unchanging. This is useful when characterizing the environments of particular populations.

These product schemes may be substituted into Eq. (3):

$$\hat{H}\left(\prod_{j=0}^{n} {}_1\omega_i^j\right) - \hat{H}\left(\prod_{j=0}^{n} {}_1\omega_i^j \,\middle|\, \prod_{j=0}^{1} {}_2\omega_i^j\right) + \hat{H}\left(\prod_{j=0}^{1} {}_2\omega_i^j \,\middle|\, \prod_{j=0}^{n} {}_1\omega_i^j\right)$$
$$\rightarrow H\left(\prod_{j=0}^{1} {}_2\omega_i^j\right). \tag{4}$$

Each of these terms can now be expanded as sums of certain characteristic entropies:

$$\sum_{j=0}^{n} \hat{H}_e({}_1\omega_i^j) - \sum_{j=0}^{n} \hat{H}_e\left({}_1\omega_i^j \,\middle|\, \prod_{j=0}^{1} {}_2\omega_i^j\right) + \sum_{j=0}^{1} \hat{H}_e\left({}_2\omega_i^j \,\middle|\, \prod_{j=0}^{n} {}_1\omega_i^j\right)$$
$$\rightarrow \sum_{j=0}^{1} H_e({}_2\omega_i^j), \tag{5}$$

where the $H_e(\cdot)$ are the characteristic entropies. These are unique linear combinations of conditional terms which specify what were previously described as the appropriate information about other levels (or compartments). These combinations are developed by using the rules governing conditional probabilities to make all possible expansions of the terms in Eq. (4), and then normalizing the sum of these expansions. For example, in the two-term expansion

$$\hat{H}(\omega_u{}^v \omega_p{}^q) = \hat{H}_e(\omega_u{}^v) + \hat{H}_e(\omega_p{}^q),$$

the characteristic entropies are given by

$$\hat{H}_e(\omega_u{}^v) \equiv \tfrac{1}{2}\{\hat{H}(\omega_u{}^v) + \hat{H}(\omega_u{}^v | \omega_p{}^q)\}, \hat{H}_e(\omega_p{}^q) \equiv \tfrac{1}{2}\{\hat{H}(\omega_p{}^q) + \hat{H}(\omega_p{}^q | \omega_u{}^v)\}.$$

In the three-term expansion

$$\hat{H}(\omega_u{}^v \omega_p{}^q \omega_r{}^s) = \hat{H}_e(\omega_u{}^v) + \hat{H}_e(\omega_p{}^q) + \hat{H}_e(\omega_r{}^s),$$

the characteristic entropies are

$$\hat{H}_e(\omega_u{}^v) \equiv \tfrac{1}{3}\{\hat{H}(\omega_u{}^v) + \tfrac{1}{2}\hat{H}(\omega_u{}^v | \omega_p{}^q) + \tfrac{1}{2}\hat{H}(\omega_u{}^v | \omega_r{}^s) + \hat{H}(\omega_u{}^v | \omega_p{}^q \omega_r{}^s)\},$$

$$\hat{H}_e(\omega_p{}^q) \equiv \tfrac{1}{3}\{\hat{H}(\omega_p{}^q) + \tfrac{1}{2}\hat{H}(\omega_p{}^q | \omega_u{}^v) + \tfrac{1}{2}\hat{H}(\omega_p{}^q | \omega_r{}^s) + \hat{H}(\omega_p{}^q | \omega_u{}^v \omega_r{}^s)\},$$

$$\hat{H}_e(\omega_r{}^s) \equiv \tfrac{1}{3}\{\hat{H}(\omega_r{}^s) + \tfrac{1}{2}\hat{H}(\omega_r{}^s | \omega_u{}^v) + \tfrac{1}{2}\hat{H}(\omega_r{}^s | \omega_p{}^q) + \hat{H}(\omega_r{}^s | \omega_u{}^v \omega_p{}^q)\}.$$

This may be continued for higher-order expansions, except that the general formula is quite complicated.

The main feature of the characteristic entropy is that the leading term represents the uncertainty associated with modifications of the level or compartment in question. The relative contributions of each of the remaining (smaller) terms describes the degree to which this modifiability is independent of the modifiability of other levels or compartments. The difference

$$\hat{H}_e({}_1\omega_i{}^j) - \hat{H}_e\left({}_1\omega_i{}^j \middle| \prod_{j=0}^{1} {}_2\omega_i{}^j\right)$$

describes the amount of environmental disturbance which the compartment can absorb. This is the adaptability of the compartment (or level).

X. Generalizations about Biological Compensation

Equation (5) expresses a general law of biological compensation. This may be stated in these words: *If the adaptability of one compartment or level is restricted this must be compensated by an increase in the adaptability of other compartments or levels or by an increase in selective (or nonselective) indifference to the environment.*

This can be reworded, using the distinction between unconditional and conditional contributions to the characteristic entropies: *If the modifiability of* *one compartment or level is restricted this must be compensated by an increase in the modifiability of other compartments or levels, by greater independence in these different modes of modifiability, by an increase in selective (or nonselective) indifference to the environment, or by better anticipation of the environment.**

Clearly, if the adaptability of some compartment (or level) is large it absorbs a large fraction of the environmental disturbance. In this case the compartment (or level) is regulating. Similarly, if the adaptability is small the compartment (or level) is regulated. In general, if adaptability is large, modifiability and independence are large. Also, if adaptability is small, modifiability is likely to be small. This is not necessarily the case, however, since a certain amount of "microscopic" modifiability may be associated with noise and reliability processes.

Naturally the unconditional (modifiability) and conditional (independence) terms may make a relatively larger or smaller contribution to the characteristic entropy. If the unconditional term decreases, but the conditional terms are constant, this means that the regulatory capabilities decrease because the compartment cannot tolerate as much modification. If the unconditional terms remain the same, but the conditional terms decrease, this means that the regulatory capabilities decrease because the disturbances have wider ramifications in the system; essentially the behavior of the compartment is coordinated to the behavior of other compartments. In general, the compartment makes the biggest contribution to regulation if both the unconditional and conditional terms are large. The compartment is completely regulated if the unconditional term (and therefore the conditional terms) are zero.

The importance of ramification within the system is particularly apparent in the special relationship between genotype and phenotype. A modification of the genotype (sequence of bases in DNA) has ramifications throughout the whole system—through the cell, the organism, even the community. However, the reverse is not true. This is the strong principle of inheritance. In statistical terms this means that the configuration of hereditary molecules (sequence of bases) is quite certain given the behavior of phenotypic compartments, but not conversely. This is because the phenotypic compartments may have many possible modes of behavior consistent with a given genotype. Essentially the hereditary compartments describe the phenotypic compartments, but only in

* This must be qualified, however, since the inherent entropies are not uniquely determined by uncertainty of the environment. Thus, the reallocation of these entropies to different levels or compartments may be associated with some change in their absolute magnitudes. This is possible if there is some change in reliability or if the new modes of adaptability require higher characteristic entropies to be effective. However, these are secondary effects.

the weak sense that they determine the possible behavior of these compartments.

The internal milieu of the genome itself—its configuration in terms of DNA and other molecules—may be quite constant. In this case the behavior of the genome compartment is highly regulated, in part because of uncertainty in the configuration of genetic molecules themselves. This is possible because these modifications have ramifications which ensure that at least some members of the population maintain the right internal milieu. Also, the direction of this ramification prevents disturbances of the environment from being communicated to the genotype.

The different possible modes of phenotypic behavior also absorb disturbances (and therefore contribute to the maintenance of this milieu). The way in which these modifications ramify within the phenotypic compartments is again important. In general, if the phenotypic compartments are more dependent on one another—for example, if the system is more centralized—the modifiability must increase if the system is to have the same behavioral uncertainty. In some cases a complicated system can decrease its overall modifiability by isolating subsystems of interdependent parts—for example, by developing specialized organ systems or levels of organization with reasonably independent behavior. This allows the system to take advantage of a complex or centralized form of organization, while at the same time localizing its modifiability within this organization.

The generalization about biological compensation gives some insight into the way populations are bound to their niches (or local environments). Ordinarily, a restriction on one mode of adaptability must be compensated by another mode of adaptability or by an increase in the population's selective indifference to the regions of space which it occupies. This means that a system which is externally specialized (has many requirements, few sources of food, greater sensitivity to physical conditions) is just as bound to its region of space as a system which is not so adaptable but which is externally unspecialized. Also a centralized system with a high degree of modifiability may be just as bound to its region of space as a decentralized system with much less modifiability. This will happen if the adaptability is the same in both cases.

XI. The Allocation of Control

A. General Considerations

The ecosystem (or a population) must fulfill the overall tendency described by Eq. (5). This is the basis for the discussion of biological compensation. The principle of compensation, however, does not determine how entropies are allocated to different populations in an ecosystem or to different levels of organization within these populations. In general, this is a problem of optimization. Each control process is more or less costly to a biological system. Also, certain mechanisms of control, or control mediated by certain levels of organization, may be more appropriate for contending with particular types of disturbances. The interplay of factors here is ordinarily quite complex. However, there are certain biological constraints which are of particular interest.

The most important factors which determine the allocation of uncertainty in a system are its size and complexity (the diversity and interdependence of its parts) and the special relationship between genotype and phenotype. Diversity and interdependence are often an advantage to a system since this means specialization of labor. This advantage has a price, however, since it increases the costs of adaptability. In particular, constitutive adaptability is always expensive. Systems which contend with environmental disturbances in this way are inefficient. This is increasingly true as any unnecessary machinery becomes more complicated, and therefore difficult to construct and maintain. In any case, if the statistical structure of the environment is such that the system can forego permanent machinery, it will do so. This is possible if the constitutive adaptability is replaced by inducible or selective adaptability. However, these also become more expensive as the system becomes more complex. This is because it is more difficult to modify a system of diverse and dependent parts than a system with a few types of relatively independent parts. In particular, the more complicated forms of organization are less amenable to alternate modes of development; essentially it takes much more information and energy to reorganize a complicated system than a simple one. Also, it is more difficult to modify the genetic description of such systems without producing lethal or subfunctional results. This is because modifications of the genetic description manifest themselves as modifications at all other levels of organization.

It has already been seen that these generalizations have certain important, although to some extent apparent, exceptions. A complex system might have special organs which allow for inexpensive inducible or selective adaptability. Such organs may be locally isolated from the genetic description. Furthermore they may make it possible to construct a really large number of viable modifications or build up and tear down temporary machinery with a smaller energy expenditure. However, these organs increase complexity by imposing new interdependencies. The system must therefore pay for this inexpensive adaptability with further restrictions on the genetic and developmental modes of adaptability. It must pay in terms of the initial costs of construction as well.

It is also possible for the system to bypass some of these adaptability mechanisms by relying on protective structures or other selective indifference. The costs of such mechanisms depend on the particular morphology of the organism and the degree to which they restrict it from using various features or resources of the environment.

B. Patterns of Control in Populations

These general relationships are very clear over the broad range of biological organizations. The unicellular organisms are adaptable at all levels of organization. Their relatively simple organization allows for considerable genetic interchange as well as mutation combined with haploidy and high growth rates. Also, such systems can exhibit considerable inducible adaptability at the physiological level: for example, inducible enzymes, dormancy, and so forth. The high potential growth rate combined with dormancy allows for culturability at the population level. Naturally, particular unicellular organisms may not utilize all these modes of adaptability, but it is possible for them to do so. This is not true for more complicated systems such as multicellular plants and animals. The multicellular plants have a relatively simple morphology which allows for both genetic and developmental plasticity. This is because the simple morphology is consistent with an open growth system and therefore with extensive alterations in cell size and number. However, since the multicellular organism must be larger than its unicellular counterpart, its potential reproduction rate must be smaller. This interferes with culturability at the population level and to some extent with genetic adaptability as well. Multicellular animals have more complicated morphologies, with concomitant closure of the growth system. This precludes the high degree of genetic and developmental plasticity found among plants. Such systems must develop special social organizations or special organs of adaptability, organs such as the neuromuscular or immunity systems, to deal with environmental variability.

C. The Ecosystem as a Whole

Now the question is, how do the different patterns of control within populations interlock with each other and with control processes at higher levels to regulate the ecosystem as a whole? Here again the interference between adaptability and efficiency requires the system to find some compromise between the independence of its parts and its overall modifiability. The stability of the ecosystem (its resistance to disturbance) increases as each population becomes more independent, or performs more of the essential ecosystem functions by itself. However, this is inefficient since greater independence means less specialization of labor (or more constitutive and inducible adaptability within each population). The system can also achieve stability, while maintaining internal specialization, if it can tolerate wider ramification of disturbances. However, this is inefficient since it means an increase in total modifiability of the ecosystem. In fact, what might be expected is a compromise situation, one with overlapping multifunctionality allowing for some degree of modifiability and independence, but organized in such a way as to achieve balance and also to rectify modifications (or correct errors) as soon as they occur. This compromise implements itself in the course of evolution and succession in three main ways.

First and most important is the development of balance itself. This contributes to regulation since it reduces disturbances which arise from modification of the physical environment by the biota. From the practical standpoint these disturbances may be described as being of environmental origin. In this sense the balanced ecosystem may be regarded as one in which the biota regulates the physical environment. In general, the development of balance requires an increase in the number of functions performed by the ecosystem and therefore increases the importance of specialization of labor.

The second main type of regulation involves the allocation of adaptability to different types of populations or trophic groups within the biota. The biota is perturbed by variations in energy input, by internal failures in the transformation of energy, and by variations in the physical conditions affecting the use of energy. This is true whether or not the ecosystem is balanced. The biota may respond to these disturbances by physiological adaptability (variations in the size or activity of individuals, or adaptability to conditions affecting the use of energy), or by variations in the size of populations. Local ecosystems (or communities) will be more efficient if they allocate the latter variations to organisms which are small and relatively simple. These construct themselves with much less energy and have higher growth rates. In short, they are culturable; that is, variations in their numbers contribute to adaptability of the ecosystem rather than to its injury. This implies that detritus pathway populations serve as a regulating element for the larger, more complex, grazing pathway organisms. This large size, in turn, facilitates physiological adaptability within these populations. It also regulates the number of grazing pathway organisms.

The third type of regulation involves the development of multifunctionalities (or selective despecializations) which refine the general patterns of allocation discussed above. Variability in the size of organisms or populations may be reduced if the system takes advantage of compensating variations in the environment. This is possible, for example, if organisms or populations have a large number of food sources, where these sources respond differently to different environmental conditions. This allows for adaptability since the utilization of such compensating variations amounts to the recognition of more statistical regularities in the environment.

Naturally the fact of compensating variations allows for a smaller amount of energy to be shunted to detritus pathways, and consequently for some reduction in the fluctuation associated with detritus populations. The amount of energy shunted may be further reduced if the system uses intra- and interspecies communication (e.g., density-dependent mechanisms) to take better advantage of these compensating variations. If such communication processes can evolve, this allows for some of the regulation associated with the fluctuation of detritus populations to be replaced by regulation at the social level; or it allows for the same amount of regulation, but with a decrease in the number of possible sources of food for each organism or population.

The recognition of compensating variations is what makes it possible for the variability of energy flow to stabilize the ecosystem (cf., Conrad, 1972b). Naturally this stability is increased if the ramification and multiplication of external disturbances is minimized, that is, if they are rectified by the first systems which are affected by them.

It is important that the regulatory processes discussed above increase the stability of the ecosystem in the sense that they make it more resistant to a certain repertoire of disturbances. These processes are based on coadaptations, that is, adaptations which confer advantages on populations or communities, in some cases at the expense of individuals (Conrad and Pattee, 1970). Such coadaptations put another restriction on the genetic variability of populations. Essentially the populations of a mature ecosystem must restrict their genetic plasticity or develop conservative mechanisms which maintain these coadaptations. However, this makes the ecosystem more unstable in the evolutionary sense, since it becomes less able to cope with disturbances which go outside of a certain range.

XII. Conclusions

The approach of this chapter has been to describe the behavior of biological systems in terms of a hierarchy of indeterminacies. This statistical description embraces all the different control characteristics of the system. Indeed, aside from thermal isolation, all these control characteristics are statistical properties appearing in different guises. In particular, they may be mediated by different mechanisms (constitutive, inducible, selective, and so forth). Or they may be mediated by different compartments or levels of organization, compartments or levels more or less coordinated to particular biological functions.

In general, if one of these guises (mechanisms or modes of control) is restricted, this must be compensated by others. These restrictions are extra conditions, conditions which must be put into the formalism by hand on the basis of biological intuition. The most important restrictions depend on the complexity of the system. In many cases this complexity may provide considerable competitive advantage. However, it cannot evolve without a concomitant development of new modes of coping with the environment: for example, modes of adaptability mediated by special organs or social organizations. This concomitant evolution is a necessity, a kind of local orthogenesis in the sense that other organizational possibilities are precluded on the grounds of efficiency.

These general patterns of regulation all fit into the larger framework of the ecosystem as a whole, characterized by the development of more stable, climax forms of organization. Sometimes this stability is discussed in dynamical terms. However, it may be more practical to treat it in statistical terms, along the lines indicated. This is because all the regulatory processes in biology contribute to stability of the ecosystem. The essential problem is to consider systematically the relations among these different processes, the rates at which they might be expected to appear in the course of succession, and how these different relations and rates are modified in different types of environments or environments of differing predictability.

XIII. Summary

This chapter has considered the question: how is the uncertainty in behavior of biological systems related to the unpredictability of their environment? Essentially the answer to this question is: the adaptability of the system plus its potential indifference to the environment must be greater than the uncertainty of this environment, and tends in the direction of equalling this uncertainty in the course of evolution.

The adaptability of the system is defined as the potential uncertainty in its behavior minus this potential uncertainty given the behavior of the environment. This adaptability may be mediated by constitutive, inducible, or trial and error mechanisms. Also it may manifest itself in terms of uncertainty at different levels of organization: the molecule, the cell, the organism, the population, or the ecosystem as a whole.

The adaptability of compartments at each of these levels is determined by their modifiability and the degree to which this modifiability is independent of other compartments, at the same or different levels. In general, if adaptability at one level is restricted, this must be compensated by adaptability at other levels or by an increase in the indifference to the environment. This generalization about biological compensation has implications for niche binding, patterns of biological control in different populations, the sequential change of these patterns in succession and evolution, and also for the ramification (or etiology) of disturbances within populations and communities.

Acknowledgment

I would like to thank Dr. W. Güttinger for helpful comments.

REFERENCES

Ashby, W. R. (1956). "An Introduction to Cybernetics." Wiley, New York.

Conrad, M. (1972a). *Curr. Mod. Biol.* **28**, 393.

Conrad, M. (1972b). *J. Theoret. Biol.* **34**, 325.

Conrad, M., and Pattee, H. H. (1970). *J. Theoret. Biol.* **28**, 393.

Kinchin, A. I. (1957). "Mathematical Foundations of Information Theory." Dover, New York.

Pattee, H. H. (1968). *In* "Towards a Theoretical Biology" (C. H. Waddington, ed.), Vol. 1, pp. 69–93. Edinburgh Univ. Press, Edinburgh, Scotland.

Shannon, C. E. (1948). *Bell Syst. Tech. J.* **27**, 379, 623.

Winograd, S., and Cowan, J. D. (1963). "Reliable Computation in the Presence of Noise." MIT Press, Cambridge, Massachusetts.

17

Propagation of Cause in Ecosystems

BERNARD C. PATTEN, ROBERT W. BOSSERMAN, JOHN T. FINN AND WILLIAM G. CALE*

DEPARTMENT OF ZOOLOGY AND INSTITUTE OF ECOLOGY
UNIVERSITY OF GEORGIA, ATHENS, GEORGIA

University of Georgia, *Contributions in Systems Ecology*, No. 30.

* Present address: Institute for Environmental Sciences, The University of Texas at Dallas, Richardson, Texas.

I. Introduction

In the search for a general systems theory of ecosystems, ecology must explore and reconsider fundamental issues of science contemplated by scientists and philosphers for centuries without resolution. One of these issues is cause and effect, and the nature of determination in concrete reality. This chapter proposes to examine the doctrine of causal determinism as a basis for or possible component of a suitable ecosystem paradigm. It will do so by reviewing the broad outlines of causalism in the context of modern state space system theory, and aligning the two into some semblance of conformity.

State theory is a mathematical model of becoming. Like all mathematical models it is at base a syntactical construct devoid of semantic content. The mathematician is interested in the former aspect; de Broglie is said to have commented that equations remain while their interpretations pass away (Bunge, 1959, p. 77). But the semantics of mathematics is more of interest to scientists than syntax, prompting Bertrand Russell (1927, p. 122) to observe that "Scientific laws can only be expressed in differential equations." The law of exponential change, $\dot{x} = kx$, giving $x(t) = x_0 \exp(kt)$ on $[0, t]$ for constant k, applies equally to radioactive decay, extinction of monochromatic light in a homogeneous medium, growth of populations, and turnover of compartments. For specific applications it is not the general law that is of interest but rather meanings attached to the elements of the mathematical description.

For example, the state space model expressed in terms of differential equations is an abstraction,

$$\dot{\mathbf{x}}(t) = \mathbf{f}[\mathbf{x}(t), \mathbf{p}(t), t], \qquad \mathbf{x}(t_0) = \mathbf{x}_0, \quad t \in [t_0, t_1], \tag{1}$$

$$\mathbf{y}(t) = \mathbf{g}[\mathbf{x}(t), \mathbf{p}(t), t], \qquad t \in [t_0, t_1], \tag{2}$$

where $\mathbf{x}$ and $\mathbf{y}$ are sets (vectors) of *state* and *output* variables, respectively, $\mathbf{p}$ is a *parameter* set, t is time, varying continuously from initial t_0 to final t_1, and **f** and **g** are *state transition* and *response functions,* respectively. A syntax is formed given **f** and **g**, with greater or lesser interest to mathematicians depending on the nature of the functions. Interpreting the elements of the model, on the other hand, endows it with a semantic content that may be both subtle and intricate scientifically.

This chapter will endeavor to explore the semantic range of Eqs. (1) and (2) in their depiction of causality. General determination and causation as one of its categories are reviewed. The causal connective and causative sequences are elaborated by a set of definitions and theorems which will serve to identify cause-and-effect aspects and implications of the state model. State is reconciled to the maxim, "Being is becoming based on what has been," in a developmental summary of a modern general system theory, that of L. A. Zadeh (Zadeh and Desoer, 1963; Zadeh, 1969). The *abstract object* concept of Zadeh, equated to Koestler's (1967) similar notion of *holon*, is made an ecological object by endowing it with an *environment.* Following Mesarovic and Takahara (1975), the holon is conceptualized as a pair of explicit objects, *genon* (output object) and *creaon* (input object), which serve different but reciprocal purposes in the propagation of cause in systems. The genon has a generative or productive role in transforming cause into effect, while the creaon serves a creative or selective function in defining object–environment consistency and compatibility. These distinguishably different aspects of the holon are illustrated by a matrix analysis of conservative, dissipative flows in several small ecosystem models. The results suggest an explanation in abstract system theory for the intuitive human perception of harmony, order, and community in nature, a perception that somehow should be incorporated into any general systems theory of the ecosystem.

II. Causal Determinism

This section is a digest, with modifications, of the philosophical synthesis of causality by Bunge (1959). It provides necessary background and context for more formal material to follow.

A. Meanings of Causality

Causality is at once a category, a principle, and a doctrine. The category, *causation,* refers to the causal bond, cause → effect ($C \rightarrow E$). The principle, which is not easy to specify (Section II. F), basically states "same cause, same effect." The doctrine, *causalism,* holds that the causal principle is exclusively valid over other principles of determination.

Causation is an epistemological category because it concerns experience and knowledge of the human observer without necessarily being otherwise a property of reality. Regarding its ontological status, Bunge concluded (1959, p. 6) "that causation is not [solely] a category of relations among ideas, but a category of connection and determination corresponding to an actual trait of the factual (external and internal) world." The causal bond can thus be taken also as an ontic category.

B. Meanings of Determination

There are three different aspects of determination, all of which are embodied in the state space model as shown later (Section III):

(i) As a property or characteristic, a thing (object) is determinate if it possesses attributes by which it can be characterized unambiguously. Mere existence or recognition of an object presumes this form of determination.

(ii) As necessary (constant and unique) connection, determination denotes relationship or association between entities. For example, in Eq. (1) the value of $\mathbf{x}(t)$ is determined by the value of t. The element of production is not present, however; t does not "produce" or "evoke" $\mathbf{x}(t)$. The function $\mathbf{f}$ only defines a constant, unique relation between t and $\mathbf{x}(t)$.

(iii) As a way of becoming, determination concerns the means of acquiring characteristics. These means are lawful rather than arbitrary, and they develop out of preexisting conditions.

C. Categories of Determination

For the purpose of identifying the range of determination as a concept, and not to provide a classification scheme with mutually disjoint categories, the following forms of determination are noted:

1. *Quantitative self-determination.* This is determination of the consequent by the antecedent. "States" that differ in quantitative respects only unfold continuously, as in the succession $[x_0, x(t)]$ on $[0, t]$ generated by $x_0 \exp(kt)$.
2. *Causal determination (causation).* A unique effect is produced by an external ("efficient") unique cause.
3. *Interaction (reciprocal causation, mutual interdependence).* Determination by mutual (internal) action.
4. *Mechanical determination.* A combination of the above three, with energy or matter serving as agents of determination.
5. *Statistical determination.* The consequent is determined by joint action of multiple independent antecedents, often large numbers of them individually determined by other types of determination.
6. *Structural (holistic) determination.* Determination of the parts by the

whole. The behavior of a part is determined by the overall structure of the collection to which it belongs.

7. *Teleological determination.* The determination of means by goals or ends.

8. *Dialectical determination (qualitative self-determination).* Determination by inner strife and eventual synthesis of essential opposite components. In biology the struggle for existence, survival of the fittest and subsequent synthesis of opposites such as autotrophs and heterotrophs into biotic communities, or resistant and resilient populations into stable ecosystems (Patten, 1974; Webster *et al.*, 1975), serve as examples.

These different, nonmutually exclusive categories of determination illustrate, as Bunge intended, that causation occupies only a small place in general determinism.

D. Principle of Determinacy

General determinism is a doctrine based on two independent but consistent principles. The *principle of lawfulness (orderliness)* states that there are laws, that every event is lawful (not unconditional, irregular, lawless, or arbitrary), and that every (epistemological) fact or (ontological) object is the locus of a set of laws. The *genetic principle,* or *principle of productivity,* holds that nothing comes out of nothing or passes into nothing (Lucretius, ca. 58 B.C.). These principles combine to form the *principle of determinacy*: Everything is lawfully produced by something else and lawfully produces something else. *Omne quod movetur ab alio movetur* (Aristotle, 1930).

E. Categories of Cause

Aristotle (1947) wrote, "There are four recognized kinds of causes. Of these we hold that one [*causa formalis*] is the essence or essential nature of the thing . . .; another [*causa materialis*] is the matter or substrate; the third [*causa efficiens*] is the source of motion; and the fourth [*causa finalis*] is the cause which is the opposite to this; namely the purpose or 'good'; for this is the end of every generative or motive process." Modern science in its concepts of input, stimulus, excitation, perturbations, etc., recognizes only efficient cause as an extrinsically acting agent (input). Material cause is taken for granted, and formal and final causes are beyond empirical demonstration.

F. Principle of Causality

Galileo (1623) defined efficient cause as the necessary and sufficient condition for the appearance of something: "that and no other is to be called cause, at the presence of which the effect always follows, and at whose removal the effect disappears." Symbolically, $C \Rightarrow E$ and also $\sim C \Rightarrow \sim E$. Bunge (1959, p. 34) criticized this Galilean concept as (i) too inclusive, admitting everything as cause that could possibly influence the effect, and (ii) too general, giving the set of conditions for occurrence of an event of whatever kind produced by any sort of determinate process. To specify the causal connection $C \rightarrow E$ more exactly, Bunge (1959, p. 35 ff.) elaborated a logical series of progressively more precise formulations, as outlined below.

1. *Constant conjunction.* Exceptions to the Galilean formulation are prevented by recognizing both C and E as sets. Then, a constant conjunction statement of cause is "If C happens, then E *always* happens." Symbolically, $\forall C$ and $\forall E$, $C \Rightarrow E$. The cause as a class is sufficient although not necessary for the effect, which is also a class. This formulation of causality is lawful ("If C" states the conditions), asymmetric (C is existentially prior to E, although not necessarily before it in time), and constant (exceptionless, as provided by the $\forall$ operator). But, it lacks uniqueness and productivity.
2. *Constant production.* Genetic linkage is outside the logic domain and must be introduced factually: "If C happens, then E is always *produced* by it." The logic is still $\forall C$ and $\forall E$, $C \Rightarrow E$.
3. *Necessary conjunction.* Uniqueness may be introduced into the constant conjunction formula by incorporating necessity: "If C happens, then *and only then E* always happens." That is, $\forall C$ and $\forall E, C \Leftrightarrow E$. The cause is now necessary (constant and unique) as well as sufficient for the effect.
4. *Necessary production.* Uniqueness and the genetic principle are both accounted for by "If C happens, then and only then E is always produced by it." Logically, $\forall C$ and $\forall E$, $C \Leftrightarrow E$. This is the working formulation of the *principle of causality* arrived at by Bunge (1959, p. 48), and will be adopted hereinafter as the meaning attached to the causal bond, $C \rightarrow E$.

G. Spatiotemporal Aspects of Cause

1. *Spatial Contiguity*

Production by contact would appear to be necessary for causation. However, spatial contiguity and causality are logically different categories, and since cause–effect at a distance cannot be empirically tested, it remains a hypothesis. Cause is consistent with contiguity, but does not require it.

2. *Temporal Antecedence*

If C and E are spatially separated, they cannot occur simultaneously; otherwise, they can. Therefore, in general no temporal priority of cause over effect is required and causality is compatible with instantaneous $C \rightarrow E$ links. Again, antecedence and causation are logically independent but consistent categories. If time precedence of C is not necessary, however, existential priority is. The cause must exist for the effect to occur. This follows from the conditional nature ("If C, . . .") of each formulation of causation noted in the preceding section, and is incorporated into the state space model by the property of "nonanticipation" (Section III. G).

H. Causality and Succession

Causation has been considered reducible or equal to uniform, invariable succession in time. Comte (1830) observed that the human mind does not seek to know the intimate causes of phenomena, but only "their effective laws, that is, their invariable relations of succession and similitude." Mill (1843) wrote that "the Law of Causation, the recognition of which is the main pillar of inductive science, is but the familiar truth that invariability of succession is found by observation to obtain between every fact in nature and some other fact which has preceded it." A causal principle based on temporal antecedence and succession can thus be framed as follows (Bunge, 1959, p. 69): "The states of a (closed) system unfold themselves in time in a unique and continuous way [and] . . . the same initial state is always followed by the same final state."

Many ecological models represented by homogeneous differential equations, including the ubiquitous Lotka–Volterra system,

$$\dot{x}_i = \left(a_i - \sum_{j=1}^{n} b_{ij} x_j\right) x_i, \qquad i = 1, \ldots, n,$$

tacitly assume this successional definition of causation. Prior states are considered sufficient to keep such systems moving through time. Later, analysis of the state space model and of cause propagation will reveal that, while states may be antecedents of later states, they are not causal. Causes are both necessary and sufficient to produce their effects and, as will be indicated, states are only necessary. They do not have a productive nature of their own, but are themselves produced by determiners from internal and external processes. They are the agent of cause, the carrier of cause into effect, but not cause–effect itself.

I. Externality of Causation

Internal causes and inward concepts of causation have been employed in the past. Bruno (1584) distinguished *principle* as "that which intrinsically concurs to the constitution of the thing and remains in the effect," from *cause* as "that which extrinsically concurs to the production of things." Self-causation, *causa sui,* occupies a diminished role in modern science although, again, it appears as an implicit assumption wherever homogeneous differential equations are used to depict causal phenomena, as in classical ecological population models. In general in science, intrinsic determinants are relegated to other categories of determination than causation, or perhaps associated with formal and material causes in which case they may bear some relationship to state. Innatism is not an accepted part of science, and in biology, for example, vitalism as inherent life force has long been put aside as a valid doctrine.

Causal determinism asserts the universal operation of efficient cause, taken by definition to be extrinsic. To sustain an effect the cause must be maintained, and removal of the cause is necessary and sufficient for disappearance of the effect: $\forall C$ and $\forall E$, $\sim C \Leftrightarrow \sim E$. *Causa cessante cessat effectus* (Aquinas, ca. 1272).

J. Causality and Functionalism

The functionalist view of causality holds that the causal bond can be replaced by mathematical relations expressing *functional dependence.* Accordingly, the arguments of $\mathbf{f}$ and $\mathbf{g}$ in Eqs. (1) and (2), for example, would be considered causes of $\dot{\mathbf{x}}$ and $\mathbf{y}$, respectively. The problem with this viewpoint is that, while functions have syntax and are therefore lawful, they lack semantics and thus the genetic quality. Functional dependencies are not a priori causal, but they can represent causation by agreement that the depicted relations are productive.

The functional model of objects and systems presented in Section III will be understood to express causal determination.

K. Causal Nexus of Systems

The causal bond $C \rightarrow E$ is only the elementary unit of influence propagation in systems, abstracted from causal chains (serial sequences), and these in turn from causal nets (with branching, convergence, and feedback). The purpose in subsequent sections will be to examine first the causal bond (Section III), then the causal sequence (Section IV), and finally the causal network (Section V) of systems.

L. Multiple Causation

If C and E are finite sets of causes and effects, then the necessary production formulation of causation, logically $\forall C$ and $\forall E$, $C \Leftrightarrow E$ (Section I. F), encompasses both plurality of causes and diversity of effects:

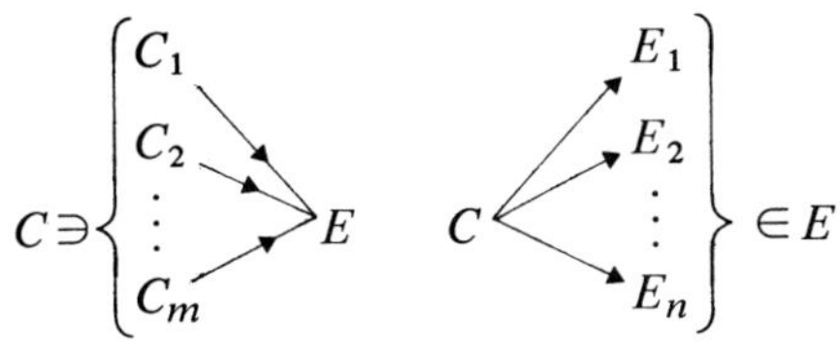

This is the reality of the causal nexus in concrete systems.

In Bunge's view, multiple causation is not strictly causal as uniqueness ($\forall C$ and $\forall E$) applies to sets and cannot be analyzed further to set elements. *Same cause, same effect,* says the ancient dictum, not *same causes, same effects.*

> As soon as the possibility of a plurality of causes (or of effects) is admitted, the picture of the causal chain ceases to be a plausible model of becoming—or at least loses its meaning after a few branchings. In the second place . . . multiple causation is nonadditive: the joint action of causes that might each alone produce the effect does not alter the result. Third, multiple causation goes over into statistical determination as soon as the set of possible causes attains a sufficient degree of complexity . . . (Bunge, 1959, pp. 124–5).

As shown in the next section, the problems of multiple causation and other obscure aspects are largely resolved when emphasis is shifted from a static to a dynamic model of causality, and the clarification that results stands as a major achievement of modern general systems theory.

III. The Causal Bond

"Being is becoming based on what has been." The *state* of a system is the link between its past and future existence. To understand causality it is necessary to understand the role of state in transforming cause into effect. The state concept occupies a central position in all modern system theories (e.g., Ashby, 1952, 1956; Bertalanffy, 1968; Gill, 1962; Kalman *et al.*, 1969; Klir, 1969, 1972; Mesarovic, 1964; Mesarovic and Takahara, 1975; Padulo and Arbib, 1974; Windeknecht, 1971; Wymore, 1967; Zadeh and Desoer, 1963; and Zadeh, 1969). One of the clearest expositions for ecological applications is that of Zadeh in the last two cited references. The relevant features of Zadeh's theory, modified in accordance with ideas of others, are summarized in this section by way of clarifying the nature of the causal bond in the framework of dynamical systems.

A. General Systems

A *system* may be defined as a collection of interacting or interdependent objects. The component objects are identified by an external observer on the basis of certain attributes. These attributes may be measurable or nonmeasurable, but together they form the object properties with which an external observer must be concerned. Thus, an object $\mathcal{H}$ is an abstract entity associated with or discriminated by a set of attributes $v_1, v_2, \ldots, v_m$, and a set of relations between these attributes. Zadeh refers to such a unit as an *abstract object*; the defining variables are termed *terminal variables* and the defining relations are *terminal relations.* Koestler (1967), in connection with hierarchical systems, introduced the term *holon* to denote subsystems at any level in a system hierarchy. Overton (1975) adopts the word holon to mean objects which are simultaneously part of a greater whole, and individualistic in their own right. The holon becomes the building block of hierarchically organized systems, and Overton formulates it in terms of two model forms based on two of Klir's (1969) five definitions of systems. Because of the correspondences between Zadeh's and Klir's theories, the holon can be usefully equated to Zadeh's abstract object (Fig. 1) even though there is no hierarchical connotation in the latter concept. In this way Zadeh's object becomes distinguished as a conceptually rich formal structure from ordinary "objects" of everyday usage.

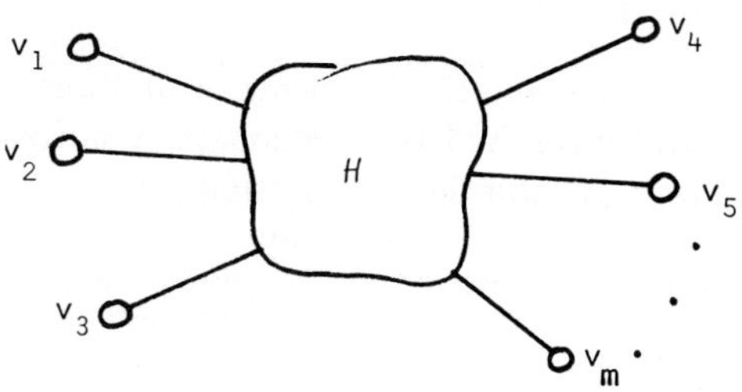

FIG. 1. The abstract object (Zadeh and Desoer, 1963) or holon (Koestler, 1967), a collection of terminal variables: $\mathcal{H} = \{v_1, \ldots, v_m\}$.

A *system* $\mathcal{H}$ can then be defined (Zadeh and Desoer, 1963, p. 65) as a partially interconnected set of holons (abstract objects) $\mathcal{H}_i$. This corresponds to the mathematical definition of a *general system* as a relation on nonempty (abstract) sets.

$$\mathcal{H} \subset \times\{\mathcal{H}_i : i \in I\}, \tag{3}$$

where $\times$ denotes Cartesian product and I is the index set (Mesarovic and Takahara, 1975, p. 11). When I is finite, Eq. (3) can be written

$$\mathcal{H} \subset \mathcal{H}_1 \times \cdots \times \mathcal{H}_n. \tag{4}$$

B. Input–Output Systems

The holon becomes *directed* or *oriented* (Fig. 2a) when its terminal variables are partitioned into inputs and outputs. That is, let $I_z \subset I$ and $I_y \subset I$ be defined

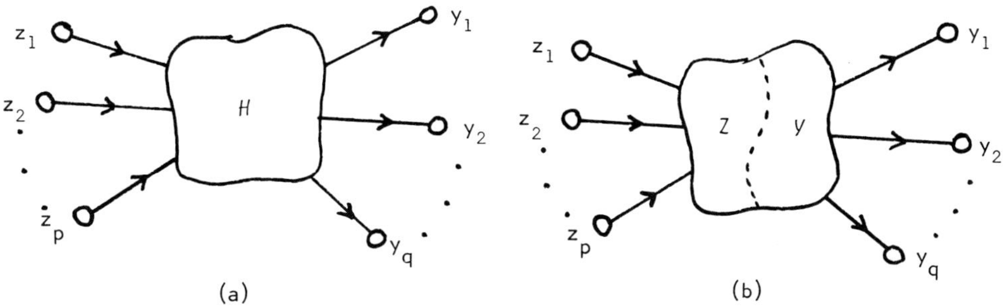

FIG. 2. (a) Oriented holon, with terminal variables $v_1, \ldots, v_m$ partitioned into inputs, $\{z_1, \ldots, z_p\}$, and outputs, $\{y_1, \ldots, y_q\}$, $p + q = m$. (b) Partition of the oriented holon into input object $\mathscr{Z}$ (creaon) and output object $\mathscr{Y}$ (genon).

such that $I_z \cap I_y = \phi$ and $I_z \cup I_y = I$. The set

$$\mathscr{Z} = \times\{\mathcal{H}_i : i \in I_z\} \tag{5}$$

is termed an *input object,* and

$$\mathscr{Y} = \times\{\mathcal{H}_i : i \in I_y\} \tag{6}$$

is an *output object* (Fig. 2b), as defined by Mesarovic and Takahara (1975, p. 11). The system $\mathcal{H}$ is then an *input–output system*:

$$\mathcal{H} \subset \mathscr{Z} \times \mathscr{Y}. \tag{7}$$

In a theory of causal objects, inputs become recognized as causes and outputs as effects. Equations (5) and (6) thus represent the beginnings of a formal depiction of the causal bond.

To emphasize significant characteristics of input and output objects, two special terms will be used. The *genon* or output object serves a generative function. If characterized by gender the genon would be considered a masculine object, responsible for broadcasting (output) attributes to the holon's environment. The *creaon* or input object performs a creative or selective role in defining a perceived environment which is specific to that particular holon. The creaon is a feminine object in its reception of (input) attributes, and serves to maintain holon–environment consistency and compatibility. These notions will be further clarified and illustrated in later sections.

C. Functional Systems

With recognition of inputs and outputs, *functional systems* can be formed by defining functions,

$$\mathcal{H} : \mathscr{Z} \to \mathscr{Y}, \tag{8}$$

on the relations (7). Object determination thereby becomes viewed as mapping of input into output, and the holon becomes formally defined as a set of input–output pairs. The domain and range of (8) are defined, respectively, by

$$D(\mathcal{H}) = \mathcal{Z} = \{z : (\exists\, y)\, ((z, y) \in \mathcal{H})\}, \tag{9}$$

$$R(\mathcal{H}) = \mathcal{Y} = \{y : (\exists\, z)\, ((z, y) \in \mathcal{H})\}. \tag{10}$$

D. Time Systems

A time set $\mathcal{T}$ is a weakly ordered abstract set, with ordering achieved by the relation $\leqslant$. The elements of $\mathcal{T}$ will be noted by t. Normally, $\mathcal{T}$ is the continuous real line $(-\infty, \infty)$, or the set of integers in the case of discrete-time systems. *Intervals* are subsets of time, denoted $\hat{t} \subset \mathcal{T}$. Time intervals for continuous systems are as follows:

$$\text{open:} \quad (t_0, t_1) = \{t \in \mathcal{T} : t_0 < t < t_1\},$$

$$\text{semiopen:} \quad [t_0, t_1) = \{t \in \mathcal{T} : t_0 \leqslant t < t_1\},$$

$$\text{semiclosed:} \quad (t_0, t_1] = \{t \in \mathcal{T} : t_0 < t \leqslant t_1\},$$

$$\text{closed:} \quad [t_0, t_1] = \{t \in \mathcal{T} : t_0 \leqslant t \leqslant t_1\}.$$

Such intervals between specified initial and final times, t_0 and t_1, are termed *observation intervals.*

Let Z and Y be arbitrary sets, $\mathcal{T}$ a time set, $Z^{\mathcal{T}}$ and $Y^{\mathcal{T}}$ sets of all maps on $\mathcal{T}$ into Z and Y, respectively, and let $\mathcal{Z} \subset Z$ and $\mathcal{Y} \subset Y$. Then, a *time system* $\mathcal{H}$ on $\mathcal{Z}$ and $\mathcal{Y}$ is a relation, $\mathcal{H} \subset \mathcal{Z} \times \mathcal{Y}$ (Mesarovic and Takahara, 1975, p. 17). Z and Y are "*alphabets*" of the input set $\mathcal{Z}$ and output set $\mathcal{Y}$, respectively, whose elements z: $\mathcal{T} \to Z$ and y: $\mathcal{T} \to Y$ are time functions. The values of $\mathcal{Z}$ and $\mathcal{Y}$ elements at time t are denoted by $z(t)$ and $y(t)$, respectively, understood to be the sets of pairs $\{t, z(t)\}$ and $\{t, y(t)\}$. When t ranges over an interval $\hat{t}$, the corresponding values $\{t, z(t)\}$ and $\{t, y(t)\}$, $\forall t \in \hat{t}$, represent input and output *segments,* respectively. Vector notation is convenient, $[\mathbf{z}(t), \mathbf{y}(t)]$ representing instantaneous values at time t of all input and output attributes of $\mathcal{H}$, and $[\mathbf{z}(\hat{t}), \mathbf{y}(\hat{t})]$ depicting corresponding segments on $\hat{t} \subset \mathcal{T}$.

Segmentation is an important concept in the formal definition of abstract objects. Let $\mathbf{v}(\hat{t})$ be segments of the unoriented attributes $v_1, \ldots, v_m$ on $\hat{t} = [t_0, t_1]$. With τ a time such that $t_0 < \tau < t_1$, the segment $\mathbf{v}(\hat{t})$ can be thought of as a *concatenation* of two segments, $\mathbf{v}^0(\hat{t})$ on $[t_0, \tau]$ and $\mathbf{v}^1(\hat{t})$ on $(\tau, t_1]$. The two segments $\mathbf{v}^0(\hat{t})$ and $\mathbf{v}^1(\hat{t})$ are *restrictions* of $\mathbf{v}(\hat{t})$, and $\mathbf{v}^1(\hat{t})$ is a *continuation* of $\mathbf{v}^0(\hat{t})$. The segment $\mathbf{v}(\hat{t})$ can be written $\mathbf{v}^0(\hat{t})\ \mathbf{v}^1(\hat{t})$ or, more succinctly, $\hat{\mathbf{v}} = \hat{\mathbf{v}}^0 \hat{\mathbf{v}}^1$. For the oriented case, the parallel representation is $[\mathbf{z}(\hat{t}), \mathbf{y}(\hat{t})] = [\mathbf{z}^0(\hat{t})\, \mathbf{z}^1(\hat{t}),\ \mathbf{y}^0(\hat{t})\, \mathbf{y}^1(\hat{t})]$, or just $(\hat{\mathbf{z}}, \hat{\mathbf{y}}) = (\hat{\mathbf{z}}^0 \hat{\mathbf{z}}^1, \hat{\mathbf{y}}^0 \hat{\mathbf{y}}^1)$.

1. *Closure under Segmentation*

Consider $\mathcal{H}$ to represent the set $\{\hat{\mathbf{v}}\}$ of segments for a holon on some observation interval $\hat{t} = [t_0, t_1]$:

$$\hat{\mathbf{v}} \in \mathcal{H}. \tag{11}$$

The family of segments, $\mathcal{H} = \{\hat{\mathbf{v}}^0 \hat{\mathbf{v}}^1\}$ is said to be *closed under segmentation* (Zadeh and Desoer, 1963; Zadeh, 1969) if for each τ in $[t_0, t_1]$, both $\hat{\mathbf{v}}^0$ and $\hat{\mathbf{v}}^1$ are elements of $\mathcal{H}$. That is,

$$\hat{\mathbf{v}}^0 \hat{\mathbf{v}}^1 \in \mathcal{H} \Rightarrow \hat{\mathbf{v}}^0 \in \mathcal{H} \quad \text{and} \quad \hat{\mathbf{v}}^1 \in \mathcal{H}; \tag{12}$$

every restriction of a segment which is a member of $\mathcal{H}$ is also a member of $\mathcal{H}$. Note that $\hat{\mathbf{v}}^0 \in \mathcal{H}$ and $\hat{\mathbf{v}}^1 \in \mathcal{H}$ does not necessarily imply $\hat{\mathbf{v}}^0 \hat{\mathbf{v}}^1 \in \mathcal{H}$. The latter is true only when $\hat{\mathbf{v}}^1$ is a continuation of $\hat{\mathbf{v}}^0$ in $\mathcal{H}$.

For the case of directed attributes, let the creaon set $\mathcal{Z}$ be the family $\{\hat{\mathbf{z}}\}$ of all input segments on $\hat{t}$ and the genon set $\mathcal{Y}$ the family $\{\hat{\mathbf{y}}\}$ of all output segments. Since $\mathcal{H} \subset \mathcal{Z} \times \mathcal{Y}$ (Fig. 2b), corresponding to (11) there is

$$(\hat{\mathbf{z}}, \hat{\mathbf{y}}) \in \mathcal{H}, \tag{13}$$

and the closure under segmentation condition becomes

$$(\hat{\mathbf{z}}^0 \hat{\mathbf{z}}^1, \hat{\mathbf{y}}^0 \hat{\mathbf{y}}^1) \in \mathcal{H} \Rightarrow (\hat{\mathbf{z}}^0, \hat{\mathbf{y}}^0) \in \mathcal{H} \quad \text{and} \quad (\hat{\mathbf{z}}^1, \hat{\mathbf{y}}^1) \in \mathcal{H}, \tag{14}$$

where $(\hat{\mathbf{z}}^1, \hat{\mathbf{y}}^1)$ is a continuation of $(\hat{\mathbf{z}}^0, \hat{\mathbf{y}}^0)$ in $\mathcal{H}$ if $(\hat{\mathbf{z}}^0 \hat{\mathbf{z}}^1, \hat{\mathbf{y}}^0 \hat{\mathbf{y}}^1) \in \mathcal{H}$.

2. *Formal Definition of Holon*

A *holon* $\mathcal{H}$ can now be defined as a set of ordered pairs of time relations (13) satisfying the condition of closure under segmentation (14). The domain and range of $\mathcal{H}$ are

$$D(\mathcal{H}) = \mathcal{Z} = \{\hat{\mathbf{z}} : (\hat{\mathbf{z}}, \hat{\mathbf{y}}) \in \mathcal{H}\}, \tag{15}$$

and

$$R(\mathcal{H}) = \mathcal{Y} = \{\hat{\mathbf{y}} : (\hat{\mathbf{z}}, \hat{\mathbf{y}}) \in \mathcal{H}\}, \tag{16}$$

representing the sets of all input and output segments that can be associated with $\mathcal{H}$.

This concept of an abstract object represents a formal expression of the fact that any interaction of an observer with a concrete object in the real world involves varying some of the attributes of the latter and observing variations which occur in other attributes. The attributes that are varied are inputs or causes, and those which respond to the variations are outputs or effects. In establishing the holon as a collection of input–output pairs, the implied assertion is that interaction with the entity takes place by observation of variations in its attributes. It is thus logical to identify the object with the totality of possible observations of attribute variations, and this totality becomes equated to the object.

Note that the formal definition does not imply that to each input segment $\hat{\mathbf{z}}$ there corresponds a unique output segment $\hat{\mathbf{y}}$. On the contrary, to each input there generally corresponds a number of possible outputs. There is thus a plurality of effects (Section II. L), each one of which corresponds to a possible *response* of $\mathcal{H}$ to $\hat{\mathbf{z}}$. Because of nonuniqueness of the dependence of output on input, an abstract object (13) is a relation rather than a function (Zadeh, 1969, p. 9). To make it a function requires an additional property, that of state. As developed in subsequent sections, the state of a holon represents an index to a subset of input–output pairs which satisfy certain consistency conditions.

Example 1

Zadeh (1969) presented an example of an abstract finite state object which lucidly clarifies the significance of closure under segmentation and also the other consistency conditions needed to establish input–output determinacy. The object is illustrated in Fig. 3 in the standard graph format of state diagrams for finite state systems (Gill, 1962). The three nodes correspond to states and the directed arcs denote state transitions. Each arc is labeled with an input–output pair. The inputs and outputs are both binary and, to distinguish them

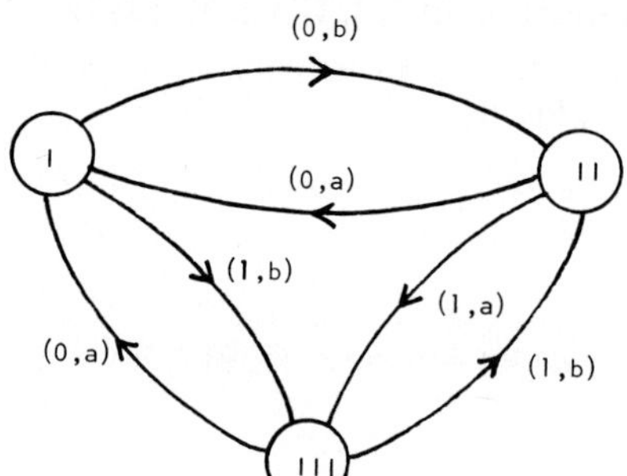

FIG. 3. State diagram for an abstract finite state holon.

notationally, the former are denoted by 0, 1 and the latter by a, b. Input and output segments are sequences of these binary symbols. The input–output pairs of lengths 1, 2, and 3 for this object are listed below; any sequence of 0's and 1's can be an input sequence.

Input–output pairs of length 1:

$$(0, \mathrm{a}) \quad (0, \mathrm{b}) \quad (1, \mathrm{a}) \quad (1, \mathrm{b}).$$

Input–output pairs of length 2:

(00, ab) (00, ba) (01, ba) (01, ab)
(10, aa) (10, ba) (11, ab) (11, ba) (11, bb).

Input–output pairs of length 3:

(000, aba) (010, baa) (100, bab) (111, aba) (000, bab) (001, bab)
(111, bba) (110, bba) (111, bbb) (001, aba) (110, baa) (011, abb)
(100, aab) (101, aab) (010, aba) (101, bab) (110, aba) (011, bab).

Note how closure under segmentation is satisfied by this system. For example, of four possible input–output segments of length 3 when the input sequence is 000, namely

(000, aba) (000, baa)

(000, abb) (000, bab),

only two (underlined), which appear in the list above, can be generated from the state diagram (Fig. 3). For these two, the restrictions (00_, ab_), (_00, _ba), (00_, ba_), and (_00, _ab) appear in the list of input–output pairs of length 2 belonging to the object. For the other two (not underlined), the possible length 2 restrictions are (00_, ab_), (_00, _bb), (00_, ba_), and (_00, _aa), of which the second and fourth do not appear in the above list. The reader can verify that all restrictions derivable from the listed input–output pairs are valid, and further, that all pairs of length other than 3 (greater or less) are derivable from those of length 3 based solely on application of the closure under segmentation condition. This object therefore qualifies as a holon according to the formal definition.

3. *Concept of Environment*

With the definition of a system as a partially interconnected set of holons, and definition of a holon as a set of input–output time segments satisfying closure under segmentation, the question of environment as a source of inputs arises. A logical difficulty is encountered because coupling of objects in interaction means constraining the terminal variables by which they are joined to be identically equal (Zadeh and Desoer, 1963, p. 52 ff). If the environment $\mathcal{H}'$ of holon $\mathcal{H}$ is taken to be the totality of interconnected abstract objects with which $\mathcal{H}$ interacts (Fig. 4a), then it is clear that both $\mathcal{H}$ and $\mathcal{H}'$ are identical as sets of input–output pairs, the only difference being reversal of what is input and what is output (Fig. 4b). $\mathcal{H}$ and $\mathcal{H}'$ are reciprocal objects, logical duals, and they differ only in orientation of terminal variable attributes. $\mathcal{H}$ and $\mathcal{H}'$ may thus be thought of as a matched pair, arising in consequence of the formal

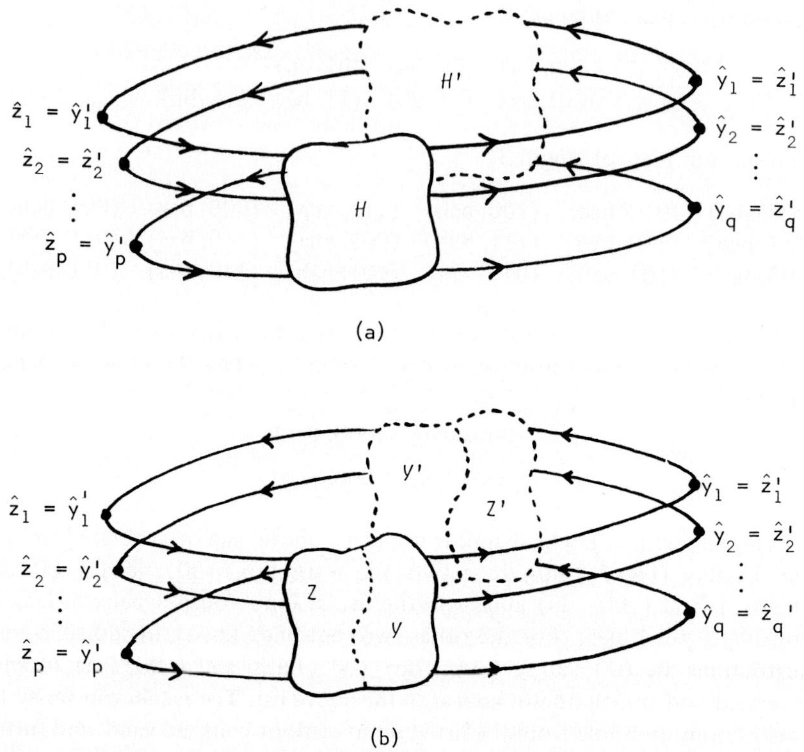

FIG. 4. (a) The holon $\mathscr{H}$ as a set of input–output time segments, $\{z_1(\hat{t}), \ldots, z_p(\hat{t}), y_1(\hat{t}), \ldots, y_q(\hat{t})\}$ coupled to an environment $\mathscr{H}'$. (b) Same as (a), but displaying creaon elements, $\mathscr{Z}$ and $\mathscr{Z}'$, and genon elements, $\mathscr{Y}$ and $\mathscr{Y}'$, of $\mathscr{H}$ and $\mathscr{H}'$.

definition of holon, and the usefulness of a concept of environment so defined remains to be investigated and clarified.

E. Holon Determinacy

The several meanings of determination were discussed previously in Section II. B. It was said first that something was determinate if it possessed attributes by which it could be characterized unambiguously. Second, determinacy signified a constant and unique relation between entities. Third, determination was concerned with the fact that the means of acquiring characteristics out of preexisting conditions be lawful. In Section II. D, a principle of determinacy was stated in terms of two other principles: lawfulness and productiveness. Assuming that input is productive of output, lawfulness of the time development of input–output segments is embodied in the following definition. A holon $\mathcal{H}$ is *determinate* if its output segment corresponding to a given input segment is unique. That is, $\forall \hat{\mathbf{z}} \in \mathcal{Z}$, $\exists$ one and only one $\hat{\mathbf{y}} \in \mathcal{Y}$. Plurality of effects (Section II. L) is precluded. If the holon is not determinate, it is *indeterminate.*

A system is said to be *past-determined* if there exists $\tau \in \mathcal{T}$ such that, for any $t \geqslant \tau$, its future evolution is determined solely by past input and output history. There is no need to refer to any auxiliary information such as states (Mesarovic and Takahara, 1975, p. 34). The system of Example 1 of Section III. D. 2 is past-determined since input–output segments of length three are sufficient, with the closure under segmentation property, to determine all segments of greater length. Although the nodes of Fig. 3 in effect represent states, this information is not needed to generate time segments once the length-three segments are available. For systems that are not past-determined to be determinate, auxiliary information is required.

1. *"Bundles" and "Tags"*

State space theory as presented by Zadeh (Zadeh and Desoer, 1963; Zadeh, 1969) serves to clarify the role of state in object determinacy better than most other available presentations. With a system defined as a set of input–output pairs, state is developed as an index attached to a subset of input–output pairs satisfying four consistency conditions. The subset of pairs so identified is termed an aggregate.

Consider holon $\mathcal{H}$ as a family of input–output pairs satisfying the closure under segmentation condition. For each initial time t_0, let $\mathcal{H}(t_0) \subset \mathcal{H}$ denote the input–output time segments that begin at t_0. These segments may be grouped according to certain as yet unspecified characteristics, and each group indexed by an identifying label $\mathbf{x}_0$. The groups are called "bundles" and the labels "tags" in Zadeh's terminology. The set of possible tags at t_0 can be denoted by $\mathcal{X}_0$, and therefore $\mathbf{x}_0 \in \mathcal{X}_0$. Thus, all the time segments $\mathcal{H}(t_0)$ of the system beginning at t_0 have been sorted into a number of bundles $\mathcal{H}_{\mathbf{x}_0}(t_0)$, each of which is indexed by a tag $\mathbf{x}_0$:

$$\mathcal{H}(t_0) = \bigcup_{\mathbf{x}_0 \in \mathcal{X}_0} \{\mathcal{H}_{\mathbf{x}_0}(t_0)\}. \tag{17}$$

Note that the bundles $\mathcal{H}_{\mathbf{x}_0}(t_0)$ are not necessarily disjoint.

If the input–output pairs beginning at t_0 are bundled in a certain way, then the bundles will be called *aggregates,* and their tags will constitute the *states* of $\mathcal{H}$. The bundling must satisfy four consistency conditions, as now set forth:

1. *Covering.* Every input–output pair in $\mathcal{H}(t_0)$ must belong to at least one bundle $\mathcal{H}_{\mathbf{x}_0}(t_0)$ in the set of bundles $\{\mathcal{H}_{\mathbf{x}_0}(t_0)\}_{\mathbf{x}_0 \in \mathcal{X}_0}$ as already specified by Eq. (17).

2. *Closure under Truncation.* Let $(\hat{z}^0\hat{z}^1, \hat{y}^0\hat{y}^1)$ be an arbitrary element of $\mathcal{H}(t_0)$, with the restrictions $(\hat{z}^0, \hat{y}^0)$ and $(\hat{z}^1, \hat{y}^1)$ being left and right truncates, respectively. Closure under truncation requires that if $(\hat{z}^0\hat{z}^1, \hat{y}^0\hat{y}^1)$ is in a bundle $\mathcal{H}_{\mathbf{x}_0}(t_0)$, then every left truncate of it is also in the same bundle:

$$(\hat{z}^0\hat{z}^1, \hat{y}^0\hat{y}^1) \in \mathcal{H}_{\mathbf{x}_0}(t_0) \Rightarrow (\hat{z}^0, \hat{y}^0) \in \mathcal{H}_{\mathbf{x}_0}(t_0)$$

$\forall t_0$, $\forall \mathbf{x}_0 \in \mathscr{X}_0$, and $\forall(\hat{z}^0\hat{z}^1, \hat{y}^1\hat{y}^1) \in \mathcal{H}_{\mathbf{x}_0}(t_0)$.

3. *Uniqueness.* The domain and range of a bundle of input–output pairs $\mathcal{H}_{\mathbf{x}_0}(t_0)$ can be defined as

$$D[\mathcal{H}_{\mathbf{x}_0}(t_0)] = \{\hat{z} : (\hat{z}, \hat{y}) \in \mathcal{H}_{\mathbf{x}_0}(t_0)\}, \tag{18}$$

and

$$R[\mathcal{H}_{\mathbf{x}_0}(t_0)] = \{\hat{y} : (\hat{z}, \hat{y}) \in \mathcal{H}_{\mathbf{x}_0}(t_0)\}. \tag{19}$$

Uniqueness requires a unique output $\hat{y} \in R[\mathcal{H}_{\mathbf{x}_0}(t_0)]$ for each input $\hat{z} \in D[\mathcal{H}_{\mathbf{x}_0}(t_0)]$:

$$(\hat{z}, \hat{y}) \in \mathcal{H}_{\mathbf{x}_0}(t_0) \quad \text{and} \quad (\hat{z}, \hat{y}') \in \mathcal{H}_{\mathbf{x}_0}(t_0) \Rightarrow \hat{y} = \hat{y}'.$$

Indeterminacy, multiple output segments corresponding to a given input segment in a bundle, is precluded; thereby, for causal holons multiple causation (Section II. L) is disallowed. Uniqueness makes it possible to model determinate objects as functions $\mathscr{X}_0 \times D[\mathcal{H}_{\mathbf{x}_0}(t_0)] \to R[\mathcal{H}_{\mathbf{x}_0}(t_0)]$, and such systems are known as *dynamical systems* (Mesarovic and Takahara, 1975, p. 21). The tags $\mathbf{x}_0 \in \mathscr{X}_0$ represent auxiliary information required for determinate mapping of input into output.

4. *Continuation.* Let $(\hat{z}^1, \hat{y}^1)$ be a continuation of $(\hat{z}^0, \hat{y}^0) \in \mathcal{H}_{\mathbf{x}_0}(t_0)$, that is a right truncate of $(\hat{z}^0\hat{z}^1, \hat{y}^0\hat{y}^1) \in \mathcal{H}_{\mathbf{x}_0}(t_0)$. This consistency condition requires, for all t_0, that the set of all continuations of $(\hat{z}^0, \hat{y}^0)$ in $\mathcal{H}_{\mathbf{x}_0}(t_0)$ be a bundle in the family $\{\mathcal{H}_{\mathbf{x}_0}(t_0)\}_{\mathbf{x}_0 t_0}$. If $\mathbf{x}_1 \in \mathscr{X}_1$ indexes this bundle of continuations in $t > \tau$, then the condition can be written

$$\mathcal{H}_{\mathbf{x}_1}(\tau) \in \{\mathcal{H}_{\mathbf{x}_0}(t_0)\}_{\mathbf{x}_0 \in \mathscr{X}_0}$$

$\forall t_0$, $\forall \tau > t_0$, $\forall \mathbf{x}_1 \in \mathscr{X}_1$, $\forall \mathbf{x}_0 \in \mathscr{X}_0$, and $\forall(\hat{z}^0, \hat{y}^0) \in \mathcal{H}_{\mathbf{x}_0}(t_0)$ on $[t_0, \tau]$.

Example 2

The finite state system of Example 1 (Section III. D. 2, Fig. 3) can be used to advantage in illustrating these four consistency conditions. Consider that for each t_0 the input–output pairs of lengths 1, 2, and 3 are bundled together into three groups indexed by the set $\mathscr{X}_0 = \{\text{I, II, III}\}$. The elements of lengths $\leqslant 3$ of the three bundles $\mathcal{H}_{\text{I}}(t_0)$, $\mathcal{H}_{\text{II}}(t_0)$, and $\mathcal{H}_{\text{III}}(t_0)$ are listed in Table I. With the bundling independent of t_0, the index set can be written $\mathscr{X}$ and the bundles $\mathcal{H}_{\text{I}}$, $\mathcal{H}_{\text{II}}$, and $\mathcal{H}_{\text{III}}$.

TABLE I
Bundles of Input–Output Time Segments of Lengths 1, 2, and 3 for the Finite State Holon of Fig. 3 and Example 1 (Section III. D. 2)

H_I	H_{II}	H_{III}
(0,b)	(0,a)	(0,a)
(1,b)	(1,a)	(1,b)
(00,ba)	(00,ab)	(00,ab)
(01,ba)	(01,ab)	(01,ab)
(10,ba)	(10,aa)	(10,ba)
(11,bb)	(11,ab)	(11,ba)
(000,bab)	(000,aba)	(000,aba)
(001,bab)	(001,aba)	(001,aba)
(010,baa)	(010,aba)	(010,aba)
(011,bab)	(011,abb)	(011,abb)
(100,bab)	(100,aab)	(100,bab)
(101,bab)	(101,aab)	(101,bab)
(110,bba)	(110,aba)	(110,baa)
(111,bba)	(111,aba)	(111,bab)

Note, in Table I, that within a bundle the output sequence associated with a given input segment is unique, and the object is therefore determinate. Plurality of static response to different inputs occurs, corresponding to multiple causation (Section II. L), but dynamically, unique response segments are produced for each unique input segment (as recorded *within* bundles). Note further, by comparing how the Table I input–output sequences are generated from Fig. 3, that all three of the tags I, II, and III are involved. For example, the pair (00, ba) in $\mathcal{H}_I$ has left truncate (0, b), one of two possibilities leading from node I in Fig. 3. The system is then subject to index II, and the two possible next behaviors are (0, a) and (1, a), the former being the right truncate of the input–output pair (00, ba) in question. The pair (01, ba) also appears in $\mathcal{H}_I$, corresponding to (1, a) as the right truncate. Exhaustive testing of this kind can be used to establish that the bundling depicted in Table I satisfies the four consistency conditions for input–output pairs of lengths 1, 2 and 3:

Covering. Comparison of Table I with the original list (Example 1, Section III, D, 2) of input–output pairs of lengths $\leqslant 3$ shows that the union of $\mathcal{H}_I$, $\mathcal{H}_{II}$, and $\mathcal{H}_{III}$ corresponds to $\mathcal{H}$ [Eq. (17)]. Note that the bundles $\mathcal{H}_I$, $\mathcal{H}_{II}$, and $\mathcal{H}_{\mathrm{III}}$ clearly are not a partition (i.e., mutually exclusive and exhaustive subsets) of $\mathcal{H}$.

Closure under Truncation. The left truncates of every input–output pair are in the same bundle. For example, the left truncates of (000, bab) in $\mathcal{H}_{\mathrm{I}}$, namely (00, ba) and (0, b) are both also in $\mathcal{H}_{\mathrm{I}}$.

Uniqueness. As already emphasized the object is determinate. Within any bundle each input sequence is listed only once, meaning that there is only the output segment corresponding to it.

Continuation. It is required that the set of all continuations of a left truncate in a bundle be itself a bundle of the system. Consider (0, b) in $\mathcal{H}_{\mathrm{I}}$ as a left truncate. Its continuations (Table I) are (0, a), (1, a), (00, ab), (01, ab), (10, aa), and (11, ab), all of which are elements of $\mathcal{H}_{\mathrm{II}}$. Further such testing verifies that the continuation condition is satisfied.

2. *Aggregates*

The concept of an aggregate can now be formally defined. Let $\mathcal{H}$ be a holon which satisfies closure under segmentation, and let the input–output time segments be grouped into bundles satisfying the further four conditions of covering, closure under truncation, uniqueness, and continuation. The set of such bundles comprises a *consistent* family, and a consistent family of bundles is termed an *aggregate.*

Each aggregate of a holon represents a law for the system's ensuing input–output behavior; an aggregate thus expresses the determinacy principle in explicit summary form. That is, the aggregate represents a catalog of future (from t_0) responses to all possible input sequences. Which catalog is to become expressed as actual system behavior is determined by the aggregate indexes, that is, by the elements of the tag set $\mathscr{X}_0$. For Example 2 this set was $\mathscr{X} = \{\mathrm{I, II, III}\}$, which tagged the aggregates $\mathcal{H}_{\mathrm{I}}$, $\mathcal{H}_{\mathrm{II}}$, and $\mathcal{H}_{\mathrm{III}}$ listed in Table I. Note again that in the example these aggregates, defined at a fixed initial time t_0, were not varied as time changed.

3. *States*

Let $\mathcal{H}_{\mathbf{x}_0}(t_0)$ be an aggregate of $\mathcal{H}$ at time t_0. The *state* of $\mathcal{H}$ at time t_0 is the identifying tag $\mathbf{x}_0$ of the aggregate. In other words it is the initial state of the system, that is, the state at t_0 that determines which aggregate is to represent the expressed behavior of the system. The set of aggregate indexes, $\mathscr{X}_0$ is the *initial state space* of the holon at time t_0, representing the range of aggregate tags at this time. When the state space is independent of initial time, as it was in Example 2 and is usually the case, it is denoted by $\mathcal{H}$ and is referred to simply as the *state space* of $\mathcal{H}$. Notationally, the state of a holon will be denoted by $\mathbf{x}$, its value at t_0 as $\mathbf{x}(t_0)$ or $\mathbf{x}_0$, its value at $t \in \mathscr{T}$ as $\mathbf{x}(t)$ or $\mathbf{x}_t$, and its time sequences on $[t_0, t_1]$ as $\mathbf{x}(\hat{t})$ or $\hat{\mathbf{x}}$. That holons may have state dynamics as well as input and output dynamics is signified by the latter notations, but what then becomes of the special significance of *initial state* in determining subsequent input–output pairs, particularly when the choice of initial time (and thus the definition of "subsequent") is arbitrary? The answer is that initial state also determines subsequent states, that is, aggregates describing input–output pairs from those times at which the new states prevail into the future. The interplay of state dynamics and input–output dynamics is conceptually difficult, but the following example will aid in clarification.

Example 3

Table II summarizes state dynamics of the Fig. 3 finite state object, beginning with all three states, I, II, and III (nodes in the Fig. 3 graph), but at different initial times. The tree diagrams of Table II were generated by reference to Fig. 3. Alongside each arrow in the table are input–output segments of appropriate length as referenced to each initial time t_0. In each case the most recent input is italicized, and in each bifurcation from a state this input *0* is on the left and *1* is on the right.

Referring to the uppermost tree in the table, the system begins in state I at $t = 0$. If it receives *0* as input it moves to state II and generates the output b. If *1* is received as input, the state transition is to III and the output is b. At the next time ($t = 1$), if the system is in state II and receives *0* it moves to I and outputs a; if it receives *1* its next state is III and the corresponding output is a. At $t = 2$, if *0* is the input the next state (from I) will be II and the corresponding output b, whereas if *1* is the input then III is the next state and b the output. The reader can verify Table II against Fig. 3. Note that the collection of all input–output pairs in the upper part of the table corresponds to aggregate $\mathcal{H}_{\mathrm{I}}$ as listed in Table I. In the next part of Table II a similar sequence is generated beginning with state II as produced at $t = 1$ above. Finally, in the lower part of the table the sequences emanating from state III beginning at $t = 2$ are tabulated. In the middle section of Table II the collection of input–output pairs corresponds to aggregate $\mathcal{H}_{\mathrm{II}}$ and in the lower part of the table the collection comprises aggregate $\mathcal{H}_{\mathrm{III}}$ (Table I).

The point of this example is that while holons do have state dynamics, the aggregate for which each state is a tag is time invariant. No matter when a state occurs, the subsequent production of input–output pairs proceeds determinately as defined by the associated aggregates. The behavior generating variable is input:

$$\mathscr{Z} : \mathscr{T} \rightarrow \mathscr{X}, \tag{20}$$

and

$$\mathscr{X} : \mathscr{Z} \rightarrow \mathscr{Y}. \tag{21}$$

TABLE II

State Dynamics of the Finite State Holon of Fig. 3, as Generated by Different Input Sequences

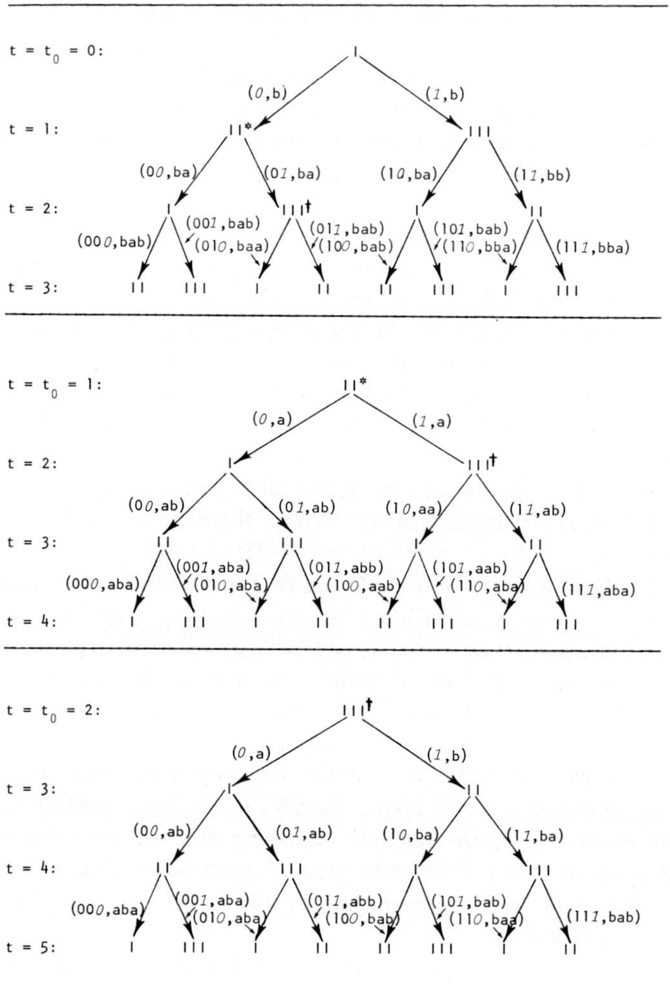

*† Identical states at identical times.

Thus, input becomes interpreted as a map from time into states (20) and states map inputs into outputs (21). This concept of the holon and its reciprocal environment is schematized in Fig. 5.

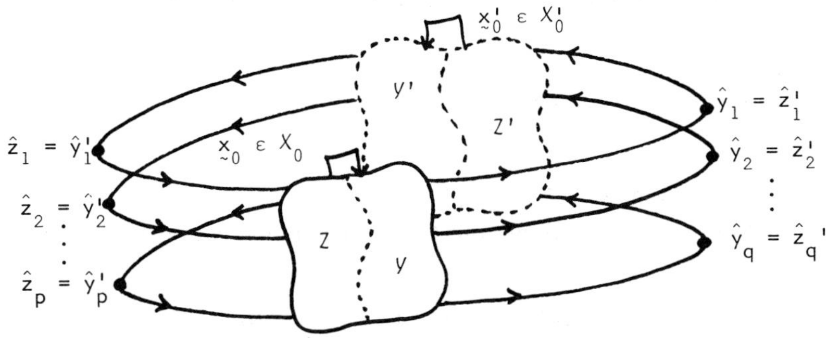

FIG. 5. Holon and corresponding environment as state determined objects.

F. Dynamical Systems

The information in Fig. 3 and Table II can be summarized in tabular form:

z_t \ x_t	I	II	III
0	II	I	I
1	III	III	II

z_t \ x_t	I	II	III
0	b	a	a
1	b	a	b

The first table is a map from states and inputs into states,

$$f\colon \mathscr{X} \times \mathscr{Z} \to \mathscr{X}, \tag{22}$$

and the second is a similar mapping into outputs,

$$g\colon \mathscr{X} \times \mathscr{Z} \to \mathscr{Y}. \tag{23}$$

Function f is termed a *state transition function,* and g is a *response* or *output function.* Together, they describe the state and output dynamics of the finite state object:

$$\mathbf{x}_{t+1} = f(\mathbf{x}_t, \mathbf{z}_t), \tag{24}$$

$$\mathbf{y}_t = g(\mathbf{x}_t, \mathbf{z}_t), \qquad t = \ldots, -1, 0, 1, \ldots . \tag{25}$$

Equation (24) can be iterated to obtain an expression for $\mathbf{x}_t$ in terms of initial state $\mathbf{x}_0$ and the input sequence $\mathbf{z}_0, \ldots, \mathbf{z}_{t-1}$:

$$\mathbf{x}_t = f(f(\ldots f(\mathbf{x}_0, \mathbf{z}_0), \ldots), \mathbf{z}_{t-1}).$$

Substituting this into Eq. (25) yields

$$\mathbf{y}_t = g(f(f(\ldots f(\mathbf{x}_0, \mathbf{z}_0), \ldots), \mathbf{z}_{t-1}), \mathbf{z}_t).$$

These last two expressions represent a *state generating function,*

$$\mathrm{x}_{t+1} = \bar{f}(\mathrm{x}_0, \hat{\mathrm{z}}), \tag{26}$$

and a *response* or *output generating function,*

$$\mathrm{y}_t = \bar{g}(\mathrm{x}_0, \hat{\mathrm{z}}) \tag{27}$$

(Mesarovic and Takahara, 1975, pp. 25, 27).

For continuous time systems the state transition function is a differential equation,

$$\dot{\mathrm{x}}_t = f(\mathrm{x}_t, \mathrm{z}_t), \qquad \mathrm{x}(t_0) = \mathrm{x}_0, \quad t \in [t_0, t_1], \tag{28}$$

and the output function is an algebraic equation,

$$\mathrm{y}_t = g(\mathrm{x}_t, \mathrm{z}_t), \qquad t \in [t_0, t]. \tag{29}$$

Corresponding generating functions are

$$\dot{\mathrm{x}}_t = \bar{f}(\mathrm{x}_0, \hat{\mathrm{z}}), \tag{30}$$

and

$$\mathrm{y}_t = \bar{g}(\mathrm{x}_0, \hat{\mathrm{z}}). \tag{31}$$

All of the equation pairs (24)–(25), (26)–(27), (28)–(29), and (30)–(31) achieve a representation of the holon, whether discrete or continuous, as a determinate function: $\mathscr{X}_0 \times \mathscr{D}[\mathcal{H}_{\mathrm{x}_0}(t_0)] \to \mathscr{R}[\mathcal{H}_{\mathrm{x}_0}(t_0)]$. Unlike past-determined systems, systems that require both state transition and response functions or the corresponding generating functions for determinate representation are referred to as *dynamical systems* (Mesarovic and Takahara, 1975, p. 21). The representations are termed *state space models* (e.g., Zadeh and Desoer, 1963).

Example 4

To illustrate how a differential equation model fits the determinacy scheme outlined in this section, consider an object described by the following simple model:

$$\dot{x}_t + x_t = z_t, \qquad x(t_0) = x_0, \quad t \in [t_0, t_1], \tag{32}$$

$$y_t = x_t, \qquad t \in [t_0, t_1]. \tag{33}$$

The output generated in response to an input segment $z(\hat{t})$, $\hat{t} = [t_0, t_1]$, is given by

$$y_t = e^{-(t-t_0)} x_0 + \int_{t_0}^{t} e^{-(t-\xi)} z(\xi)\, d\xi, \qquad t \in [t_0, t_1]. \tag{34}$$

The initial state x_0 parameterizes the response, and each value of x_0 in the state space (say all real numbers) identifies a bundle of input–output pairs:

$$\mathcal{H}_{x_0}(t_0) = \left\{ z_t, e^{-(t-t_0)} x_0 + \int_{t_0}^{t} e^{-(t-\xi)} z(\xi)\, d\xi \right\}, \qquad t \in [t_0, t_1]. \quad (35)$$

Note that input $z(t)$ is the generating variable.

System (32)–(33) is a holon. That is, it is a family of input–output time segments, and it is closed under segmentation. Equation (35) denotes an aggregate of input–output pairs as follows. Covering, closure under truncation, and uniqueness are satisfied by the definition of $\mathcal{H}$ [Eqs. (32)–(33)] and obvious properties of differential equations. To verify continuation, consider an arbitrary input–output pair (z^0, y^0) in bundle $\mathcal{H}_{x_0}(t_0)$. It must be shown that the set of all continuations in $\mathcal{H}_{x_0}(t_0)$ corresponds to another aggregate $\mathcal{H}_{x_\tau}(\tau)$ for some x_τ in the state space $\mathcal{X} = (-\infty, \infty)$.

Let (z^1, y^1) be a continuation of (z^0, y^0) in $\mathcal{H}_{x_0}(t_0)$. Then, since $(z^0 z^1, y^0 y^1) \in \mathcal{H}_{x_0}(t_0)$, it follows that

$$y^1(t) = e^{-(t-t_0)} x_0 + \int_{t_0}^{t} e^{-(t-\xi)} z(\xi) d\xi, \; t \geqslant \tau.$$

This can be written

$$y^1(t) = e^{-(t-t_0)} x_0 + \int_{t_0}^{\tau} e^{-(t-\xi)} z(\xi)\, d\xi + \int_{\tau}^{t} e^{-(t-\xi)} z(\xi)\, d\xi$$

$$= e^{-(t-t_0)} x_0 + e^{-(t-\tau)} \int_{t_0}^{\tau} e^{-(\tau-\xi)} z^0(\xi)\, d\xi + \int_{\tau}^{t} e^{-(t-\xi)} z^1(\xi)\, d\xi.$$

Letting

$$x_\tau = e^{-(t-t_0)} x_0 + \int_{t_0}^{\tau} e^{-(\tau-\xi)} z^0(\xi)\, d\xi,$$

then

$$y^1(t) = e^{-(t-\tau)} x_\tau + \int_{\tau}^{t} e^{-(t-\xi)} z^1(\xi)\, d\xi, \qquad t \geqslant \tau.$$

Thus, the set (z^1, y^1) of all continuations of (z^0, y^0) in $\mathcal{H}_{x_0}(t_0)$ corresponds to the aggregate

$$\mathcal{H}_{x_\tau}(\tau) = \left\{ z_t, e^{-(t-\tau)} x_\tau + \int_{\tau}^{t} e^{-(t-\xi)} z(\xi)\, d\xi \right\}, \qquad t \in [\tau, t_1], \quad (36)$$

and the continuation condition is satisfied by the input–output pairs defined by (35).

G. Holon Nonanticipation

Causality implies an ability to forecast future events. It requires a causal representation of a phenomenon consisting of recognition of associated causes and effects and descriptions of how causes are converted to effects. If a holon is determinate it would clearly seem to meet these criteria, but there is an exception.

Let $(\hat{\mathbf{z}}^0\hat{\mathbf{z}}^1, \hat{\mathbf{y}}^0\hat{\mathbf{y}}^1) \in \mathcal{H}_{\mathbf{x}_0}(t_0)$ be an input–output pair of determinate holon $\mathcal{H}$. The left truncate occurs on the interval $[t_0, \tau]$ and the right on $(\tau, t_1]$. Let $\hat{\mathbf{z}}^0$ remain fixed while $\hat{\mathbf{z}}^1$ is varied. The resulting input–output pairs can be denoted $(\hat{\mathbf{z}}^0\hat{\mathbf{z}}^{1*}, \hat{\mathbf{y}}^{0*}\hat{\mathbf{y}}^1)$ where asterisks indicate segments of interest. The experiment is such that $\hat{\mathbf{z}}^{1*} \neq \hat{\mathbf{z}}^1$. Only two outcomes are possible: (i) $\hat{\mathbf{y}}^{0*} \neq \hat{\mathbf{y}}^0$, or (ii) $\hat{\mathbf{y}}^{0*} = \hat{\mathbf{y}}^0$. In the first case the object's response on $[t_0, \tau]$ differs from the original even though $\hat{\mathbf{z}}^1$ was held constant. Therefore, either $\mathcal{H}$ is indeterminate, or it must be responding to future inputs $\hat{\mathbf{z}}^{1*}$ on $(\tau, t_1]$. Determinate objects capable of such behavior are termed *anticipatory*. In the second case the new input $\hat{\mathbf{z}}^{1*}$ on $(\tau, t_1]$ had no effect on the output during $[t_0, \tau]$, and such an object is *nonanticipatory*. In such a system changes in output cannot precede or anticipate changes in the input.

H. The Causal Holon

A holon is defined to be *causal* if it is both determinate and nonanticipatory. The causal holon $\mathcal{H}$ will be taken as the elementary or canonical form of the causal bond $C \rightarrow E$ (Fig. 6). This is the fundamental hypothesis of the theory.

Subsequent sections will consider the synthesis of causal holons into causal systems.

IV. The Causal Sequence

Hutchinson (1948) was first to introduce the explicit concept of ecosystems as circular causal systems. The basis for such a view was biogeochemical cycling and the closing of causal loops by nutrient regeneration. The causal "ecosystem" depicted in Fig. 6d displays this characteristic loop closure. In consequence of it, all components of ecosystems are mutually dependent via complexly interconnected causal chains which branch, converge, overlap, and feed back. An ecosystem, as a partially interconnected set of causal components, may best be considered a web or network of interactions (Fig. 6c). The webs are decomposable into serial sequences (Fig. 6b) and ultimately to individual causal bonds (Fig. 6a). The rich network of complex interactions within ecosystems is here considered a fundamental characteristic of systems at this level of

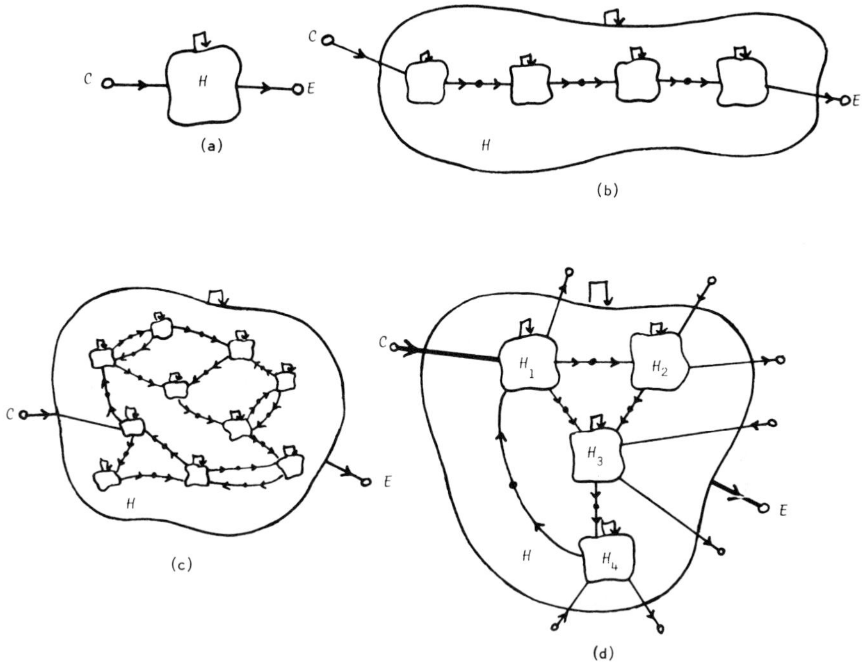

FIG. 6. Causal systems, $\mathcal{H}$: (a) causal bond; (b) causal sequence; (c) causal network; (d) causal ecosystem, with (say) $\mathcal{H}_1$, producers; $\mathcal{H}_2$, consumers; $\mathcal{H}_3$, decomposers; and $\mathcal{H}_4$, nutrients.

organization. To understand such systems it will be necessary to understand their interactive structure, which provides the means for propagation of efficient cause.

A. Hierarchies and Boundaries

Efficient cause is *external* cause (Section II. E), and intrinsic causation is generally not admitted to the causal concept (Section II. I). However, in endeavoring to apply the causal principle to large abstract systems like ecosystems, the meanings of "external" and "internal" must be closely examined. Placing boundaries around the ecosystem by definition has always been a difficult matter in ecology.

A hierarchical view of systems is useful in distinguishing exogenous and endogenous features. Koestler's (1967) holon was originally intended to apply to hierarchical systems, but this aspect has been suppressed in equating the holon to Zadeh's abstract object. In the present view of a holon as depicted in Fig. 5, contact with the external world ("environment") is provided by inputs and outputs. The state of the object is a more intrinsic characteristic than input and output although, as represented in Zadeh's theory, state is only a parameter labeling a set of input–output time segments beginning at some arbitrary initial time. There is thus nothing particularly basic about state that gives it a high heuristic or intuitive value, and yet, the state of a causal object represents its total memory about past input–output behavior and provides the necessary and sufficient information for a determinate and nonanticipatory mapping of cause into effect.

Simon (1973), in considering the *decomposability* of hierarchical systems, suggested a plausible argument that levels of organization within such systems are separated from one another because they operate in different time frames. The basis of hierarchy is that most interactions decrease in strength with distance. Components in a network have their strongest interactions with nearby components, and systems typically behave as collections of localized subsystems with graded bond strengths at successive levels of organization. The behavioral modes at each level can be decomposed into a set of elementary frequencies, each frequency proportional to an associated bond energy. At microscale levels bond strengths are greatest and dynamic frequencies highest. At macroscale levels bond strengths are weakest and behavioral frequencies lowest. Medium frequencies are associated with middle levels of the hierarchy. Systems are nearly always defined in terms of three levels: the level of interest (system), a level within (components), and a level without (environment). Simon refers to such organizations as "nearly decomposable" because each level is in effect segregated from others, as follows.

Behaviors of the system determined by high-frequency (component) modes will control the internal interactions of components of lower-level subsystems in a hierarchy, but will not be involved in interactions between those subsystems. These motions will be so rapid that the system, operating in the middle band of frequencies, is blind to them. The subsystems appear always to the system to be in steady state as behavioral variability and detail are filtered out, and endogenous parameters driving system states are, to the system, correspondingly smooth and well behaved. At the other end of the frequency spectrum, system behaviors determined by low-frequency (environmental) modes will be so slow relative to the median frequencies of the middle level that they appear to the system as virtual constants. The exogenous parameters driving states from without are thereby also regular and well behaved as the system sees them. The middle frequencies, then, determine the observable dynamics of the focal level of organization essentially independently of either the fast dynamics of internal structure or the slow dynamics of external environment.

Thus, in formulating a state space model of a system, a model whose principal quantities are input, output, and state, "input" to the level of interest can be construed as impinging both from within (i.e., below) as well as from without (above). The system, bounded by temporal isolation as a level in a hierarchy, has an endogenous environment beneath it and an exogenous environment above it, and both contribute "input," which maps time into state. The philosophical position against innate cause (Section II. I) is in effect rejected by viewing "within" as hierarchical "beneath."

B. Parameters

In state models of systems, inputs have the property that processes generating them are not explicitly modeled. They are taken as time-dependent drives on the system defined by state transition or generating functions. For example, in modeling photosynthetic primary production in ecology, solar processes responsible for generating insolation are not modeled, but the solar radiation which results is taken as measured as a time-dependent input. Similarly, processes generating endogenous "inputs" are likewise not explicitly modeled. They appear in mathematical representations as *coefficients,* which may be either constant or time varying. Coefficients define relationships between dependent variables of the model, and in state models they serve the role of coupling holons together to form systems. Coefficients, like inputs, are terminal variables, but at a lower level of organization. Thus, in Fig. 6 open circles denote terminal variables at the system level, $\mathcal{H}$, whereas closed circles depict coefficients as terminal variables of the component level. With the interpretation of coefficients as endogenous inputs, the mapping of time into state from below in a hierarchy emerges as an additional insight about *coupling.*

For the purpose of clarifying the propagation of cause in state space models, inputs $\mathbf{z}(t)$ and coefficients $\mathbf{c}(t)$ can be combined into a single class of model variables, *parameters* $\mathbf{p}(t)$, $t \in \mathcal{T}$. The state space model for continuous systems, for example [Eqs. (28)–(29)], can then be written as it was in Eqs. (1) and (2), where $\mathbf{p}(t) = [\mathbf{z}(t), \mathbf{c}(t)]$. With these concepts, roles of the different classes of system variables in transmitting cause and effect along causal chains can now be investigated.

C. Genetic Productivity

Let $\mathcal{H}$ be a system defined by m undifferentiated *system variables* $q_1, \ldots, q_m$. These variables can be partitioned into parameters, state variables, and outputs based on their contribution to dependency relationships within the system. Causation is transmitted in systems along dependency paths which constitute the causal structure.

Let $P_i(q_1, \ldots, q_m)$, $i = 1, \ldots, m$, be a set of functions expressing how each undifferentiated system variable is produced by the others according to the genetic principle (Section II. D). Then, productive determination can be defined as follows.

1. *Direct Dependence*

System variable q_i is defined to be *directly dependent* on system variable q_j if and only if

$$\partial P_i / \partial q_j \neq 0. \tag{37}$$

Direct dependency, expressing the philosophical concept of genetic productivity, will be symbolized in this section by a single-headed arrow, $q_j \to q_i$, read q_j "produces" q_i or q_i "is directly dependent upon" q_j. The different categories of system variables can be distinguished on the basis of this definition as follows.

A system variable q_p is a *parameter* if and only if

$$\sum_{j=1}^{m} |\partial P_p / \partial q_j| = 0, \tag{38}$$

and

$$\sum_{j=1}^{m} |\partial P_j / \partial q_p| \neq 0. \tag{39}$$

That is, q_p a parameter means that it is not directly dependent on any system variables [Eq. (38)], and that at least one system variable is directly dependent on it [Eq. (39)]. If the parameter is an input q_z, then there exists at least one state variable which is directly dependent on it. If the parameter is a coefficient q_c, then there exists at least one state variable or output which is produced by it.

A system variable q_x is a *state variable* if and only if

$$\sum_{j=1}^{m} |\partial P_x / \partial q_j| \neq 0, \tag{40}$$

and

$$\sum_{j=1}^{m} |\partial P_j / \partial q_x| \neq 0. \tag{41}$$

The variable q_x is directly dependent upon at least one system variable [Eq. (40)], and at least one system variable is directly dependent on it [Eq. (41)].

A system variable q_y is an *output* if and only if

$$\sum_{j=1}^{m} |\partial P_y / \partial q_j| \neq 0, \tag{42}$$

and

$$\sum_{j=1}^{m} |\partial P_j/\partial q_y| = 0. \tag{43}$$

That is, q_y is directly dependent upon at least one system variable q_j [Eq. (42)], and there are no system variables produced by q_y [Eq. (43)]. The system variables q_j upon which output variables depend are state variables; by the state space theory of input–output determinacy, an output of dynamical systems cannot be directly dependent on input.

Example 5

Several small compartment models of energy or material flows in ecosystems are described in the Appendix. A linear version of Eqs. (A.1) representing the marine coprophagy model can be used to illustrate correspondences between model variables and production functions. State transition and response equations are

$$\begin{array}{ll} \text{State transition equations} & \left\{ \begin{aligned} \dot{x}_1 &= z_{10} - a_{01}x_1 - a_{21}x_1, \\ \dot{x}_2 &= a_{21}x_1 - a_{02}x_2 - a_{32}x_2, \\ \dot{x}_3 &= z_{30} + a_{32}x_2 + a_{34}x_4 - a_{03}x_3 - a_{43}x_3, \\ \dot{x}_4 &= a_{43}x_3 - a_{04}x_4 - a_{34}x_4; \end{aligned} \right. \\ \text{Output equations} & \left\{ \begin{aligned} y_1 &= a_{01}x_1, \\ y_2 &= a_{02}x_2, \\ y_3 &= a_{03}x_3, \\ y_4 &= a_{04}x_4, \\ y_5 &= x_1 + x_2 + x_3 + x_4. \end{aligned} \right. \end{array} \tag{44}$$

Note that outputs $y_1, \ldots, y_4$ represent actual *outflows* of energy from the system, while y_5 is an attribute defined for its general interest. Production functions for the system variables are

$q_1 = z_{10}$:	$P_1 = 0$	$q_7 = a_{32}$:	$P_7 = 0$
$q_2 = a_{01}$:	$P_2 = 0$	$q_8 = z_{30}$:	$P_8 = 0$
$q_3 = x_1$:	$P_3 = z_{10}$	$q_9 = a_{34}$:	$P_9 = 0$
$q_4 = a_{21}$:	$P_4 = 0$	$q_{10} = x_4$:	$P_{10} = a_{43}x_3$
$q_5 = a_{02}$:	$P_5 = 0$	$q_{11} = a_{03}$:	$P_{11} = 0$
$q_6 = x_2$:	$P_6 = a_{21}x_1$	$q_{12} = x_3$:	$P_{12} = z_{30} + a_{32}x_2 + a_{34}x_4$

$$q_{13} = a_{43}: \quad P_{13} = 0 \qquad q_{17} = y_3: \quad P_{17} = a_{03}x_3$$

$$q_{14} = a_{04}: \quad P_{14} = 0 \qquad q_{18} = y_4: \quad P_{18} = a_{04}x_4$$

$$q_{15} = y_1: \quad P_{15} = a_{01}x_1 \qquad q_{19} = y_5: \quad P_{19} = x_1 + x_2 + x_3 + x_4.$$

$$q_{16} = y_2: \quad P_{16} = a_{02}x_2$$

System variable q_1 is an input because $\partial P_1/\partial q_j = 0$, for all j [Eq. (38)], and $\partial P_3/\partial q_1 \neq 0$ [Eq. (39)]; also, q_2 is a coefficient as $\partial P_2/\partial q_j = 0$, for all j, and $\partial P_{15}/\partial q_2 \neq 0$. Variable q_3 is a state variable because $\partial P_3/\partial z_{10} \neq 0$ [Eq. (40)] and $\partial P_i/\partial q_3 \neq 0$ [Eq. (41)] for $i = 6, 15, 19$. Variable q_{15} is an output because $\partial P_{15}/\partial q_j \neq 0$ for $j = 2, 3$ [Eq. (42)] and $\partial P_i/\partial q_{15} = 0$ for $i = 1, \ldots, 19$ [Eq. (43)]. The reader can verify that the remaining system variables also conform to the definitions for parameters, state variables, and outputs, as given in Eqs. (38)–(39), (40)–(41), and (42)–(43), respectively.

2. *Direct Dependency Matrices*

The binary matrix shown in Table III represents a *direct dependency matrix* for the marine coprophagy model discussed in Example 5. The matrix is square, $m \times m$ dimensional, with the ith row and ith column corresponding to system variables q_i, $i = 1, \ldots, m$. The matrix entries are binary, 0 in the ijth position if $\partial P_i/\partial q_j = 0$, and 1 if $\partial P_i/\partial q_j \neq 0$. Each nonzero element indicates the presence of a direct dependency link from system variable q_j to system variable q_i, $q_j \rightarrow q_i$. The Table III matrix is termed an *incidency matrix* in graph theory (Busaker and Saaty, 1965). A family of matrices can be generated from the direct dependency matrix by substituting various quantities for the binary 1's.

An *adjacency matrix* is formed by replacing each nonzero element in the incidency matrix with the number of direct dependency links between corresponding system variables. This matrix has significance in other contexts where the number of dependency bonds is a relevant consideration. A related matrix is the *Boolean adjacency matrix* in which each direct dependency link is indicated by a Boolean 1 in the appropriate position. A *Jacobian matrix* results when the binary 1's of the direct dependency matrix are replaced by values of the nonzero partial derivatives $\partial P_i/\partial q_j$. In ecology such matrices are known as *community matrices* (Levins, 1968).*

In a model describing flows of conserved quantities (energy or matter), a *flow matrix* can be formed by substituting amounts of flow from q_j to q_i. Such matrices have also been called *transaction* (Chenery and Clark, 1959) and *production matrices* (Hannon, 1973). A *fractional inflow* (or *outflow*) *matrix* is formed by replacing nonzero elements of the direct dependency matrix with the

* Levins uses both positive and negative terms of the state transition equations for his production functions. This does not affect the theorems that follow.

TABLE III
Direct Dependency Matrix for the Marine Coprophagy Model[a]

q_i \ q_j (*to* \ *from*)	x_1	x_2	x_3	x_4	z_{10}	z_{30}	a_{01}	a_{21}	a_{02}	a_{32}	a_{03}	a_{43}	a_{04}	a_{34}	y_1	y_2	y_3	y_4	y_5
x_1	0	0	0	0	1	0	0	0	0	0	0	0	0	0	0	0	0	0	0
x_2	1	0	0	0	0	0	0	1	0	0	0	0	0	0	0	0	0	0	0
x_3	0	1	0	1	0	1	0	0	0	1	0	0	0	1	0	0	0	0	0
x_4	0	0	1	0	0	0	0	0	0	0	0	1	0	0	0	0	0	0	0
z_{10}	0	0	0	0	0	0	0	0	0	0	0	0	0	0	0	0	0	0	0
z_{30}	0	0	0	0	0	0	0	0	0	0	0	0	0	0	0	0	0	0	0
a_{01}	0	0	0	0	0	0	0	0	0	0	0	0	0	0	0	0	0	0	0
a_{21}	0	0	0	0	0	0	0	0	0	0	0	0	0	0	0	0	0	0	0
a_{02}	0	0	0	0	0	0	0	0	0	0	0	0	0	0	0	0	0	0	0
a_{32}	0	0	0	0	0	0	0	0	0	0	0	0	0	0	0	0	0	0	0
a_{03}	0	0	0	0	0	0	0	0	0	0	0	0	0	0	0	0	0	0	0
a_{43}	0	0	0	0	0	0	0	0	0	0	0	0	0	0	0	0	0	0	0
a_{04}	0	0	0	0	0	0	0	0	0	0	0	0	0	0	0	0	0	0	0
a_{34}	0	0	0	0	0	0	0	0	0	0	0	0	0	0	0	0	0	0	0
y_1	1	0	0	0	0	0	1	0	0	0	0	0	0	0	0	0	0	0	0
y_2	0	1	0	0	0	0	0	0	1	0	0	0	0	0	0	0	0	0	0
y_3	0	0	1	0	0	0	0	0	0	0	1	0	0	0	0	0	0	0	0
y_4	0	0	0	1	0	0	0	0	0	0	0	0	1	0	0	0	0	0	0
y_5	1	1	1	1	0	0	0	0	0	0	0	0	0	0	0	0	0	0	0

[a] A 1 indicates presence of a direct dependency link *from* q_j (columns) *to* q_i (rows), and 0 denotes absence of such a link.

fraction of the total inflow (outflow) of conserved substance to (from) q_i that comes from (goes to) each of the other system variables q_j. These types of flow matrices will figure significantly in later analysis of causality in complexly connected systems (Section V. B).

D. The Causal Chain

In addition to direct dependence, system variables are also indirectly influenced by one another as cause and effect are propagated along serial dependency sequences. Such sequences are known as causal chains (Bunge, 1959).

1. *Sequential Dependence*

A system variable q_i is defined to be *sequentially dependent* on system variable q_j if and only if there exists a finite sequence of system variables between q_j and q_i such that

$$q_j \rightarrow q_{j+1} \rightarrow q_{j+2} \rightarrow \cdots \rightarrow q_i, \tag{45}$$

where the subscripts serve to number the variables of the sequence. Each pair $q_k \rightarrow q_{k+1}$, $k = j, \ldots, i-1$, denotes a direct dependency link as defined in Eq. (37). The sequence (45) will be termed a *sequential dependency path* from q_j to q_i, symbolized by $q_j \dashrightarrow q_i$. The number of direct dependency links in such a path will be referred to as the *length* of the path.

The dependency associations of system variable q_i are represented by the following sets:

1. The *direct dependency set* D_i of system variable q_i is the set of variables q_j for which a direct dependency bond from q_j to q_i exists:

$$D_i = \{q_j : q_j \rightarrow q_i; j = 1, \ldots, m\}, \qquad i = 1, \ldots, m. \tag{46}$$

2. The *sequential dependency set* S_i of q_i is the set of system variables q_j for which a sequential dependency path from q_j to q_i exists:

$$S_i = \{q_j : q_j \dashrightarrow q_i; j = 1, \ldots, m\}, \qquad i = 1, \ldots, m. \tag{47}$$

Theorem 1

Given a set of undifferentiated system variables q_i, $i = 1, \ldots, m$, the following statements are true:

(i) Every sequential dependency path $q_j \dashrightarrow q_i$ contains an element of D_i.
(ii) $D_i \subset S_i$.

Proof:

(i) Every sequential dependency path $q_j \dashrightarrow q_i$ has the form $q_j \rightarrow q_{j+1} \rightarrow \cdots \rightarrow q_{i-1} \rightarrow q_i, \forall q_j \in S_i$. The penultimate system variable q_{i-1} is obviously a member of the direct dependency set of q_i.

(ii) Let $q_{i-1} \in D_i$ and $q_j \in S_i$. Every sequential dependency path originating from the set of variables q_j terminates at one and only one q_{i-1}. That is, in the dependency sequence $q_j \rightarrow q_{j+1} \rightarrow \cdots \rightarrow q_{i-1} \rightarrow q_i$, $q_j, q_{j+1}, \ldots, q_{i-1} \in S_i$, but only $q_{i-1} \in D_i$. In general, then, $\forall q_{i-1} \in D_i$, $q_{i-1} \in S_i$; but $\forall q_j \in S_i$, $q_j \notin D_i$. Hence, $D_i \subset S_i$.

Corollary 1

The sequential dependency set S_p of parameter q_p is the empty set.

Proof:

Suppose $S_p \neq \phi$. Theorem 1(i) asserts that every sequential dependency path to q_p must contain an element of D_p. However, from the definition of a parameter, Eq. (38), q_p is known to be directly dependent on no system variable. Therefore, $D_p = \phi$, which contradicts Theorem 1(i). Thus, $S_p = \phi$.

The following definitions pertain. Let $q_j \dashrightarrow q_i$ be a sequential dependency path from q_j to q_i. The former system variable is said to be *originating* and the latter is *terminating.* A *complete dependency path* is one which originates with a parameter q_p and terminates with an output q_y. An *incomplete* or *partial dependency path* is one that originates and/or terminates with a state variable q_x.

Theorem 2

Given a set of undifferentiated system variables q_i, $i = 1, \ldots, m$, the following statements are true:

(i) Every parameter q_p originates at least one sequential dependency path.

(ii) Every output variable q_y terminates at least one sequential dependency path.

(iii) Every system variable between the terminating and originating variables of a sequential dependency path is a state variable q_x.

Proof:

(i) Every parameter q_p has at least one system variable that is directly dependent upon it [Eq. (39)]. Therefore, there exists at least one q_j such that $\partial P_j/\partial q_p \neq 0$. That is, q_j is directly dependent on q_p, which thereby originates at least one direct dependency bond. By Theorem 1(ii), q_p thus originates at least one sequential dependency path.

(ii) Every output variable q_y is directly dependent upon some other system variable [Eq. (42)]. That is, at least one system variable q_j exists such that $\partial P_y/\partial q_j \neq 0$, and there is a direct dependency link $q_j \rightarrow q_y$ from q_j to q_y. By Theorem 1(ii), there is thus also a sequential dependency path $q_j \dashrightarrow q_y$ from q_j to q_y. Hence, q_y is the terminating variable of at least one sequential dependency path.

(iii) Every system variable q_j between the originating and terminating variables of a sequential dependency path must be directly dependent on the preceding variable q_{j-1}. Similarly, the system variable q_{j+1} that directly succeeds q_j must be directly dependent on the latter. Therefore, $\partial P_j/\partial q_{j-1} \neq 0$ and $\partial P_{j+1}/\partial q_j \neq 0$. By definitions (40) and (41), q_j must be a state variable. Hence, any system variable between the originating and terminating variables of a sequential dependency path is a state variable.

Theorem 2 is consistent with the state theory of the causal bond as developed in Section III. In the single holon case (Fig. 6a) input maps time into holon state, and state maps input causally into output. In the sequential dependency case (Fig. 6b) input q_p (i.e., both exogenous and endogenous, Section III. B) is construed as taking time into a sequence of directly dependent states, and these then map inputs into outputs q_y.

Theorem 3

Given a set of undifferentiated system variables q_i, $i = 1, \ldots, m$, the following statements are true in a causally determined model:

(i) Every state variable q_x is sequentially dependent on some system parameter q_p. That is, $\forall q_x$, $\exists q_p \in S_x$, where S_x is the sequential dependency set of q_x.

(ii) Every state variable q_x is in the sequential dependency set S_y of some output q_y. That is, $\forall q_x$, $\exists q_y$ such that $q_x \in S_y$, where S_y is the sequential dependency set of q_y.

(iii) Every state variable q_x is contained in at least one complete sequential dependency path. That is, $\forall q_x$, $\exists q_p, q_y$ such that $q_p \dashrightarrow q_x \dashrightarrow q_y$.

(iv) Every system output q_y is sequentially dependent on some system parameter q_p. That is, $\forall q_y$, $\exists q_p$ such that $q_p \dashrightarrow q_y$. Every system output terminates at least one complete sequential dependency path.

(v) Every system parameter q_p is in the sequential dependency set S_y of some system output q_y. That is, $\forall q_p$, $\exists q_y$ such that $q_p \in S_y$. Every system parameter originates at least one complete sequential dependency path.

Proof:

(i) As established in Section III, the principal role of state in causal determination is to map parameters (i.e., exogenous or endogenous "inputs") into outputs. Any state variable q_x of a system model which did not do this would be superfluous. Therefore, $\forall q_x$, $\exists q_p$ such that $q_p \in S_x$. Every state variable has at least one parameter in its sequential dependency set.

(ii) By the same reasoning, if a state variable q_x did not serve to map some parameter q_p into some output q_y, it would be superfluous in a causally determinate model. Therefore, $\forall q_x$, $\exists q_y$ such that $q_x \in S_y$. Every state variable is in the sequential dependency set of at least one system output.

To prove Theorem 3(iii) a lemma is needed.

Lemma

Sequential dependence is transitive: $q_j \dashrightarrow q_k$ and $q_k \dashrightarrow q_i \Rightarrow q_j \dashrightarrow q_i$.

Proof:

Since $q_j \dashrightarrow q_k, \exists q_{j+1}, \ldots, q_{k-1}$ such that $q_j \to q_{j+1} \to \cdots \to q_{k-1} \to q_k$. Also, since $q_k \dashrightarrow q_i, \exists q_{k+1}, \ldots, q_{i-1}$ such that $q_k \to q_{k+1} \to \cdots \to q_{i-1} \to q_i$. Therefore, $q_j \to q_{j+1} \to \cdots \to q_k \to q_{k+1} \to \cdots q_{i-1} \to q_i$, or $q_j \dashrightarrow q_i$.

(iii) Let q_x be any state variable. By Theorem 3(i), q_x is sequentially dependent on at least one system parameter q_p, that is, $q_p \dashrightarrow q_x$. By Theorem 3(ii), q_x is in the sequential dependency set of some output q_y, that is, $q_x \dashrightarrow q_y$. Therefore, by the above lemma a complete sequential dependency path containing q_x exists that originates at q_p and terminates at q_y: $q_p \dashrightarrow q_x \dashrightarrow q_y$.

(iv) Each output q_y is directly dependent on at least one state variable [Eq. (44)]. Therefore, there exists a state variable q_x such that $q_x \to q_y$. From Theorem 1(ii), it is known that $q_x \dashrightarrow q_y$, and, from Theorem 3(i), q_x is sequentially dependent upon some parameter q_p. Hence, $q_p \dashrightarrow q_x$, and the lemma establishing transitivity of sequential dependence gives $q_p \dashrightarrow q_y$. The latter is a complete sequential dependency path, and thus q_y terminates at least one complete dependency sequence.

(v) Every parameter q_p is in the direct dependency set of either a state variable q_x or an output q_y [Eq. (39)]. Suppose q_p is in the direct dependency set of q_y. Then there exists a complete sequential dependency path from q_p to q_y since $q_p \to q_y$, which by Theorem 1(ii) implies $q_p \dashrightarrow q_y$. On the other hand, suppose q_p is in the direct dependency set of a state variable q_x. Then $q_p \to q_x$ which, by Theorem 1(ii), implies $q_p \dashrightarrow q_x$. From Theorem 3(i), q_x is in the sequential dependency set of q_y, $q_x \dashrightarrow q_y$. Applying the transitivity lemma, $q_p \dashrightarrow q_y$. This is a complete dependency path, and hence q_p originates at least one complete dependency sequence.

The foregoing theorem establishes that there are no superfluous state variables in a system. Theorem 3(i) asserts that every state variable relates causally to some parameter, illustrating the role of input in determining the system state space [i.e., $\mathscr{Z}: \mathscr{T} \to \mathscr{X}$, Eq. (20)]. Theorem 3(ii) establishes that every state variable relates to some output, this result corresponding in the context of sequential dependence to the notion of states taking inputs into outputs [$\mathscr{X}: \mathscr{Z} \to \mathscr{Y}$, Eq. (21)]. Theorem 3(iii) makes every state variable of a system relevant to the propagation of cause. Later, the significance of inputs (parameters) q_p and outputs q_y in determining (via the genon) and verifying (via the creaon) the state space of causal objects and systems will be examined in some greater detail.

Two undifferentiated system variables q_i and q_j are defined to be *mutually independent* if and only if $q_i \notin S_j$ and $q_j \notin S_i$. Conversely, two variables q_i and q_j are *mutually dependent* if and only if $q_i \in S_j$ and $q_j \in S_i$.

Theorem 4

Given a set of undifferentiated system variables q_i, $i = 1, \ldots, m$, the following statements are true:

(i) Every set of mutually dependent system variables contains only state variables.

(ii) If system variables q_i and q_j are mutually dependent, then their sequential dependency sets are identical.

Proof:

(i) Mutual dependence of q_i and q_j means $q_i \in S_j$ and $q_j \in S_i$. Thus, $q_i \dashrightarrow q_j$ and $q_j \dashrightarrow q_i$. Applying the transitivity lemma, this signifies the existence of a sequential dependency path, $q_i \dashrightarrow q_j \dashrightarrow q_i$, for which q_i serves as both originating and terminating variable. By Theorem 2(iii), q_j within the sequence must be a state variable. Similarly, $q_i \dashrightarrow q_j$ and $q_j \dashrightarrow q_i$ also implies $q_j \dashrightarrow q_i \dashrightarrow q_j$, and q_i is therefore a state variable as a member of the dependency sequence originating and terminating on q_j. In addition, by definition neither a parameter [Eq. (38)] nor an output [Eq. (43)] can be mutually dependent.

(ii) Suppose q_i and q_j are mutually dependent system variables such that $q_i \in S_j$ and $q_j \in S_i$. Then, as above $q_i \dashrightarrow q_j \dashrightarrow q_i$. Let q_k be a system variable such that $q_k \in S_i$, that is, $q_k \dashrightarrow q_i$. Since $q_k \dashrightarrow q_i$ and $q_i \dashrightarrow q_j$, it follows from the previous lemma that $q_k \dashrightarrow q_j$. Therefore, $q_k \in S_j$ and $S_i \subset S_j$. By the same reasoning any element of S_j is also an element of S_i, so $S_j \subset S_i$. Hence, $S_i = S_j$, and the sequential dependency sets of mutually dependent system variables are identical.

2. *Causal Closure of Ecosystems*

As indicated earlier, Hutchinson (1948) introduced the concept, based on considerations of biogeochemical cycling, that an ecosystem is circularly causal. Modern ecological research in nutrient cycling has only served to confirm that matter recycling is a fundamental attribute of ecosystems. These facts, together with Theorem 4, lead to the following:

Conjecture. The state variables of ecosystems are all mutually dependent and thus subject to the same sets of causes.

This property of *causal closure* is illustrated in Fig. 6d. Every system variable that is a parameter is an element of the sequential dependency set of every state or output variable, and every output is sequentially dependent upon every state variable and parameter. In philosophical terms, according to the conjecture, the ecosystem is a monistic "block universe" (Bunge, 1959).

3. *Sequential Dependency Matrices*

Multiplication of direct dependency matrices can be used to trace sequential dependency paths through interactive networks of complex systems. Let $A = (a_{ij})$ be an $m \times m$ adjacency matrix (Section IV. C. 2) whose entries denote the position and number of direct dependency links from system variable q_j to variable q_i. Elements of A^2 represent the number of 2-link sequential dependency paths from q_j to q_i, and in general elements of A^l indicate the number of sequential dependency paths of length l (l direct dependency links) from q_j to q_i. The following property is known from linear algebra and network theory: With A the adjacency matrix describing the position and number of

TABLE IV
Transitive Closure Matrix B^* for the Marine Coprophagy Model[a]

to q_i \ *from* q_j	x_1	x_2	x_3	x_4	z_{10}	z_{30}	a_{01}	a_{21}	a_{02}	a_{32}	a_{03}	a_{43}	a_{04}	a_{34}	y_1	y_2	y_3	y_4	y_5
x_1	0	0	0	0	1	0	0	0	0	0	0	0	0	0	0	0	0	0	0
x_2	1	0	0	0	1	0	0	1	0	0	0	0	0	0	0	0	0	0	0
x_3	1	1	1	1	1	1	0	1	0	1	0	1	0	1	0	0	0	0	0
x_4	1	1	1	1	1	1	0	1	0	1	0	1	0	1	0	0	0	0	0
z_{10}	0	0	0	0	0	0	0	0	0	0	0	0	0	0	0	0	0	0	0
z_{30}	0	0	0	0	0	0	0	0	0	0	0	0	0	0	0	0	0	0	0
a_{01}	0	0	0	0	0	0	0	0	0	0	0	0	0	0	0	0	0	0	0
a_{21}	0	0	0	0	0	0	0	0	0	0	0	0	0	0	0	0	0	0	0
a_{02}	0	0	0	0	0	0	0	0	0	0	0	0	0	0	0	0	0	0	0
a_{32}	0	0	0	0	0	0	0	0	0	0	0	0	0	0	0	0	0	0	0
a_{03}	0	0	0	0	0	0	0	0	0	0	0	0	0	0	0	0	0	0	0
a_{43}	0	0	0	0	0	0	0	0	0	0	0	0	0	0	0	0	0	0	0
a_{04}	0	0	0	0	0	0	0	0	0	0	0	0	0	0	0	0	0	0	0
a_{34}	0	0	0	0	0	0	0	0	0	0	0	0	0	0	0	0	0	0	0
y_1	1	0	0	0	1	0	1	0	0	0	0	0	0	0	0	0	0	0	0
y_2	1	1	0	0	1	0	0	1	1	0	0	0	0	0	0	0	0	0	0
y_3	1	1	1	1	1	1	0	1	0	1	1	1	0	1	0	0	0	0	0
y_4	1	1	1	1	1	1	0	1	0	1	0	1	1	1	0	0	0	0	0
y_5	1	1	1	1	1	1	0	1	0	1	0	1	0	1	0	0	0	0	0

[a] Element $b^*_{ij} = 1$ denotes the presence of a sequential dependency path of some length *from* system variable q_j (columns) *to* variable q_i (rows); $b^*_{ij} = 0$ indicates absence of sequential dependency.

direct dependency links between system variables, the series $\Sigma_{k=1}^{l} A^k$ is convergent to a matrix $A^{(l)}$ whose entries describe the position and number of all sequential dependency paths of length $\leqslant l$ between system variables q_j and q_i, $i = 1, \ldots, m$.

If the adjacency matrix is Boolean, then only the positions of sequential dependency paths are indicated. Let $B = (b_{ij})$ be an $m \times m$ Boolean adjacency matrix, that is, an incidency matrix (Table III) whose zero and nonzero entries are subject to Boolean operations. Then, B^2 defines the positions of 2-link sequential dependency paths, B^l the positions of l-link paths, and $B^{(l)} = \Sigma_{k=1}^{l} B^k$ the positions of dependency sequences of all lengths $\leqslant l$ (Ponstein, 1966). The infinite series

$$\lim_{l \to \infty} \sum_{k=1}^{l} B^k = B^* \tag{48}$$

represents the positions of all possible sequential dependency paths of any length. B^* is termed the *transitive closure matrix* (Ore, 1962), and is illustrated for the marine coprophagy model in Table IV.

Theorem 5

The infinite series (48) converges to the transitive closure matrix B^* after at most $n + 1$ terms, where n is the order (number of state variables) of the system. That is,

$$B^* = \sum_{k=1}^{n+1} B^k, \tag{49}$$

and with $b_{ij}^* \in B^*$, (i) $b_{ij}^* = 1 \Rightarrow q_j \dashrightarrow q_i$, whereas (ii) $b_{ij}^* = 0 \Rightarrow q_j \not\dashrightarrow q_i$, $i = 1, \ldots, m$.

Proof:

Consider the sequential dependency path $x_j \dashrightarrow x_i$ from state variable x_j to state variable x_i. By Theorem 2(iii) the propagating variables between x_j and x_i are all state variables. The shortest path $\min[x_j \dashrightarrow x_i]$ from x_j to x_i will contain each intermediary state variable only once, otherwise a cycle or closed path $(x_j \dashrightarrow x_k \dashrightarrow x_k \dashrightarrow x_i)$ would be indicated and the path would not be shortest. Considering all possible pairs of state variables in the system, the longest possible shortest path between them, that is, $\max\{\min[x_j \dashrightarrow x_i], \forall x_j, \forall x_i\}$, can therefore contain at most only n state variables, where n is the order of the system. The length of such a path would be $n - 1$ direct dependency links. By Theorem 2(i) and 2(iii), every system parameter q_p is only one direct dependency link away from its nearest state variable, say x_j. Therefore, the longest possible shortest path between an originating parameter and the farthest state variable x_i from it, that is, $\max\{\min[q_p \dashrightarrow x_i], \forall q_p, \forall x_i\}$, will be composed of no more than n direct dependency links. Similarly, by Theorem 2(ii) and 2(iii), every system output variable q_y is only one direct dependency link away from its nearest state variable, say x_i. The longest possible shortest path between a parameter and the farthest terminating output variable from it, that is, $\max\{\min[q_p \dashrightarrow q_y], \forall q_p, \forall q_y\}$, will thus consist of at most $n+1$ direct dependency links. Therefore, the truncated series $\Sigma_{k=1}^{n+1} B^k$ is sufficient to identify the positions of all sequential dependency paths of all possible lengths in the system, and hence

$$\lim_{l\to\infty} \sum_{k=1}^{l} B^k = \sum_{k=1}^{n+1} B^k. \tag{50}$$

Elements b_{ij}^* of matrix (50), that is, B^*, have a value of 1 when a sequential dependency path of some length exists between q_j and q_i, and a value of 0 otherwise.

Corollary 5a

The sequential dependency set S_i of system variable q_i consists of the variables in row i of the transitive closure matrix B^* for which nonzero entries are registered.

Proof:

This follows directly from the definitions of S_i [Eq. (47)] and B^* [Eq. (48)]. As an example, in Table IV the sequential dependency set of state variable x_3 is seen to be $\{x_1, x_2, x_3, x_4, z_{10}, z_{30}, a_{21}, a_{32}, a_{43}, a_{34}\}$. Note that x_4 also has this same sequential dependency set, in accordance with Theorem 4(ii).

Corollary 5b

The system variables q_i that are sequentially dependent on q_j, that is, whose sequential dependency sets S_i contain q_j, consist of the variables in column j of B^* for which entries are nonzero.

Proof:

This also follows directly from the definition of B^*. In Table IV, for example, x_3 and x_4 are shown to be productive of the elements in $\{x_3, x_4, y_3, y_4, y_5\}$.

Example 6

To illustrate the developments of this section, consider the submatrix of state variables for the marine coprophagy model in Table III:

	$q_i \backslash q_j$	from x_1	x_2	x_3	x_4
	x_1	0	0	0	0
	x_2	1	0	0	0
to	x_3	0	1	0	1
	x_4	0	0	1	0

$= B.$

This is a Boolean adjacency matrix whose nonzero entries denote direct dependency links from column variables q_j to row variables q_i. The system state structure as described in terms of component holons in the Appendix (Fig. A.1) is reflected in the matrix, that is,

$$x_1 \xrightarrow{[1]} x_2 \xrightarrow{[1]} x_3 \underset{[1]}{\overset{[1]}{\rightleftarrows}} x_4.$$

The bracketed numbers alongside the arrows indicate lengths of the associated dependency paths, in this case all 1 since direct dependencies are involved.

The matrix B^2 is

	$q_i \backslash q_j$	from x_1	x_2	x_3	x_4
	x_1	0	0	0	0
	x_2	0	0	0	0
to	x_3	1	0	1	0
	x_4	0	1	0	1

$= B^2.$

The nonzero entries identify sequential dependency paths of length 2 from q_j to q_i:

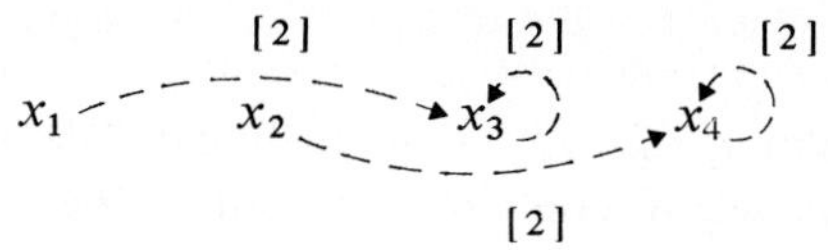

These paths are easily traced in the direct dependency diagram above. The matrix $B^{(2)}$ is

	$q_i \backslash q_j$	from x_1	x_2	x_3	x_4
	x_1	0	0	0	0
	x_2	1	0	0	0
to	x_3	1	1	1	1
	x_4	0	1	1	1

$= B + B^2.$

Sequential dependency paths of lengths ⩽2 are identified:

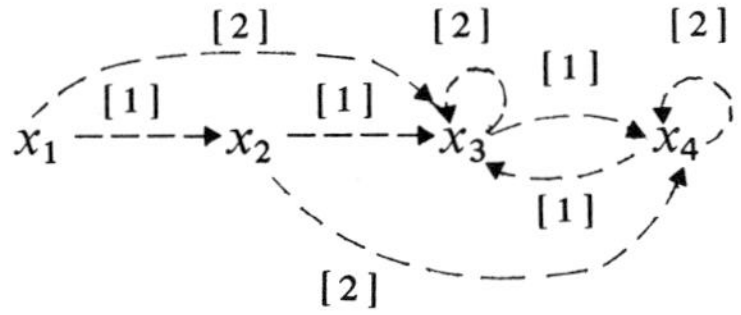

The matrix defining dependency sequences of length 3 is

q_i \ q_j (to \ from)	x_1	x_2	x_3	x_4
x_1	0	0	0	0
x_2	0	0	0	0
x_3	0	1	0	1
x_4	1	0	1	0

$= B^3$,

indicating the following sequential dependency paths:

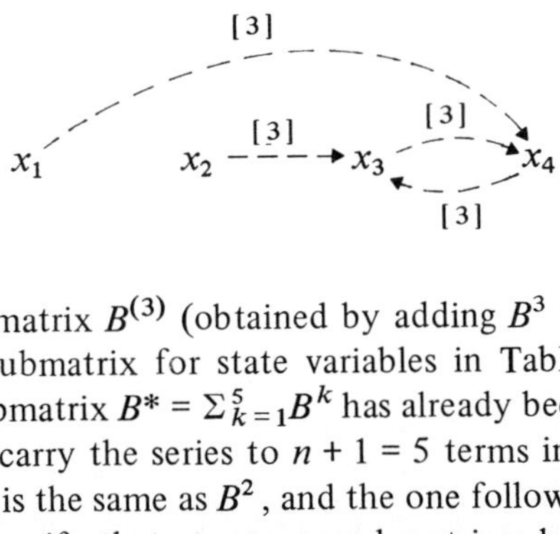

Comparison of the matrix $B^{(3)}$ (obtained by adding B^3 to $B^{(2)}$ preceding) with the corresponding submatrix for state variables in Table IV indicates that the transitive closure submatrix $B^* = \Sigma_{k=1}^{5} B^k$ has already been reached; therefore, it is not necessary to carry the series to $n + 1 = 5$ terms in order to compute B^*. The next matrix B^4 is the same as B^2, and the one following that, B^5, is equal to B^3. The reader can verify that even-powered matrices beginning with B^2 are all identical, as are odd-powered matrices beginning with B^3. The infinite series [Eq. (48)] is thus harmonic for the marine coprophagy system beginning with the second term B^2.

E. Flow Paths

Phagy—feeding—is the basis for the archetypal causal chain in ecology, the food chain. It was the study of food chains and feeding relations that led to modern trophodynamic (Lindeman, 1942) considerations and emphasis on ecosystem energy flow and material cycling. The flow of conserved quantities represents also propagation of cause, and the causal theme is implicit in ecological treatments of energy–matter. Conservative flows in systems correspond to Aristotle's material cause (Section II. E), but the extrinsic element of efficient cause is also present. That is, in Section IV. A, discussion of the decomposability of hierarchical systems led to the hypothesis of internal as well as external sources of causation. In Section IV. B inputs **z** (external) and coefficients **c** (internal) were combined into parameters **p** for the purpose of developing properties of causal sequences. When conserved physical quantities are the agents of genetic productivity, a distinction between inputs and coefficients becomes apparent.

A direct dependency link $q_j \rightarrow q_i$ which represents transfer of material or energy from system variable q_j to variable q_i will be termed a *direct flow path.* A dependency sequence $q_j \dashrightarrow q_i$ based on exchange of conserved quantities is

TABLE V
Direct Flow Matrix for the Marine Coprophagy Model[a]

to q_i \ *from* q_j	x_1	x_2	x_3	x_4	z_{10}	z_{30}	a_{01}	a_{21}	a_{02}	a_{32}	a_{03}	a_{43}	a_{04}	a_{34}	y_1	y_2	y_3	y_4	y_5
x_1	0	0	0	0	1	0	0	0	0	0	0	0	0	0	0	0	0	0	0
x_2	1	0	0	0	0	0	0	0	0	0	0	0	0	0	0	0	0	0	0
x_3	0	1	0	1	0	1	0	0	0	0	0	0	0	0	0	0	0	0	0
x_4	0	0	1	0	0	0	0	0	0	0	0	0	0	0	0	0	0	0	0
z_{10}	0	0	0	0	0	0	0	0	0	0	0	0	0	0	0	0	0	0	0
z_{30}	0	0	0	0	0	0	0	0	0	0	0	0	0	0	0	0	0	0	0
a_{01}	0	0	0	0	0	0	0	0	0	0	0	0	0	0	0	0	0	0	0
a_{21}	0	0	0	0	0	0	0	0	0	0	0	0	0	0	0	0	0	0	0
a_{02}	0	0	0	0	0	0	0	0	0	0	0	0	0	0	0	0	0	0	0
a_{32}	0	0	0	0	0	0	0	0	0	0	0	0	0	0	0	0	0	0	0
a_{03}	0	0	0	0	0	0	0	0	0	0	0	0	0	0	0	0	0	0	0
a_{43}	0	0	0	0	0	0	0	0	0	0	0	0	0	0	0	0	0	0	0
a_{04}	0	0	0	0	0	0	0	0	0	0	0	0	0	0	0	0	0	0	0
a_{34}	0	0	0	0	0	0	0	0	0	0	0	0	0	0	0	0	0	0	0
y_1	1	0	0	0	0	0	0	0	0	0	0	0	0	0	0	0	0	0	0
y_2	0	1	0	0	0	0	0	0	0	0	0	0	0	0	0	0	0	0	0
y_3	0	0	1	0	0	0	0	0	0	0	0	0	0	0	0	0	0	0	0
y_4	0	0	0	1	0	0	0	0	0	0	0	0	0	0	0	0	0	0	0
y_5	0	0	0	0	0	0	0	0	0	0	0	0	0	0	0	0	0	0	0

[a] A 1 indicates presence of a direct flow link *from* system variable q_j (columns) *to* variable q_i (rows), and 0 denotes absence of such a flow.

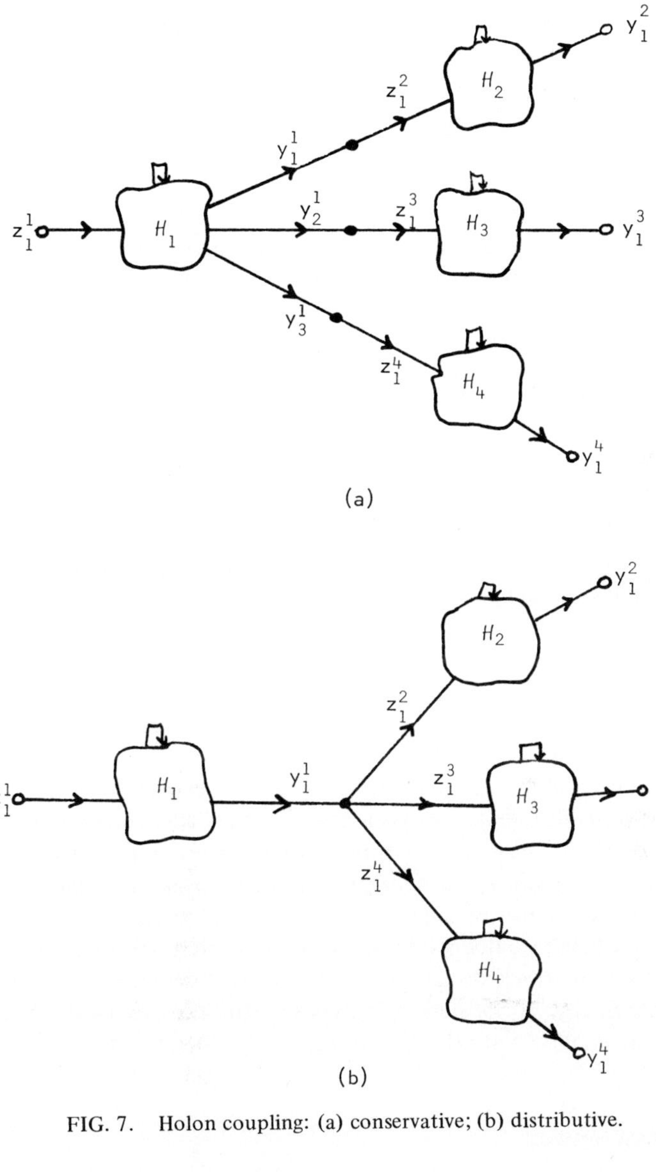

FIG. 7. Holon coupling: (a) conservative; (b) distributive.

a *sequential flow path.* An incidency matrix can be constructed to represent system dependency relations based on physical flows. Table V depicts such a matrix for the marine coprophagy system. Comparison of this matrix with the direct dependency matrix of Table III illustrates the distinction between inputs and coefficients. No flows originate from the latter, only from inputs. Physical contact of a holon with its environment occurs only through inputs or outputs. When inputs and outputs represent energy or material flows, they will be referred to as *inflows* and *outflows,* respectively. In the marine coprophagy model the inputs z_{10} and z_{30} are also inflows, whereas only the outputs $y_1, \ldots, y_4$ are outflows [Eq. (44)]. Output y_5 does not involve an energy loss from the system, but only a nonconservative transmission of information.

Thus, causality comes in two forms: (i) *conservative,* when energy or matter are transferred along causal paths, and (ii) *distributive* or *nonconservative,* when informational flows only are involved. The distinction is illustrated in Fig. 7. In Fig. 7a the total output from $\mathcal{H}_1$ is allocated as inputs to $\mathcal{H}_2$, $\mathcal{H}_3$, and $\mathcal{H}_4$. The coupling constraints are

$$y_1{}^1 = z_1{}^2, \qquad y_2{}^1 = z_1{}^3, \qquad y_3{}^1 = z_1{}^4,$$

and, from the conservation law,

$$y_1{}^1 + y_2{}^1 + y_3{}^1 = z_1{}^2 + z_1{}^3 + z_1{}^4. \tag{51}$$

On the other hand, when information is exchanged the amounts need not be conserved. In Fig. 7b the total output from $\mathcal{H}_1$ again affects $\mathcal{H}_2$, $\mathcal{H}_3$, and $\mathcal{H}_4$. The interaction constraints hold,

$$y_1{}^1 = z_1{}^2, \qquad y_1{}^1 = z_1{}^3, \qquad y_1{}^1 = z_1{}^4,$$

but this time the conservation principle is violated:

$$y_1{}^1 \neq z_1{}^2 + z_1{}^3 + z_1{}^4. \tag{52}$$

Leopold (1949) comprehended the distinction between energy–matter and information as a difference between consumption and perception. He wrote, "The same dilution and damage [from hunting] is not apparent in the yield of 'indirect' trophies such as photographs. Broadly speaking, a piece of scenery snapped by a dozen tourist cameras daily is not physically impaired, thereby, nor does any other resource suffer when the rate increases to a hundred." The camera, unlike the gun, is a harmless collector of distributive output from the environment. Energy is transferred in exposing the film, certainly, but it is not relevant to the context since (by day at least) it is superabundant.

V. The Causal Network

The causal bond and causal sequence are appealing models of becoming because of their basic simplicity and reliance on the intuitive concept of state. In Section III it was shown that determinate and nonanticipatory mappings of causes into effects required memory about past behavior, and that if a system was not past-determined this memory, expressed in terms of a set of consistency conditions, was provided by state. The causal holon as the canonical unit of causation was thus formulated as a functional relation on input and output sets:

$$y = \mathcal{H}(z), \qquad z \in \mathcal{Z},\ y \in \mathcal{Y} \tag{53}$$

[i.e., Eqs. (8)–(10), (13)–(16), (22)–(31)]. In Section IV serial sequences of holons were shown to have the property [Theorem 2(iii)] that internal input–output relations were suppressed in the propagation of cause and effect. That is, sequential dependency of a system $\mathcal{H}$ consisting of holons $\mathcal{H}_1, \ldots, \mathcal{H}_n$ in cascades was transmitted by state variables of the component holons, preserving the basic input–output formulation [Eq. (53)]; that is,

$$y = \mathcal{H}(z) = \mathcal{H}_n(\mathcal{H}_{n-1}(\cdots \mathcal{H}_1(z)), \qquad z \in \mathcal{Z}, \quad y \in \mathcal{Y}, \tag{54}$$

corresponding to the dependency sequence $z \to x_1 \to x_2 \to \cdots \to x_n \to y$. Just as the causal bond was the module for the causal sequence, in this section the latter becomes the module for propagation in networks of series and parallel paths with branching, convergence, and feedback. As with food chains and causal sequences, the food web in ecology is the type concept for representation of ecosystems as interactive networks.

A. Transmittance

An algebra for representing causal relations in networks is provided by signal flow graph theory (e.g., Huggins and Entwisle, 1968). Signal flow graphs are directed graphs in which "signals" appear at the *nodes* and (conceptually) flow over and are quantitatively transformed by the *branches,* which represent input–output operators. Signal flow graphs corresponding to Eqs. (53) and (54) are depicted in Figs. 8a and 8b, respectively, where $\mathcal{H}$ and $\mathcal{H}_1, \ldots, \mathcal{H}_n$ are construed as operators. The elementary graph unit is as shown in Fig. 8a, corresponding to the causal holon. The operator $\mathcal{H}$ is termed the branch *transmittance.*

Two properties of signal flow graphs are (i) the *addition rule,* which states that the signal at a node is the algebraic sum of all signals flowing into the node, and (ii) the *transmission rule,* namely that the signal at a node is operated upon independently by each outgoing branch. The signal content of a node is thus not affected by outgoing signals, that is it is distributive (nonconservative). Figures 8c and 8d, respectively, illustrate these two features. The transmittance of multiple direct dependency links in parallel is shown in Fig. 8e.

1. *Graph Reduction*

An *input* or *originating* node in a signal flow graph is one with no incoming branches. All other nodes are *dependent,* having one or more incoming branches.

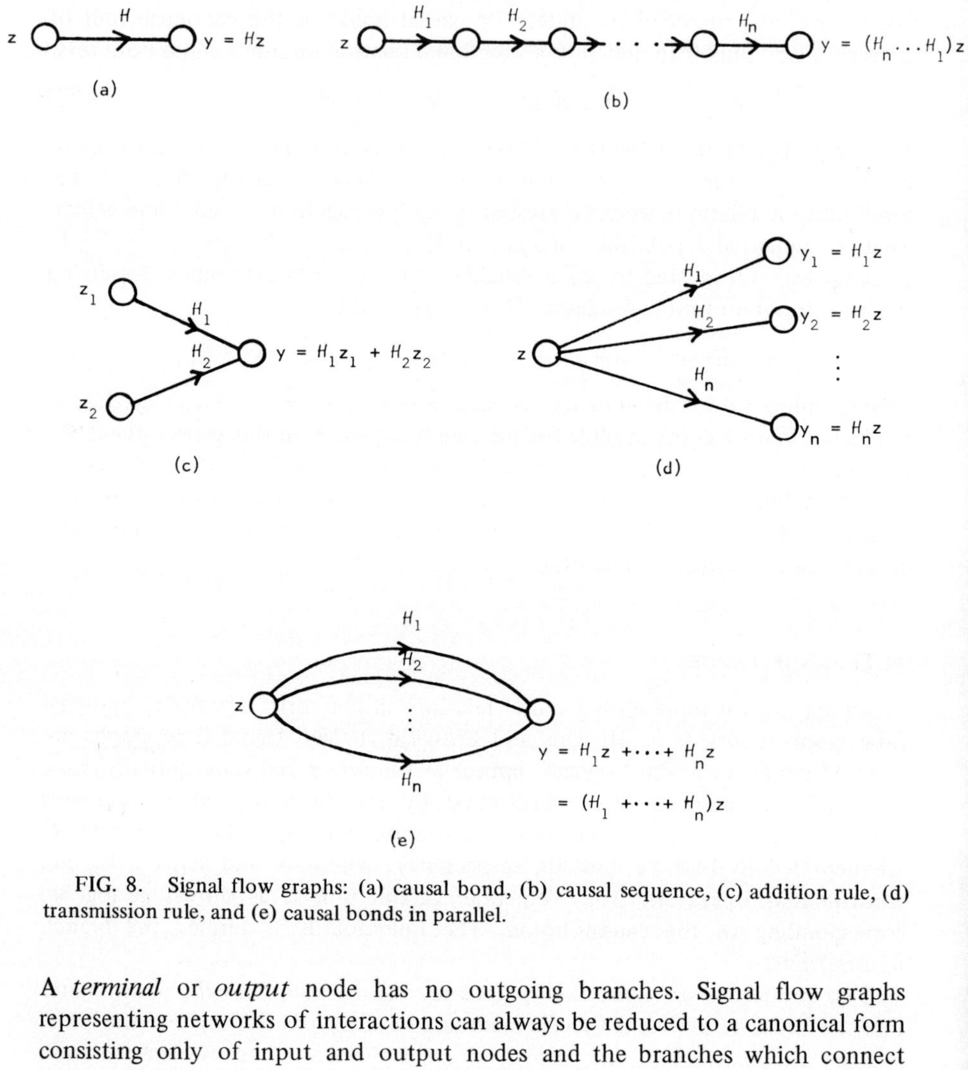

FIG. 8. Signal flow graphs: (a) causal bond, (b) causal sequence, (c) addition rule, (d) transmission rule, and (e) causal bonds in parallel.

A *terminal* or *output* node has no outgoing branches. Signal flow graphs representing networks of interactions can always be reduced to a canonical form consisting only of input and output nodes and the branches which connect them. All dependent nodes in the network representation have been eliminated, and the branches represent the total transmittance of the entire system in propagating each indicated input to corresponding indicated outputs. Elementary graph reductions are illustrated in Fig. 9.

Example 7

Suppose it is desired to reduce the graph of Fig. 10a by eliminating nodes x_1, x_2, and x_3. In Fig. 10b, nodes x_2 and x_3 have been eliminated by application of the rule for transmittances in series (Fig. 9a). The two parallel branches labeled

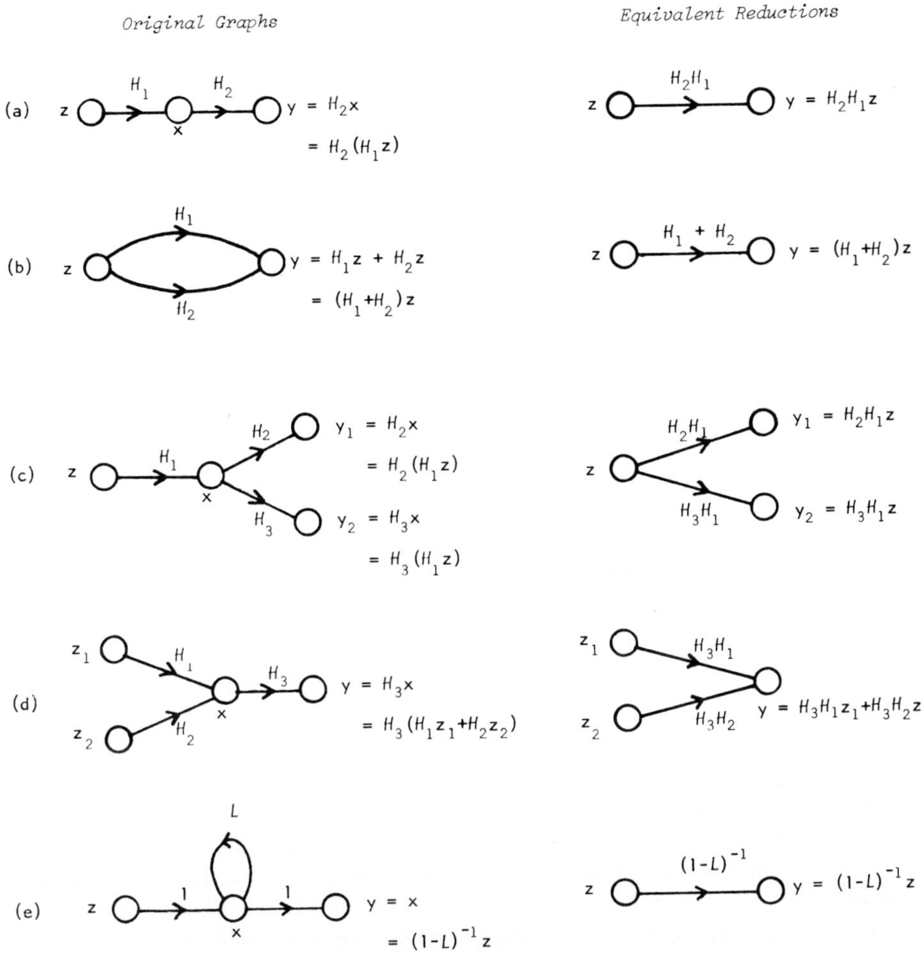

FIG. 9. Elementary signal flow graph reductions: (a) serial sequences, (b) parallel paths, (c) divergence, (d) convergence, and (e) feedback.

$\mathscr{L}_1$ and $\mathscr{L}_2$ are next combined in Fig. 10c by application of the rule for operators in parallel (Fig. 9b). Then finally, node x_1 is "absorbed" into y by application of the cascade rule (Fig. 9a) again. The final reduced graph is depicted in Fig. 10d.

2. *Feedback*

The graph reduction in the preceding example concluded with a loop on node y. Feedback occurs in a signal flow graph whenever a sequential dependency

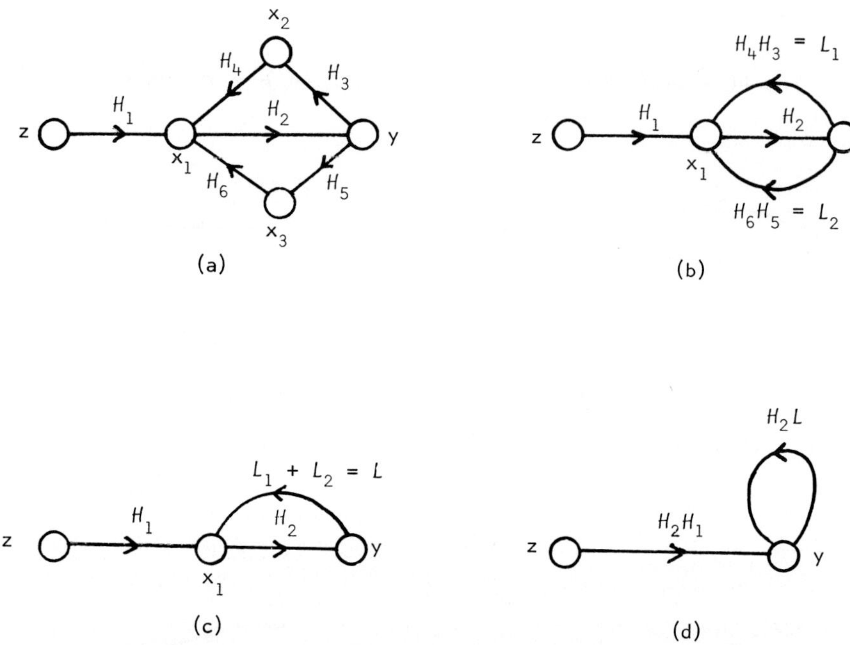

FIG. 10. Steps in reducing graph (a) by eliminating x_1, x_2, and x_3.

path (serial sequence of nodes and branches) returns to a node already in the path. Such feedback paths are termed *circuits* or *loops.* A *self-loop* consists of a branch originating and terminating on the same node as in Fig. 10d. The effect of such loops on graph transmittance is clarified in Fig. 9e. The two forward branches in the left graph of Fig. 9e each have unit transmittances. Since the signal at a node is the sum of entering signals, the signal at node x is

$$x = z + \mathscr{L}x. \tag{55}$$

Therefore,

$$(1-\mathscr{L})x = z, \qquad \text{and} \qquad x = (1-\mathscr{L})^{-1}z.$$

The output signal is thus

$$y = x = (1-\mathscr{L})^{-1}z. \tag{56}$$

Hence, node x is absorbed and the transmittance of the effective branch between z and y is $(1-\mathscr{L})^{-1}$, as shown in the right graph of Fig. 9e. Since at node x the self-loop furnishes the signal $\mathscr{L}x$, only the difference $(1-\mathscr{L})x$ is supplied from the input node. For this reason the quantity $(1-\mathscr{L})$ is termed the *return difference.* If the loop has unit transmittance, $\mathscr{L} = 1$ and the return difference is 0, meaning that no external signal source (efficient cause) is required to sustain a signal at node x. To interpret the effect of a self-loop further, consider in the left diagram of Fig. 9e the number of different routes by which the input signal can find its way to the terminal node. This number of paths is infinite since there is no specified limit to the number of times the self-loop may be traversed. These paths have respective transmittances $1, \mathscr{L}, \mathscr{L}^2, \ldots$, and hence the output from the graph can be written

$$y = \left(\sum_{k=0}^{\infty} \mathscr{L}^k \right) z. \tag{57}$$

The indicated transmittance is an infinite series describing a geometric progression, and it is well known that the series converges to $1/(1 - \mathscr{L})$, if $0 < \mathscr{L} < 1$. The input–output relation of the Fig. 9e graph is thus

$$y = (1 - \mathscr{L})^{-1} z, \tag{58}$$

as obtained previously.

Example 8

Figure 11 is a signal flow graph for the marine coprophagy model $\mathscr{H}$ as described by Eqs. (44). Note that the coefficient notations a_{ij} here represent branch transmittances and are used only for convenience in identifying correspondences with the state space model (44). Steps in the reduction of this graph to canonical form are shown in Fig. 12. Node x_1 has been eliminated in Fig. 12a, leaving the residual graph as indicated. The darkened node x_2 is

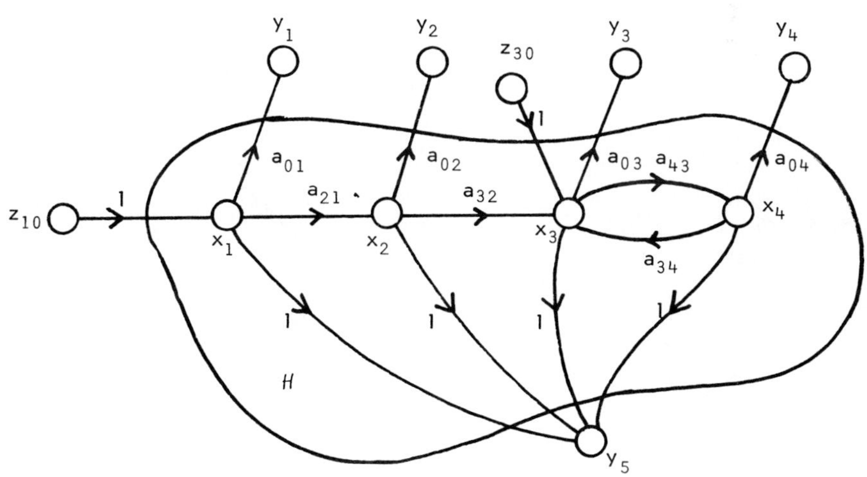

FIG. 11. Signal flow graph for the marine coprophagy model [Eq. (44)].

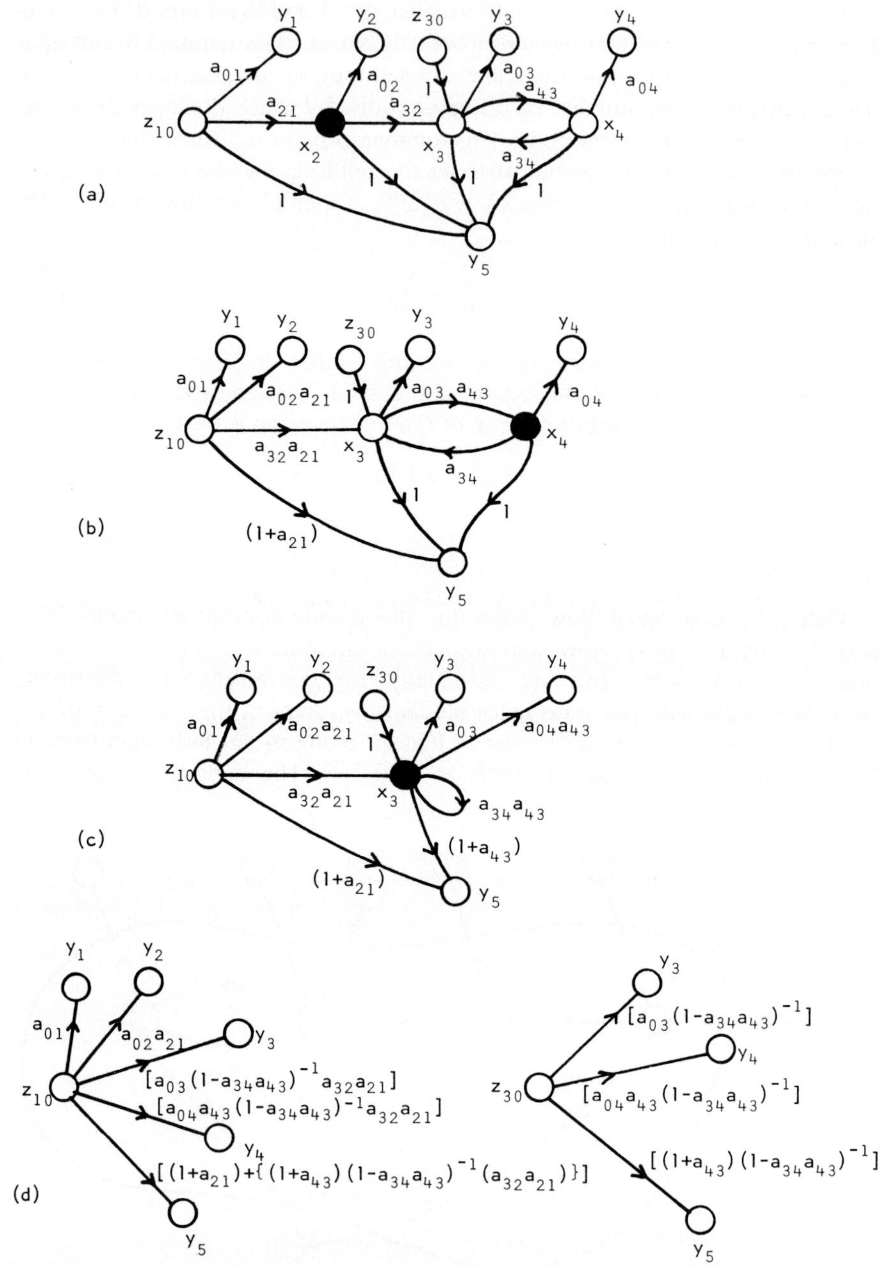

FIG. 12. Steps in reduction of the marine coprophagy model signal flow graph (Fig. 11).

eliminated next, giving Fig. 12b. In Fig. 12c, the next darkened node x_4 has been eliminated, leaving a self-loop on node x_3 as shown. In the last diagram, Fig. 12d, all intermediary nodes have been removed and only input and output nodes remain. The transmittances from each input z_{10} and z_{30} to each of the five outputs are as indicated adjacent to the remaining branches. Expressions summarizing the system input–output relations in terms of propagation of cause are as follows:

$$\begin{aligned}
y_1 &= [a_{01}]z_{10}, \\
y_2 &= [a_{02}a_{21}]z_{10}, \\
y_3 &= [a_{03}(1-a_{34}a_{43})^{-1}a_{32}a_{21}]z_{10} + [a_{03}(1-a_{34}a_{43})^{-1}]z_{30}, \\
y_4 &= [a_{04}a_{43}(1-a_{34}a_{43})^{-1}a_{32}a_{21}]z_{10} + [a_{04}a_{43}(1-a_{34}a_{43})^{-1}]z_{30}, \\
y_5 &= [(1+a_{21})+\{(1+a_{43})(1-a_{34}a_{43})^{-1}(a_{32}a_{21})\}]z_{10} \\
&\quad + [(1+a_{43})(1-a_{34}a_{43})^{-1}]z_{30}.
\end{aligned} \tag{59}$$

These equations indicate in an operator sense which endogenous parameters of the system are involved in taking exogenous causes into effects. That is, $y_1 = [a_{01}]z_{10}$ means that a respiration operator a_{01} combines input z_{10} and state x_1 into an output mapping y_1. The actual process is specifically defined in Eqs. (44) in accordance with the state space model [Eqs. (22–23)]; that is,

$$\begin{aligned}
f_1&: \{x_1\} \times \{z_{10}, a_{01}, a_{21}\} \to \{x_1\}, \\
g_1&: \{x_1\} \times \{a_{01}\} \to \{y_1\}.
\end{aligned}$$

3. *Graph Determinant*

It is evident from the foregoing example for a simple model that graph reduction of complex systems would be an immensely complicated problem without an algorithm. Such an algorithm requires the concept of a graph determinant. Let node q_i of a signal flow graph representing an interactive network be sequentially dependent upon node q_j, $i, j = 1, \ldots, m$. In general, there may be numerous paths from q_j to q_i, with or without loops. A *forward path* is a dependency sequence with no feedbacks, that is, passing through a given node only once. A *reverse path* is a loop such that signals flow away from q_i and toward q_j. Loops and paths are said to be *touching* if they share a node, and *nontouching* otherwise.

Consider the graph reduction illustrated in Fig. 13. Node q_k in Fig. 13a is eliminated in 13b. This elimination involves formation of a composite operator which takes into account the forward path transmittance $\mathcal{H}_{kj}\mathcal{H}_{ik}$ as well as the effect of the loop, $(1-\mathcal{H}_{kk})^{-1}$. In Fig. 13c the two parallel forward paths from q_j to q_i are combined to form the composite forward path transmittance $\mathcal{P}_{ij}$.

The loop $\mathscr{P}_{ij}\mathcal{H}_{ji}$ is then expressed (Fig. 13d) as a self-loop on node q_i, and finally in Fig. 13e the paths from q_j to q_i are all combined in a single operator whose numerator represents the forward paths and whose denominator represents feedback. Expressing this transmittance in terms of the individual branch transmittances (commutativity of operators is assumed for algebraic convenience with some loss of generality), the subgraph transmittance $\mathscr{G}_{ij}$ from node q_j to node q_i is obtained as follows:

$$\mathscr{G}_{ij} = (1 - \mathscr{P}_{ij}\mathcal{H}_{ji})^{-1}\mathscr{P}_{ij} = \mathscr{P}_{ij}/(1 - \mathscr{P}_{ij}\mathcal{H}_{ji})$$

$$= \frac{\mathcal{H}_{ij} + \mathcal{H}_{ik}(1 - \mathcal{H}_{kk})^{-1}\mathcal{H}_{kj}}{1 - (\mathcal{H}_{ij} + \mathcal{H}_{ik}(1 - \mathcal{H}_{kk})^{-1}\mathcal{H}_{kj})\mathcal{H}_{ji}}$$

$$= \frac{\mathcal{H}_{ij} + \mathcal{H}_{ik}(1 - \mathcal{H}_{kk})^{-1}\mathcal{H}_{kj}}{1 - \mathcal{H}_{ij}\mathcal{H}_{ji} - \mathcal{H}_{ik}(1 - \mathcal{H}_{kk})^{-1}\mathcal{H}_{kj}\mathcal{H}_{ji}}$$

$$= \frac{[\mathcal{H}_{ij} + \mathcal{H}_{ik}(1 - \mathcal{H}_{kk})^{-1}\mathcal{H}_{kj}](1 - \mathcal{H}_{kk})}{[1 - \mathcal{H}_{ij}\mathcal{H}_{ji} - \mathcal{H}_{ik}(1 - \mathcal{H}_{kk})^{-1}\mathcal{H}_{kj}\mathcal{H}_{ji}](1 - \mathcal{H}_{kk})}$$

$$= \frac{\mathcal{H}_{ij}(1 - \mathcal{H}_{kk}) + \mathcal{H}_{ik}\mathcal{H}_{kj}}{(1 - \mathcal{H}_{kk}) - \mathcal{H}_{ij}\mathcal{H}_{ji}(1 - \mathcal{H}_{kk}) - \mathcal{H}_{ik}\mathcal{H}_{kj}\mathcal{H}_{ji}}$$

$$= \frac{\mathcal{H}_{ij}(1 - \mathcal{H}_{kk}) + \mathcal{H}_{ik}\mathcal{H}_{kj}}{1 - \mathcal{H}_{kk} - \mathcal{H}_{ij}\mathcal{H}_{ji} + \mathcal{H}_{ij}\mathcal{H}_{ji}\mathcal{H}_{kk} - \mathcal{H}_{ik}\mathcal{H}_{kj}\mathcal{H}_{ji}}.$$

Denoting the three loops in the original graph segment (Fig. 13a) as $\mathscr{L}_1 = \mathcal{H}_{kk}$, $\mathscr{L}_2 = \mathcal{H}_{ij}\mathcal{H}_{ji}$, and $\mathscr{L}_3 = \mathcal{H}_{ik}\mathcal{H}_{kj}\mathcal{H}_{ji}$, the final expression becomes

$$\mathscr{G}_{ij} = [\mathcal{H}_{ij}(1 - \mathscr{L}_1) + \mathcal{H}_{ik}\mathcal{H}_{kj}]/(1 - \mathscr{L}_1 - \mathscr{L}_2 - \mathscr{L}_3 + \mathcal{H}_{ij}\mathcal{H}_{ji}\mathcal{H}_{kk}). \quad (60)$$

The structure of cause-and-effect propagation from q_j to q_i is now apparent. The numerator of Eq. (60) represents the two forward paths. The first term consists of the direct dependency transmittance $\mathcal{H}_{ij}$ augmented by an expression representing the only loop in the system ($\mathscr{L}_1 = \mathcal{H}_{kk}$, Fig. 13a) that does not touch it. The second term consists of the forward path transmittance along the second path, unaugmented by any loop expression since $\mathscr{L}_1$, $\mathscr{L}_2$, and $\mathscr{L}_3$ all touch this path. The denominator consists of (i) the identity operator minus (ii) the operators denoting feedback paths plus (iii) the product of the transmittances of nontouching loops. The denominator is called the *determinant* of the subgraph formed by selecting q_j as originating and q_i as terminating node. When the signal flow graph for an entire system is considered, the denominator quantity represents the *graph determinant.*

(a)

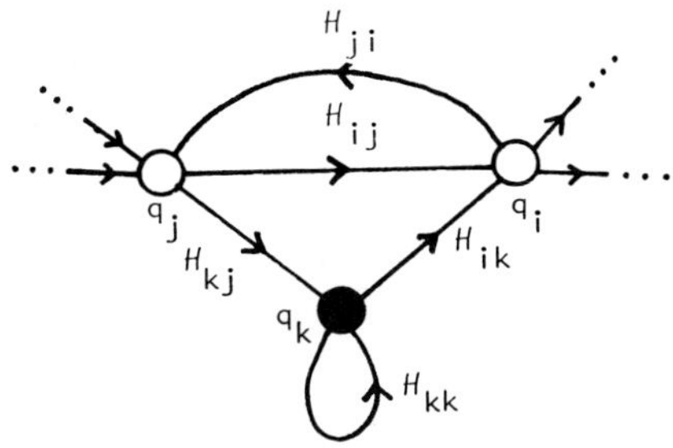

(b)

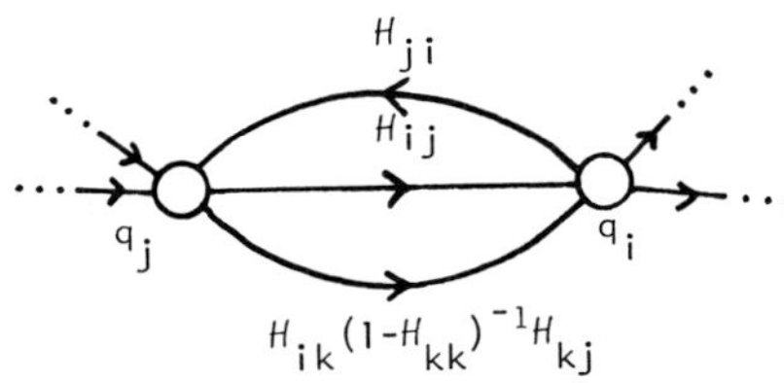

(c)

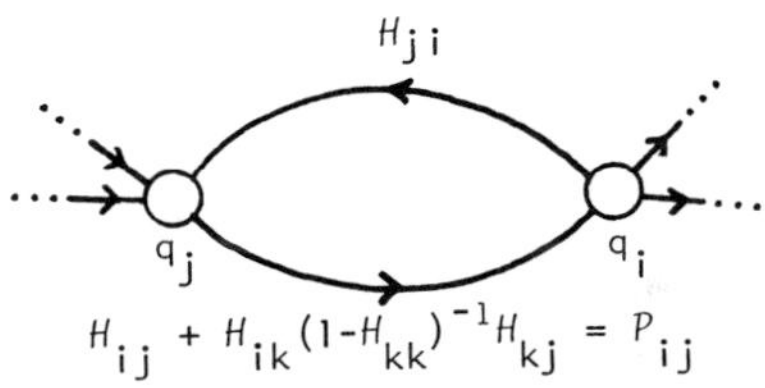

(d)

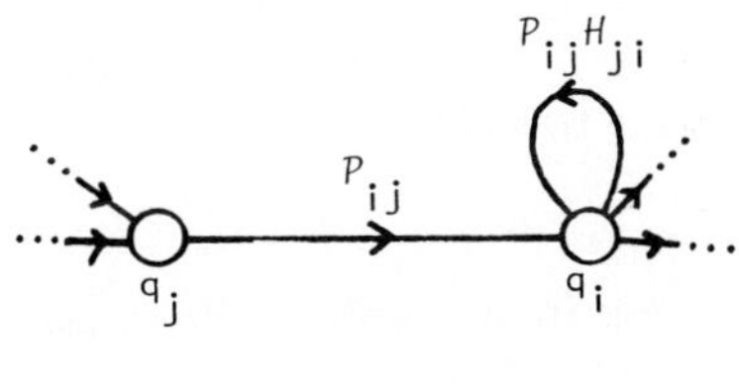

(e)

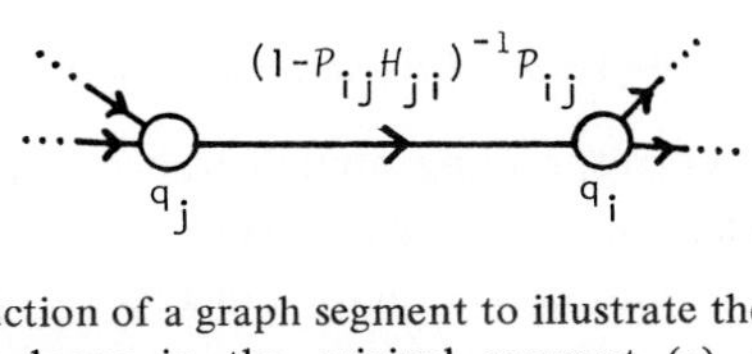

FIG. 13. Steps in the reduction of a graph segment to illustrate the nature of the graph determinant. There are three loops in the original segment (a), with transmittances $\mathcal{L}_1 = \mathcal{H}_{kk}$, $\mathcal{L}_2 = \mathcal{H}_{ij}\,\mathcal{H}_{ji}$, and $\mathcal{L}_3 = \mathcal{H}_{ik}\mathcal{H}_{kj}\mathcal{H}_{ji}$, respectively.

4. *Mason's Theorem*

The algorithm for determining system transmittance between specified inputs and outputs without explicit node examination and graph reduction is based on a theorem due to Mason (1956). Let $\mathscr{G}_{ij}$ be the graph transmittance from node q_j to q_i in a network, $\mathscr{P}_{ij}^k$ the transmittance of the kth forward path from q_j to q_i, Δ_{ij}^k the kth *path cofactor,* that is, the graph determinant when transmittances for all feedback loops touching the kth path are deleted, and Δ the graph determinant. Then, Mason's theorem gives

$$\mathscr{G}_{ij} = \frac{\sum_{k=1}^{p} \mathscr{P}_{ij}^k \Delta_{ij}^k}{\Delta}, \qquad i, j = 1, \ldots, m. \tag{61}$$

Or, in expanded form,

$$\mathscr{G}_{ij} = \left[\frac{(\mathscr{P}_{ij}^1 + \cdots + \mathscr{P}_{ij}^k + \cdots + \mathscr{P}_{ij}^p)(1 - \mathscr{L}_1)(1 - \mathscr{L}_2) \cdots (1 - \mathscr{L}_l)}{(1 - \mathscr{L}_1)(1 - \mathscr{L}_2) \cdots (1 - \mathscr{L}_l)}\right]^*, \tag{62}$$

where p is the number of forward paths from q_j to q_i, l is the number of loops in the system, and the asterisk signifies that (i) products of touching loop transmittances are deleted to form Δ, and (ii) any terms of Δ that represent loops touching the kth forward path are deleted to form Δ_{ij}^k.

Example 9

Graph transmittances for the signal flow graph of the marine coprophagy model (Fig. 11) can readily be determined by use of Mason's rule. For example, consider the signal flow from input z_{10} to output y_5. There are four forward paths with the following transmittances:

$$\mathscr{P}_{y_5 z_{10}}^1 = 1, \qquad \mathscr{P}_{y_5 z_{10}}^2 = a_{21}, \qquad \mathscr{P}_{y_5 z_{10}}^3 = a_{32} a_{21}, \qquad \mathscr{P}_{y_5 z_{10}}^4 = a_{43} a_{32} a_{21}.$$

There is only one loop in the system, $\mathscr{L}_1 = a_{34} a_{43}$. The graph determinant is therefore simply $\Delta = (1 - \mathscr{L}_1) = (1 - a_{34} a_{43})$. Inserting these quantities into Eqs. (61)–(62) the graph transmittance from z_{10} to y_5 is

$$\mathscr{G}_{y_5 z_{10}} = (1 + a_{21} + a_{32} a_{21} + a_{43} a_{32} a_{21})/(1 - a_{34} a_{43}).$$

This is exactly the corresponding branch transmittance as labeled in Fig. 12d. It is left as an exercise for the reader to verify by applying Mason's theorem the input–output transmittances as obtained in Fig. 12 by graph reduction.

B. Input–Output Flow Analysis

For systems in which causality is transmitted by conservative energy–matter flows, input–output analysis as developed in economics permits the tracing of cause and effect within the interactive network. The focus on conservative flows restricts the domain of the causality concept so considered, but what significant classes or characteristics of causation as a general concept are ruled out is uncertain. Later, classes that are ruled in will be stated in the explicit form of a theorem, and for the moment the admissible set can be taken to include at the least conservative flows which by themselves justify ecological attention.

Economics has a much longer history than ecology in attempting to understand causal relationships in complex systems. Quesnay (1758) devised a table ("Tableau Économique") which displayed the interdependence of economic activities on a farm. Successive "rounds" of wealth producing activities were graphically presented. Walras (1877) developed a theory to determine all prices in an economy simultaneously. The production demands of each industry for raw materials and the flexibility of production requirements in various industries determined the interdependence between industries. Leontief (1936) simplified the Walrasian system and devised a way to determine direct and indirect requirements of each industry for a unit of production. This type of analysis assumed that demand, and therefore outflow, was the originator of cause in the system, an interesting inversion of Aristotelian efficient cause. Phillips (1955) interpreted Quesnay's "Tableau Économique" as a static, closed Leontief system. The derivation of Leontief's static, steady-state, open system which follows parallels Leontief (1966) and Chenery and Clark (1959). For other applications of the methodology to ecological systems, see Hannon (1973), Finn (1976), and Cale (1975).

1. *Production Matrix*

Given a partition of the economic system into sectors (corresponding to holons) a matrix of flows is constructed, the transaction (Chenery and Clark, 1959) or production (Hannon, 1973) matrix. The term *production matrix* will be used here to reflect the fact that genetic productivity (Section II. D) in the form of flows is represented. For steady-state systems the production matrix P is a modified direct flow matrix (i.e., Section IV. D. 3, Table V) in which (i) binary 1's are replaced by static flow values, (ii) inflow and outflow submatrices are de-diagonalized, and (iii) an additional row of column sums and an additional column of row sums are added. With n the order of the system the structure of the square $n + 3$ production matrix P is as shown in (63).

(63)

q_i \ q_j	*from* x_1	x_2	$\dots$	x_n	Inflow	Outflow	Row Σ
x_1	ϕ_{11}	ϕ_{12}	$\dots$	ϕ_{1n}	$\phi_{10} = z_{10}$	0	T_1
x_2	ϕ_{21}	ϕ_{22}	$\dots$	ϕ_{2n}	$\phi_{20} = z_{20}$	0	T_2
$\vdots$	$\vdots$	$\vdots$		$\vdots$	$\vdots$	$\vdots$	$\vdots$
to x_n	ϕ_{n1}	ϕ_{n2}	$\dots$	ϕ_{nn}	$\phi_{n0} = z_{n0}$	0	T_n
Inflow	0	0	$\dots$	0	0	0	0
Outflow	$\phi_{01} = y_{01}$	$\phi_{02} = y_{02}$	$\dots$	$\phi_{0n} = y_{0n}$	0	0	Σy
Column Σ	T_1	T_2	$\dots$	T_n	Σz	0	$TST + (\Sigma z = \Sigma y)$

Intrasystem direct flows from state variables x_j to x_i are denoted ϕ_{ij}, $i, j = 0, \dots, n$. Total inflow to the system is

$$\Sigma z = \sum_{i=1}^{n} z_{i0}, \tag{64}$$

and total outflow

$$\Sigma y = \sum_{j=1}^{n} y_{0j}. \tag{65}$$

Throughflow of the kth holon is defined equivalently (in the steady state) as the sum either of inflows

$$T_k = \sum_{j=1}^{n} \phi_{kj} + z_{k0}, \qquad k = 1, \dots, n, \tag{66}$$

or of outflows

$$T_k = \sum_{i=1}^{n} \phi_{ik} + y_{0k}, \qquad k = 1, \dots, n. \tag{67}$$

Total system throughflow is the sum of component throughflows

$$TST = \sum_{k=1}^{n} T_k. \tag{68}$$

Table VI is a production matrix for the marine coprophagy model based on steady-state carbon flows as described in the Appendix, Fig. A.1. Note that this matrix has the same structure as the corresponding direct flow matrix of Table V. Furthermore, the entries in the Table VI matrix represent quantification of the original production function used in Section IV. C. 1 to define direct dependence [Eq. (39)]. The coprophagy model production functions $P_i(q_1, \dots, q_m)$, $i = 1, \dots, m$, actually employed in Example 5 correspond to entries in Table VI as shown in the accompanying matrix.

q_i \ q_j (from →)	x_1	x_2	x_3	x_4	Inflow	Outflow	Row Σ
x_1	0	0	0	0	P_3	0	P_3
x_2	P_6	0	0	0	0	0	P_6
x_3	0	ϵP_{12}	0	ϵP_{12}	ϵP_{12}	0	P_{12}
to x_4	0	0	P_{10}	0	0	0	P_{10}
Inflow	0	0	0	0	0	0	0
Outflow	P_{15}	P_{16}	P_{17}	P_{18}	0	0	$\Sigma_{i=15}^{18} P_i$
Column Σ	$P_6 + P_{15}$	$\epsilon P_{12} + P_{16}$	$P_{10} + P_{17}$	$\epsilon P_{12} + P_{18}$	$P_3 + \epsilon P_{12}$	0	–

TABLE VI
Production Matrix P for the Steady-State Marine Coprophagy Model[a]

q_i \ q_j (from →)	x_1	x_2	x_3	x_4	inflow	outflow	row Σ
x_1	0	0	0	0	124.1	0	124.1
x_2	21.9	0	0	0	0	0	21.9
x_3	0	15.3	0	52.9	255.5	0	323.7
to x_4	0	0	79.6	0	0	0	79.6
inflow	0	0	0	0	0	0	0
outflow	102.2	6.6	244.1	26.7	0	0	379.6
column Σ	124.1	21.9	323.7	79.6	379.6	0	928.9
CE_j	0.18	0.70	0.25	0.66			TST = 549.3

[a] Nonzero entries are carbon flows, in gC m^{-2} y^{-1}; *TST* is total system throughflow [Eq. (68)]; and CE_j is cycling efficiency as discussed in Example 15.

2. *Creaon Inflow Analysis*

To analyze the dependency structure implied by network flows, Leontief assumed that each direct flow ϕ_{ij} from holon $\mathcal{H}_j$ to holon $\mathcal{H}_i$ [$i = 1, \ldots, n$; $j = 0, \ldots, n$, where $j = 0$ denotes environment $\mathcal{H}'$ (see Fig. 4a) of the system $\mathcal{H}$ defined to include $\mathcal{H}_i$ and $\mathcal{H}_j$ as components] could be expressed as a fraction q'_{ij} of throughflow T_i into $\mathcal{H}_i$ (Fig. 14a):

$$\phi_{ij} = q'_{ij} T_i, \qquad j = 0, \ldots, n. \tag{69}$$

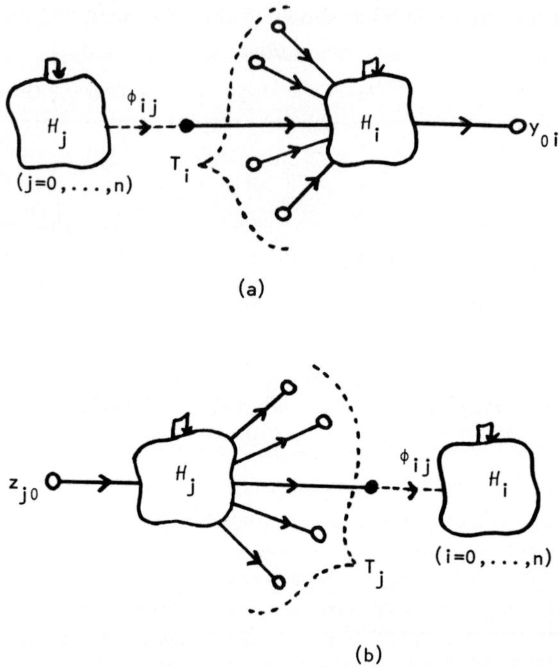

FIG. 14. Derivation of fractional inflow and outflow matrices: (a) input object (creaon); (b) output object (genon).

The expression for throughflow of $\mathcal{H}_j$ can thus [from Eq. (67)] be written

$$T_j = \sum_{i=1}^{n} \phi_{ij} + y_{0j} = \sum_{i=1}^{n} q'_{ij} T_i + y_{0j}, \qquad j = 0, \ldots, n. \tag{70}$$

Letting **T** and **y** be row vectors $[1 \times (n+1)]$ of throughflows and outflows, respectively, and Q' be an $(n+1) \times (n+1)$ square *fractional inflow matrix*, Eq. (70) can be written

$$\mathbf{T} = \mathbf{T}Q' + \mathbf{y}.$$

Rearranging, and solving for **T**,

$$\mathbf{T}(I - Q') = \mathbf{y},$$

and

$$\mathbf{T} = \mathbf{y}(I - Q')^{-1}. \tag{71}$$

That is, the throughflow at each $\mathcal{H}_j$ is partitioned into portions responsible for the maintenance of each system output y_{0i}:

TABLE VII Creaon Inflow Analysis Matrices for the Marine Coprophagy Model[a]

(i) *Q' matrix:*

q_i \ q_j (to \ from)	x_1	x_2	x_3	x_4	z_{10}	z_{30}
x_1	0	0	0	0	1.0	0
x_2	1.0	0	0	0	0	0
x_3	0	0.047	0	0.163	0	0.789
x_4	0	0	1.0	0	0	0
z_{10}	0	0	0	0	0	0
z_{30}	0	0	0	0	0	0

(ii) *(I-Q') matrix:*

q_i \ q_j (to \ from)	x_1	x_2	x_3	x_4	z_{10}	z_{30}
x_1	1.0	0	0	0	-1.0	0
x_2	-1.0	1.0	0	0	0	0
x_3	0	-0.047	1.0	-0.163	0	-0.789
x_4	0	0	-1.0	1.0	0	0
z_{10}	0	0	0	0	1.0	0
z_{30}	0	0	0	0	0	1.0

(iii) *$(I-Q')^{-1}$ matrix:*

q_i \ q_j (to \ from)	x_1	x_2	x_3	x_4	z_{10}	z_{30}	YPL_i
x_1	1.0	0	0	0	1.0	0	1.0
x_2	1.0	1.0	0	0	1.0	0	2.0
x_3	0.057	0.057	1.195	0.195	0.057	0.943	1.504
x_4	0.057	0.057	1.195	1.195	0.057	0.943	2.504
z_{10}	0	0	0	0	1.0	0	
z_{30}	0	0	0	0	0	1.0	

(iv) *T matrix, $Y(I-Q')^{-1}$*

q_i \ q_j (to \ from)	x_1	x_2	x_3	x_4	z_{10}	z_{30}
x_1	102.2	0	0	0	102.2	0
x_2	6.6	6.6	0	0	6.6	0
x_3	13.8	13.8	291.8	47.7	13.8	230.3
x_4	1.5	1.5	31.9	31.9	1.5	25.2
z_{10}	0	0	0	0	0	0
z_{30}	0	0	0	0	0	0
T_j	124.1	21.9	323.7	79.6	-	-
inflow Σ	-	-	-	-	124.1	255.5

[a] Top to bottom: (i) fractional inflow matrix Q'; (ii) $[I - Q']$ matrix; (iii) transitive closure inflow matrix, $(I - Q')^{-1}$; (iv) throughflow matrix, $Y(I - Q')^{-1}$, in gC m^{-2} y^{-1}. YPL_i in (iii) is outflow path length, discussed in Example 14.

$$T_j = \sum_{k=1}^{n} y_{0k}(I - Q')_{kj}^{-1}$$

$$= y_{01}(I - Q')_{1j}^{-1} + \cdots + y_{0i}(I - Q')_{ij}^{-1} + \cdots + y_{0n}(I - Q')_{nj}^{-1},$$

or, letting $\Phi_{ij} = y_{0i}(I - Q')_{ij}^{-1}$,

$$T_j = \Phi_{1j} + \cdots + \Phi_{ij} + \cdots + \Phi_{nj}. \tag{72}$$

Just as ϕ_{ij} [Eq. (63)] represents direct dependence of $\mathcal{H}_i$ on $\mathcal{H}_j$, each quantity Φ_{ij} denotes the causal flow from $\mathcal{H}_j$ (i.e., the fraction of T_j) sequentially propagated over all paths of all lengths to $\mathcal{H}_i$ to maintain y_{0i}. Thus,

$$y_{0i}(I - Q')_{ij}^{-1} = \Phi_{ij} = T_j - \sum_{\substack{k=1 \\ k \neq i}}^{n} \Phi_{kj},$$

and

$$(I - Q')_{ij}^{-1} = \Phi_{ij}/y_{0i} = \left[T_j - \sum_{\substack{k=1 \\ k \neq i}}^{n} \Phi_{kj}\right] \Big/ y_{0i}. \tag{73}$$

The matrix $(I - Q')^{-1}$ is a transitive closure inflow matrix whose elements $(I - Q')_{ij}^{-1}$ represent [Eq. (73)] the flow of cause from $\mathcal{H}_j$ to $\mathcal{H}_i$, or the fraction of throughflow T_j, required to produce one unit of outflow y_{0i} from $\mathcal{H}_i$. When $\mathcal{H}_j$ and $\mathcal{H}_i$ are directly dependent as in Fig. 14a, element $(I - Q')_{ij}^{-1}$ in effect quantifies the causal bond. However, the matrix $(I - Q')^{-1}$ ranges over all system variables and thus its elements collectively quantify sequential dependencies within the system. As shown in Section V. B. 4, all dependency paths of all lengths are accounted for, which is why $(I - Q')^{-1}$ is a "transitive closure" matrix.

Example 10

Consider the marine coprophagy model production matrix (Table VI). In accordance with Eq. (69), divide each flow ϕ_{ij} ($j = 0, \ldots, n$ so the inflows are included, but $i = 1, \ldots, n$ so the outflows are excluded) by the throughflow T_i of the ith row (row sums in Table VI). The two inflows z_{10} and z_{30} are put into separate vectors in order to identify their effects individually. Table VII(i) displays the resulting $(n + 2) \times (n + 2)$ Q' matrix. Note that the state variable rows sum to 1, signifying normalization of inflows. Subtracting Q' from the identity matrix I gives $(I - Q')$ as shown in Table VII(ii). The inverse of this matrix, $(I - Q')^{-1}$, as obtained by computer appears in Table VII(iii). Elements of this matrix [Eq. (73)] represent all direct and indirect (sequential) flows in the system, which is the transitive closure property. Premultiplying $(I - Q')^{-1}$ by the $\mathbf{y}$ vector in accordance with Eq. (71) would yield a 1×4 row vector $\mathbf{T}$. More information can be retained if instead the y vector is diagonalized to a 6 x 6 matrix:

$$Y = \begin{bmatrix} 102.2 & 0 & 0 & 0 & 0 & 0 \\ 0 & 6.6 & 0 & 0 & 0 & 0 \\ 0 & 0 & 244.1 & 0 & 0 & 0 \\ 0 & 0 & 0 & 26.7 & 0 & 0 \\ 0 & 0 & 0 & 0 & 0 & 0 \\ 0 & 0 & 0 & 0 & 0 & 0 \end{bmatrix}.$$

Postmultiplying this matrix by $(I - Q')^{-1}$, analogously to Eq. (71), gives a matrix T of throughflows, Table VII(iv). Columns of this matrix sum to the throughflows T_j associated with $\mathcal{H}_j$ (compare with Table VI row sums). The ith row of T presents actual flows (in $\mathrm{gCm^{-2}\,y^{-1}}$) required to productively cause outflow y_{0i} from component $\mathcal{H}_i$.

Thus, comparing Table VI outflows with components of inflows and throughflows as listed in Table VII(iv), the following results may be summarized (values in $\mathrm{gCm^{-2}\,y^{-1}}$):

(i) Outflow $y_{01} = 102.2$ from $\mathcal{H}_1$ requires a contribution of 102.2 to $\mathcal{H}_1$ inflow, the total value of which is $z_{10} = 124.1$ (Table VI).

(ii) Outflow of $y_{02} = 6.6$ from $\mathcal{H}_2$ requires 6.6 units from z_{10}, seen as a contribution of 6.6 units to the throughflows of both $\mathcal{H}_1$ and $\mathcal{H}_2$.

(iii) Outflow of $y_{03} = 244.1$ from $\mathcal{H}_3$ requires inflow components of 13.8 into $\mathcal{H}_1$ as z_{10} and 230.3 into $\mathcal{H}_3$ as z_{30}, and throughflow components of 13.8, 13.8, 291.8, and 47.7 at $\mathcal{H}_1, \ldots, \mathcal{H}_4$, respectively.

(iv) Outflow $y_{04} = 26.7$ requires inflow components of 1.5 and 25.2 into $\mathcal{H}_1$ and $\mathcal{H}_3$, and throughflow components of 1.5, 1.5, 31.9, and 31.9 at $\mathcal{H}_1, \ldots, \mathcal{H}_4$, respectively. Further interpretation of these matrices will be deferred until Section V. D. 1.

3. *Genon Outflow Analysis*

Referring to Fig. 14b, an equally valid expression to Eq. (69) for the steady-state direct flow ϕ_{ij} from holon $\mathcal{H}_j$ to holon $\mathcal{H}_i$, $i = 0, \ldots, n$, is

$$\phi_{ij} = q''_{ij} T_j, \qquad i = 0, \ldots, n, \tag{74}$$

where q''_{ij} is the fraction of throughflow T_j destined for $\mathcal{H}_i$. Then, from Eq. (67), the steady-state throughflow of $\mathcal{H}_i$ can be formulated as

$$T_i = \sum_{j=1}^{n} \phi_{ij} + z_{i0} = \sum_{j=1}^{n} q''_{ij} T_j + z_{i0}, \qquad i = 0, \ldots, n. \tag{75}$$

Letting **T** and **z** now be column vectors $[(n+1) \times 1]$ of throughflows and inflows, respectively, and Q'' be an $(n+1) \times (n+1)$ square *fractional outflow matrix.* Eq. (75) can be written

$$\mathbf{T} = Q''\mathbf{T} + \mathbf{z}.$$

Rearranging, and solving for **T** as before,

$$(I - Q'')\mathbf{T} = \mathbf{z},$$

and

$$\mathbf{T} = (I - Q'')^{-1}\mathbf{z}. \tag{76}$$

That is, the throughflow at each $\mathcal{H}_i$ is apportioned into fractions produced by each system input z_{j0}:

$$T_i = \sum_{k=1}^{n} (I - Q'')_{ik}^{-1} z_{k0}$$

$$= (I - Q'')_{i1}^{-1} z_{10} + \cdots + (I - Q'')_{ij}^{-1} z_{j0} + \cdots + (I - Q'')_{in}^{-1} z_{n0},$$

or, letting $\Phi_{ij} = (I - Q'')_{ij}^{-1} z_{j0}$,

$$T_i = \Phi_{i1} + \cdots + \Phi_{ij} + \cdots + \Phi_{in}. \tag{77}$$

In this case, Φ_{ij} represents the causal flow at $\mathcal{H}_i$ (i.e., the portion of T_i) sequentially propagated over all paths of all lengths from $\mathcal{H}_j$ in response to input z_{j0}. Thus,

$$(I - Q'')_{ij}^{-1} z_{i0} = \Phi_{ij} = T_i - \sum_{\substack{k=1 \\ k \neq j}}^{n} \Phi_{ik},$$

and

$$(I - Q'')_{ij}^{-1} = \Phi_{ij}/z_{i0} = \left[T_i - \sum_{\substack{k=1 \\ k \neq j}}^{n} \Phi_{ik}\right] \Big/ z_{i0}. \tag{78}$$

The matrix $(I - Q'')^{-1}$ is a transitive closure outflow matrix whose elements $(I - Q'')_{ij}^{-1}$ represent [Eq. (78)] the flow of cause from $\mathcal{H}_j$ to $\mathcal{H}_i$, or the fraction of throughflow T_i, generated by one unit of inflow z_{j0} to $\mathcal{H}_j$. The matrix $(I - Q'')^{-1}$ thus partitions throughflows of each component in an interactive network into fractions productively caused by each system inflow. It has, as with $(I - Q')^{-1}$ for the creaon analysis, the transitive closure property of accounting for all direct and indirect pathways through the network from any system inflow.

Example 11

Table VIII presents matrices for genon outflow analysis of the marine coprophagy model. Following Eq. (74), each flow ϕ_{ij} from Table VI ($j = 1, \ldots, n$ so inflows are excluded, but $i = 0, \ldots, n$ so outflows are included) is divided by the throughflow T_j of the jth column. The four outflows $y_{01}, \ldots, y_{04}$ are diagonalized into separate vectors in order to identify their individual dependencies. Table VIII(i) gives the resulting Q'' matrix; note that the state variable columns sum to 1. The $(I - Q'')$ matrix is obtained by subtracting from I, and it is then inverted to give $(I - Q'')^{-1}$, which appears in Table VIII(ii). Then, the inflow vector (Table VI) is diagonalized to match the dimensions of $(I - Q'')^{-1}$, and the resulting matrix Z used to postmultiply $(I - Q'')^{-1}$ [per Eq. (76)]. This yields a throughflow matrix [Table VIII(iii)], the jth column of which denotes contributions to throughflows and outflows caused by inflows z_{0j}. The state variable rows in this matrix sum to total throughflow of the corresponding holon, as comparison with Table VI verifies, and the outflow rows sum to total outputs signifying that transitive closure has been achieved.

The allocation of inflows is evident by comparing Table VI inflows against Table VIII(iii) components of throughflow and outflow, all figures in $\text{g C m}^{-2}\,\text{y}^{-1}$:

1. Inflow $z_{10} = 124.1$ is the total throughflow at $\mathcal{H}_1$; 102.2 appears as outflow y_{01} and 21.9 is propagated to $\mathcal{H}_2$. Of this latter amount 6.6 becomes outflow y_{02} and the remainder, 15.3, passes to $\mathcal{H}_3$. Another 3.0 flows to $\mathcal{H}_3$ from $\mathcal{H}_4$, making $T_3 = 18.3$. Of this, 13.8 is lost as y_{03} and the remaining 4.5 is transmitted to $\mathcal{H}_4$ where 1.5 appears as y_{04}, and 3.0 flows to $\mathcal{H}_3$.
2. Inflow $z_{30} = 255.5$ enters at the point of the system feedback loop, and so is responsible for maintaining a throughflow of 305.4 at $\mathcal{H}_3$. Of this, 230.3 is contributed to outflow y_{03}. The remainder contributes 75.1 to $\mathcal{H}_4$ throughflow totaling 79.6, and of the former amount 25.2 flows out as y_{04}.

Further discussion of these relationships will be undertaken in Section V. D. 1.

4. *Causality and Transitive Closure*

In Section IV. D. 3 it was discussed how powers B^k, $k = 1, 2, \ldots$, of Boolean adjacency matrices described the positions of sequential dependency paths of length k, and how the sum $\Sigma_{k=1}^{l} B^k$ of successive powers denotes positions of dependency sequences of all lengths up to and including l. In Theorem 5 it was proved that the series $\Sigma_{k=1}^{l} B^k$ converges to B^*, a "transitive closure" matrix which specifies all sequential dependency paths of all lengths, after at most only $n + 1$ terms, where n is the order of the system. There is thus something

TABLE VIII

Genon Outflow Analysis Matrices for the Marine Coprophagy Model[a]

(i) *Q″ matrix:*

from (columns), *to* (rows)

q_i \ q_j	x_1	x_2	x_3	x_4	y_{01}	y_{02}	y_{03}	y_{04}
x_1	0	0	0	0	0	0	0	0
x_2	0.176	0	0	0	0	0	0	0
x_3	0	0.700	0	0.664	0	0	0	0
x_4	0	0	0.246	0	0	0	0	0
y_{01}	0.824	0	0	0	0	0	0	0
y_{02}	0	0.300	0	0	0	0	0	0
y_{03}	0	0	0.754	0	0	0	0	0
y_{04}	0	0	0	0.336	0	0	0	0

(ii) *$(I-Q'')^{-1}$ matrix*

from (columns), *to* (rows)

q_i \ q_j	x_1	x_2	x_3	x_4	y_{01}	y_{02}	y_{03}	y_{04}
x_1	1.0	0	0	0	0	0	0	0
x_2	0.177	1.0	0	0	0	0	0	0
x_3	0.148	0.837	1.195	0.794	0	0	0	0
x_4	0.036	0.206	0.294	1.195	0	0	0	0
y_{01}	0.824	0	0	0	1.0	0	0	0
y_{02}	0.053	0.300	0	0	0	1.0	0	0
y_{03}	0.111	0.631	0.901	0.598	0	0	1.0	0
y_{04}	0.012	0.069	0.099	0.402	0	0	0	1.0
ZPL_j	1.361	2.043	1.489	1.989				

(iii) *T matrix, $(I-Q'')^{-1}z$*

from (columns), *to* (rows)

q_i \ q_j	x_1	x_2	x_3	x_4	T_i	outflow Σ
x_1	124.1	0	0	0	124.1	-
x_2	21.9	0	0	0	21.9	-
x_3	18.3	0	305.4	0	323.7	-
x_4	4.5	0	75.1	0	79.6	-
y_{01}	102.2	0	0	0	-	102.2
y_{02}	6.6	0	0	0	-	6.6
y_{03}	13.8	0	230.3	0	-	244.1
y_{04}	1.5	0	25.2	0	-	26.7

[a] Top to bottom: (i) fractional outflow matrix Q''; (ii) transitive closure outflow matrix, $(I-Q'')^{-1}$; (iii) throughflow matrix, $(I-Q'')^{-1}Z$, in gC m^{-2} y^{-1}; ZPL_j in (ii) is inflow path length, discussed in Example 14.

fundamental in networks about the variety of paths over which cause may be propagated, and all pathways must be considered.

In Section V. A. 2 it was shown, in discussion of the effect of feedback loops in networks, that the transmittance of a signal flow graph (Fig. 9e) consisting only of a loop between input and output could be expressed as the sum of transmittances of the different paths, which are infinite in number around a loop. Hence, with $\mathscr{L}^k$ the transmittance of the kth path, the transmittance between input and output was $\Sigma_{k=0}^{\infty}\mathscr{L}^k$ [Eq. (57)], which converged to $(1-\mathscr{L})^{-1}$ [Eq. (58)]. There is thus also something basic about expressions of the form $(1-\mathscr{L})^{-1}$ in systems with feedback. This is reflected in Mason's theorem which, as a formula for propagation, must account for all dependency paths between cause and effect. The inverse graph determinant which appears in Mason's expression for graph transmittance has the same general form: $\Delta^{-1} = [\Pi_{k=1}^{l}(1-\mathscr{L}_k)]^{-1}$ [Eqs. (61)–(62)].

The form is apparent again in the input–output flow analysis inverse matrices $(I-Q')^{-1}$ and $(I-Q'')^{-1}$, the former of which Leontief (1936) and Hannon (1973) have called the "structure" of the system. All dependency paths are accounted for in these matrices, otherwise the flow analysis methodology would be invalid. Therefore, the matrices implicitly satisfy the transitive closure property [Eq. (48)], as has already been indicated. That is,

$$\lim_{l\to\infty}\sum_{k=0}^{l}(Q')^k = (I-Q')^{-1} \tag{79}$$

and

$$\lim_{l\to\infty}\sum_{k=0}^{l}(Q'')^k = (I-Q'')^{-1}. \tag{80}$$

The question now is what restrictions are imposed on the philosophically wide ranging concept of causal determination (Bunge, 1959) by the requirement of convergence of infinite series to inverse matrices? What are the properties of a system requisite to such convergence and to existence of the matrices, and how far beyond the stricture of flows obeying the conservation law does the causal concept amenable to the flow analysis methodology extend?

That geometric progressions $\Sigma_{k=0}^{\infty}Q^k$, if they converge, converge to $(I-Q)^{-1}$ is well-known (Faddeev and Faddeeva, 1963, p. 113). The relationship is readily apparent through the following equalities:

$$(I+Q+Q^2+\cdots+Q^k+\cdots)(I-Q) = I, \tag{81}$$

and therefore,

$$(I+Q+Q^2+\cdots+Q^k+\cdots) = (I-Q)^{-1}. \tag{82}$$

For the series to converge it is necessary and sufficient that $Q^k \to 0$ as $k\to\infty$.

Mathematical conditions for convergence include: (i) all eigenvalues of Q must be less than one in modulus, or (ii) at least one norm of Q must be less than one (Faddeev and Faddeeva, 1963, pp. 111–112). Ortega (1972), cited by Hannon (1973), gives conditions somewhat more amenable to physical interpretation, as detailed below for input and output objects.

Creaon. Elements q'_{ij} of Q', now construed as a generalized (i.e., not necessarily conservative) input matrix, denote fractions of measures of throughput T_i at holon $\mathcal{H}_i$ received as direct input ϕ_{ij} from $\mathcal{H}_j$ (Fig. 14a). If there are multiple direct dependency links from $\mathcal{H}_j$ to $\mathcal{H}_i$, their effect is nevertheless summarized in the single link represented by ϕ_{ij} based on the relationship for parallel coupling (Fig. 8e). There is thus implicit in ϕ_{ij} possible many-to-one mapping of direct dependencies. $\mathcal{H}_j$ also generates outputs to other destinations than $\mathcal{H}_i$ (Fig. 14b), upon some or all of which $\mathcal{H}_i$ may be sequentially dependent. It is clear that elements $(q'_{ij})^k$ of matrices $(Q')^k$, $k = 1, 2, \ldots$, represent fractions of measures of T_i at $\mathcal{H}_i$ propagated over sequential dependency paths of length k from $\mathcal{H}_j$. In a system with feedback $k \to \infty$ and hence the need [Eq. (79)] to establish existence of the $(I - Q')^{-1}$ matrix.

Let

$$Q' = \begin{bmatrix} Q_1' & 0 & \cdots & 0 \\ 0 & Q_2' & \cdots & 0 \\ \vdots & \vdots & \ddots & \vdots \\ 0 & 0 & \cdots & Q_m' \end{bmatrix}, \tag{83}$$

where Q_k', $k = 1, \ldots, m$, are irreducible block diagonal submatrices such that $\det Q_1' \cdot \det Q_2' \cdot \ldots \cdot \det Q_m' = \det Q'$. For each block matrix, let $(q'_{ij})_k \in Q_k'$ be the ith row element directly dependent upon the jth state variable, and $(q'_{ij})_k \in Q_k'$ be the ith row element directly dependent upon the jth input variable. Then, for convergence (79) it is sufficient that

$$\sum_{j=1}^{n} (q'_{ij})_k \leqslant 1, \qquad k = 1, \ldots, m, \tag{84}$$

with strict inequality holding for at least one row i. The physical significance of this condition for existence of $(I - Q')^{-1}$ will be clearer after an example.

Example 12

The marine coprophagy model Q' matrix [Table VII(i)] can be block diagonalized as shown in the accompanying matrix.

q_i \ q_j (from)	z_{10}	x_1	z_{30}	x_2	x_4	x_3	State variable row Σ
x_1	1.0	0	0	0	0	0	0
x_2	0	1.0	0	0	0	0	1.0
to x_3	0	0	0.789	0.047	0.163	0	0.2
x_4	0	0	0	0	0	1.0	1.0
z_{10}	0	0	0	0	0	0	0
z_{30}	0	0	0	0	0	0	0

For the 2 x 2 submatrix the row sums of state variable columns are 0 for row 1 and 1.0 for row 2. The inequality condition is satisfied by the first row. For the 4 x 4 submatrix, row sums of state variable columns are 0.2, 1.0, 0, 0 for rows 3, . . ., 6, respectively. Rows 3, 5, and 6 satisfy strict inequality. Therefore, the power series $\Sigma_{k=0}^{\infty} (Q')^k$ converges to $(I - Q')^{-1}$, and transitive closure (all paths of all lengths accounted for) is guaranteed.

Table IX illustrates the sequence of matrices $Q', (Q')^2, \ldots, (Q')^{10}$ and their sum $(Q')^{(10)}$ for the marine coprophagy model. The entries $(q'_{ij})^k$, $k = 1, 2, \ldots, 10$, denote fractional energy contributions of sequential dependency paths of length k to throughflow T_i. For example, 0.047 of T_3 of $\mathcal{H}_3$ is contributed by $\mathcal{H}_2$ in the direct link $x_2 \rightarrow x_3$, 0.008 in the path of length 3 ($x_2 \rightarrow x_3 \rightarrow x_4 \rightarrow x_3$), 0.001 in the length 5 path ($x_2 \rightarrow x_3 \rightarrow x_4 \rightarrow x_3 \rightarrow x_4 \rightarrow x_3$), and essentially 0 in subsequent turns around the $x_3 \leftrightarrows x_4$ loop. Also, z_{30} contributes 0.789 of T_3 in direct flow ($z_{30} \rightarrow x_3$), 0.129 in the length 3 path ($z_{30} \rightarrow x_3 \rightarrow x_4 \rightarrow x_3$), 0.021 in the length 5 path, 0.003 for length 7, 0.001 for length 9, and essentially 0 thereafter. The reader can investigate other examples. Note that the last matrix of Table IX, representing $\Sigma_{k=0}^{10} (Q')^k$, corresponds to within computational error to $(I - Q')^{-1}$ [Table VII(iii)]. The series converges to final values effectively after $k = 10$ terms as the value of $(Q')^{10}$ is zero. Note also that, in conformity with Theorem 5 (Section IV. D. 3), all paths of all lengths have been identified by the fourth term $(Q')^4$ of the series $\Sigma_{k=0}^{10} (Q')^k$, one term before the theorem guarantees (and the fifth term of $\Sigma_{k=0}^{10} (Q')^k$).

The physical significance of condition (84) for convergence (79) of the geometric progression of fractional input matrices is as follows. Block diagonalization associates subsets of state and input variables. Each irreducible block submatrix contains at least one system input column. Therefore, the sum of state variable column entries in the row for the state variable which receives the input must be <1. The other rows will have either all inputs from state variables (row sums equal 1) or no entries at all (row sums equal 0). In example 12, the 2 x 2 submatrix associated with input z_{10} has the row sum of state variable entries to the x_1 row less than 1 as required. Similarly the 4 x 4 submatrix associated with input z_{30} has the row sum of state variable entries to the x_3 row less than 1. Thus, the only effect of the Ortega (1972) condition for convergence of $\Sigma_{k=0}^{\infty} (Q')^k$ to $(I - Q')^{-1}$ is the existence of input! There is specifically no requirement that the signals flowing through the system be conservative, and it is now obvious, referring to Fig. 14a, that the focal flow ϕ_{ij}

TABLE IX
Matrices in the Geometric Progression $\Sigma_{k=1}^{10}(Q')^k$ for the Marine Coprophagy Model, with $\Sigma_{k=0}^{10}(Q')^k = (Q')^{(10)}$ Converging to $(I - Q')^{-1}$

Q' matrix:

	q_i \ q_j (from)	x_1	x_2	x_3	x_4	z_{10}	z_{30}
	x_1	0	0	0	0	1.0	0
	x_2	1.0	0	0	0	0	0
to	x_3	0	0.047	0	0.163	0	0.789
	x_4	0	0	1.0	0	0	0
	z_{10}	0	0	0	0	0	0
	z_{30}	0	0	0	0	0	0

$(Q')^2$ matrix:

	q_i \ q_j (from)	x_1	x_2	x_3	x_4	z_{10}	z_{30}
	x_1	0	0	0	0	0	0
	x_2	0	0	0	0	1.0	0
to	x_3	0.047	0	0.163	0	0	0
	x_4	0	0.047	0	0.163	0	0.789
	z_{10}	0	0	0	0	0	0
	z_{30}	0	0	0	0	0	0

$(Q')^3$ matrix:

	q_i \ q_j (from)	x_1	x_2	x_3	x_4	z_{10}	z_{30}
	x_1	0	0	0	0	0	0
	x_2	0	0	0	0	0	0
	x_3	0	0.008	0	0.027	0.047	0.129
to	x_4	0.047	0	0.163	0	0	0
	z_{10}	0	0	0	0	0	0
	z_{30}	0	0	0	0	0	0

TABLE IX (*continued*)

$(Q')^4$ *matrix*: (*from* q_j, *to* q_i)

q_i \ q_j	x_1	x_2	x_3	x_4	z_{10}	z_{30}
x_1	0	0	0	0	0	0
x_2	0	0	0	0	0	0
x_3	0.008	0	0.027	0	0	0
x_4	0	0.008	0	0.027	0.047	0.129
z_{10}	0	0	0	0	0	0
z_{30}	0	0	0	0	0	0

$(Q')^5$ *matrix*: (*from* q_j, *to* q_i)

q_i \ q_j	x_1	x_2	x_3	x_4	z_{10}	z_{30}
x_1	0	0	0	0	0	0
x_2	0	0	0	0	0	0
x_3	0	0.001	0	0.004	0.008	0.021
x_4	0.008	0	0.027	0	0	0
z_{10}	0	0	0	0	0	0
z_{30}	0	0	0	0	0	0

$(Q')^6$ *matrix*: (*from* q_j, *to* q_i)

q_i \ q_j	x_1	x_2	x_3	x_4	z_{10}	z_{30}
x_1	0	0	0	0	0	0
x_2	0	0	0	0	0	0
x_3	0.001	0	0.004	0	0	0
x_4	0	0.001	0	0.004	0.008	0.021
z_{10}	0	0	0	0	0	0
z_{30}	0	0	0	0	0	0

$(Q')^7$ *matrix*: (*from* q_j, *to* q_i)

q_i \ q_j	x_1	x_2	x_3	x_4	z_{10}	z_{30}
x_1	0	0	0	0	0	0
x_2	0	0	0	0	0	0
x_3	0	0	0	0.001	0.001	0.003
x_4	0.001	0	0.004	0	0	0
z_{10}	0	0	0	0	0	0
z_{30}	0	0	0	0	0	0

(*continued*)

TABLE IX (*continued*)

$(Q')^8$ matrix:

to	q_i \ q_j (from)	x_1	x_2	x_3	x_4	z_{10}	z_{30}
	x_1	0	0	0	0	0	0
	x_2	0	0	0	0	0	0
	x_3	0	0	0.001	0	0	0
	x_4	0	0	0	0.001	0.001	0.003
	z_{10}	0	0	0	0	0	0
	z_{30}	0	0	0	0	0	0

$(Q')^9$ matrix:

to	q_i \ q_j (from)	x_1	x_2	x_3	x_4	z_{10}	z_{30}
	x_1	0	0	0	0	0	0
	x_2	0	0	0	0	0	0
	x_3	0	0	0	0	0	0.001
	x_4	0	0	0.001	0	0	0
	z_{10}	0	0	0	0	0	0
	z_{30}	0	0	0	0	0	0

$(Q')^{10}$ matrix:

to	q_i \ q_j (from)	x_1	x_2	x_3	x_4	z_{10}	z_{30}
	x_1	0	0	0	0	0	0
	x_2	0	0	0	0	0	0
	x_3	0	0	0	0	0	0
	x_4	0	0	0	0	0	0
	z_{10}	0	0	0	0	0	0
	z_{30}	0	0	0	0	0	0

$(Q')^{(10)}$ matrix:

to	q_i \ q_j (from)	x_1	x_2	x_3	x_4	z_{10}	z_{30}
	x_1	1.0	0	0	0	1.0	0
	x_2	1.0	1.0	0	0	1.0	0
	x_3	0.057	0.057	1.195	0.195	0.057	0.943
	x_4	0.057	0.057	1.195	1.195	0.057	0.943
	z_{10}	0	0	0	0	0	0
	z_{30}	0	0	0	0	0	0

from $\mathcal{H}_j$ could be distributed nonconservatively (as illustrated in Fig. 7b) to many different holons other than $\mathcal{H}_i$ without affecting the flow analysis derivation [Eqs. (69)–(72)].

Therefore, it is concluded that input–output analysis applies not only to conservative energy–matter flows in systems, but to distributive informational signal flows as well, and therefore places no restrictions on the type of causality to which it applies except that the cause be measurable and quantifiable. The existence condition for $(I - Q')^{-1}$ serves only to guarantee that a system cannot exist in isolation but must be driven by an external environment. This, of course, is the epitome of causal determination, "Everything is lawfully produced by something else . . ." (Section II. D), and that transitive closure and implied endogenous path structure of systems are so inextricably related to contact with an external world (Figs. 4 and 5) evidences the power and sweep of Aristotle's ancient principle.

Genon. Elements q''_{ij} [Eq. (75)] of Q'', now considered a generalized fractional output rather than outflow matrix, denote fractions of throughput T_j at $\mathcal{H}_j$ generated as direct output ϕ_{ij} to $\mathcal{H}_i$ (Fig. 14b). $\mathcal{H}_i$ may also receive inputs from other sources than $\mathcal{H}_j$ (Fig. 14a), some of which may make $\mathcal{H}_i$ sequentially dependent on $\mathcal{H}_j$. Elements $(q''_{ij})^k$ of $(Q'')^k$, $k = 1, 2, \ldots$, signify fractions of T_j at $\mathcal{H}_j$ propagated over sequential dependency paths of length k to $\mathcal{H}_i$. To establish existence of the $(I - Q'')^{-1}$ matrix, block submatrices are formed as before:

$$Q'' = \begin{bmatrix} Q''_1 & 0 \cdots & 0 \\ 0 & Q''_2 \cdots & 0 \\ \vdots & \vdots \ddots & \vdots \\ 0 & 0 \cdots & Q''_m \end{bmatrix}, \tag{85}$$

where $\det Q''_1 \cdot \det Q''_2 \cdot \ldots \cdot \det Q''_m = \det Q''$. For each such block matrix, let $(q''_{ij})_k \in Q''_k$, $k = 1, \ldots, m$, be the jth column variable directly producing the ith row state variable, and $(q''_{ij})_k \in Q''_k$ be the jth column variable directly causing the ith row output variable. Then, for convergence (80) it is sufficient that

$$\sum_{i=1}^{n} (q''_{ij})_k \leqslant 1, \qquad k = 1, \ldots, m, \tag{86}$$

with the inequality holding strictly for at least one column j.

Example 13

The marine coprophagy model Q'' matrix [Table VIII(i)] can be block diagonalized as shown in the accompanying matrix. There are three block *from*

	$q_i \backslash q_j$	x_1	y_{01}	x_2	x_4	y_{02}	x_3	y_{03}	y_{04}
	y_{01}	0.834	0	0	0	0	0	0	0
	x_2	0.176	0	0	0	0	0	0	0
	y_{02}	0	0	0.300	0	0	0	0	0
	x_3	0	0	0.700	0.664	0	0	0	0
to	y_{04}	0	0	0	0.336	0	0	0	0
	x_4	0	0	0	0	0	0.246	0	0
	y_{03}	0	0	0	0	0	0.754	0	0
	x_1	0	0	0	0	0	0	0	0
	State variable column Σ	0.176	0	0.700	0.664	0	0.246	0	0

submatrices, and in each of them the column sums of state variable rows only are all less than 1. Condition (86) is thereby satisfied, and transitive closure (80) of the model is established.

In the genon case, then, the formation of block diagonal submatrices [Eq. (85)] associates subsets of state and output variables. Each irreducible submatrix contains at least one output row with a nonzero entry. The corresponding column sum of state variable rows must then be less than 1 since the sum over all nonzero rows is unity. Therefore, the condition for convergence of $\Sigma_{k=0}^{\infty} (Q'')^k$ to $(I - Q'')^{-1}$ serves to guarantee system output. Systems must drive exogenous environments, and this is the other half of the principle of determinacy, "[Everything] . . . lawfully produces something else" (Section II. D). That transitive closure of internal dependency structure of systems cannot be established without reference to an environment again verifies the fundamental validity of the causal principle.*

* Note that this same conclusion can be reached independently from Markov chain theory. Partition Q' into submatrices of absorbing "states" (inputs) Q'_z and nonabsorbing "states" (state and output variables) $Q'_{z'}$, and likewise partition Q'' into submatrices of absorbing "states" (outputs) Q''_y and nonabsorbing "states" (input and state variables) $Q''_{y'}$. A theorem for absorbing Markov chains (i.e., chains whose nontransient stages are sources or sinks) states that $(I - Q'_{z'})^{-1}$ and $(I - Q''_{y'})^{-1}$ exist if and only if such submatrices can be formed. In each case the inverses are $\Sigma_{k=0}^{\infty}(Q'_{z'})^k$ and $\Sigma_{k=0}^{\infty}(Q''_{y'})^k$, respectively (Kemeny and Snell, 1960). Thus, contact with an environment, as represented by the "absorbing states," is necessary and sufficient for transitive closure. These results and the causal principle itself may be a manifestation of Goedel's theorem (e.g., Nagel and Newman, 1956) specifying that the logical consistency of any deductive system cannot be established without reference to an external system whose own internal consistency is in question without reference to a further external system, and so forth. That is, logical systems also have logical "environments," which must be consulted to demonstrate the internal consistency of the former.

In conclusion, this discussion of causality and transitive closure of fractional inflow and outflow matrices reinforces the concept, introduced in Section III. D. 3, of holon $\mathcal{H}$ and environment $\mathcal{H}'$ as an inseparable, reciprocal pair (Figs. 4 and 5). To consider one without the other is utterly meaningless. Although the applicability of input–output flow analysis to nonconservative signals has been established, in subsequent sections which consider applications to ecosystem energy and material flows, conservative flow terminology (inflow, outflow, and throughflow versus input, output, and throughput) will be employed.

C. Flow Analysis Measures

Several measures will facilitate discussion of steady-state ecosystem model flow analysis characteristics. These include path length, cycling efficiency, and cycling index, as defined below.

1. *Path Length*

Referring to Fig. 14a and Eq. (73), each element $(I - Q')^{-1}_{ij}$ of the transitive closure matrix $(I - Q')^{-1}$ denotes the throughflow T_j at $\mathcal{H}_j$ required to generate one unit of output y_{0i} from $\mathcal{H}_i$. These fractions, summed over all j, represent a measure of the magnitude of endogenous propagation required to produce y_{0i}, and can be referred to as *outflow path length*:

$$YPL_i = \sum_{j=1}^{n} (I - Q')^{-1}_{ij}, \qquad i = 1, \ldots, n. \tag{87}$$

Similarly, with reference to Fig. 14b and Eq. (78), each entry $(I - Q'')^{-1}_{ij}$ of $(I - Q'')^{-1}$ specifies the fraction of throughflow T_i at $\mathcal{H}_i$ caused by inflow z_{j0} at $\mathcal{H}_j$. Summed over all i, *inflow path length* measures the relative causal strength of z_{j0}:

$$ZPL_j = \sum_{i=1}^{n} (I - Q'')^{-1}_{ij}, \qquad j = 1, \ldots, n. \tag{88}$$

Mean path length $\overline{PL}$ can also be computed for outflows or inflows as the weighted mean of all YPL_i or ZPL_j. For the former,

$$\overline{YPL} = \sum_{i=1}^{n} \frac{y_{0i}}{\Sigma y} YPL_i, \tag{89}$$

and for the latter,

$$\overline{ZPL} = \sum_{j=1}^{n} \frac{z_{j0}}{\Sigma z} ZPL_j, \tag{90}$$

where Σz and Σy are total system inflow and outflow [Eqs. (64) and (65)], respectively. Alternatively, applying Eqs. (87), (73), (72), and (68) to (89):

$$\overline{YPL} = \sum_{i=1}^{n} \frac{1}{\Sigma y}\left[y_{0i} \sum_{j=1}^{n} (I-Q')_{ij}^{-1}\right]$$

$$= \frac{1}{\Sigma y} \sum_{j=1}^{n}\left[\sum_{i=1}^{n} \Phi_{ij}\right]$$

$$= \frac{1}{\Sigma y}\left(\sum_{j=1}^{n} T_j\right),$$

and therefore

$$\overline{YPL} = TST/\Sigma y, \tag{91}$$

where TST is total system throughflow [Eq. (68)]. Similarly for input path length, apply Eqs. (88), (78), (77), and (68) to (90):

$$\overline{ZPL} = \sum_{j=1}^{n} \frac{1}{\Sigma z}\left[z_{j0} \sum_{i=1}^{n} (I-Q'')_{ij}^{-1}\right]$$

$$= \frac{1}{\Sigma z} \sum_{i=1}^{n}\left[\sum_{j=1}^{n} \Phi_{ij}\right]$$

$$= \frac{1}{\Sigma z}\left(\sum_{i=1}^{n} T_i\right),$$

and hence

$$\overline{ZPL} = TST/\Sigma z. \tag{92}$$

Since $\Sigma z = \Sigma y$ at steady state, $\overline{YPL} = \overline{ZPL}$ [Eqs. (91) and (92)], and mean path length $\overline{PL}$ is computable from any of Eqs. (89)–(92). $\overline{PL}$ indicates combined effects of component holons, and will tend to increase with more components, greater throughflows, and more feedback in the system. When TST is partitioned into sequential or straight through (TST_s) as opposed to cyclic or feedback (TST_c) components (Section V. C. 3), $\overline{PL}$ can similarly be apportioned into feedback $\overline{PL}_c$ and nonfeedback $\overline{PL}_s$ portions.

Example 14

For the marine coprophagy model outflow path lengths YPL_i are shown in Table VII(iii). $YPL_1 = 1.0$ denotes that outflow y_{01} is only one direct dependency link, $x_1 \rightarrow y_{01}$, away from the state variable that carries cause z_{10} into effect y_{01}. Similarly, $YPL_2 = 2.0$ indicates that y_{02} is two direct dependency links, $x_1 \rightarrow x_2 \rightarrow y_{02}$, away in the sequential dependency path to its source, also z_{10}. These are straightforward examples because only simple serial sequences are involved. They do illustrate that in pure form YPL_i does measure path length within a system, and hence represents how "deeply" the internal propagating structure is involved in generating each indicated output. $YPL_3 = 1.504$ quantitatively expresses the resultant of y_{03} being produced by two inflows z_{10} and z_{30} through sequential dependency paths that involve a loop. Qualitatively the paths to inputs are

$$\circlearrowleft x_3 \rightarrow y_{03} \quad \text{and} \quad x_1 \rightarrow x_2 \rightarrow \circlearrowleft x_3 \rightarrow y_{03}.$$

Similarly, $YPL_4 = 2.504$ quantitatively reflects longer paths to the two inflows z_{10} and z_{30} which both drive y_{04} through dependency sequences that qualitatively are

$$x_3 \rightleftarrows x_4 \rightarrow y_{04} \quad \text{and} \quad x_1 \rightarrow x_2 \rightarrow x_3 \rightleftarrows x_4 \rightarrow y_{04}.$$

Inflow path lengths ZPL_j are given in Table VIII(ii). $ZPL_1 = 1.361$ is a quantitative measure of the qualitative sequence

$$z_{10} \rightarrow x_1 \rightarrow x_2 \rightarrow x_3 \rightleftarrows x_4.$$

$ZPL_2 = 2.043$, reflecting the dependency structure

$$z_{02} \rightarrow x_2 \rightarrow x_3 \rightleftarrows x_4,$$

signifies that inflows at x_2 would be propagated further into the loop than those introduced at x_1. There are no such inflows in the model, of course, but ZPL_2 nevertheless indicates the path length of endogenously propagated inflow to x_2. $ZPL_3 = 1.489$ quantifies the dependency sequence

$$z_{03} \rightarrow x_3 \rightleftarrows x_4,$$

and $ZPL_4 = 1.989$ quantitatively reflects

$$z_{04} \rightarrow x_4 \rightleftarrows x_3,$$

even though again the only inflow to x_4 in the model is endogenous.

Mean path length $\overline{PL}$ for the coprophagy system is 1.447, as calculated, for example, by Eq. (90),

$$\overline{PL} = \frac{124.1}{379.6}(1.361) + \frac{0}{379.6}(2.043) + \frac{255.5}{379.6}(1.489) + \frac{0}{379.6}(1.989),$$

or by Eq. (92),

$$\overline{PL} = 549.3/379.6.$$

2. *Cycling Efficiency*

The *component cycling efficiency* CE_j of $\mathcal{H}_j$ is defined as the ratio of endogenous outflow to total outflow (throughflow):

$$CE_j = \left(\sum_{k=1}^{n} \phi_{jk}\right)/T_j, \qquad j = 1, \ldots, n. \tag{93}$$

The information required for this computation is apparent in the system production matrix P (63).

A feedback loop is a sequential dependency path which closes on itself, that is, originates and terminates on the state variable of the same holon, say $\mathcal{H}_j$. In Sections V. A. 3 and V. A. 4 the significance of loops touching forward paths and other loops was detailed. The *loop cycling efficiency* of $\mathcal{H}_j$ is its fractional throughflow propagated to other loops, or, alternatively, its fractional throughflow arriving from other loops within the system. Consider the principal diagonal elements of $(I - Q')^{-1}$ and $(I - Q'')^{-1}$. $(I - Q')_{jj}^{-1} = 1$ implies that all the throughflow of $\mathcal{H}_j$ is propagated out and never returns. Similarly, $(I - Q'')_{jj}^{-1} = 1$ if $\mathcal{H}_j$ receives only inflow that it never generated. If $(I - Q')_{jj}^{-1} > 1$, the difference $(I - Q')_{jj}^{-1} - 1 > 0$ represents flow once generated by $\mathcal{H}_j$ and returned to it around loops. If $(I - Q'')_{jj}^{-1} > 1$, the difference $(I - Q'')_{jj}^{-1} - 1 > 0$ represents flow received by $\mathcal{H}_j$ that it once generated. Therefore, loop return flows R_j are a component of the total contribution of $\mathcal{H}_j$ to intrasystem cycling. They may be computed in terms of output or input aspects directly from the principal diagonal entries in $(I - Q')^{-1}$ and $(I - Q'')^{-1}$, respectively:

$$YR_j = [(I - Q')_{jj}^{-1} - 1]y_{0j}, \tag{94}$$

$$ZR_j = [(I - Q'')_{jj}^{-1} - 1]z_{j0}. \tag{95}$$

In addition to return flows, causes generated by $\mathcal{H}_j$ and propagated to other cycles in the system whose components are not in the sequential dependency set of $\mathcal{H}_j$, or flows propagated from other loops to $\mathcal{H}_j$ upon which the originating loops are not sequentially dependent, must be accounted for. This will be achieved graphically in Section V. D to determine loop cycling flow due to $\mathcal{H}_j$:

$$YC_j = YR_j + YR_j', \qquad \text{and} \qquad ZC_j = ZR_j + ZR_j',$$

where R terms represent flow contributions of return or reachable loops, and R' terms contributions of nonreturn and nonreachable loops. Loop cycling efficiency due to $\mathcal{H}_j$ is calculated by dividing loop cycling flow by total loop flow:

$$YCE_j = \frac{YC_j}{[(I - Q')_{jj}^{-1}]y_{0j} + YR_j'}, \tag{96}$$

and

$$ZCE_j = \frac{ZC_j}{[(I - Q'')_{jj}^{-1}]z_{j0} + ZR_j'}. \tag{97}$$

System cycling efficiency CE is then the weighted sum of these loop cycling efficiencies:

$$CE = \sum_{j=1}^{n} \frac{y_{0j}}{\Sigma y} YCE_j = \sum_{j=1}^{n} \frac{z_{j0}}{\Sigma z} ZCE_j. \tag{98}$$

Example 15

Component cycling efficiencies CE_j for the marine coprophagy model are listed in Table VI. State variables x_1 and x_3 are shown to be inefficient contributors since they lose most of their throughflow to noncyclic outflows. On the other hand, x_2 and x_4 are efficient contributors to loop cycling.

Loop return flows (Eqs. (94) and (95)] are zero for x_1 and x_2 as the principal diagonal entries for these components are one in both $(I - Q')^{-1}$ and $(I - Q'')^{-1}$ [Tables VII(iii) and VIII(ii)]. However, endogenous flows propagated from both x_1 and x_2 to x_3 cycle in the loop $x_3 \rightleftarrows x_4$ and cannot return to their origin. Therefore, while system cycling efficiency computed from Eq. (96) will be correct,

$$CE = \frac{102.2}{379.6}(0) + \frac{6.6}{379.6}(0) + \frac{244.1}{379.6}\left(\frac{0.195}{1.195}\right) + \frac{26.7}{379.6}\left(\frac{0.195}{1.195}\right) = 0.116, \tag{99}$$

that calculated from Eq. (97) will be an underestimate,

$$CE = \frac{124.1}{379.6}(0) + 0(0) + \frac{255.5}{379.6}\left(\frac{0.195}{1.195}\right) + 0\left(\frac{0.195}{1.195}\right) = 0.109. \tag{100}$$

The missing value $ZR'_1 = 0.024\ (124.1) = 2.978$ is obtainable by a graphic procedure, described later.

3. *Cycling Index*

The principal diagonal entries of $(I - Q')^{-1}$ and $(I - Q'')^{-1}$ differ from other elements in that noncyclic flow is immediately apparent whereas for off-diagonal elements it is not. However, beginning with the main diagonal entries, it is possible to reveal cycled portions of $(I - Q')_{ij}^{-1}$ and $(I - Q'')_{ij}^{-1}$ elements, $i \neq j$. The procedure is described below in Example 16, and yields matrices $(I - Q')_c^{-1}$ and $(I - Q'')_c^{-1}$ of cycled fractions of throughflow causing each outflow y_{0i} or caused by each inflow z_{j0}, $i, j = 1, \ldots, n$. When these matrices are multiplied by **y** and **z**, respectively [Eqs. (72) and (77)], a vector $\mathbf{T}_c$ of cycled throughflow results. This is the fraction of **T** contributed by cycled flow, and $\mathbf{T}_s$ which remains is the straight through or noncycled portion:

$$\mathbf{T} = \mathbf{T}_c + \mathbf{T}_s. \tag{101}$$

Total system throughflow [Eq. (68)] can then be partitioned into feedback and nonfeedback portions:

$$TST = TST_c + TST_s. \tag{102}$$

Mean path length [Eqs. (91)–(92)] can also be apportioned into cyclic and acyclic components:

$$\overline{PL} = \overline{PL_c} + \overline{PL_s}. \tag{103}$$

The ratio of cycled to straight through TST or $\overline{PL}$ defines the system *cycling index*:

$$CI = TST_c/TST_s = \overline{PL_c}/\overline{PL_s}. \tag{104}$$

This measure reflects, in its representation of feedback versus nonfeedback characteristics, how far in terms of path length inflow will tend to penetrate and from how deeply within the endogenous structure outflow has emanated. It is a measure of the importance of cycling to a system.

Example 16

The procedure for calculating the cycling index from $(I - Q')^{-1}$ for the marine coprophagy model [Table VII(iii)] is as follows. Uncycled portions of $(I - Q')^{-1}$ are identified by first eliminating outflows from the diagonal elements by subtracting 1 from each. First-row entries of the resulting matrix are then all 0 implying no contribution of $x_1, \ldots, x_4$ to the system cycling index through x_1. Next, since x_2 receives all the cause of its outflow y_{02} from x_1, 1 is subtracted from element 21, making elements of the second row all 0 also. Variables $x_1, \ldots, x_4$ contribute nothing to the cycling index through x_2. The only state variables involved in cycling are x_3 and x_4. From Table VII(iii), x_3 receives 0.057 of its outflow from z_{10} over the sequential dependency path

$z_{10} \to x_1 \to x_2 \to x_3$. Therefore, 0.057 is subtracted from entries 31 and 32. The remaining fraction (0.943) of outflow y_{03} from x_3 originates with z_{30}, and has already been eliminated. The final state variable x_4 has 0.057 of its outflow y_{04} generated from z_{10} as propagated over the sequence $z_{10} \dashrightarrow x_3 \rightleftarrows x_4$. The remaining fraction (0.943) comes from z_{30} via the dependency sequence $z_{30} \to x_3 \rightleftarrows x_4$. Therefore, 0.057 should be subtracted from entries 41, 42, and 43, and then an additional 0.943 subtracted from entry 43 to give 0.195. The resulting matrix $(I - Q')_c^{-1}$ represents fractions of the jth cycled throughflow responsible for generating a unit of outflow y_{0i}. This matrix is shown in Table X. Premultiplying it by $\mathbf{y}$ in accordance with Eq. (71) gives the vector $\mathbf{T}_c$

TABLE X

Cycling Index *CI* Data for the Marine Coprophagy Model[a]

	q_i \ q_j (*from*)	x_1	x_2	x_3	x_4
to	x_1	0	0	0	0
	x_2	0	0	0	0
	x_3	0	0	0.195	0.195
	x_4	0	0	0.195	0.195
	$(T_c)_j$	0	0	52.81	52.81

$TST_c = 105.61 \qquad TST_s = 549.30 - 105.61 = 443.69$

$\overline{PL}_c = 0.28 \qquad \overline{PL}_s = 1.17$

$CI = \dfrac{105.61}{443.69} = 0.24$

[a] The matrix is $(I - Q')_c^{-1}$, representing normalized cycled flows; $(T_c)_j$ is the jth element of the cycled throughflow vector $\mathbf{T}_c$ [Eq. (101)]; TST_c, TST_s, $\overline{PL}_c$, $\overline{PL}_s$, and *CI* are as defined in Eqs. (102)–(104).

of cycled throughflows, whose elements are also listed in the table. $TST_c = 105.6$ and $\overline{PL}_c = 0.28$ which, compared to total system throughflow and mean path length, give a system cycling index of $CI = 0.24$ (Table X). Thus, one quarter as much carbon causally propagated through the model is cycled as flows straight through.

D. Static Flow Analysis of Ecosystem Models

In this section causal flow diagrams are introduced to aid interpretation of the transitive closure matrices $(I - Q')^{-1}$ and $(I - Q'')^{-1}$. Two additional simple ecosystem models described in the Appendix are analyzed and interpretations offered to indicate the kind of ecological insight that flow analysis can provide. The models are to be considered as examples. They are too simple to be realistic or to demonstrate the full potential of the methodology in ecosystem analysis.

1. *Marine Coprophagy Model*

Figure 15 depicts for the coprophagy model four causal flow diagrams derived from the $(I - Q')^{-1}$ creaon matrix [Table VII(iii)]. The diagrams illustrate for each matrix row i ($i = 1, \ldots, n$) normalized: (i) inflows z_{j0}^i to holon $\mathcal{H}_j$, (ii) throughflows $T_j^{\,i}$ at $\mathcal{H}_j$, and (iii) intrasystem flows ϕ_{jk}^i from $\mathcal{H}_k$ to $\mathcal{H}_j$ ($j, k = 1, \ldots, n$) required to generate one unit of outflow y_{0i}^i from $\mathcal{H}_i$, as derived from the matrix elements $(I - Q')_{ij}^{-1}$. The focal unit outflow y_{0i}^i is depicted in each case by a heavy arrow. The procedure for constructing the diagram for the ith row is as follows: (i) column j *entries, representing* $T_j^{\,i}$, are placed inside the jth holon symbol; (ii) normalized inflows z_{j0}^i to each appropriate $\mathcal{H}_j$ are indicated; (iii) normalized outflow y_{0i}^i from $\mathcal{H}_i$ is given a value of 1, and all other outflows a value of 0; and (iv) direct dependency flows ϕ_{jk}^i between $\mathcal{H}_k$ and $\mathcal{H}_j$ are calculated by difference based on the conservation principle. Note in step (ii) that $\Sigma_{j=1}^n z_{j0}^i = 1$ (the sum of all normalized inflows is one), and in step (iii) $\Sigma_{j=1}^n y_{0j}^i = 1$ (the sum of all normalized outflows is also one). In computing interholon normalized flows, $\Sigma_{k=1}^n \phi_{jk}^i + z_{j0}^i = \Sigma_{k=1}^n \phi_{kj}^i + y_{0j}^i$ (endogenous and exogenous inflows and outflows to and from each holon $\mathcal{H}_j$ must be conserved).

Figure 15a illustrates normalized flows, derived from the first row of Table VII(iii), required to generate $y_{01}^1 = 1$ as outflow $\mathcal{H}_1$. The row has only two nonzero entries. The first column element, representing $T_1^{\,1} = 1.0$, is placed in the symbol for $\mathcal{H}_1$; the entry indicating $z_{10}^1 = 1.0$ is also diagrammed. Remaining elements are all zero and it is understood that $y_{01}^1 = 1.0$, which is entered by the heavy arrow. Flows between holons in this case are all zero. Row two (Fig. 15b) is similar, showing one unit of normalized throughflow ($T_1^{\,2} = T_2^{\,2} = 1.0$) at both $\mathcal{H}_1$ and $\mathcal{H}_2$ required to yield $y_{02}^2 = 1.0$. From this information, normalized flow $\phi_{21}^2 = 1.0$ from $\mathcal{H}_1$ to $\mathcal{H}_2$ is constructed. For the third row (Fig. 15c) of Table VII(iii), normalized inflows $z_{10}^3 = 0.057$ and $z_{30}^3 = 0.943$ and the unit outflow $y_{03}^3 = 1.0$ are entered. Remaining outflows are zero. Intrasystem flows to support the unit outflow are calculated by mass balance. Thus, $\phi_{21}^3 = 0.057$ since $z_{10}^3 = T_1^{\,3} = 0.057$ and $y_{01}^3 = 0$. Similarly, $\phi_{32}^3 = 0.057$ as $T_2^{\,3} = 0.057$ and $y_{02}^3 = 0$. For the flow from $\mathcal{H}_3$ to $\mathcal{H}_4$, $T_3^{\,3} = 1.195$ and $y_{03}^3 = 1.0$ giving by difference $\phi_{43}^3 = 0.195$. This produces $T_4^{\,3} = 0.195$ at $\mathcal{H}_4$ which, since $y_{04}^3 = 0$, returns to $\mathcal{H}_3$ as $\phi_{34}^3 = 0.195$. The fourth row of Table VII(iii), as diagrammed in Fig. 15d, indicates the following characteristics of the marine coprophagy model.

To observe one unit of outflow $y_{04}^4 = 1.0$ from the benthic invertebrate feces

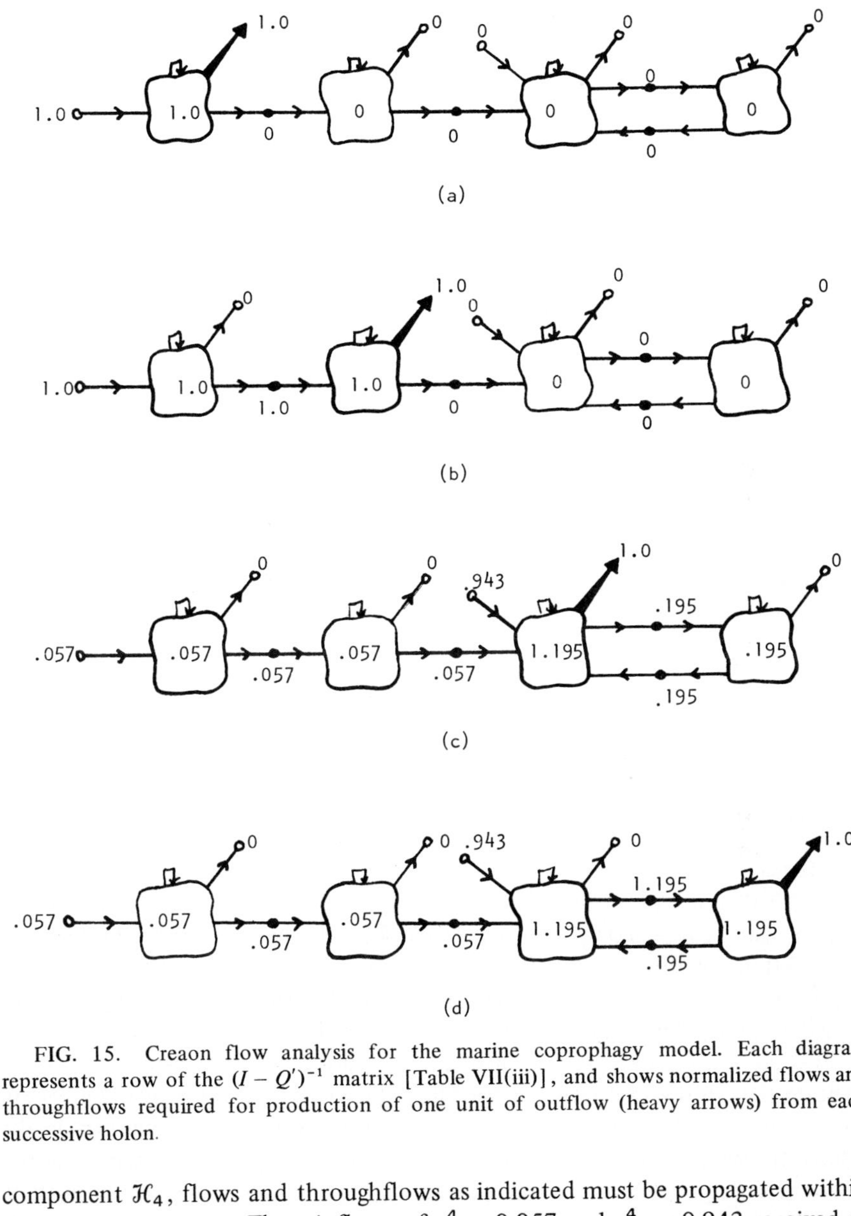

FIG. 15. Creaon flow analysis for the marine coprophagy model. Each diagram represents a row of the $(I - Q')^{-1}$ matrix [Table VII(iii)], and shows normalized flows and throughflows required for production of one unit of outflow (heavy arrows) from each successive holon.

component $\mathcal{H}_4$, flows and throughflows as indicated must be propagated within the system structure. Thus, inflows of $z_{10}^4 = 0.057$ and $z_{30}^4 = 0.943$ received at $\mathcal{H}_1$ and $\mathcal{H}_3$ are implied by observation of one unit of outflow from $\mathcal{H}_4$. These inflows generate the normalized throughflows T_j^4 and interholon flows ϕ_{jk}^4 shown. With $\mathcal{H}_4$ taken as reference and with everything else depicted in Fig. 15d considered to be its environment, the environment-defining nature of the $\mathcal{H}_4$ creaon becomes apparent. Similarly for the other holons within the system. A unit normalized outflow from any one of them implicitly signifies the existence of a complementary environmental structure, details of which are apparent in each case as displayed in Fig. 15. This perspective illustrates that it is in the nature of the creaon as an input selector and receptor to verify the structure of its holon's environment. This creative verification of the causal nexus leading to output is the key property which the name "creaon" is intended to denote.

The corresponding genon input analysis for the marine coprophagy model is depicted in Fig. 16, which presents flow diagrams based on the $(I - Q'')^{-1}$ matrix as listed in Table VIII(ii). These diagrams are constructed in an analogous way to those for the outflow analysis. Each diagram in Fig. 16 represents a column of the Table VIII(ii) matrix. For the jth column inflow is normalized to $z_{j0}^j = 1.0$ and all other inflows are set to zero. The inflow of interest in each case is indicated by a heavy arrow. The entry in the ith row of column j is the normalized throughflow T_i^j of the ith holon as generated by $z_{j0}^j = 1.0$ $(i, j = 1, \ldots, n)$. These values appear in the holon symbols in the flow diagram. Normalized outflows y_{0i}^j which appear in the jth column are entered in the jth diagram. Flows between holons ϕ_{ik}^j $(i, k = 1, \ldots, n)$ are calculated by mass balance as before.

Thus, each number associated with an arrow in Fig. 16 represents a carbon flow normalized to one unit of inflow at the holon indicated by the bold arrows. In Fig. 16a, for example, $z_{10}^1 = 1.0$ enters $\mathcal{H}_1$ giving throughflow $T_1^1 = 1.0$. Of this, $y_{01}^1 = 0.824$, representing respiration and other losses from *Callianassa*, $\mathcal{H}_1$, and the remainder $\phi_{21}^1 = 0.176$ is excreted. Of the throughflow $T_2^1 = 0.177$ at $\mathcal{H}_2$, representing fecal material generated by *Callianassa*, $y_{02}^1 = 0.053$ is lost from the system and $\phi_{32}^1 = 0.124$ constitutes feeding by benthic invertebrates. The latter holon $\mathcal{H}_3$ also receives $\phi_{34}^1 = 0.024$ units of flow from $\mathcal{H}_4$. Of the $T_3^1 = 0.148$ units of throughflow in $\mathcal{H}_3$, $y_{03}^1 = 0.111$ represents loss from the system and $\phi_{43}^1 = 0.036$ is passed to $\mathcal{H}_4$ to become the throughflow $T_4^1 = 0.036$. Of this amount, $y_{04}^1 = 0.012$ leaves the system and the remainder $\phi_{34}^1 = 0.024$ cycles back to $\mathcal{H}_3$. These are the relationships generated by one unit of inflow at $\mathcal{H}_1$. The other flow diagrams in Fig. 16 demonstrate the same kind of information for unit inflows into $\mathcal{H}_2$, $\mathcal{H}_3$, and $\mathcal{H}_4$, respectively. The results indicate that inflows (external causes) introduced at different points in a system have quantitatively different effects at any given point of reference within the interconnection. Also, the relative contributions of the different inputs at any point in the system can be evaluated. As an example, for the carbon outflow from $\mathcal{H}_4$ the system causal structure is (Fig. 16) such that one unit of inflow at $\mathcal{H}_1, \ldots, \mathcal{H}_4$, respectively, would contribute 0.012, 0.069, 0.099, and 0.402 units of observed outflow. The genon analysis summarizes causal propagation within a system from a generative perspective.

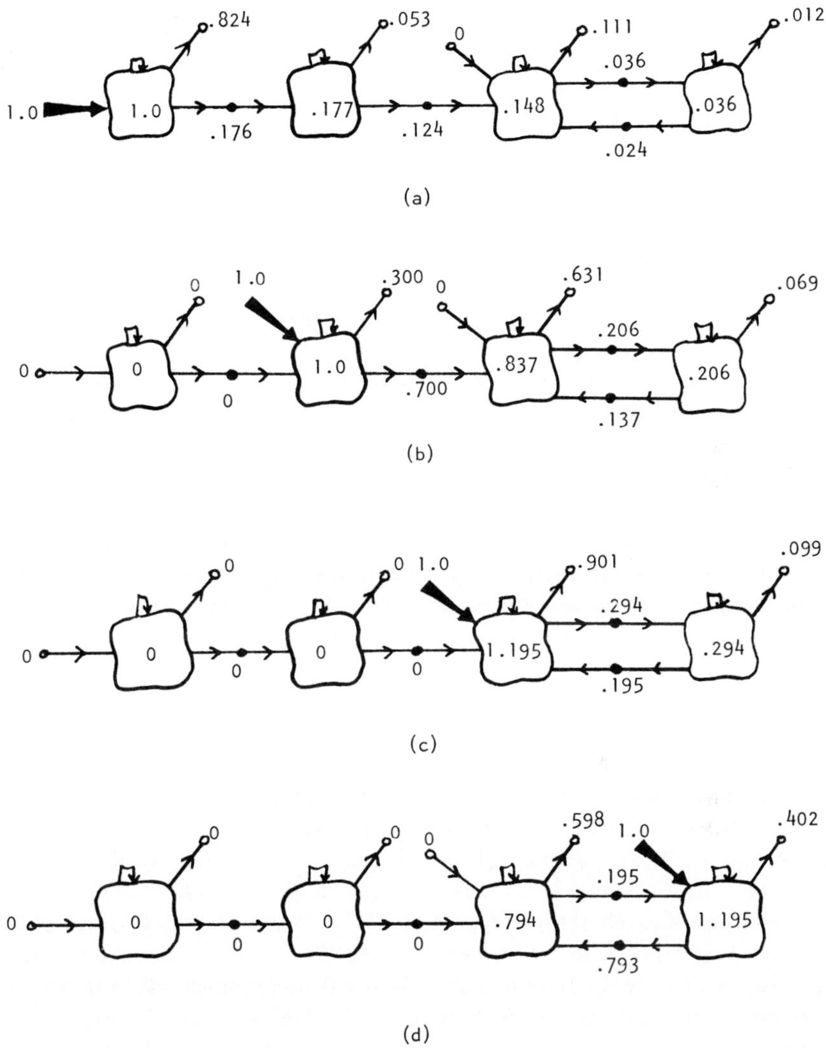

FIG. 16. Genon flow analysis for the marine coprophagy model. Each diagram represents a column of the $(I - Q'')^{-1}$ matrix [Table VIII(ii)], and shows normalized flows and throughflows generated by one unit of inflow (heavy arrows) to each successive holon.

Path length characteristics of this model were discussed in Example 14. The global value of $\overline{PL} = 1.447$ [Eqs. (91)–(92)] indicates that the system loses most of its inflow from the holon at which it enters. That is, the mean path length is less than two, indicating that carbon does not flow very deeply into the system structure before it exits. Cycling efficiency measures were discussed in Example 15, and it was indicated that the system cycling efficiency *CE* [Eq. (98)] was underestimated when calculated from ZCE_j [Eq. (97)]. Component $ZR'_1 = 2.978$ could not be obtained conveniently from the $(I - Q'')^{-1}$ matrix, but it was indicated that this value could be revealed by graphic analysis. Figure 16a indicates that a cycled flow $\phi^1_{34} = 0.024$ is generated by unit inflow $z^1_{10} = 1.0$ to $\mathcal{H}_1$. Therefore, the calculation for ZC_1 is $ZC_1 = 0.024 = ZR'_1\,(124.1) = 2.978$. Then [Eq. (97)], $ZCE_1 = 2.978/(124.1 + 2.978) = 0.023$. The contribution of z_{10} to *CE* is then [Eq. (98)]

$$\frac{124.1}{379.6}\,(0.023) = 0.007,$$

as needed to correct the underestimated value of *CE* in Eq. (100), $CE = 0.116$. Finally, the cycling index $CI = 0.24$, as discussed in Example 16, indicates that approximately one quarter as much carbon cycles within the model as flows straight through.

In concluding the static flow analysis discussion of this simple model, which has served to illustrate many of the developments in this chapter, it is instructive to compare diagrams in Figs. 15 and 16 with the diagrammed description of the system as presented in the Appendix (Fig. A.1). As examples of the kinds of observations that can be made, consider the following.

1. *Creaon.* In Fig. A.1 an outflow $y_{04} = 26.7$ gC m^{-2} y^{-1} is indicated from $\mathcal{H}_4$. Figure 15d shows that $z^4_{10} = 0.057$ and $z^4_{30} = 0.943$ units of input are required to support $y^4_{40} = 1.0$ unit of output. Thus, $z_{10} = 1.5$ and $z_{30} = 25.2$ gC m^{-2} y^{-1} are sources of the observed outflow, $y_{04} = 26.7$ gC m^{-2} y^{-1}, representing 1.2% of measured inflow to $\mathcal{H}_1$ and 9.9% of inflow to $\mathcal{H}_3$.

Genon. In Fig. 16a, $z^1_{10} = 1.0$ is indicated to generate $y^1_{04} = 0.012$ units of outflow from $\mathcal{H}_4$. Therefore, $z_{10} = 124.1$ gC m^{-2} y^{-1} (Fig. A.1) productively generates $y_{04} = 1.5$ gC m^{-2} y^{-1}, representing 5.6% of the observed outflow $y_{04} = 26.7$ gC m^{-2} y^{-1}. In Fig. 16c, $z^3_{30} = 1.0$ determines $y^3_{04} = 0.099$ corresponding units. So, $z_{30} = 255.5$ gC m^{-2} y^{-1} (Fig. A.1) can be expected to produce $y_{04} = 25.3$ gC m^{-2} y^{-1}, or 94.4% of the observed value $y_{04} = 26.7$ gC m^{-2} y^{-1}.

2. *Creaon.* In the system feedback loop $\phi_{34} = 52.9$ gC m^{-2} y^{-1} (Fig. A.1) represents carbon flow returned to $\mathcal{H}_3$ from $\mathcal{H}_4$. Observed outflows from these holons are $y_{03} = 244.1$ and $y_{04} = 26.7$ gC m^{-2} y^{-1}, respectively. From Fig. 15c, $y^3_{03} = 1.0$ corresponds to ϕ^3_{34} 0.195 in comparable units. Therefore, $y_{03} = 244.1$ gC m^{-2} y^{-1} observed as outflow from $\mathcal{H}_3$ implies existence of $\phi_{34} = 47.6$ gC m^{-2} y^{-1} of feedback flow between $\mathcal{H}_4$ and $\mathcal{H}_3$. In Fig. 15d,

$y_{04}^4 = 1.0$ corresponds to $\phi_{34}^4 = 0.195$ in the same units. Hence, $y_{04} = 26.7$ gC m^{-2} y^{-1} mandates $\phi_{34} = 5.2$ gC m^{-2} y^{-1} as the feedback flow. Consequently, observation of both $y_{03} = 244.1$ and $y_{04} = 26.7$ gC m^{-2} y^{-1} verifies existence of the total observed flow from $\mathcal{H}_4$ to $\mathcal{H}_3$, $\phi_{34} = 47.6 + 5.2 = 52.8$ gC m^{-2} y^{-1}.

Genon. Figure 16a relates $z_{10}^1 = 1.0$ to $\phi_{34}^1 = 0.024$ flow units; therefore, $z_{10} = 124.1$ gC m^{-2} y^{-1} (Fig. A.1) generates $\phi_{34} = 3.0$ gC m^{-2} y^{-1}. Figure 16c indicates that $z_{30}^3 = 1.0$ produces in the same units $\phi_{34}^3 = 0.195$; hence, $z_{30} = 255.5$ gC m^{-2} y^{-1} (Fig A.1) causes the associated feedback flow $\phi_{34} = 49.8$ gC m^{-2} y^{-1}. Both observed inflows together (Fig. A.1) account for $\phi_{34} = 3.0 + 49.8 = 52.8$ gC m^{-2} y^{-1}, the indicated recycling by benthic invertebrates of their fecal carbon.

The reader is free to examine numerous additional relationships to illustrate the fact that both modes of input–output flow analysis correctly reveal identical causal relationships in the model, but in reciprocal perspectives, one passive-receptive (creaon) and the other active-generative (genon).

2. *Cone Spring*

Cone Spring is a cold water spring ecosystem studied by Tilly (1968) and modeled in terms of energy flow (kcal m^{-2} y^{-1}) as described in the Appendix. Figure A.2 in the Appendix depicts the structure of the model which consists of five component holons $\mathcal{H}_1, \ldots, \mathcal{H}_5$ represented by state variables $x_1, \ldots, x_5$, two inflows z_{10} and z_{20}, five outflows $y_{01}, \ldots, y_{05}$, and five feedback loops $x_2 \rightleftarrows x_3$, $x_2 \rightleftarrows x_4$, $x_2 \dashrightarrow x_3 \dashrightarrow x_4 \dashrightarrow x_2$, $x_2 \dashrightarrow x_4 \dashrightarrow x_5 \dashrightarrow x_2$, and $x_2 \dashrightarrow x_3 \dashrightarrow x_4 \dashrightarrow x_5 \dashrightarrow x_2$.

The system production matrix P is shown in Table XI (i). As modeled the spring is very close to steady state. The fractional inflow matrix Q' is given in Table XII(i), and the transitive closure matrix $(I - Q')^{-1}$ in Table XIII(i). The fractional outflow matrix Q'' appears in Table XIV(i) and the transitive closure matrix in Table XV(i). Flow diagrams derived from the $(I - Q')^{-1}$ matrix for outflow analysis are shown in Fig. 17, and diagrams derived from the $(I - Q'')^{-1}$ matrix for inflow analysis appear in Fig. 18.

Total system throughflow [Table XI(i)] is $TST = 30626.5$ kcal m^{-2} y^{-1} based on total outflow $\Sigma y = 11822.9$, which slightly exceeds total inflow $\Sigma z = 11819.0$. Mean path length $\overline{PL}$, as given by Eq. (91), is therefore $30626.5/11822.9 = 2.590$, compared to a value of 1.447 for the marine coprophagy model. Considering the number of loops, $\overline{PL}$ is smaller than expected, most inflow exiting the system before it passes through three components. Outflow path lengths YPL_i [Eq. (87)] are listed in Table XIII(i), and inflow path lengths ZPL_j [Eq. (88)] in Table XV(i). Both sets of values are

TABLE XI
Production Matrices P for (i) Cone Spring and (ii) Tropical Rainforest Models[a]

(i) Cone Spring

to \ from (q_i \ q_j)	x_1	x_2	x_3	x_4	x_5	inflow	outflow	row Σ
x_1	0	0	0	0	0	11184.0	0	11184.0
x_2	8881.4	0	1600.2	200.0	167.0	635.0	0	11483.6
x_3	0	5204.5	0	0	0	0	0	5204.5
x_4	0	2309.2	75.1	0	0	0	0	2384.3
x_5	0	0	0	370.1	0	0	0	370.1
inflow	0	0	0	0	0	0	0	0
outflow	2302.6	3973.4	3529.9	1814.0	203.1	0	0	11822.9
column Σ	11184.0	11487.1	5205.2	2384.1	370.1	11819.0	0	42449.4
CE_j	0.79	0.65	0.32	0.24	0.45	TST = 30626.5		

(ii) Tropical rainforest

to \ from (q_i \ q_j)	x_1	x_2	x_3	x_4	x_5	inflow	outflow	row Σ
x_1	0	0	0	0	10.2	7.5	0	17.7
x_2	16.0	0	0	0	0	0	0	16.0
x_3	0	16.0	0	0	0	2.7	0	18.7
x_4	0	0	9.2	0	0	0	0	9.2
x_5	0	0	3.6	6.6	0	0	0	10.2
inflow	0	0	0	0	0	0	0	0
outflow	1.7	0	5.9	2.6	0	0	0	10.2
column Σ	17.7	16.0	18.7	9.2	10.2	10.2	0	82.0
CE_j	0.90	1.00	0.68	0.72	1.00	TST = 71.8		

[a] Nonzero entries are energy flows (kcal m^{-2} y^{-1}) for Cone Spring, and nitrogen flows (gN m^{-2} y^{-1}) for the tropical rainforest.

larger than those for the coprophagy model [Tables VII(iii) and VIII(ii)]. From Eq. (90) for $\overline{PL}$ in terms of input components,

$$\overline{PL} = \frac{11184.0}{11819.0}(2.622) + \frac{635.0}{11819.0}(2.060) = 2.48 + 0.11,$$

indicating that most of the model path length is contributed by plant inflow z_{10} rather than by detritus inflow z_{20}.

TABLE XII
Fractional Inflow Matrices Q' for (i) Cone Spring and (ii) Tropical Rainforest Models

(i) Cone Spring

	$q_i \backslash q_j$	x_1	x_2	x_3	*from* x_4	x_5	z_{10}	z_{20}
	x_1	0	0	0	0	0	1.0	0
	x_2	0.773	0	0.139	0.017	0.015	0	0.055
	x_3	0	1.0	0	0	0	0	0
to	x_4	0	0.969	0.031	0	0	0	0
	x_5	0	0	0	1.0	0	0	0
	z_{10}	0	0	0	0	0	0	0
	z_{20}	0	0	0	0	0	0	0

(ii) Tropical rainforest

	$q_i \backslash q_j$	x_1	x_2	x_3	*from* x_4	x_5	z_{10}	z_{30}
	x_1	0	0	0	0	0.576	0.424	0
	x_2	1.0	0	0	0	0	0	0
	x_3	0	0.856	0	0	0	0	0.144
to	x_4	0	0	1.0	0	0	0	0
	x_5	0	0	0.352	0.647	0	0	0
	z_{10}	0	0	0	0	0	0	0
	z_{30}	0	0	0	0	0	0	0

Component cycling efficiencies CE_j [Eq. (93)] are listed in Table XI(i), and the values may be compared with those in Table VI for the marine coprophagy model. $\mathcal{H}_1$ and $\mathcal{H}_2$ in Cone Spring have higher ratios of endogenous to total flow than the other components, and $\mathcal{H}_4$ has the lowest. This is not apparent by positions of these holons in the network structure (Fig. A.2, Appendix).

Calculations for loop cycling efficiencies may be made with the $(I - Q')^{-1}$ and $(I - Q'')^{-1}$ matrices together with corresponding flow diagrams. Referring to Eqs. (94) and (96), $YCE_1 = 0$ [Table XIII(i)]. No cycled flows contribute to outflow y_{01} from the system (Fig. 17a). Table XIII(i) and Fig. 17b both indicate that $YCE_2 = 0.207\ (3973.4)/1.207\ (3973.4) = 0.171$. This is all due to the YR_2 component of YC_2 since no system loops are nonreachable or do not return to state variable x_2. In fact, since x_2 originates every loop in the system, cycled flow associated with the remaining outflows is equal to that for y_{02}. $YCE_j = 0.171$, $j = 2, \ldots, 5$. System cycling efficiency CE [Eq. (98)], as computed based on outflow data, is

TABLE XIII
Transitive Closure Inflow Matrices $(I - Q')^{-1}$ for (i) Cone Spring and (ii) Tropical Rainforest Models

(i) Cone Spring

	$q_i \backslash q_j$	from x_1	x_2	x_3	x_4	x_5	z_{10}	z_{20}	YPL_i
	x_1	1.0	0	0	0	0	1.0	0	1.0
	x_2	0.933	1.207	0.169	0.039	0.018	0.933	0.067	3.37
	x_3	0.933	1.207	1.169	0.039	0.018	0.933	0.067	4.37
to	x_4	0.933	1.207	0.200	1.039	0.018	0.933	0.067	3.40
	x_5	0.933	1.207	0.200	1.039	1.018	0.933	0.067	4.40
	z_{10}	0	0	0	0	0	1.0	0	
	z_{20}	0	0	0	0	0	0	1.0	

(ii) Tropical rainforest

	$q_i \backslash q_j$	from x_1	x_2	x_3	x_4	x_5	z_{10}	z_{30}	YPL_i
	x_1	1.97	0.97	1.14	0.74	1.14	0.84	0.16	5.96
	x_2	1.97	1.97	1.14	0.74	1.14	0.84	0.16	6.96
	x_3	1.69	1.69	1.97	0.63	0.97	0.72	0.28	6.95
to	x_4	1.69	1.69	1.97	1.63	0.97	0.72	0.28	7.95
	x_5	1.69	1.69	1.97	1.28	1.97	0.72	0.28	8.60
	z_{10}	0	0	0	0	0	1.0	0	
	z_{30}	0	0	0	0	0	0	1.0	

$$CE = \frac{2302.6}{11822.9}(0) + \frac{3973.4}{11822.9}(0.171) + \frac{3529.9}{11822.9}(0.171) + \frac{1814.0}{11822.9}(0.171) + \frac{203.1}{11822.9}(0.171) = 0.138.$$

For loop cycling efficiencies related to inflows, reference to Eq. (95) and Table XV(i) indicates $ZR_1 = 0$. However, it is apparent in Fig. 18a that flows propagaged from x_1 contribute to all system loops, but their influence cannot feed back to $\mathcal{H}_1$. For inflow z_{10} the cycled flow generated is the difference between normalized throughflow $T_2^1 = 0.958$ at $\mathcal{H}_2$ and flow $\phi_{21}^1 = 0.794$ from $\mathcal{H}_1$ to $\mathcal{H}_2$. Thus, the normalized cycled flow is 0.164 for one unit of inflow at z_{10}, and $ZR'_1 = 0.164\ (11184.0)$. Hence, $ZCE_1 = 0.164\ (11184.0/(1.0 + 0.164)$

TABLE XIV
Fractional Outflow Matrices Q'' for (i) Cone Spring and (ii) Tropical Rainforest Models

(i) Cone Spring

	q_i \ q_j (*from*)	x_1	x_2	x_3	x_4	x_5	y_{01}	y_{02}	y_{03}	y_{04}	y_{05}
	x_1	0	0	0	0	0	0	0	0	0	0
	x_2	0.794	0	0.307	0.084	0.451	0	0	0	0	0
	x_3	0	0.453	0	0	0	0	0	0	0	0
	x_4	0	0.201	0.615	0	0	0	0	0	0	0
to	x_5	0	0	0	0.155	0	0	0	0	0	0
	y_{01}	0.206	0	0	0	0	0	0	0	0	0
	y_{02}	0	0.346	0	0	0	0	0	0	0	0
	y_{03}	0	0	0.678	0	0	0	0	0	0	0
	y_{04}	0	0	0	0.761	0	0	0	0	0	0
	y_{05}	0	0	0	0	0.549	0	0	0	0	0

(ii) Tropical rainforest

	q_i \ q_j (*from*)	x_1	x_2	x_3	x_4	x_5	y_{01}	y_{03}	y_{04}
	x_1	0	0	0	0	1.0	0	0	0
	x_2	0.904	0	0	0	0	0	0	0
	x_3	0	1.0	0	0	0	0	0	0
to	x_4	0	0	0.492	0	0	0	0	0
	x_5	0	0	0.193	0.717	0	0	0	0
	y_{01}	0.096	0	0	0	0	0	0	0
	y_{03}	0	0	0.315	0	0	0	0	0
	y_{04}	0	0	0	0.283	0	0	0	0

(11184.0) = 0.141 [Eq. (97)]. For inflow z_{20}, ZCE_2 is obtained from Table XV(i) since x_2 originates every loop in the system. That is, $ZR'_2 = 0$, and $ZCE_2 = 0.206\,(635.0)/1.206\,(635.0) = 0.171$. The product of this value with the througflow at $\mathcal{H}_2$ ($T_2^1 = 0.958$) equals the cycled fraction 0.164 as obtained above. Since there are no inflows to $\mathcal{H}_3$, $\mathcal{H}_4$, and $\mathcal{H}_5$, $ZCE_j = 0$, $j = 3, 4, 5$. System cycling efficiency computed based on inflow data is as follows:

$$CE = \frac{11184.0}{11819.0}\,(0.141) + \frac{635.0}{11819.0}\,(0.171) + 0 + 0 + 0 = 0.143.$$

TABLE XV

Transitive Closure Outflow Matrices $(I - Q'')^{-1}$ for (i) Cone Spring and (ii) Tropical Rainforest Models

(i) Cone Spring

q_i \ q_j (to \ *from*)	x_1	x_2	x_3	x_4	x_5
x_1	1.0	0	0	0	0
x_2	0.958	1.206	0.373	0.186	0.544
x_3	0.434	0.564	1.169	0.084	0.246
x_4	0.199	0.251	0.092	1.039	0.113
x_5	0.031	0.039	0.014	0.161	1.018
y_{01}	0.206	0	0	0	0
y_{02}	0.331	0.417	0.129	0.064	0.188
y_{03}	0.294	0.371	0.793	0.057	0.167
y_{04}	0.151	0.191	0.070	0.790	0.086
y_{05}	0.017	0.021	0.008	0.088	0.559
ZPL_j	2.622	2.060	1.648	1.470	1.921

(ii) Tropical rainforest

q_i \ q_j (to \ *from*)	x_1	x_2	x_3	x_4	x_5
x_1	1.97	1.08	1.08	1.42	1.97
x_2	1.78	1.97	0.97	1.28	1.78
x_3	1.78	1.97	1.97	1.28	1.78
x_4	0.80	.97	0.97	1.63	0.88
x_5	0.97	1.08	1.08	1.42	1.97
y_{01}	0.19	0.10	0.10	0.14	0.19
y_{03}	0.56	0.62	0.62	0.40	0.56
y_{04}	0.25	0.27	0.27	0.46	0.25
ZPL_j	7.38	7.07	6.07	7.03	8.38

The difference between this value and that computed above based on outflow data is due to the fact that the model is not quite in steady-state balance.

Derivation of the matrix $(I - Q')_c^{-1}$ [Table XVI] of cycled throughflows is facilitated by flow diagrams (Fig. 17) of $(I - Q')^{-1}$ [Table XIII(i)]. No cycling occurs in row 1 since $(I - Q')^{-1} = 1.0$. This is verified by reference to Fig. 17a where the dependency sequence $z_{10} \rightarrow x_1 \rightarrow y_{01}$ is indicated. Therefore, in Table XVI the first column can be eliminated as well as the first row. Figure 17b shows the normalized flow diagram required to generate $y_{02}^2 = 1.0$ unit of outflow. The noncyclic flows from inputs to y_{02}^2 in this diagram pass only

TABLE XVI
Cycling Index *CI* Data for the Cone Spring Model[a]

				from		
	q_i \ q_j	x_1	x_2	x_3	x_4	x_5
	x_1	0	0	0	0	0
	x_2	0	0.207	0.169	0.039	0.018
to	x_3	0	0.207	0.169	0.039	0.018
	x_4	0	0.207	0.169	0.039	0.018
	x_5	0	0.207	0.169	0.039	0.018
	$(T_c)_j$	0	1970.7	1608.9	371.3	171.4

$TST_c = 4122.3$ $\overline{PL}_c = 0.35$

$TST_s = 26504.2$ $\overline{PL}_s = 2.24$

$CI = 0.16$

[a] The matrix is $(I - Q')_c^{-1}$, representing normalized cycled flows; $(T_c)_j$ is the jth element of the $\mathbf{T}_c$ vector [Eq. (101)], TST_c, TST_s, $\overline{PL}_c$, $\overline{PL}_s$, and *CI* are as defined in Eqs. (102)–(104).

through $\mathcal{H}_1$ and $\mathcal{H}_2$ (i.e., $z_{10} \to x_1 \to x_2 \to y_{02}$). All other flows are cycled around the five system loops. Subtracting 1 from the second-row diagonal element $(I - Q')_{22}^{-1}$ in Table XIII(i) yields the amount of cycled flow returned through $\mathcal{H}_2$. Since $\mathcal{H}_2$ originates every feedback loop, there are no nonreturning portions of cycled flow. The third row (Fig. 17c) and Table XIII(i) are much the same except that $\mathcal{H}_3$ is not reachable from several loops to which it may contribute flow. Subtracting 1 from the entries for x_2 and x_3 in the third row results in the same numbers as for row 2 (Table XVI). In row 4 (Fig. 17d) flow is propagated from $\mathcal{H}_2$ to $\mathcal{H}_4$ both directly and via x_3 ($x_2 \dashrightarrow x_4 \to y_{04}$ or $x_2 \dashrightarrow x_3 \to x_4 \to y_{04}$). The uncycled portion of flow through $\mathcal{H}_3$ can be determined by comparing $\phi_{43}^2 = \phi_{43}^3 = 0.001$ (Fig. 17b, c) with $\phi_{43}^4 = 0.032$ (Fig. 17d). If cycled flow is the same, then straight throughflow is 0.031 units relative to $y_{04}^4 = 1.0$ unit of outflow. The cycled flows are the same for several reasons: (i) ϕ_{23}^k, ϕ_{24}^k, ϕ_{25}^k, and ϕ_{54}^k, $k = 2, 3, 4$, are loop flows which do not have uncycled components superimposed upon them; they have the same values in Fig. 17d as in Fig. 17b, c. (ii) After subtracting noncyclic flow from x_2 and x_4, the cycled portions of these components and $\mathcal{H}_5$ are the same as before. (iii) Since all cycling loops touch $\mathcal{H}_2$ and since inflows enter these loops only through state variable x_2, the distribution of cycled flows will be the same for any outflow in the cycling loops. Thus, $\mathcal{H}_3$ contributes 0.031 units of outflow

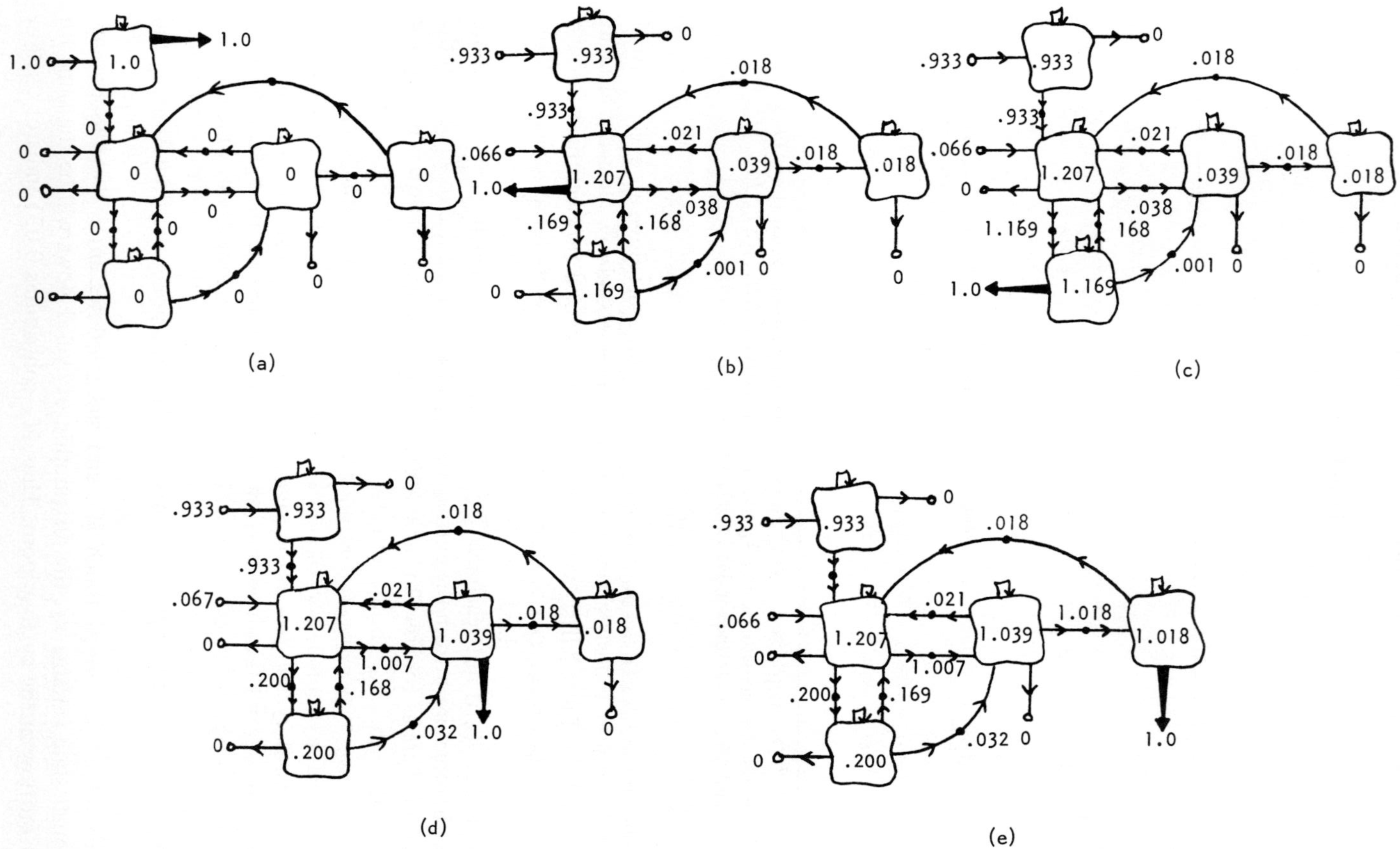

FIG. 17. Creaon flow analysis for the Cone Spring energy flow model. Each diagram represents a row of the $(I-Q')^{-1}$ matrix [Table XIII(i)], and shows normalized flows and throughflows required to generate one unit of outflow (heavy arrows) from each holon.

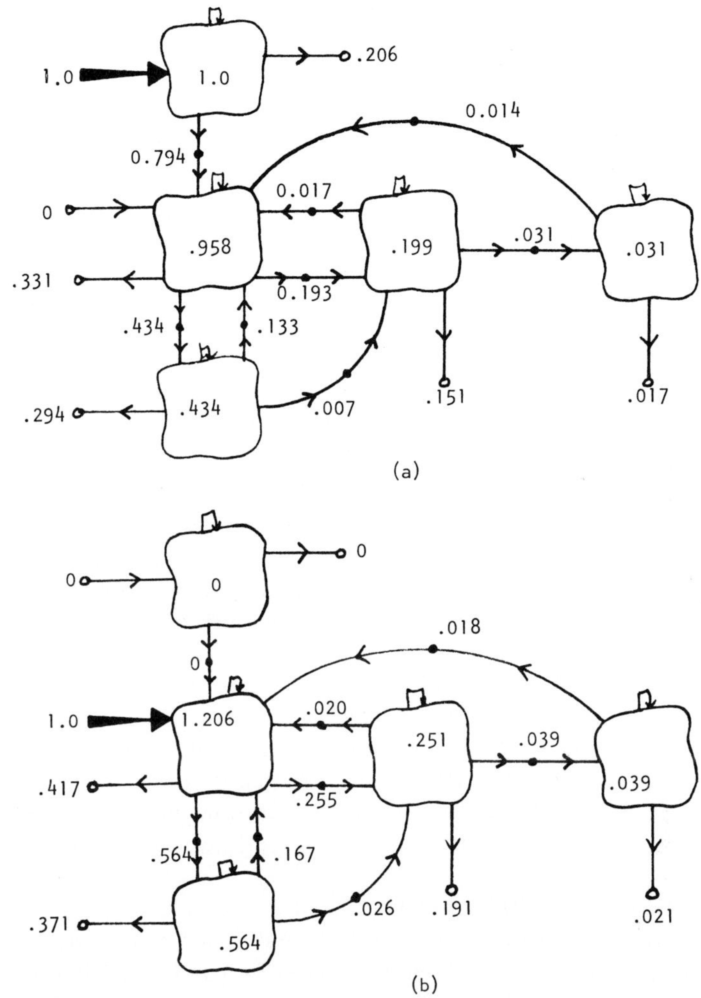

FIG. 18. Genon flow analysis for the Cone Spring model. The diagrams represent the first two columns of the $(I - Q'')^{-1}$ matrix [Table XV(i)], corresponding to the two system inputs: (a) z_{10}, and (b) z_{20}. Normalized flows and throughflows generated by the unit inflows in each case (heavy arrows) are depicted.

directly to y_{04}, and this amount is subtracted from the throughflow in $\mathcal{H}_3$ to obtain the cycled flow. The cycled flows in the fifth row of Table XIII(i) are also the same as before and the resulting matrix of normalized cycled flows $(I - Q')_c^{-1}$ is obtained (Table XVI). Premultiplying this matrix by the **y** vector according to Eq. (71) gives the vector $\mathbf{T}_c$ of cycled throughflows, also indicated in Table XVI. TST_c is then the sum of the elements of $\mathbf{T}_c$, and equals 4122.3 kcal m^{-2} y^{-1}. Applying Eqs. (102)–(104), the measures TST_s, $\overline{PL_c}$, $\overline{PL_s}$, and CI can be calculated. These values are listed in Table XVI. Despite the relatively complex qualitative feedback structure of this system, the cycling index $CI = 0.16$ is only two thirds of the value (Table X) for the coprophagy model. This is partly due to the fact that the model for Cone Spring represents energy flow while in the coprophagy model carbon is propagated. However, the importance of quantitative relationships as well as qualitative ones in propagation of cause in networks is indicated.

3. *Tropical Rainforest*

The tropical rainforest nitrogen cycling model described in the Appendix was based on an original model developed by Edmisten (1970). Causality in the model is expressed in terms of nitrogen flows, gN m^{-2} y^{-1}. Figure A.3 in the Appendix illustrates the model structure consisting of five holons $\mathcal{H}_1, \ldots, \mathcal{H}_5$, two inflows of nitrogen z_{10} and z_{30}, three outflows y_{01}, y_{03}, and y_{04}, and two feedback loops $x_1 \rightarrow x_2 \rightarrow x_3 \rightarrow x_5 \rightarrow x_1$ and $x_1 \rightarrow x_2 \rightarrow x_3 \rightarrow x_4 \rightarrow x_5 \rightarrow x_1$. Of the three ecosystem models considered herein, this is the only one that meets the requirement of the causal closure conjecture put forth in Section IV. D. 2. The state variables of the rainforest model are all mutually dependent and they each therefore possess identical sequential dependency sets.

The production matrix P for the model is shown in Table XI(ii). Inflows and outflows balance and so the model depicts steady state. The fractional inflow matrix Q' is given in Table XII(ii), and the transitive closure inflow matrix $(I - Q')^{-1}$ in Table XIII(ii). The fractional outflow matrix Q'' appears in Table XIV(ii), and the transitive closure outflow matrix $(I - Q'')^{-1}$ in Table XV(ii).

Referring to Table XIII(ii) and Fig. 19, it is observed that the normalized inflows in each row change between rows 1 and 2, and rows 3, 4, and 5 after flow from z_{30} has entered the system cycles. This inflow is almost twice as important [z_{30} column, Table XIII(ii)] for outflows y_{03}, y_{04}, and y_{05} as for outflows y_{01} and y_{02}. In Table XV(ii) and Fig. 20 the same effect is present although not as marked. The outflow ratios change between columns, and the columns are grouped somewhat as x_2, x_3; x_4; and x_5, x_1. The closer the inflow (column) is to an outflow, the larger its contribution to that outflow tends to be, which confirms the above observation.

Path lengths to outflows YPL_i are twice as great in this model as in Cone Spring (Table XIII), and approximately four times greater than those in the marine coprophagy system [Table VII(iii)]. Input path lengths ZPL_j are of the same order as those for outputs [Table XV(ii)], but considerably greater than those observed in the two preceding models [Tables XV(i) and VIII(ii)]. The

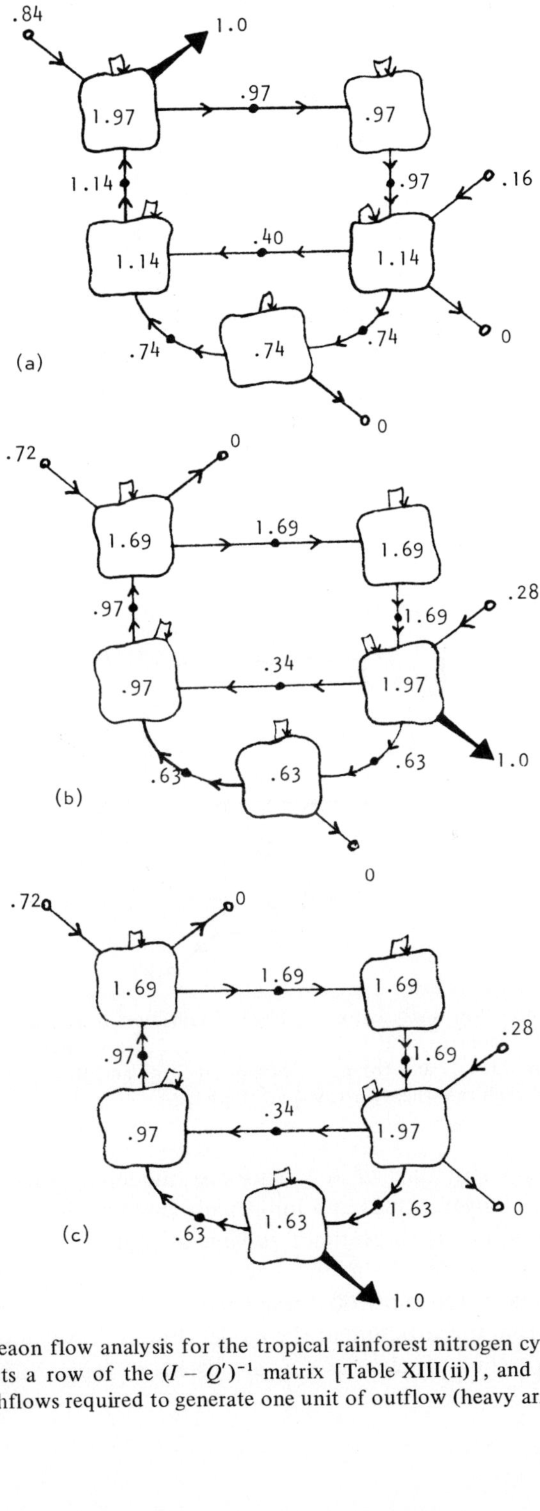

FIG. 19. Creaon flow analysis for the tropical rainforest nitrogen cycling model. Each diagram represents a row of the $(I - Q')^{-1}$ matrix [Table XIII(ii)], and shows normalized flows and throughflows required to generate one unit of outflow (heavy arrows): (a) y_{01}, (b) y_{03}, and (c) y_{04}.

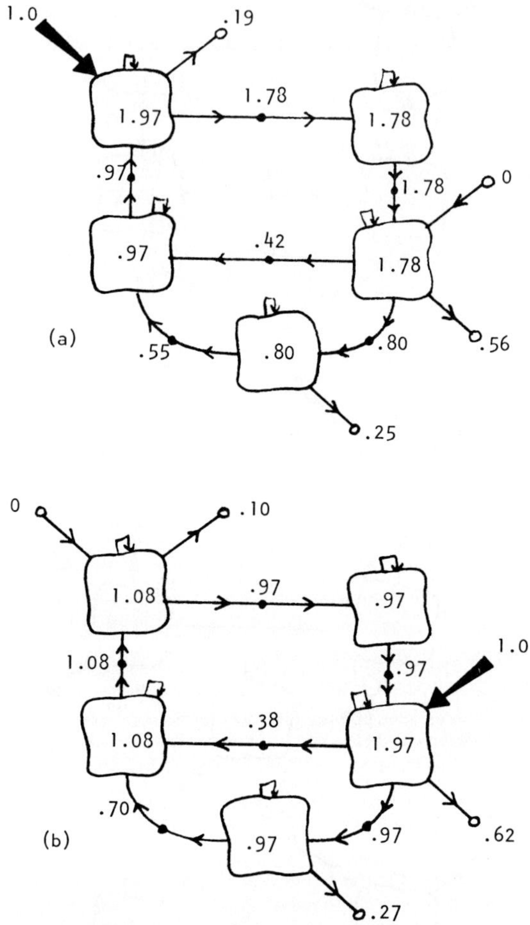

FIG. 20. Genon flow analysis for the tropical rainforest nitrogen cycling model. Each diagram represents a column of the $(I - Q'')^{-1}$ matrix [Table XV(ii)], corresponding to the two system inputs: (a) z_{10}, and (b) z_{30}. Normalized flows and throughflows generated by the unit inflows in each case (heavy arrows) are depicted.

global closed-loop structure of the rainforest model appears to account for similar as well as longer output and input path lengths. Mean path length $\overline{PL}$ is 71.8/10.2 = 7.039, about three times as large as that for Cone Spring and five times larger than that for the marine coprophagy system. The two inflows z_{10} and z_{30} contribute 5.426 and 1.607, respectively, to the system $\overline{PL}$, which is not at all apparent from the structure of the model (Fig. A.3, Appendix).

Component cycling efficiencies CE_j are listed in Table XI(ii), and may be compared to those [Tables XI(i) and VI] for the previous models. The efficiencies are much higher and in two instances, $\mathcal{H}_2$ (loose litter) and $\mathcal{H}_5$ (wood), components are indicated to be processing exclusively nitrogen already within the system. Leaves $\mathcal{H}_1$ which receive input z_{10} from rainfall and fixation, still have a cycling efficiency of 90%. Fibrous roots $\mathcal{H}_3$, which receive z_{30} through fixation, are indicated along with soil $\mathcal{H}_4$ to be processing exogenously received nitrogen to a greater extent. These observations nevertheless support the general concept of closed nutrient cycles in mature ecosystems, particularly the tropical rainforest. System cycling efficiency CE [Eq. (98)], as computed based on inflow [Table XI(ii)], is

$$CE = \frac{7.5}{10.2}(0.492) + 0(0.492) + \frac{2.7}{10.2}(0.492) + 0(0.492) + 0(0.492) = 0.492.$$

Here, the loop cycling efficiencies were computed [Eq.(97)] as

$$ZCE_j = 0.97z_{j0}/1.97\, z_{j0} = 0.492, j = 1, 2, 3, 5;$$

and $ZCE_4 = (0.63 + 0.34)z_{40}/(1.63 + 0.34)z_{40} = 0.492$ [Table XV(ii)]. The rainforest system cycling efficiency of 0.492 is 3.5 and 4.2 times greater than the corresponding values for Cone Spring and the marine coprophagy model, respectively.

TABLE XVII
Cycling Index CI Data for the Tropical Rainforest Model[a]

	$q_i \backslash q_j$	from x_1	x_2	x_3	x_4	x_5
	x_1	0.97	0.97	0.97	0.63	0.97
	x_2	0.97	0.97	0.97	0.63	0.97
to	x_3	0.97	0.97	0.97	0.63	0.97
	x_4	0.97	0.97	0.97	0.63	0.97
	x_5	0.97	0.97	0.97	0.63	0.97
	$(T_c)_j$	9.89	9.89	9.89	6.43	9.89

$TST_c = 46.0$ $\overline{PL}_c = 4.45$

$TST_s = 25.8$ $\overline{PL}_s = 2.59$

$CI = 1.78$

[a] The matrix is $(I - Q')_c^{-1}$ and represents normalized cycled flows; $(T_c)_j$ is the jth element of the cycled throughflow vector $\mathbf{T}_c$ [Eq. (101)]; TST_c, TST_s, $\overline{PL_c}$, $\overline{PL_s}$, and CI are as defined in Eqs. (102)–(104).

Information pertaining to the cycling index CI [Eq. (104)] is presented in Table XVII. The matrix of cycled throughflows $(I - Q')_c^{-1}$ was straightforwardly computed from the principal diagonal entries of $(I - Q')^{-1}$. Premultiplication of the cycled flow matrix by the output vector $\mathbf{y}$ yields [Eq. (71)] the vector $\mathbf{T}_c$ [Eq. (101)] of cycled throughflows. These values $(T_c)_j$, $j = 1, \ldots, 5$, are given in Table XVII and they clearly indicate predominance of cyclical flows in this model. Values for feedback and nonfeedback portions of total system throughflow and path length are also listed in the table as is the system cycling index ($CI = 1.78$) computed from them [Eq. (104)]. This index is an order of magnitude greater than those for the other two systems.

Table XVIII presents selected summary data for quick comparison of the three demonstration models of this section.

TABLE XVIII

Summary of Total System Throughflows (TST) and Intermediate Measures Leading to the Cycling Index (CI) for Comparison of the Marine, Cold Spring, and Rainforest Ecosystem Models

Model	TST	Σz	CE	$\overline{PL}$	TST_c	TST_s	$\overline{PL}_c$	$\overline{PL}_s$	CI
Marine coprophagy	549.3 gC $m^{-2}y^{-1}$	379.6	0.116	1.45	105.6	443.7	0.28	1.17	0.24
Cone Spring	30626.5 kcal $m^{-2}y^{-1}$	11822.9	0.138	2.59	4122.2	26504.2	0.35	2.24	0.16
Tropical rainforest	71.8 gN $m^{-2}y^{-1}$	10.2	0.492	7.04	46.0	25.8	4.45	2.59	1.78

E. Dynamic Flow Analysis

The causal bond was modeled in Section III as a dynamic concept in response to the problem raised in philosophy (Section II. L) of associating unique effects with unique causes. It is clear that causal relationships revealed by static input–output analysis may be useful for representing nondynamic steady states of systems, but for the resolution of causal dynamics further development is required. In this section a preliminary approach to dynamic input–output analysis of conservative flow systems is described.

To obtain a dynamic formulation throughflow must be redefined. Four alternative definitions will be considered, and the last one applied to derive instantaneous transitive closure matrices. For continuous time systems with conserved flows, a state transition function [Eq. (28)] can be constructed based on the conservation principle:

$$\dot{x}_k(t) = \sum_{j=1}^{n} \phi_{kj}(t) + z_{k0}(t) - \sum_{i=1}^{n} \phi_{ik}(t) - y_{0k}(t), \qquad (105)$$

$$k = 1, \ldots, n, \quad t \in [t_0, t_1],$$

where $\Sigma_{j=1}^{n} \phi_{kj}(t)$ and $z_{k0}(t)$ are endogenous and exogenous inflows to holon $\mathcal{H}_k$, and $\Sigma_{i=1}^{n} \phi_{ik}(t)$ and $y_{0k}(t)$ are endogenous and exogenous outflows, respectively.

1. Referring to Eq. (66), a dynamic throughflow for $\mathcal{H}_k$ can be defined as the sum of inflows:

$$ {}_zT_k(t) = \sum_{j=1}^{n} \phi_{kj}(t) + z_{k0}(t), \qquad k = 1, \ldots, n, \quad t \in [t_0, t_1]. \tag{106} $$

Combining Eqs. (106) and (105) this throughflow can be expressed as

$$ {}_zT_k(t) = \sum_{i=1}^{n} \phi_{ik}(t) + y_{0k}(t) + \dot{x}_k(t), \qquad k = 1, \ldots, n, \quad t \in [t_0, t_1]. \tag{107} $$

2. Similarly, referring to Eq. (67) a dynamic throughflow for $\mathcal{H}_k$ can be defined as the sum of outflows:

$$ {}_yT_k(t) = \sum_{i=1}^{n} \phi_{ik}(t) + y_{0k}(t), \qquad k = 1, \ldots, n, \quad t \in [t_0, t_1]. \tag{108} $$

Substituting from Eq. (105), this becomes

$$ {}_yT_k(t) = \sum_{j=1}^{n} \phi_{kj}(t) + z_{k0}(t) - \dot{x}_k(t), \qquad k = 1, \ldots, n, \quad t \in [t_0, t_1]. \tag{109} $$

3. A third formulation can be defined as the instantaneous minimums of (107) and (109):

$$ \min_z T_k(t) = \sum_{i=1}^{n} \phi_{ik}(t) + y_{0k}(t) + \dot{x}_k(t)_- \tag{110} $$

and

$$ \min_y T_k(t) = \sum_{j=1}^{n} \phi_{kj}(t) + z_{k0}(t) - \dot{x}_k(t)_+, \tag{111} $$

where

$$ \dot{x}_k(t)_- = \begin{cases} \dot{x}_k(t) & \text{if } \dot{x}_k(t) < 0 \\ 0 & \text{otherwise,} \end{cases} \qquad \dot{x}_k(t)_+ = \begin{cases} \dot{x}_k(t) & \text{if } \dot{x}_k(t) > 0 \\ 0 & \text{otherwise.} \end{cases} \tag{112} $$

4. Finally, a fourth formulation can be defined as the instantaneous maximums of Eqs. (107) and (109):

$$ \max_z T_k(t) = \sum_{i=1}^{n} \phi_{ik}(t) + y_{0k}(t) + \dot{x}_k(t)_+, \tag{113} $$

and

$$\max_y T_k(t) = \sum_{j=1}^{n} \phi_{kj}(t) + z_{k0}(t) - \dot{x}_k(t)_-, \qquad (114)$$

$$k = 1, \ldots, n, \quad t \in [t_0, t_1],$$

where $\dot{x}_k(t)_+$ and $\dot{x}_k(t)_-$ are according to Eqs. (112). This formulation represents the actual flow of substance through $\mathcal{H}_k$, as follows. First, it is observed that in dynamical systems (Section III. F) the dependency sequence $z_k(t) \to y_k(t)$ is precluded and transfer of inflow to outflow must always involve some storage [i.e., some state variable $x_k(t)$, in accordance with Theorem 2(iii), Section IV. D. 1], however momentary: $z_k(t) \to x_k(t) \to y_k(t)$. If $\dot{x}_k(t) > 0$, then there will be an increment to storage $x_k(t)$, and $\max_z T_k(t)$ [Eq. (113)] represents a maximum inflow. On the other hand, if $\dot{x}_k(t) < 0$, then storage $x_k(t)$ will be decremented and $\max_y T_k(t)$ [Eq. (114)] represents a maximum outflow. Hence, Eqs. (113)–(114) in effect define the actual transfer of inflow to outflow: that is, they represent *throughflow*. Expressions (113) and (114) will therefore be taken as creaon and genon formulations, respectively, appropriate for dynamic flow analysis. To minimize cumbersome notation, both formulations will be written simply $T_k(t)$ hereafter, and the context will make it apparent which is in use. It should be noted that in steady state, $\dot{x}_k(t) = 0$, the four throughflow definitions above reduce to those originally given in Eqs. (66)–(67). Note also that $\max_z T_k(t)$ is identically equal to $\max_y T_k(t)$ [from Eqs. (105) and (112)].

The dynamic production matrix $P(t)$ for a nonsteady-state system has the form shown in the accompanying matrix. Derivation of dynamic transitive

to \ *from* (q_i \ q_j)	x_1	x_2	...	x_n	Inflow	$-\dot{x}_i(t)_-$	Outflow	Row Σ
x_1	$\phi_{11}(t)$	$\phi_{12}(t)$	...	$\phi_{1n}(t)$	$z_{10}(t)$	$-\dot{x}_1(t)_-$	0	$T_1(t)$
x_2	$\phi_{21}(t)$	$\phi_{22}(t)$	...	$\phi_{2n}(t)$	$z_{20}(t)$	$-\dot{x}_2(t)_-$	0	$T_2(t)$
⋮	⋮	⋮		⋮	⋮	⋮	⋮	⋮
x_n	$\phi_{n1}(t)$	$\phi_{n2}(t)$	...	$\phi_{nn}(t)$	$z_{n0}(t)$	$-\dot{x}_n(t)$	0	$T_n(t)$
Inflow	0	0	...	0	0	0	0	0
$+\dot{x}_j(t)_+$	$+\dot{x}_1(t)_+$	$+\dot{x}_2(t)_+$	...	$+\dot{x}_n(t)_+$	0	0	0	$+\Sigma\dot{x}(t)_+$
Outflow	$y_{01}(t)$	$y_{02}(t)$	...	$y_{0n}(t)$	0	0	0	$\Sigma y(t)$
Column Σ	$T_1(t)$	$T_2(t)$	...	$T_n(t)$	$\Sigma z(t)$	$-\Sigma\dot{x}(t)_-$	0	$TST+[\Sigma z(t) - \Sigma\dot{x}(t)_-$ or $\Sigma y(t) + \Sigma\dot{x}(t)_+]$

closure matrices parallels the static case (Sections V. B. 2 and V. B. 3) as follows:

Creaon. Paralleling Eqs. (69)–(72), each instantaneous flow $\phi_{ij}(t)$ from $\mathcal{H}_j$ to $\mathcal{H}_i$ is expressed as a fraction $q'_{ij}(t)$ of instantaneous throughflow $T_i(t)$ into $\mathcal{H}_i$ (Fig. 14a):

$$\phi_{ij}(t) = q'_{ij}(t)\, T_i(t), \qquad j = 0, \ldots, n, \quad t \in [t_0, t_1]. \tag{115}$$

For throughflow of $\mathcal{H}_j$ substitute Eq. (115) into Eq. (113):

$$\begin{aligned} T_j(t) &= \sum_{i=1}^{n} \phi_{ij}(t) + y_{0j}(t) + \dot{x}_j(t)_+ \\ &= \sum_{i=1}^{n} q'_{ij}(t)\, T_i(t) + y_{0j}(t) + \dot{x}_j(t)_+, \\ &\qquad j = 0, \ldots, n, \quad t \in [t_0, t_1]. \end{aligned} \tag{116}$$

In matrix form,

$$\mathbf{T}(t) = \mathbf{T}(t)\, Q'(t) + \mathbf{y}(t) + \dot{\mathbf{x}}(t)_+, \qquad t \in [t_0, t_1]. \tag{117}$$

Rearranging and solving for $\mathbf{T}(t)$,

$$\mathbf{T}(t)\, [I - Q'(t)] = \mathbf{y}(t) + \dot{\mathbf{x}}(t)_+,$$

and

$$\mathbf{T}(t) = [\mathbf{y}(t) + \dot{\mathbf{x}}(t)_+]\, [I - Q'(t)]^{-1}, \qquad t \in [t_0, t_1]. \tag{118}$$

The only difference in this from the static formulation [Eq. (71)] is that, wherever Σy in P (Section V. B. 1) is used in the steady-state case, $[\Sigma y(t) + \Sigma \dot{x}(t)_+]$ in $P(t)$ applies for the nonsteady state.

Genon. Paralleling Eqs. (74)–(77), each instantaneous flow $\phi_{ij}(t)$ from $\mathcal{H}_j$ to $\mathcal{H}_i$ is expressed as a fraction $q''_{ij}(t)$ of the instantaneous throughflow $T_j(t)$ destined for $\mathcal{H}_i$ (Fig. 14b):

$$\phi_{ij}(t) = q''_{ij}(t)\, T_j(t), \qquad i = 0, \ldots, n, \quad t \in [t_0, t_1]. \tag{119}$$

Throughflow of $\mathcal{H}_i$ can be formulated by substituting Eq. (119) into Eq. (114):

$$\begin{aligned} T_i(t) &= \sum_{j=1}^{n} \phi_{ij}(t) + z_{i0}(t) - \dot{x}_i(t)_- \\ &= \sum_{j=1}^{n} q''_{ij}(t)\, T_j(t) + z_{i0}(t) - \dot{x}_i(t)_-, \\ &\qquad i = 0, \ldots, n, \quad t \in [t_0, t_1]. \end{aligned} \tag{120}$$

In matrix form,

$$\mathbf{T}(t) = Q''(t)\,\mathbf{T}(t) + \mathbf{z}(t) - \dot{\mathbf{x}}(t)_-, \qquad t \in [t_0, t_1]. \tag{121}$$

Rearranging and solving for $\mathbf{T}$,

$$[I - Q''(t)]\,\mathbf{T}(t) = \mathbf{z}(t) - \dot{\mathbf{x}}(t)_-,$$

and

$$\mathbf{T}(t) = [I - Q''(t)]^{-1}[\mathbf{z}(t) - \dot{\mathbf{x}}(t)_-], \qquad t \in [t_0, t_1]. \tag{122}$$

This differs from the static formulation [Eq. (77)] in that, wherever Σz in P (Section V. B. 1) was used formerly, $[\Sigma z(t) - \Sigma \dot{x}(t)_-]$ in $P(t)$ is employed in the nonsteady-state case.

Throughflow formulations (118) and (122) and their associated inflow and outflow transitive closure matrices permit the same operations to be carried out dynamically as accomplished previously for static systems. The matrices defined instantaneous causal structure with applicability limited to conservative flow systems [Eq. (105)]. Extension to discrete time systems is straightforward. The causality relations depicted by the dynamic transitive closure matrices are only momentary, and in the next instant of time they will change according to the inflow–outflow dynamics of the system. The next and concluding example provides an illustration.

Example 17

Consider the marine coprophagy model defined by a linear, stationary, constant inflow state transition function as obtained by setting Eq. (44)

TABLE XIX

Initial Production Matrix $P(0)$ for the Marine Coprophagy Model[a]

	q_i \ q_j	from x_1	x_2	x_3	x_4	inflow	$-\dot{x}_i(t)_-$	out-flow	row Σ
to	x_1	0	0	0	0	136.5	0	0	136.5
	x_2	21.9	0	0	0	0	0	0	21.9
	x_3	0	15.3	0	52.9	281.1	0	0	349.3
	x_4	0	0	79.6	0	0	0	0	79.6
	inflow	0	0	0	0	0	0	0	0
	$+\dot{x}_i(t)_+$	12.4	0	25.6	0	0	0	0	38.0
	outflow	102.2	6.6	244.1	26.7	0	0	0	379.6
	column Σ	136.5	21.9	349.3	79.6	417.6	0	0	1004.9

TST = 587.3

[a] Same as Table VI except for 10% increases in $z_{10}(0)$ and $z_{30}(0)$, resulting in derivative increases for $\dot{x}_1(0)_+$ and $\dot{x}_3(0)_+$.

TABLE XX

Initial (i) Fractional Inflow Matrix $Q'(0)$, and (ii) Transitive Closure Inflow Matrix $[I - Q'(0)]^{-1}$ for the Marine Coprophagy Model upon 10% Step Increases in $z_{10}(0)$ and $z_{30}(0)$

(i) *Q' matrix:*

	q_i \ q_j	from x_1	x_2	x_3	x_4	z_{10}	z_{30}
to	x_1	0	0	0	0	1.0	0
	x_2	1.0	0	0	0	0	0
	x_3	0	0.044	0	0.152	0	0.805
	x_4	0	0	1.0	0	0	0
	z_{10}	0	0	0	0	0	0
	z_{30}	0	0	0	0	0	0

(ii) *(I-Q') matrix:*

	q_i \ q_j	from x_1	x_2	x_3	x_4	z_{10}	z_{30}
to	x_1	1.0	0	0	0	1.0	0
	x_2	1.0	1.0	0	0	1.0	0
	x_3	0.052	0.052	1.178	0.178	0.052	0.943
	x_4	0.052	0.052	1.178	1.178	0.052	0.943
	z_{10}	0	0	0	0	1.0	0
	z_{30}	0	0	0	0	0	1.0

TABLE XXI

Time Dynamics of Nonzero Matrix Elements of Marine Coprophagy Model after 10% Step Increases in $z_{10}(0)$ and $z_{30}(0)$, as Percent Deviations from Steady-State Values

t (years)	Fractional Inflow Matrix Elements $q'_{32}(t)$	$q'_{34}(t)$	$q'_{36}(t)$	Transitive Closure Inflow Matrix Elements $[I-Q'(t)]^{-1}_{13}$	$[I-Q'(t)]^{-1}_{33}$	$[I-Q'(t)]^{-1}_{43}$	$[I-Q'(t)]^{-1}_{63}$
< 0	0	0	0	0	0	0	0
0.0	-6.4	-7.4	2.0	-8.7	-1.4	-8.6	0.5
0.1	-6.4	-2.5	1.0	-8.6	-0.5	-2.8	0.5
1.0	-2.2	0	0.3	-3.0	-0.1	-0.1	0.2
2.0	0	0	0.1	-1.0	0	0	0.1
3.0	0	0	0	-0.2	0	0	0

TABLE XXII

Initial (i) Fractional Outflow Matrix $Q''(0)$, and (ii) Transitive Closure Outflow Matrix $[I - Q''(0)]^{-1}$ for the Marine Coprophagy Model upon 10% Step Increases in $z_{10}(0)$ and $z_{30}(0)$

(i) *Q' matrix*

to q_i \ *from* q_j	x_1	x_2	x_3	x_4	y_{01}	y_{02}	y_{03}	y_{04}	$\dot{x}_1$	$\dot{x}_3$
x_1	0	0	0	0	0	0	0	0	0	0
x_2	0.160	0	0	0	0	0	0	0	0	0
x_3	0	0.699	0	0.665	0	0	0	0	0	0
x_4	0	0	.228	0	0	0	0	0	0	0
y_{01}	0.749	0	0	0	0	0	0	0	0	0
y_{02}	0	0.301	0	0	0	0	0	0	0	0
y_{03}	0	0	0.699	0	0	0	0	0	0	0
y_{04}	0	0	0	0.335	0	0	0	0	0	0
$\dot{x}_1$	0.091	0	0	0	0	0	0	0	0	0
$\dot{x}_3$	0	0	0.073	0	0	0	0	0	0	0

(ii) *(I-Q') matrix:*

to q_i \ *from* q_j	x_1	x_2	x_3	x_4	y_{01}	y_{02}	y_{03}	y_{04}	$\dot{x}_1$	$\dot{x}_3$
x_1	1.0	0	0	0	0	0	0	0	0	0
x_2	0.160	1.0	0	0	0	0	0	0	0	0
x_3	0.132	0.823	1.179	0.783	0	0	0	0	0	0
x_4	0.030	0.188	0.269	1.179	0	0	0	0	0	0
y_{01}	0.749	0	0	0	1.0	0	0	0	0	0
y_{02}	0.048	0.301	0	0	0	1.0	0	0	0	0
y_{03}	0.092	0.575	0.824	0.547	0	0	1.0	0	0	0
y_{04}	0.010	0.062	0.090	0.395	0	0	0	1.0	0	0
$\dot{x}_1$	0.091	0	0	0	0	0	0	0	1.0	0
$\dot{x}_3$	0.010	0.060	0.086	0.057	0	0	0	0	0	1.0

coefficients and exogenous inputs constant on $(-\infty, t_0)$. If both inputs z_{10} and z_{30} are step-increased 10% at time t_0, the model will during $[t_0, t_1]$ undergo transient dynamics toward a new steady state. To demonstrate dynamic flow relationships, this was simulated on the interval [0, 3] years, numerically integrating the equations. By three years the state variable derivatives $\dot{x}_k(t)$, $k = 1, \ldots, 4$, were virtually 0.

Table XIX shows the initial production matrix $P(0)$. The matrix is the same as that in Table VI except for the 10% increases in $z_{10}(0)$ and $z_{30}(0)$, resulting in

TABLE XXIII
Time Dynamics of Flow Analysis Measures for the Marine Coprophagy Model during the Interval [0, 3] Years Following an Initial 10% Step Increase in $z_{10}(0)$ and $z_{30}(0)$

t (years)	ZPL_1	ZPL_3	$\overline{PL}$	TST	CE	TST_c	$\overline{PL}_c$	CI
steady state	1.361	1.489	1.447	549.3	0.116	105.6	0.278	0.238
0.0	1.323	1.447	1.406	587.3	0.107	105.7	0.253	0.220
0.1	1.328	1.477	1.428	596.4	0.113	112.5	0.269	0.232
0.2	1.333	1.486	1.435	599.4	0.115	114.9	0.275	0.237
0.3	1.335	1.488	1.438	600.4	0.116	115.4	0.276	0.238
0.4	1.337	1.489	1.439	601.0	0.116	115.7	0.277	0.238
0.5	1.339	1.489	1.440	601.3	0.116	115.8	0.277	0.238
1.0	1.348	1.489	1.443	602.5	0.116	115.9	0.278	0.238
1.5	1.355	1.489	1.445	603.4	0.116	116.1	0.278	0.238
2.0	1.357	1.489	1.446	603.7	0.116	116.2	0.278	0.238
2.5	1.359	1.489	1.447	604.0	0.116	116.3	0.278	0.238
3.0	1.360	1.489	1.447	604.1	0.116	116.3	0.278	0.238

corresponding derivative increases for $\dot{x}_1(0)_+$ and $\dot{x}_3(0)_+$. Note that $TST(0) = 587.3$ gC m^{-2} y^{-1}, representing an increase of $\Sigma z(0) = 38.0$ compared to the steady-state value $TST = 549.3$ gC m^{-2} y^{-1} (Table VI). None of the endogenous flows has yet changed in Table XIX as the derivative increases have not had time to be propagated to the state variables.

Table XX presents initial $Q'(0)$ and $[I - Q'(0)]^{-1}$, which can be compared with Tables VII(i) and VII(iii), respectively. The dynamic matrices $Q'(t)$ and $[I - Q'(t)]^{-1}$ were virtually identical with their steady-state counterparts by three years. Table XXI summarizes the dynamics of nonzero $Q'(t)$ and $[I - Q'(t)]^{-1}$ elements during [0, 3] years. Numbers show percent deviations (±) from corresponding steady-state matrix entries. No dynamic element changes by as much as 10%, the level of input increases, and after two years only one element $[I - Q'(t)]^{-1}_{13}$ deviates by as much as 1%. Table XXII shows initial $Q''(0)$ and $[I - Q''(0)]^{-1}$ matrices for comparison with the corresponding steady-state matrices [Tables VIII(i) and VIII(ii)]. The nonzero state variable derivatives contribute to these matrices, adding two rows as indicated. Time dynamics of the elements of the two matrices were similar to that indicated in Table XXI for the $Q'(t)$ and $[I - Q'(t)]^{-1}$ matrices.

Finally, Table XXIII illustrates the dynamics of selected flow analysis measures introduced in Section V. C. The behavior is very simple because the model and input perturbations are simple. All measures except TST and TST_c

decreased initially and then monotonically increased to their steady-state values. The throughflow measures TST and TST_c rose monotonically to final values which were 10% greater than the corresponding steady-state figures.

In concluding this section on the causal network, it may be observed that input–output flow analysis in general and its dynamic aspects in particular are new to ecological applications. Cale (1975) only recently has applied the static methodology to a nontrivial, large-scale, causally closed ecosystem model of shortgrass prairie. It is clear that the methods complement and with certain restrictions implement the present causal theory of ecology. These techniques, described here to illuminate causation in webs, and future methods that they and the problem give rise to, will be the instruments that enable ecologists to dissect and tease apart the structural dynamics of complex natural systems. The difficulties are great, and progress is only beginning, but it is just as imperative now as ever that, "to understand ecosystems ultimately will be to understand networks" (Patten and Witkamp, 1967).

VI. Summary and Conclusions

In a complex and dangerous biological world of eating and being eaten, of pursuit by and escape from the multifarious forces of natural selection, contemporary ecology portrays an existence based in struggle, conflict, and competition. A deeper theme of harmony and order is provided by naturalists, vaguely expressed in ecological concepts like niche, biocoenose, and ecosystem. Ecological theory is a system theory, but a dialectical one in which order of the whole arises from the evolutionary shakedown of parts.

The objective of this study has been to capture the intuitive perception that nature is unified, and make it integral to a proposed formal blueprint for an emerging systems paradigm about ecosystems. The approach has been to express causality, as a scientific principle of first moment, in terms of state space system theory, then to examine in causal chains and networks the role of state as internuncial propagator of causes into effects.

A. Summary

In the Introduction, Section I, it was pointed out that state theory was a mathematical model of becoming, and that like all mathematical constructs it had automatic syntax but semantic content had to be provided. To do so in an ecological context would be the undertaking of the chapter.

Section II, Causal Determinism, reviewed the broad philosophical outlines of causality as presented by Bunge (1959). A variety of general determination categories were identified to emphasize that causation is only one kind of determination. Determinacy was formulated in terms of two separate principles, orderliness (lawfulness) and genetic productivity (producing or evoking effects): "Everything is lawfully produced by something else and lawfully produces something else." Aristotle's four kinds of cause were then identified—formal, material, efficient, and final—and the observation made that only efficient (external) cause was important in science. The logical structure of causation was then investigated and the "necessary production" formulation of Bunge adopted as a working meaning of the causal bond, $C \rightarrow E$: "If C occurs, then and only then E is always produced by it." The problem of associating unique causes with unique effects in multiple causation was raised, and went unsolved in the philosophical context.

In Section III, The Causal Bond, the uniqueness problem is solved by equating the causal bond, $C \rightarrow E$, to state theoretic objects of system theory. This implies that between every cause and effect of any kind there exists some entity which uniquely takes the cause into the effect. The entity is termed a *holon,* consisting of a receptor object, *creaon*, and an effector object, *genon.* The properties of holons which permit causal production of effects are then investigated according to the state space system theory of Zadeh (Zadeh and Desoer, 1963; Zadeh, 1969). They are (i) closure under segmentation, (ii) covering, (iii) closure under truncation, (iv) uniqueness, (v) continuation, and (vi) nonanticipation. These formalistic properties are embodied in the concept of *state,* which is sufficient to uniquely carry causes (*inputs*) into effects (*outputs*). Inputs are received by the creaon and serve to map time into states; this aspect of the holon is formulated mathematically by a state transition function. Then, by action of the genon states map inputs into outputs; this characteristic of the holon is described by a response function. The state transition and response functions together comprise a state space model of the causal bond. So conceived, the holon $\mathcal{H}$ is converted to an ecological object by endowing it with an *environment,* $\mathcal{H}'$. This new object, also seen as causal, represents the specific observed universe of the holon as perceived by its creaon. The name "creaon" connotes the creative act of environment definition implied by reception of input. The holon–environment pair are taken as reciprocal and inseparable (e.g., Figs. 4 and 5), and in the next two sections propagation of received causes within the holon "structure" is investigated.

Section IV, The Causal Sequence, examines properties of serial sequences of causal bonds, taken as modules for causal chains. The issue of boundaries is raised in response to the concept of environment, wherein external causes originate. In a hierarchical view of the holon, environment is defined at both sides of the levels spectrum. Thus, causes emanate from both lower-level and higher-level processes. This leads to a distinction between exogenous and endogenous causes—*inputs* and *coefficients,* respectively, in mathematical representations—but both can be combined into one class of system variable (*parameter*) for the purpose of developing properties of causal chains. The development begins by defining $\mathcal{H}$ in terms of undifferentiated system variables that are to become differentiated into parameters, states, and outputs. The process of making distinctions begins with *production functions* expressing, according to the determinacy principle (Section II), genetic productivity between the variables. If one variable is caused by another, it is said to be *directly dependent* on the latter. Parameters (causes), state variables, and outputs (effects) are differentiated solely by their direct dependency characteristics. Parameters are not directly dependent on any system variables, and have at least one system variable directly depedent on them. State variables are directly dependent on at least one system variable, and at least one system variable is directly dependent on them. Outputs are directly dependent on at least one system variable, but have no system variables directly dependent on them. If a variable is caused by another through a chain of intermediary variables, it is *sequentially dependent* on the originator of the chain. It is shown that every parameter of a system originates at least one causal sequence, every output variable terminates at least one such sequence, and every system variable between the originating and terminating variables of a causal chain is a state variable. The role of state in endogenously propagating causes into effects is thus demonstrated in the context of serial dependency sequences. It is then further shown that no system variable is superfluous. Every parameter originates a dependency sequence to output, every output variable terminates a dependency sequence traceable back to some parameter, and every state variable exists in some sequence from a cause to an effect. System variables may be mutually (sequentially) dependent or mutually independent, but only state variables can have the former property, which implies the same set of sequential causes. Based on the propagation of material cause through ecosystems, as represented by biogeochemical cycling, it is conjectured that ecosystem objects consist of mutually dependent state variables, and thus that ecosystems are causally closed. In considering alternate sequential dependency routes through a system, dependency matrices are discussed. A direct dependency matrix defines the direct dependency links (chain length equal to 1) between system variables. Successive powers of such matrices identify sequences of length 2, 3, . . ., and eventually all causal chains of all lengths are accounted for and represented by the sum of all the matrices. This property, *transitive closure,* figures significantly in the relationship of internal propagating structure of systems to their external environment, as shown in the succeeding section. Finally, in considering physical aspects of causal chains, *flow paths* are defined over which conserved quantities (energy and matter) are transferred, and parallel terminology for material causation (*inflow* and *outflow*) is introduced.

The Causal Network, in which causal sequences serve as modules, is investigated in Section V. Consequences of branching, convergence, and feedback of causal chains are developed through use of signal flow graph algebra. Feedback loops generate causal sequences of infinite length, and the graph algebra expressing propagation between any two system variables is shown to account for all dependency paths of all lengths. This is the transitive closure property. Leontief (1936) input–output analysis, requiring transitive closure, is then developed in both its original creaon form and also in its complementary genon aspect. *Transitive closure matrices,* representing inflows to each component creaon required to generate unit outflows, or outflows from each component genon produced by unit inflows, are derived. Conditions for existence of these matrices establish that all sequential dependency paths of all lengths have been accounted for. The conditions are, for the creaon, existence of input and, for the genon, existence of output. Thus, for the internal propagating structure of causal systems to have consistent meaning, the system cannot exist in isolation but must be driven by ("Everything is lawfully produced by something else, . . .") and must drive (". . . and lawfully produces something else") an external environment. That holon $\mathcal{H}$ and environment $\mathcal{H}'$ are an inseparable, reciprocal pair, each requiring for existence of its own internal propagating structure causal contact with the other is the profoundest but at the same time most truistic conclusion of the theory. The remainder of Section V develops flow analysis measures and applies them to several small ecosystem models, in both static and dynamic cases, to illustrate how input–output analysis reveals the internal propagating structure of causal networks.

B. Conclusions

1. *The Holon and Aristotelian Causality*

The state theoretic model of the causal bond seems to accommodate Aristotle's four categories of cause quite naturally. Efficient cause is input and has a creaon association. If the agent of efficient cause is energy or matter, then material cause is involved. Both efficient and material causes are "external" in the hierarchical sense of impinging from either a higher or lower level of organization. That is, they may be represented in a mathematical model as parameters. However, while efficient cause may take the form of either exogenous inputs or endogenous coefficients, material cause can only impact a holon from a higher hierarchical level, that is, can only be exogenous. Final cause in Aristotle's doctrine is the teleological end or goal toward which a process is directed in development. In the context of causal chains and webs with propagating paths of very great or even infinite length, final cause as the opposite of efficient cause may reasonably be associated with output and the genon function. Formal cause as the essential nature of a thing aligns naturally with the concept of state.

While these correspondences are tentative, they suggest how the state model of the causal bond can encompass the earliest historical ideas about causation.

2. *Holon–Environment Reciprocity*

The flow analyses of Section V. D served to illustrate propagation of cause within a system, and these same analyses can be used to illustrate the mutually defining nature of the holon–environment pair. Consider the marine coprophagy model (Section V. D. 1). The creaon aspect of a holon's relationship to its environment is shown in Fig. 21a. In Fig. 15c, consider $\mathcal{H}_3$ to be the holon illustrated in Fig. 21a. The remaining holons and their contact with the

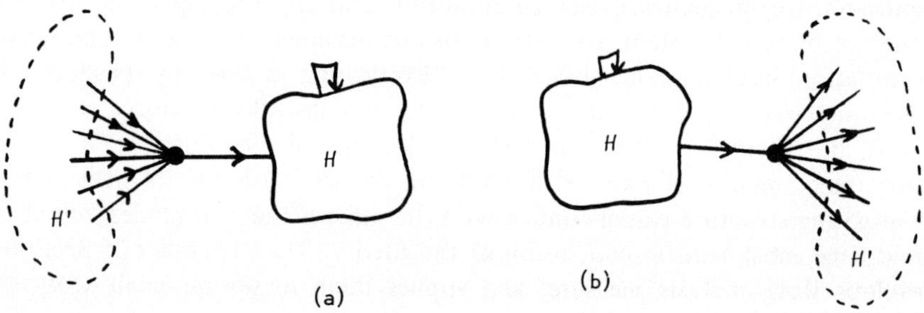

FIG. 21. (a) The creaon: "Everything is lawfully produced by something else." (b) The genon: "Everything lawfully produces something else."

environment $\mathcal{H}'$ of the entire model $\mathcal{H}$ represent the environment $\mathcal{H}_3'$ as perceived by the creaon of $\mathcal{H}_3$. It is apparent from Fig. 15c that observation of one unit of output from $\mathcal{H}_3$ defines flow components of $\mathcal{H}_3'$ as indicated. That is, the creaon element serves to define, by the inputs it admits, the environment of the object. This is a perceived environment (implicit), specific to $\mathcal{H}_3$, and its relation to the absolute environment of $\mathcal{H}_3$ is evident through comparison of Fig. 15c (with its normalized flows multiplied by 244.1 to give gC m^{-2} y^{-1}) with modeled flows in gC m^{-2} y^{-1} as depicted in Fig. A.1 of the Appendix. Hence, all flows corresponding to 0.057 and 0.195 in Fig. 15c are 13.9 and 47.6 gC m^{-2} y^{-1}, respectively. These values represent different fractions of the flows 124.1, 21.9, 15.3, 79.6, and 52.9 in Fig. A.1. Each creaon in the system defines some portion of the measured flows, and together all portions constitute the system as modeled.

Thus, the nature of the creaon as an input selector and receptor is to "create" the environment of its holon and, thereby, (i) to define a portion of the propagating state structure of the system $\mathcal{H}$ of which the holon is a member, and (ii) to contribute also to definition of the system environment $\mathcal{H}'$. The creaon's role is exactly analogous to that of a radio receiver which verifies the existence of a distant transmitter and, implicitly, the entire causal nexus of the transmitter's being in place and outputting signals. If the receiver happens to be a radiotelescope, its environment-defining potential becomes immediately and dramatically apparent.

The genon role is opposite, but similar (Fig. 21b). Consider, in Fig. 16c, $\mathcal{H}_3$ to be the holon illustrated in Fig. 21b. The remaining holons (actually only $\mathcal{H}_4$ in this case) constitute the environment of $\mathcal{H}_3'$ as influenced by its genon. The capacity of $\mathcal{H}_3'$ to be productively moved by $\mathcal{H}_3$ in effect validates the former as the latter's environment. As before, all genons in the system collectively (i) produce the internal propagating state structure of $\mathcal{H}$, and (ii) determine the system environment $\mathcal{H}'$. A holon, however defined, is always matched to a reciprocal environment produced by the genon and verified by the creaon.

3. *Ecosystem as a Coevolutionary Unit*

A holon and its environment are a mutually consistent reciprocal pair. For existence of an internally consistent causal propagating structure input or output contact with the environment is required for transitive closure. Endogenous propagation of input to output is achieved by state variables, and a system boundary exists at points where cause is transferred from input to state variable or state variable to output. The state variables of ecosystems are mutually dependent, by conjecture, so all ecosystem states are sequentially determined by the same set of efficient causes. How do these observations from the present theory bear on the issue of the unity of nature?

Dunbar (1960, 1972) has suggested that ecosystems act as units of natural selection, and that alternative ecosystem structures compete for survival. Emerson (1960) wrote about interspecific adjustments that they have had a "long evolutionary history, and in some instances the species have so modified the environment in a favorable direction that the interspecies system and its physical and biotic environment seem to have evolved as a unit—an ecosystem." The state model of causation gives a strong imperative favoring the view that all evolution, including coevolution above the species level, is an undeniable expression of intra- or intersystem consistency. The question "Do species coevolve?" becomes a special case of the more general phenomenon of system coevolution, which seems an inescapable conclusion of the causal model.

Figure 22 schematizes the holon–environment concept. System $\mathcal{H}$, at whatever level of organization, transmits potential causation to the environmental object depicted at the right. At the same time the environment, now as shown on the left, transmits cause to $\mathcal{H}$. The holon simultaneously generates (as genon) an environment influencing potential, and defines or verifies (as creaon) some portion of the absolute physical environment. The latter perceived environment and the holon together comprise a consistent coevolving pair.

For added perspective, let $\mathcal{H}'$ in Fig. 22 be a species (genotype) undergoing natural selection by its environment, $\mathcal{H}$. The environment is now construed as a holon whose genon is generating selective pressures as potential inputs to $\mathcal{H}'$, the species, and whose creaon is simultaneously verifying through success or failure of $\mathcal{H}'$ the compatibility or consistency of the latter with $\mathcal{H}$. A successful species and its environment must always be a consistent reciprocal pair, and this is a special case of the generalization that a holon and its environment are a coevolutionary unit.

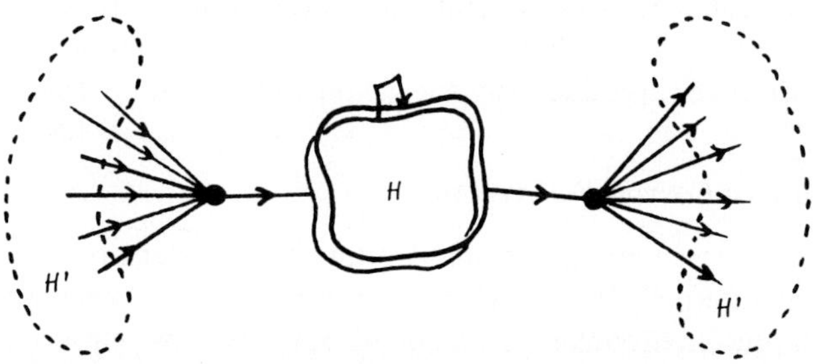

FIG. 22. Coevolution of holon–environment pair: "Nothing comes out of nothing, and nothing passes into nothing."

The relation of perceived environments to the concrete, absolute environment is implicit in the input–output analysis derivations of transitive closure matrices [Eqs. (69)–(72) and (74)–(77)]. This relationship was illustrated in Section VI. B. 2 above in reference to synthesis of Fig. A.1 from combined creaon or genon flow contributions as depicted (normalized) in Figs. 15 and 16. The absolute environment is a composition of all perceptive or productive contributions to it by all holons present. In the causally closed ecosystem (Fig. 6d), moreover, all component holons are mutually dependent. Each genon nexus propagated forward is generative of every state variable, and each creaon nexus projected backward observes every state variable. Thus, each component holon $\mathcal{H}_i$, $i = 1, \ldots, n$, both causes and is caused by every other one in the ecosystem, all of which are members of the internally consistent environment $\mathcal{H}_i'$. The whole system must coevolve in a unitary way because no alternative exists under the causal principle. The community of nature is as naturalists have seen it, unified, whole, and true.

Acknowledgments

Nancy R. Payne prepared the typescript.

Appendix. Description of Small Ecosystem Compartment Models

Descriptions of the three ecosystem models used in the text to illustrate the discussion are provided below. These models were each prepared by graduate students in a systems ecology course at the University of Georgia. They were not formulated so much as exercises in realistic ecosystem modeling but rather to provide small-scale demonstration models to which various techniques of simulation and systems analysis could be applied. While the ecological content of these models is not an issue here, they should not be assumed to have no such value. Several such course models have been published to demonstrate systems ecology methods and modes of thinking (Rykiel and Kuenzel, 1971; Williams, 1971; Brylinsky, 1972; Child and Shugart, 1972), and one (Waide *et al.*, 1974) has been extensively examined in context of the realistic ecology of its subject system, the Hubbard Brook watershed ecosystem.

A.I. Marine Coprophagy Model

Cale and Ramsey (1970) examined the role of detritus feeding in marine benthic ecosystems. Noting Johannes and Satomi's (1966) suggestion that combined feces and bacteria could be treated as one compartment which displays singular respiration and other dissipation, and following a conceptual model (Haven and Morales-Alamo, 1966) of fecal deposition by an intertidal filter feeder, the oyster *Crassostrea virginica,* the model illustrated in Fig. A.1 was constructed. The model incorporates descriptions of feces production by an intertidal crustacean, *Callianassa major,* and coprophagy rates of benthic organisms as reported by Frankenberg *et al.* (1967) and Frankenberg and Smith (1967).

In Fig. A.1, $\mathcal{H}_1$ represents *Callianassa major,* $\mathcal{H}_2$ is *C. major* feces, $\mathcal{H}_3$ is an aggregate of other benthic invertebrates, and $\mathcal{H}_4$ is fecal matter generated by $\mathcal{H}_3$. The model describes carbon storage and flow. The state variables, $x_i(t) \in \mathcal{X}_i$, $t \in [t_0, t_1]$, of each holon $\mathcal{H}_i$, $i = 1, \ldots, 4$, represent mean standing crop, gC m^{-2}. The state space $\mathcal{X} = \mathcal{X}_1 \times \cdots \times \mathcal{X}_4$ is $R_+{}^4$, where R_+ is the set of nonnegative real numbers. Coupling between holons and to the system environment is achieved by carbon flows, gC m^{-2} y^{-1}. These flows and initial states are denoted in Fig. A.1.

Mass balance state transition equations are

$$\begin{aligned}
\dot{x}_1 &= z_{10} - \phi_{21} - y_{01},\\
\dot{x}_2 &= \phi_{21} - \phi_{32} - y_{02},\\
\dot{x}_3 &= z_{30} + \phi_{32} + \phi_{34} - \phi_{43} - y_{03},\\
\dot{x}_4 &= \phi_{43} - \phi_{34} - y_{04},
\end{aligned} \tag{A.1}$$

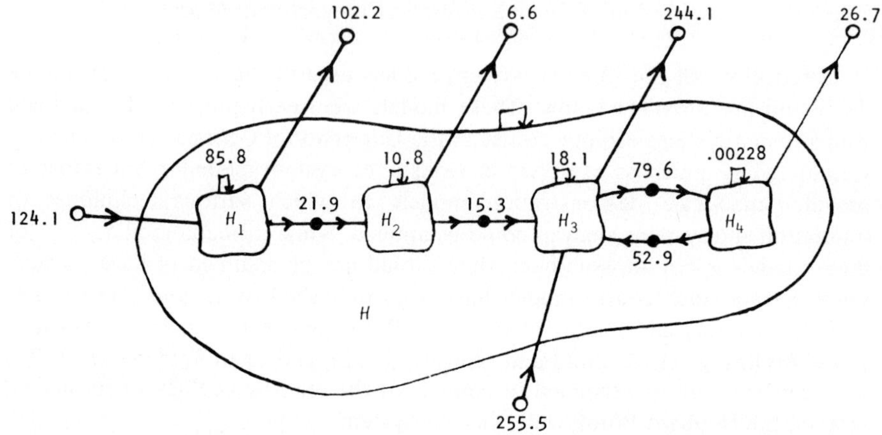

FIG. A.1. Marine coprophagy carbon model of Cale and Ramsey (1970): $\mathcal{H}_1$, *Callianassa major*; $\mathcal{H}_2$, *C. major* feces; $\mathcal{H}_3$, other benthic invertebrates; $\mathcal{H}_4$, benthic invertebrate feces. State variables in gC m^{-2}, and flows in gC m^{-2} y^{-1} (Finn, 1976).

where $\dot{x}_i$ $(i = 1, \ldots, 4)$ are time derivatives of state variables, z_{10} and z_{30} are inflows, ϕ_{ij} $(i,j = 1, \ldots, 4)$ are internal coupling flows, and $y_{01}, \ldots, y_{04}$ are outflows. Inflows to $\mathcal{H}_1$ and $\mathcal{H}_3$ represent feeding imports of organic matter, immigration, etc. Outflows represent carbon losses due to metabolism, export, and emigration. Intrasystem flows ϕ_{21} and ϕ_{43} denote excretory fecal production and ϕ_{32} and ϕ_{34} represent coprophagy. Sources of data for phytoplankton availability and coprophagy rates, represented by z_{30} and ϕ_{43}, were Pomeroy (1959), Williams (1962), and Frankenberg *et al.* (1967). Weight to metabolic rate conversions for y_{01} and y_{03} computations were obtained from Zeuthen (1947) and Sanders *et al.* (1966). The export loss component of y_{01} was estimated based on the percentage of uninhabited *Callianassa* burrows (Frankenberg and Smith, 1967). Biomass carbon content of $\mathcal{H}_1$ was based on a mean of densities, and that of $\mathcal{H}_2$ on an average of pellet densities, both as reported by Frankenberg *et al.* (1967). The mass of $\mathcal{H}_3$ was based on data of Beers (1966) and Sanders *et al.* (1966).

A.II. Cone Spring

Williams and Crouthamel (1972) modeled a cold water spring, Cone Spring, based on an original study by Tilly (1968). Tilly partitioned the system into four compartments and determined their energy storage (kcal m^{-2}) and energy fluxes (kcal m^{-2} y^{-1}) between them. The four compartments of Tilly were (i) plants, (ii) detritus and decomposers, (iii) detritus feeders, and (iv) carnivores. Based on average estimated bacterial populations over a year in Cone Spring, Williams and Crouthamel formed a fifth compartment by separating bacteria out from the original detritus–decomposer group. Their five-compartment system is diagrammed in Fig. A.2.

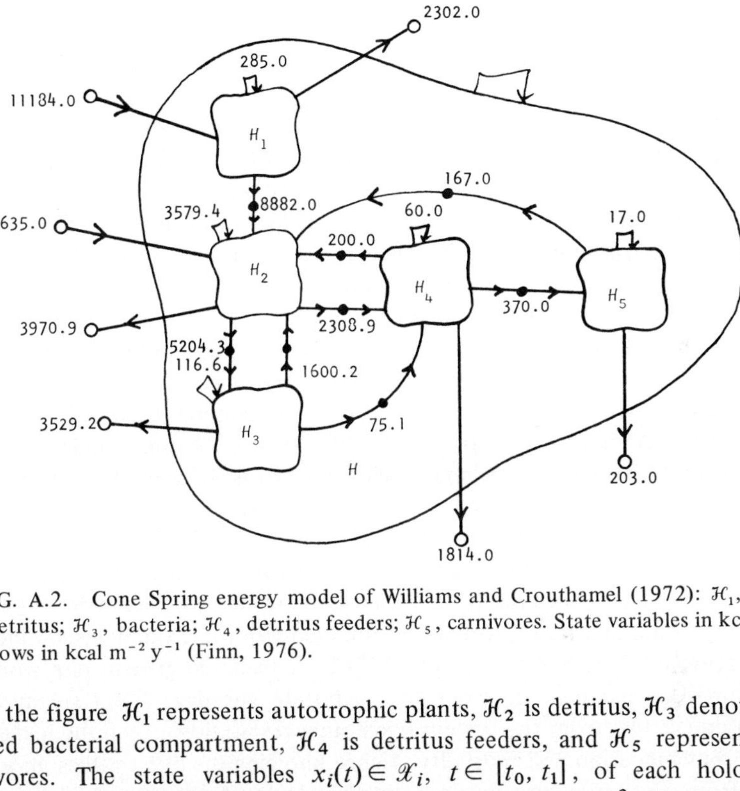

FIG. A.2. Cone Spring energy model of Williams and Crouthamel (1972): $\mathcal{H}_1$, plants; $\mathcal{H}_2$, detritus; $\mathcal{H}_3$, bacteria; $\mathcal{H}_4$, detritus feeders; $\mathcal{H}_5$, carnivores. State variables in kcal m^{-2}, and flows in kcal $m^{-2}\ y^{-1}$ (Finn, 1976).

In the figure $\mathcal{H}_1$ represents autotrophic plants, $\mathcal{H}_2$ is detritus, $\mathcal{H}_3$ denotes the derived bacterial compartment, $\mathcal{H}_4$ is detritus feeders, and $\mathcal{H}_5$ represents the carnivores. The state variables $x_i(t) \in \mathscr{X}_i$, $t \in [t_0, t_1]$, of each holon $\mathcal{H}_i$, $i = 1, \ldots, 5$, represent mean standing crops of energy, kcal m^{-2}. The state space $\mathscr{X} = \mathscr{X}_1 \times \cdots \times \mathscr{X}_5$ is R_+^5. Coupling between the holons and to the system environment, as illustrated in Fig. A.2, is achieved by energy flows, kcal $m^{-2}\ y^{-1}$.

State transition equations based on the energy conservation principle are

$$
\begin{aligned}
\dot{x}_1 &= z_{10} - \phi_{21} - y_{01}, \\
\dot{x}_2 &= z_{20} + \phi_{21} + \phi_{23} + \phi_{24} + \phi_{25} - \phi_{32} - \phi_{42} - y_{02}, \\
\dot{x}_3 &= \phi_{32} - \phi_{23} - \phi_{43} - y_{03}, \\
\dot{x}_4 &= \phi_{42} + \phi_{43} - \phi_{24} - \phi_{54} - y_{04}, \\
\dot{x}_5 &= \phi_{54} - \phi_{25} - y_{05},
\end{aligned}
\tag{A.2}
$$

where $\dot{x}_i$ ($i = 1, \ldots, 5$) are time derivatives of state variables, z_{10} and z_{20} are inflows, ϕ_{ij} ($i,j = 1, \ldots, 5$) are endogenous flows, and $y_{01}, \ldots, y_{05}$ are outflows. Inflow z_{10} to plants represents gross primary production. This value was not provided by Tilly, but was estimated by assuming steady-state inflow–outflow balance. Inflow z_{20} represents allochthonous import of organic matter into the spring (leaf fall, litterfall, ground water, etc.). Coupling between holons is provided by two basic processes, feeding (ϕ_{32}, ϕ_{42}, ϕ_{43}, and ϕ_{54}) and mortality (ϕ_{21}, ϕ_{23}, ϕ_{24}, and ϕ_{25}). These flows, except ϕ_{23} and ϕ_{32}, were all directly calculable from Tilly's data. In some cases small adjustments were made to satisfy the steady-state assumption and such changes were based on the author's evaluation that some flows were probably either underestimated or overestimated. Standing crops, $x_1, \ldots, x_5$, were provided in the original paper and only the bacteria compartment had to be separated from the detritus–decomposer group. The estimated standing crop of bacteria was 22×10^9 cells m^{-2}. It has been approximated that one gram of bacteria equals 10^9 cells (Pelczar and Reid, 1965). Therefore, Cone Spring contained an average estimated 22 g m^{-2} of bacterial biomass. Mayberry *et al.* (1967) determined that the average heat of combustion of one gram of bacterial cells is 5.3 kcal m^{-2}. Thus, the bacteria compartment in the Cone Spring model contains 116.6 kcal m^{-2}, taken as initial state. This represents 3.2% of the energy in the original compartment composed of both detritus and bacteria. This percentage was used to calculate the separate fluxes of detritus and bacteria from the total flow given both to and from the original detritus–decomposer compartment. The flux ϕ_{23} from bacteria to detritus as mortality was estimated based on personally communicated information obtained from microbiologists. Since the temperature of Cone Spring averaged 12° C, the bacterial growth rate would be approximately one half an order of magnitude per day. If, at steady state, growth equals loss from the system (excepting respiration), then the loss equals export plus predation and mortality. Hence knowing the other values, mortality flow from bacteria to detritus was calculated to be approximately 1600 kcal m^{-2} y^{-1}. The reverse feeding flow ϕ_{32} was obtained by difference since system inflow equals outflow. Two processes were responsible for generating outflows, respiration, and downstream export. Respiration contributed to all of the outflows and export to y_{02} and y_{03}.

A.III. Tropical Rainforest

Auble *et al.* (1973) modeled the nitrogen cycle in a Puerto Rican montane rainforest based on studies of Edmisten (1970). The area investigated in the Luquillo Experimental Forest (El Verde Site) was characterized by stable temperature and moisture conditions, and had been essentially undisturbed for half a century. Edmisten's original five compartments were (i) leaves, (ii) loose litter, (iii) fibrous roots, (iv) mineral soil, and (v) wood. This scheme encompasses three compartments of nitrogen in living tissues, one in dead organic matter, and one in soil. Storages ($gN\,m^{-2}$) were calculated by multiplying nitrogen levels in the compartment materials (Edmisten, 1970, Table 2, p. H213) by their biomass density (Ovington and Olson, 1970). The compartments were assumed to contain steady-state nitrogen levels. Flows ($gN\,m^{-2}\,y^{-1}$) were given by the products of nitrogen concentration (ppm) in water samples collected at seven levels [above canopy, throughfall, stemflow, soil water below litter, soil water at 6-in. depth, soil water at 24 in., and river water (Edmisten, 1970, Table 1, p. H213)] and flow rates between the indicated levels. The sources of waterflow data were not explicitly provided. Field measurements span only a three-month period and accordingly results were considered preliminary.

The model by Auble *et al.* (1973), illustrated in Fig. A.3, retains the same basic structure as diagrammed originally (Edmisten, 1970, Fig. 2, p. H212). In another version flows were added from loose litter to soil and wood to loose litter, but these alterations are not included here. The leaf compartment $\mathcal{H}_1$ also

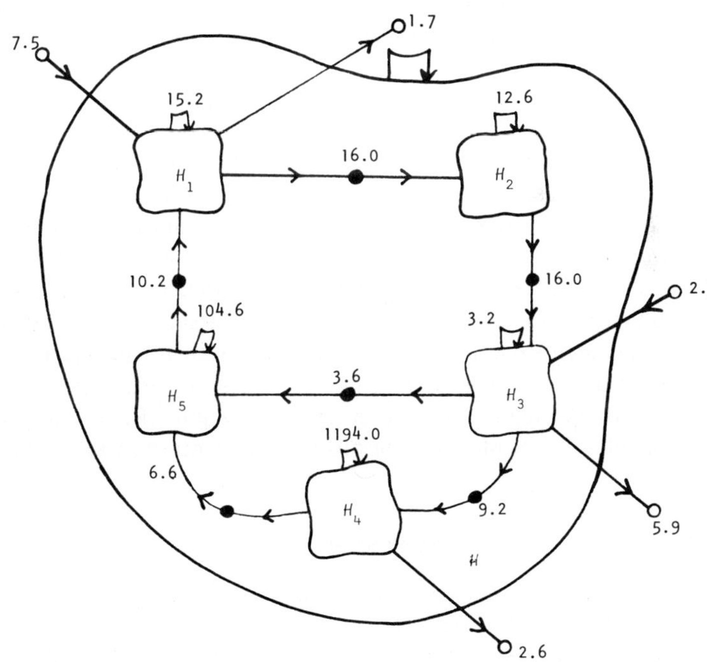

FIG. A.3. Tropical rainforest nitrogen model of Auble *et al.* (1973): $\mathcal{H}_1$, leaves and epiphyllae; $\mathcal{H}_2$, loose litter; $\mathcal{H}_3$, fibrous roots; $\mathcal{H}_4$, soil; $\mathcal{H}_5$, wood. State variables in $gN\,m^{-2}$, and flows in $gN\,m^{-2}\,y^{-1}$ (Finn, 1976).

includes leaf algae, lichens, liverworts, and other epiphyllae. Loose litter $\mathcal{H}_2$ represents litter exclusively, thereby excluding organic residues associated with soil. Fibrous roots $\mathcal{H}_3$ consists of the small root–mycorrhiza complex including associated bacteria and algae. All other roots are included in the wood component $\mathcal{H}_5$. The state-variables $x_i(t) \in \mathscr{X}_i$, $t \in [t_0, t_1]$, of the system represent nitrogen storages, gN m^{-2}. The state space $\mathscr{X} = \mathscr{X}_1 \times \cdots \times \mathscr{X}_5$ is R_+^5. External and internal coupling are provided by nitrogen flows, $\text{gN m}^{-2}\ \text{y}^{-1}$. These flows together with the nitrogen levels taken as initial states are shown in Fig. A.3.

State transition equations accounting for mass conservation are

$$
\begin{aligned}
\dot{x}_1 &= z_{10} + \phi_{15} - \phi_{21} - y_{01}, \\
\dot{x}_2 &= \phi_{21} - \phi_{32}, \\
\dot{x}_3 &= z_{30} + \phi_{32} - \phi_{43} - \phi_{53} - y_{03}, \\
\dot{x}_4 &= \phi_{43} - \phi_{54} - y_{04}, \\
\dot{x}_5 &= \phi_{53} + \phi_{54} - \phi_{15},
\end{aligned}
\tag{A.3}
$$

where $\dot{x}_i$ $(i = 1, \ldots, 5)$ are time derivatives of the state variables, z_{10} and z_{30} are inflows, ϕ_{ij} $(i,j = 1, \ldots, 5)$ are intrasystem flows, and y_{01}, y_{03} and y_{04} are outflows. Inflow z_{10} represents two sources: nitrogen dissolved in rainwater and withdrawn by components of $\mathcal{H}_1$, and biological fixation of atmospheric nitrogen by the epiphyllous complex. The first of these flows depends upon precipitation and nitrogen concentration in rain (0.56 ppm, assumed constant). Inflow z_{30} is nitrogen fixation by algae and bacteria associated with fibrous roots. Outflows represent nitrogen losses to the environment due to free nitrogen released by denitrifying bacteria (y_{01}), and denitrification and runoff associated with fibrous roots (y_{03}) and soil (y_{04}). Flow processes within the system include litterfall (ϕ_{21}), throughfall and stemflow (ϕ_{21}), runoff (ϕ_{32} and ϕ_{43}), and biological uptake (ϕ_{15}, ϕ_{53}, and ϕ_{54}). Data for the model were as provided by Edmisten except that no values were given for nitrogen density in the soil compartment $\mathcal{H}_4$. This storage was therefore assessed from soil nitrogen studies of Roberts (1942) and Edmisten (1970) in Puerto Rico. Using data for the top 61 cm of soil, the steady-state nitrogen concentration was calculated to be 1194.0 gN m^{-2}.

REFERENCES

Aquinas, T. (c. 1272). "Summa theologica" (Fathers of the English Dominican Province, transl.), Vol. 1, 96, 3, ob. 3. Burns, Oates and Washbourne, London.

Aristotle (1930). "Physics," Bk. 7, Ch. 1, 214b, 242a; Bk. 8, Ch. 4, 254b, Ch. 5, 257, *In* "Works" (W. D. Ross, ed.) Oxford Univ. Press (Clarendon), London and New York.

Aristotle (1947). "Metaphysics" (H. Tredennick, transl.), Bk. 1, Ch. 3, 983a, b. Heinemann, London.

Ashby, W. R. (1952). "An Introduction to Cybernetics." Wiley, New York.

Ashby, W. R. (1956). "Design for a Brain." Chapman and Hall, London.

Auble, G. T., Finn, J. T., Hamilton, D. B., Ojasti, J., and Rodman, D. (1973). Dynamics of nitrogen in a tropical rain forest. Unpublished Systems Ecology course report. Univ. of Georgia, Athens, Georgia

Beers, J. R. (1966). *Limnol. Oceanogr.* **11**, 520.

Bertalanffy, L. von (1968). "General System Theory. Foundations, Development, Applications." Braziller, New York.

Bruno, G. (1584). *In* "Opere italiane" (G. Gentile, ed.), Vol. 1, p. 178. Laterza, Bari (1925–1927).

Brylinsky, M. (1972). *In* "Systems Analysis and Simulation in Ecology" (B. C. Patten, ed.), Vol. II, pp. 81–101. Academic Press, New York.

Bunge, M. (1959). "Causality." Harvard Univ. Press, Cambridge, Massachusetts.

Busaker, R. G., and Saaty, T. L. (1965). "Finite Graphs and Networks." McGraw-Hill, New York.

Cale, W. G. (1975). Simulation and Systems Analysis of a Shortgrass Prairie Ecosystem. Ph.D. thesis, Univ. of Georgia, Athens, Georgia.

Cale, W. G., and Ramsey, P. R. (1970). Trophic significance of coprophagy by benthic organisms. Unpublished Systems Ecology course report. Univ. of Georgia, Athens, Georgia.

Chenery, H. B., and Clark, P. G. (1959). "Interindustry Economics." Wiley, New York.

Child, G. I., and Shugart, H. H. (1972). *In* "Systems Analysis and Simulation in Ecology" (B. C. Patten, ed.), Vol. II, pp. 103–135. Academic Press, New York.

Comte, A. (1830). "Cours de philosophie positive," *l.* I, Vol. I, p. 3. Costes, Paris (1934).

Dunbar, M. J. (1960). *Amer. Natur.* **94**, 129.

Dunbar, M. J. (1972). *Trans. Connecticut Acad. Arts Sci.* **44**, 113.

Edmisten, J. A. (1970). *In* "A Tropical Rain Forest" (H. T. Odum and R. F. Pigeon, eds.), H79–87, H211–215. U.S. At. Energy Comm., Oak Ridge, Tennessee.

Emerson, A. E. (1960). *In* "Evolution after Darwin," Vol. I, pp. 307–348. Univ. Chicago Press, Chicago, Illinois.

Faddeev, D. K., and Faddeeva, V. N. (1963). "Computational Methods of Linear Algebra" (R. C. Williams, transl.). Freeman, San Francisco, California.

Finn, J. T. (1976). *J. Theoret. Biol.* **56**, 363.

Frankenberg, D., and Smith, K. L. (1967). *Limnol. Oceanogr.* **12**, 443.

Frankenberg, D., Coles, S. L., and Johannes, R. E. (1967). *Limnol. Oceanogr.* **12**, 113.

Galileo, G. (1623). *In* "Opere" Vol. 6, p. 265. Ricciardi, Milan (1953).

Gill, A. (1962). "Introduction to the Theory of Finite State Machines." McGraw-Hill, New York.

Hannon, B. (1973). *J. Theoret. Biol.* **41**, 535.

Haven, D. S., and Morales-Alamo, R. (1966). *Limnol. Oceanogr.* **11**, 487.

Huggins, W. H., and Entwisle, D. R. (1968). "Introductory Systems and Design." Ginn (Blaisdell), Boston, Massachusetts.

Hutchinson, G. E. (1948). *Ann. N.Y. Acad. Sci.* **50**, 221.

Johannes, R. E., and Satomi, M. (1966). *Limnol. Oceanogr.* **11**, 191.

Kalman, R. E., Falk, P. L., and Arbib, M. A. (1969). "Topics in Mathematical System Theory." McGraw-Hill, New York.

Kemeny, J. G., and Snell, J. L. (1960). "Finite Markov Chains." Von Nostrand-Reinhold, Princeton, New Jersey.

Klir, G. J. (1969). "An Approach to General Systems Theory." Van Nostrand-Reinhold, Princeton, New Jersey.

Klir, G. J., ed. (1972). "Trends in General Systems Theory." Wiley (Interscience), New York.

Koestler, A. (1967). "The Ghost in the Machine." MacMillan, New York.

Leontief, W. W. (1936). *Rev. Econ. Statist.* **18,** 105.

Leontief, W. W. (1966). "Input–Output Economics." Oxford Univ. Press, London and New York.

Leopold, A. (1949). "A Sand County Almanac, and Sketches Here and There." Oxford Univ. Press, London and New York.

Levins, R. (1968). "Evolution in Changing Environments." Princeton Univ. Press, Princeton, New Jersey.

Lindeman, R. L. (1942). *Ecology* **23,** 399.

Lucretius (ca. 58 B.C.). "Of the Nature of Things" (W. E. Leonard, transl.), Ch. 1, pp. 150, 206, 249–250. Dutton, New York (1921).

Mason, S. J. (1956). *Proc. IRE* **44,** 920.

Mayberry, W. R., Prochazka, G. J., and Payne, W. J. (1967). *Appl. Microbiol.* **15,** 1332.

Mesarovic, M. D. (ed.) (1964). "Views on General Systems Theory." Wiley, New York.

Mesarovic, M. D., and Takahara, Y. (1975). "General Systems Theory: Mathematical Foundations." Academic Press, New York.

Mill, J. (1843). "A System of Logic," Bk. 3, Ch. 5, Sect. 2. Longmans, Green, London (1936).

Nagel, E., and Newman, J. R. (1956). *Scient. Amer.* **206,** 71.

Ore, O. (1962). "Theory of Graphs," Colloq. Publ. 38. Amer. Math. Soc., Providence, Rhode Island.

Ortega, J. M. (1972). "Numerical Analysis." Academic Press, New York.

Overton, W. S. (1975). *In* "Systems Analysis and Simulation in Ecology" (B. C. Patten, ed.), Vol. III, pp. 117–138. Academic Press, New York.

Ovington, J. D., and Olson J. S. (1970). *In* "A Tropical Rain Forest" (H. T. Odum and R. F. Pigeon eds.), H53–77. U.S. At. Energy Comm., Oak Ridge, Tennessee.

Padulo, L., and Arbib, M. A. (1974). "System Theory, A Unified State-Space Approach to Continuous and Discrete Systems." Saunders, Philadelphia, Pennsylvania.

Patten, B. C. (1974). The zero state and ecosystem stability, *Proc. Int. Congr. Ecol., 1st, The Hague, The Netherlands, September 8–14, 1974 (Suppl.).*

Patten, B. C., and Witkamp, M. (1967). *Ecology* **48,** 813.

Pelczar, M. J., and Reid, R. D. (1965). "Microbiology." McGraw-Hill, New York.

Phillips, A. (1955). *Quart. J. Econ.* **69,** 137.

Pomeroy, L. R. (1959). *Limnol. Oceanogr.* **4,** 386.

Ponstein, J. (1966). "Matrices in Graph and Network Theory." Van Gorcum, Assen, Netherlands.

Quesnay, F. (1758). "Tableau Économique" (1759 edition, published in 1894), Brit. Econ. Ass., London.

Roberts, R. C. (1942). Soil Survey of Puerto Rico, Soil Surv. Ser. No. 8, U.S. Dep. of Agriculture, Washington D.C.

Russell, B. (1927). *In* "An Outline of Philosophy," p. 122. Allen and Unwin, London.

Rykiel, E. J., and Kuenzel, N. T. (1971). *In* "Systems Analysis and Simulation in Ecology" (B. C. Patten, ed.), Vol. I, pp. 513–541. Academic Press, New York.

Sanders, H. L., Goudsmit, E. M., Mills, E. L., and Hampson, G. E. (1966). *Limnol. Oceanogr.* **7,** 63.

Simon, H. A. (1973). *In* "Hierarchy Theory, The Challenge of Complex Systems" (H. H. Pattee, ed.), pp. 1–27. Braziller, New York.

Tilly, L. J. (1968). *Ecol. Monogr.* **38,** 169.

Waide, J. B., Krebs, J. E., Clarkson, S. P., and Setzler, E. M. (1974). *In* "Progress in Theoretical Biology " (R. Rosen and F. M. Snell, eds.), Vol. III, pp. 261–345. Academic Press, New York.

Walras, L. (1877). "Elements of Pure Economics" (W. Jaffe, transl., republished in 1954). Irwin, Homewood, Illinois.

Webster, J. R., Waide, J. B., and Patten, B. C. (1975). *In* "Mineral Cycling in Southeastern Ecosystems" (F. G. Howell, J. B. Gentry, and M. H. Smith, eds.), pp. 1–27. ERDA Symposium Series (Conf-740513), Springfield, Virginia.

Williams, M., and Crouthamel, D. (1972). Systems analysis of Cone Spring. Unpublished Systems Ecology course report, Univ. of Georgia, Athens, Georgia.

Williams, R. B. (1962). The Ecology of Diatom Populations in a Georgia Salt Marsh. Ph.D. thesis, Harvard Univ., Cambridge, Massachusetts.

Williams, R. B. (1971). *In* "Systems Analysis and Simulation in Ecology" (B. C. Patten, ed.), Vol. I, pp. 543–582. Academic Press, New York.

Windeknecht, T. G. (1971). "General Dynamic Processes, A Mathematical Introduction." Academic Press, New York.

Wymore, A. W. (1967). "A Mathematical Theory of Systems Engineering: The Elements." Wiley, New York.

Zadeh, L. A. (1969). *In* "System Theory" (L. A. Zadeh and E. Polak, eds.), pp. 3–42. McGraw-Hill, New York.

Zadeh, L. A., and Desoer, C. A. (1963). "Linear System Theory, The State Space Approach." McGraw-Hill, New York.

Zeuthen, E. (1947). *C. R. Trav. Lab. Carlsberg* **26,** 17.

Index

F

I

J

K

L

A 6
B 7
C 8
D 9
E 0
F 1
G 2
H 3
I 4
J 5